Radar Meteorology

ATMOSPHERIC AND OCEANOGRAPHIC SCIENCES LIBRARY

VOLUME 27

The titles published in this series are listed at the end of this volume.

Radar Meteorology

by

S. Raghavan

Former Deputy Director-General of Meteorology,
India Meteorological Department, India

KLUWER ACADEMIC PUBLISHERS
DORDRECHT / BOSTON / LONDON

A C.I.P. Catalogue record for this book is available from the Library of Congress.

ISBN 1-4020-1604-2

Published by Kluwer Academic Publishers,
P.O. Box 17, 3300 AA Dordrecht, The Netherlands.

Sold and distributed in North, Central and South America
by Kluwer Academic Publishers,
101 Philip Drive, Norwell, MA 02061, U.S.A.

In all other countries, sold and distributed
by Kluwer Academic Publishers,
P.O. Box 322, 3300 AH Dordrecht, The Netherlands.

Printed on acid-free paper

CONTENTS

FOREWORD

As we all know, weather radar came into existence during the Second World War when aircraft detection radars had their vision limited by echoes from rain bearing clouds. What was often considered to be of nuisance value by the air force personnel trying to locate enemy aircraft was seen as an opportunity by the weather men. Thus adversity in one field was converted into an opportunity in another. Since then weather radar has found myriad applications with the increased sophistication of technology and processing systems. It has now become an indispensable tool for the operational forecasters, cloud physicists and atmospheric scientists.

The current generation radar is but a distant echo of the radars of the 1940s. As a result, its operation and maintenance have become very complex, like the technology it uses. Therefore, there is a definite requirement of focussing our special attention not only on the science of radar meteorology but also on its operational aspects. The present book, as pointed out by the author, attempts to fill this gap. The author has presented the subject with a balanced blend of science, technology and practice. The canvas is indeed very broad. Starting with the history of weather radar development the book goes on to discuss in a lucid style the physics of the atmosphere related to radar observation, radar technology, echo interpretation, different applications and finally attempts to look into the future to indicate potential new opportunities in this field.

S. Raghavan has the right credential to write about the subject he has chosen. He is one of the foremost exponents of radar meteorology in India and has made significant scientific contribution to the radar meteorology of the tropics where India is located. He has seen the subject in the light of his own hands-on experience. I am sure this book will be found useful as a source material, by scientists, technologists and practitioners.

New Delhi
20 January 2003

(N. SEN ROY)
Director-General of Meteorology (Retired),
India Meteorological Department,
and
Former Vice-President,
World Meteorological Organization.

PREFACE

"The worst miser is the learned man that will not write" was the quotation used by one of my colleagues while suggesting that I should write a book on Radar Meteorology. While the implication that I am a learned man is debatable, I think he was right in pointing out that there is not enough documentation of the practical aspects of the subject.

Research in radar meteorology has made enormous strides. There are numerous research weather radars including "clear air" radars. In many countries there are networks of operational weather radars, which have produced substantial improvement in areas such as short period forecasting, aviation services, hydrometeorology and tropical cyclone warnings. Yet there has been an insufficient dialogue between the radar specialist and the general meteorologist or forecaster leading to an inadequate appreciation of how the radar can help the latter. When I tracked a tropical cyclone by radar some thirty years ago, I realised for the first time its importance for saving lives and mitigating economic losses. I hold the view that whatever be the advances in equipment, the human meteorologist has the most important role to play in getting the best out of the equipment. It is generally accepted that education and training in radar meteorology have not kept pace with developments in technology. The need for a comprehensive book is therefore apparent.

But, a book for whom? The graduate student in meteorology or the operational or research meteorologist or the specialist in weather radar? There are several excellent books and compilations published in the USA and Europe, in English, covering various aspects [e.g. Atlas (1990), Battan (1973), Bringi and Chandrasekar (2001), Collier (1989), Collinge and Kirby (1987), Doviak and Zrnić (1984, 1993), Gossard and Strauch (1983) and Sauvageot (1991), see bibliography]. But, as Rinehart (1999) points out in his book, these books are aimed at the research community with little practical information for "frontline" meteorologists. Rinehart's (1999) own book and some publications of the World Meteorological Organization (WMO) fill the latter need to some extent.

In this context, I have discussed the plan of this book with experts in India and abroad whose advice I gratefully acknowledge. They advised that in a single volume it will not be possible to cover all aspects of the subject and that I should focus on some areas and give references for other aspects for the interested reader to explore further. Hence this book has an emphasis on operational applications and the results obtainable from radar. At the same time, the entire range of the subject is introduced. Theoretical background including details of signal processing is kept brief and references are given to existing work on these aspects. Practical aspects of classification of weather radar echoes, observation of precipitating systems, quantitative estimation of precipitation, observation of tropical cyclones and application to aviation, including the latest advances in these areas are dealt with in greater detail. The treatment of tropical cyclones is a major highlight of the book. An extensive bibliography is provided to assist the interested reader to pursue any topic further.

Chapter one gives an elementary introduction to the subject. Chapter two deals with propagation, scattering, and absorption of microwaves in the lower atmosphere. Chapter three surveys signal processing and display, and also deals with clutter suppression, airborne radar and polarimetric radar. In Chapter four, various phenomena observable on weather radar are introduced. Chapter five is devoted to operational aspects of weather radar including calibration, maintenance and networking. Chapter six deals in further detail with observation of precipitating weather systems. Chapter seven discusses methods of precipitation estimation by ground-based and space-borne radar. Chapter eight has a comprehensive discussion of radar observation of tropical cyclones. Chapter nine deals with aviation. In Chapter ten we deal with clear air (VHF/UHF) radars and wind profilers. Recent developments and some special applications are discussed in Chapter 11 and an attempt is made to foresee likely progress in the coming years.

I have drawn in large measure on my operational experience, my previous publications, my lectures at various levels to meteorological personnel, to university students and at various WMO gatherings in India and abroad and what I have seen during visits to meteorological institutions in other countries. Though manual and traditional techniques are being rapidly replaced by automated techniques, I have dealt in some detail with the former for the reason that these give an insight into basic principles. It cannot be denied that even with the most modern computer technology, the judgement of the human meteorologist is vital in both operations and research.

I hope that the book will serve as guidance and reference material for field meteorologists and at the same time stimulate the interest of students and academics.

Chennai, India
January 2003

S.Raghavan

ACKNOWLEDGEMENTS

This book is made possible by an initial grant from the Department of Science and Technology (DST), Government of India. Library facilities and permission to reproduce photographs, drawings and other material from the collections and publications of the India Meteorological Department (IMD) have been graciously extended by the Director-General of Meteorology through various offices of the IMD, particularly the Regional Meteorological Centre, Chennai. I am thankful to the DST as well as to the IMD and to my numerous colleagues in the IMD who helped me actively. In particular, my thanks are due to Dr. N.Sen Roy, former Director-General of Meteorology and former Vice-President of WMO, who has written the foreword for this book and to the present Director-General Dr. R.R. Kelkar.

Other organisations, notably the American Meteorological Society (AMS), the Meteorological Society of Japan (MSJ), Her Majesty's Stationery Office (HMSO), UK, the American Geophysical Union (AGU), the World Meteorological Organization (WMO), the Indian Space Research Organisation (ISRO), the Council of Scientific and Industrial Research (CSIR), New Delhi and several individual authors have permitted reproduction of photographs and diagrams from their publications. Due credit has been indicated in the figure captions. I express my gratitude to these organisations and authors.

I am specially indebted for their advice, critical comments, discussions, encouragement, good wishes, supply of relevant material and various other assistance to: Drs. T.N.Krishnamurti of Florida State University, USA; J. Kruus of Environnement-Canada and former President of the Commission for Instruments and Methods of Observation of the WMO; W-C. Lee, T.W. Schlatter, R.J. Serafin, J.Wilson and J. Vivekanandan of the National Center for Atmospheric Research (NCAR), USA; R.A.Brown, R.J.Doviak, J.J. Gourley, V.T.Wood and D.S.Zrnić of the National Severe Storms Laboratory (NSSL), USA; E.E. Gossard, R. J. Lataitis, F.M. Ralph and B.B. Stankov of Environmental Technology Laboratory, USA; G.E.Klazura of the WSR-88D Operational Support Facility, USA; R.A. Houze Jr. of the University of Washington, USA; R.J. Donaldson Jr., and F.I. Harris of Hughes STX Corporation, USA; V. Chandrasekar and S.A . Rutledge and of Colorado State University, USA; J.T. Schaefer of the National Weather Service Training Center (NWSTC), USA; J.H. Golden of National Oceanic and Atmospheric Administration (NOAA), USA; C.R. Williams of

NOAA Aeronomy Lab., USA; R. Meneghini of National Aeronautics and Space Administration (NASA) USA; P. Kollias of the University of Miami, USA; W. Krajewski of the University of Iowa, USA, J. McCarthy of Aviation Weather Associates, USA; A.L. Pazmany of the University of Massachusetts, USA; C. Ulbrich of Clemson University, USA; E.E. Clothiaux of Pennsylvania State University, USA; D.W. Martin of the University of Wisconsin, USA; M. Steiner of Princeton University, USA; W.K. Hocking of the University of Western Ontario, Canada; F. Bouttier of the European Centre for Medium range Weather Forecasting (ECMWF); C.G. Collier and P. J. Hardaker of Meteorological Office, UK; V. Klaus and J. Pilon of Météo-France; J. Dibbern of Deutscher Wetterdienst, Germany; Yumao Xu of Nanjing University, China; J. Aoyagi of the Remote Sensing Technology Center of Japan; T. Muramatsu of the Japan Meteorological Agency; T. Sato of Kyoto University, Japan; Y-H. Chu of National Central University, Taiwan; P.T. May and P.J.Meighen of the Australian Bureau of Meteorology; (the late) R.L.Southern of Australia; D. Rosenfeld of the institute of Earth Sciences, Israel; former Directors-General of Meteorology, IMD, P.K. Das, S.M.Kulshrestha, (the late) P. Koteswaram and (the late) L.S.Mathur; K.R.Saha of New Delhi; A.K. Sen Sarma of West Bengal; S.M.Bhandari, O.P.N. Calla, S.C. Chatterjee, P.S.Desai, B.V. Krishna Murthy, M.S. Narayanan, B.M.Rao, and G.Viswanathan of the Indian Space Research Organisation (ISRO); A.R. Jain and P.B. Rao of the National MST Radar Facility; P.R.Mahapatra of the Indian Institute of Science, Bangalore; D.K. Sinha of the University of Calcutta; B.M. Reddy of the National Physical Laboratory, New Delhi; D.K.Rakshit of Jadhavpur University, Calcutta; K. Krishna Reddy, K. Mohan, D. Narayana Rao, T. Narayana Rao and S. Vijaya Bhaskara Rao of Sri Venkateswara University, Tirupati; P.V. Joseph and C.K. Rajan of Cochin University of Science and Technology, Kochi; A.B. Bhattacharya of Kalyani University, Kalyani and K.G. Vernekar and A.A. Shiralkar of the Indian Institute of Tropical Meteorology (IITM), Pune.

My special thanks are due to my former colleague Mr. V. Ramaswami and my friend Mr. R.Jagannathan, who spared their time for preparing some of the figures for the book and to Mr. V. Thyagarajan and Ms. Bharathy of SRA Systems, Drs. T.N. Jegdish and S. Srinivasan of Quads, Chennai and Mr. G. Raguraman and Mr. S. Subramaniam of Umayal Print House who assisted with the formatting of the manuscript.

I also take this opportunity to thank Dr. R. Sadourney, co-Chief Editor of the Atmospheric and Oceanic Science Library at Kluwer, for accepting this book for their Series and Ms. Marie Johnson of Kluwer Academic Publishers for guiding me regarding the publication procedures.

Finally, I thank my family members Anuradha, Lakshmi, Manmatha, and Shankar for their active assistance, understanding and support during the preparation of the book.

Chennai, India
January 2003

S.Raghavan

Chapter 1

RADAR METEOROLOGY – HISTORY, PRINCIPLES AND TECHNOLOGY

In section 1.1 named History, we survey the origin, history and evolution of meteorological radar. Then in section 1.2 we shall have a discussion of the general principles of radar and a preliminary view of scattering of electromagnetic radiation by meteorological targets. In section 1.3 on technology, we deal with radar equipment and circuitry to the extent needed to facilitate understanding of its operation, capabilities and limitations.

1.1 History

1.1.1 Remote Detection by Radio

RADAR (RAdio Detection And Ranging) may be considered to be the first of the modern remote sensing techniques to be used in meteorology. The possibility of using radio waves to detect or locate a distant object appears to have been thought of from the time Hertz detected radio waves for the first time. In 1904 a German Engineer Hülsmeyer obtained a patent for a device to detect radio waves reflected by ships. In 1922 Marconi proposed what we would now call bistatic CW (Continuous Wave) radar to detect ships in fog or "thick" weather. In the same decade a pulse modulated device was developed for sounding the ionosphere. FM-CW (Frequency Modulated Continuous Wave) radio altimeter was developed in the thirties of the last century. Watson-Watt in Britain had made some experiments in what was then called RDF (Radio Direction Finding) and is regarded as the inventor of radar. But pulse modulated radar did not make much progress probably because the technology for producing adequate power at the high frequencies needed to get sufficient directivity, was not available.

1.1.2 World War II

The impetus for rapid development was provided by World War II or the prospect of it. In Britain, a Committee for the scientific study of Air Defence with Prof. Henry Tizard as chairman, identified RDF as the best option for Air Defence and mobilised every available scientist including biologists like Harrison-Mathews to work on it (Snow 1971, Harrison-Mathews 1982). Earlier a Telecommunication Research Establishment – a camouflage for radar – had been set up. The "resonant cavity magnetron", the crucial transmitting tube was developed by Randall and Boot in Britain and manufactured in the USA where work on radar had been started at the Radiation Laboratory of the Massachusetts Institute of Technology. The effort was so successful that it has been said that radar was the single most important factor contributing to the success of the western powers in the War. It can equally well be said that success in radar development was due to the War.

1.1.3 Radar vs. Infrared Remote Sensing

There was no lack of sceptics, however. F.A. Lindemann (later Lord Cherwell), advisor to Winston Churchill thought that infra-red remote sensing was a better bet for immediate development in the military context than radar. There is an anecdote that when he visited the radar laboratory he said, "It stinks" (Wears 1982). The scientists promptly labelled radar as H_2S, for Hydrogen Sulphide. When Lindemann visited again and wanted to know what H_2S was, they improvised the explanation "Height to Surface". Probably "Height to Surface" was a reference to the mapping of the ground from airborne radar from the clutter observation, which had been achieved by then. It is interesting to note that now both radar and infrared remote sensing have become two of the most important observational tools of the meteorologist.

1.1.4 Radar and the Atmosphere

If the War hastened the development of military radar it also meant that most research was classified and therefore known only to a few. The military users of radar found radar scopes cluttered with precipitation return. Radar return from long ranges due to anomalous propagation was also encountered. These meteorological effects no doubt stimulated further, the theoretical work on the scattering, absorption and propagation of microwaves in the lower atmosphere started in the earlier decade. They also encouraged experimental work and field observations in these areas. The first observation of precipitation by radar was probably made in Britain in 1941. The first major post-war symposium on the subject was organised in London in 1946 by the (British) Physical Society and the Royal Meteorological Society (see e.g. Booker 1948, Durst 1946, Hatcher and Sawyer 1947, Ryde 1946, Saxton and Lane 1946). The monumental work of Kerr (1951) on the "Propagation of Short Radio Waves" had been published soon after. The first account of radar observation of a tropical cyclone was published in 1945 (Maynard 1945). Meteorological use of radar was made by the US armed services in support of aviation during the war and radar was deployed immediately afterwards in the thunderstorm project (Byers and Braham 1949), the first of the major field experiments in meteorology.

1.1.5 Early Equipment

Early radar meteorological work was carried out with ground-based or airborne equipment designed primarily to detect military targets. The very word "target" has obvious military connotations and military use has been, over the years, the principal use of radar. Also of military origin is the coding of wave bands as "L" for 23 cm wavelength, "S" 10 cm, "C" 5 cm, "X" 3 cm, "K" 1 cm and so on. (A classification of radiofrequency bands of interest in radar meteorology is given in section 1.2) The military radars were not ideal for meteorological use but early radar meteorologists were quite innovative in adapting them to their needs. The radars were used for two purposes in meteorology – (1) for tracking balloons (and later, "chaff") to determine upper winds, and (2) for detection of precipitating cloud systems. For instance, the British "Baby Maggie" radars operating at a frequency of 204 MHz were in use for some years in India for radiosonde balloon tracking until replaced by more sophisticated devices. The early radars were mostly in the X band but finding that this wavelength suffers from heavy attenuation by rainfall, many switched over to longer wavelengths, 5 and 10 cm. The AN/CPS-9, 3-cm radar produced in the US around 1949 was probably the first radar designed specifically for meteorological use. Progress both in hardware and in techniques was rapid thereafter. In Britain, aviation and marine radars were modified for operational meteorological use in the early fifties. The first meteorological observation with Doppler radar was made in Britain in 1953 (Barratt and Browne 1953).

1.1.6 Work in other Countries

In other countries too, radar meteorology was taken up during or soon after the war. In Canada, a "Project Stormy Weather" was started in 1943 under the Army and later moved to McGill University. This was headed by Marshall who can be called the father of radar meteorology, if any one can be so described. With this group was associated some of the best-known pioneers in this field – Palmer, Hitschfeld and Gunn among others – and they were responsible for some of the most important studies and techniques – the Marshall-Palmer drop size distribution, the definition of reflectivity factor, melting band studies, time-lapse photography and the Constant Altitude Plan Position Indicator (CAPPI), to name a few.

In Japan, a committee on weather radar was formed in 1950 and by 1954 the first 3 cm weather radar made in Japan, was installed at the Meteorological Research Institute near Tokyo. Research on drop size distribution in different types of rainfall, estimation of precipitation, and tropical cyclone observation followed while a dense operational network was established. A remote-controlled weather radar – probably the world's highest – was set up on Mt. Fuji (3776 m above sea level) in 1965, primarily to detect typhoons. Digitisation and development of MTI (Moving Target Indicator) and Doppler facilities followed in the seventies and eighties. An interesting feature in Japan is the close collaboration between the Japan Meteorological Agency (JMA) and the radar manufacturers at every stage of design, field testing and operation.

In China, work was started in the fifties with imported radars. Around 1970 China started making its own radars and later built also dual wavelength and dual polarisation systems. China has 237 operational weather radars (Zou Jingmeng 1993). Doppler weather radars have also been installed. Principal applications have been in observation of tropical cyclones and mesoscale weather systems and in heavy rainfall forecasting and flood management.

The (erstwhile) Soviet Union carried out hardware development as well as fundamental work in cloud physics and precipitation estimation. Dual wavelength systems were successfully used for hail studies and operational hail suppression (see Kostarev et al. 1968). Other countries where notable work on radar meteorology was carried out from the early days, include Australia, France, Hong Kong, Israel, Switzerland and Taiwan. The RONSARD (Recherche sur les Orages et les Nuages par un Système A deux Radars Doppler) dual-Doppler radar system developed in France has been used successfully for thermodynamic and microphysical retrievals (see Chapter 6). An overview of the operational network in France may be seen in Cheze et al. (1991).

1.1.7 Early Reviews

Among reviews of early work on the subject may be mentioned those of Atlas (1964), Battan (1959, 1973), Kostarev et al. (1968), Ligda (1951), Marshall et al. (1955) and Wexler (1951). A very interesting review of the development – or "invention" as he calls it – has been given by Hitschfeld (1986).

1.1.8 The Indian Experience

In India, much information of meteorological interest was collected from military radars and aircraft measurements by the Royal Air Force in the war years. The most remarkable result was the observation of anomalous propagation over very long distances across the Bay of Bengal and the Arabian sea (Durst 1946, Hatcher and Sawyer 1947, also Chapters 2 and 4 of this book). The India Meteorological Department (IMD) acquired war disposal radars immediately after the war. These were used to gain experience. Qualitative observations of convective cells (e.g. Mathur 1958) as well as the melting band phenomenon in stratiform monsoon rainfall (Gupta et al. 1961) were made. Drop size distribution of rain was studied (Sivaramakrishnan 1961). Two operational networks, one of 10 cm wavelength primarily for detection of tropical cyclones and another of 3 cm wavelength, mainly for aviation weather observations have been set up. Manufacture of weather radars within the country started in the late sixties and currently some of the operational radars are of Indian manufacture. Doppler weather radars have been introduced in the operational network recently. The Indian Space Research Organisation (ISRO) has designed a Doppler Weather Radar for the IMD. A Mesosphere Stratosphere Troposphere (MST) radar has also been designed and operates as a research facility.

1.1.9 Progress in the Seventies and Later

From about 1975 rapid development has taken place in many parts of the world in automation of radar operation, digital processing of data, use of radar in conjunction with satellites and other devices for short period forecasting and nowcasting of severe weather, precipitation estimation, use of Doppler systems for determining velocity fields and turbulence and for thermodynamic and microphysical retrievals, use of airborne and ground-based systems for observation of tropical cyclones, cloud physics studies, support for weather modification efforts, development and application of dual polarisation radars, exploration of the capabilities of radars at various wavelengths, and in co-ordinated field experiments to study phenomena on various scales. In the last two decades, several countries in Europe have come together under the COST (CO-operation in Science and Technology) Programme to network their radars (Collier 1989b). They have thus achieved an almost continuous radar coverage of much of that continent, yielding data in real time for various applications. Space-borne radar for precipitation estimation has become a reality recently.

A recent and comprehensive review of Radar in Meteorology is that published by the American Meteorological Society (Atlas 1990).

1.1.10 Clear Air Radar

Echoes from the clear atmosphere were reported from VHF equipment intended to sound the ionosphere even in the thirties by, among others, Prof. S.K. Mitra's group at Calcutta University (Rakshit and Bhar 1936). Clear air phenomena were observed with microwave radars also subsequently, but it was not until the seventies that the use of "clear air" radars in the VHF and UHF bands for wind profiling as well as for study of phenomena not directly associated with precipitation, assumed prominence. MST (Mesosphere, Stratosphere, Troposphere) Radars operating in the VHF band have been established in a number of places around the world and are in use for meteorological and upper atmospheric research. Wind profilers based on the same principle have come into operational use in several European countries and in the USA. The capability of clear air radars to study precipitation has also been recognised and exploited in the last few years, in the research mode.

1.1.11 Concept of Total Observation

A very interesting feature of the history of meteorological radar is that it not only developed into a versatile tool but gave rise to new concepts and attitudes. Thus the term "mesoscale" was coined by Ligda (1951) to describe phenomena observable by weather radar. The term has come to stay in meteorology with a more basic meaning. Indian meteorologists found confirmation from radar observations that convective clouds in the monsoon season could overshoot the high summer tropopause far more often than was thought to be the case earlier (see Chapter 6). Findings of early Indian meteorologists on rainfall distribution in tropical cyclones have been confirmed by radar (Chapter 8). New thinking on convective and mesoscale cloud growth processes was stimulated by radar observations (e.g. Houze and Hobbs 1982). Or again, the concept of areally integrated measurement of rainfall as distinct from point measurement gained ground through the attempted application of radar to precipitation measurement. This concept has now found wider application in respect of several meteorological quantities in the context of satellite observation and numerical modelling. The concept of "total observation" or an "integrated observing system" using a variety of remote sensing and in situ observation techniques and use of the data for short period forecasting and nowcasting has its origin in the continuous observation made possible by radar.

1.2 Principles of Radar

1.2.1 Echo-sounding

Radar works on the echo-sounding principle. When a radio wave emitted by a transmitter meets any object, a part of the energy contained in the radiation is reflected, a part is transmitted through, a part is scattered in all directions and a part absorbed by the material of the object. The relative proportions of the energy disposed of in various ways depend on the properties of the object and its position and orientation relative to the incident radiation. A suitably placed radio receiver receives the reflected or scattered radiation. From the time of travel of the radiation and the direction from which it is received, the position of the object (referred to hereafter as the target[1]) is found. The measurement of the received energy and monitoring of various characteristics of the received radiation enable determination of several characteristics of the target. If the transmitter and receiver are located at two different points, the system is known as bistatic radar. The more common arrangement, known as monostatic radar, is one in which the transmitter and receiver are at the same site and usually share a common antenna. In this case the radiation reflected or scattered back towards the radar is received as an "echo".

1.2.2 Continuous Wave (CW) Radar

If the radar transmits a continuous sine wave with no modulation, the echo received can be distinguished from the transmitted radiation because of the phase

[1] As mentioned in section 1.1, the word target is of military origin. In meteorology, the object of interest (usually an ensemble of hydrometeors, or a region of varying refractive index) may be more appropriately called a scatterer.

shift, which occurs. If the distance to the target is r, the number of wavelengths contained in the two-way path is,

$$\frac{2r}{\lambda} = \frac{2rf}{c} \tag{1.1}$$

where λ is the operating wavelength, f is the corresponding frequency of the radar and c is the velocity of light. The corresponding phase shift ϕ is

$$\phi = 2\pi \frac{2rf}{c} = \frac{4\pi rf}{c} \tag{1.2}$$

Hence r can be determined as in Fig. 1.1. This will give an unambiguous range only if the phase shift is less than 2π radians i.e. range is less than half a wavelength. In the case illustrated in the figure, the value is 2000 metres. If the range exceeds this figure. an integral multiple of half a wavelength has to be added to the computed range. This ambiguity is acceptable if the range is a continuously varying quantity e.g. the range of a radiosonde balloon. The frequency of 75 kHz as in the figure is too low for use as a radar frequency. It is used as a modulation on a VHF transmitting frequency and is returned as a phase-shifted signal by a transponder in the radiosonde. This is an example of a "secondary" radar i.e. one in which the target co-operates by giving a response signal on receipt of the radiation from the radar.

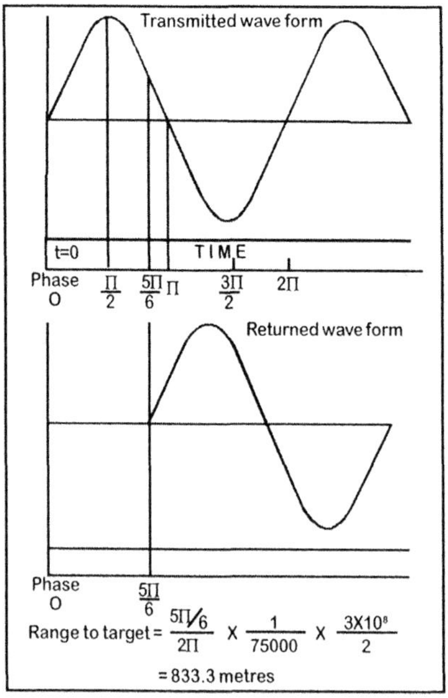

Fig. 1.1. Range finding by continuous wave of frequency 75 kHz.

1.2.3 Doppler Shift

If, in the above example, the target is moving with a radial (i.e. away from or towards the radar) component of velocity v, r and ϕ change continuously. A change in ϕ with respect to time is equal to a frequency. This is the "Doppler"[2] angular frequency given by,

$$\omega = 2\pi F = \frac{\mathrm{d}\phi}{\mathrm{d}t} = \left(\frac{4\pi}{\lambda}\frac{\mathrm{d}r}{\mathrm{d}t}\right) = \frac{4\pi v}{\lambda} \tag{1.3}$$

where F = Doppler frequency shift, or,

$$F = \frac{2v}{\lambda} = \frac{2vf}{c} \tag{1.4}$$

and

$$v = \frac{Fc}{2f} = \frac{F\lambda}{2} \tag{1.5}$$

E.g. for $\lambda = 10$ cm, the Doppler shift for a velocity of 10 m s^{-1} is 200 Hz. By measuring the small Doppler shift in frequency the radial component of the velocity of the target can be determined. The shift can also be used to distinguish moving targets from overlapping stationary echoes such as ground or sea clutter. This is equally true of pulsed radar and as we shall see later, the use of this technique has added a new dimension to weather radar capability in recent years.

1.2.4 Pulsed Radar

CW radar at microwave frequencies will obviously have an unacceptable range ambiguity for most applications. Transmission of radiation in short duration pulses with relatively long intervals of silence enables the echo from one pulse to be received before the next is transmitted. This is illustrated in Fig. 1.2. If the time of receipt of the echo measured from the time of transmission is t, the range of the target is:

$$r = \frac{ct}{2} \tag{1.6}$$

The duration τ of the transmission is known as the pulse duration and is expressed in microseconds. It corresponds to a pulse width h expressed in metres,

---

[2] Doppler with a capital letter because the "Doppler effect" is named after its discoverer in 1842, the Austrian physicist Christian Doppler (1803 – 1853). See also Balsley (1998).

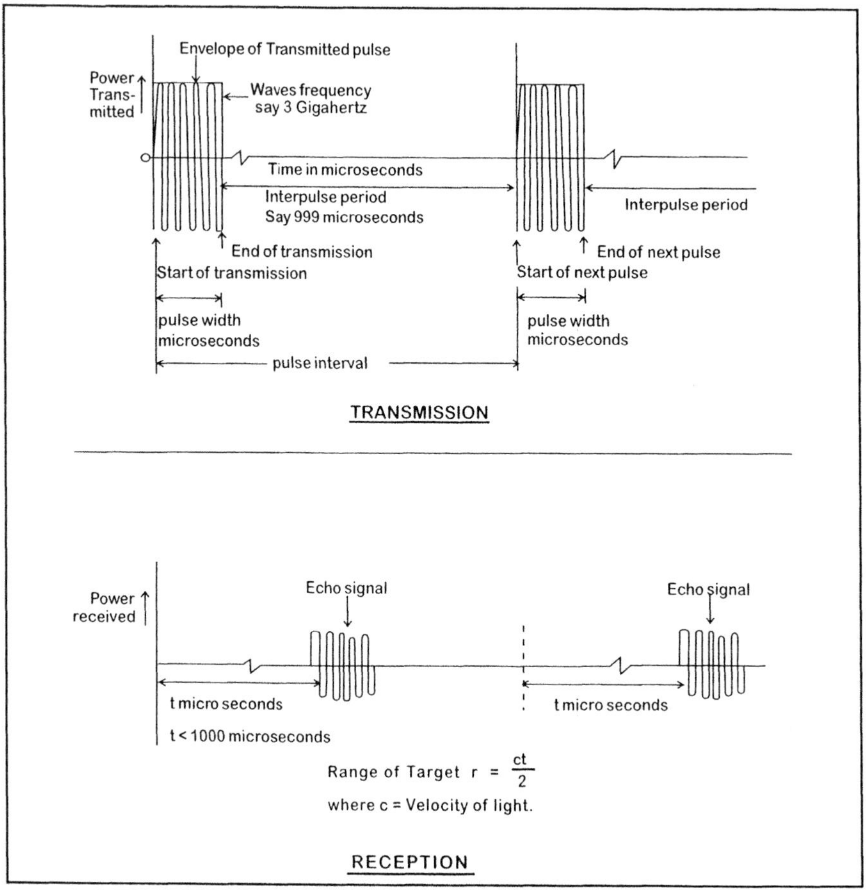

Fig. 1.2. Principle of pulsed radar.

being the distance travelled during one pulse duration. The number of pulses transmitted per second is known as pulse repetition frequency (p.r.f.) and is the reciprocal of the time interval T_s between pulses. In Fig. 1.2 the interval is 1 millisecond and the p.r.f. is 1000. Therefore echoes from all targets within 150 km will arrive before the subsequent pulse is transmitted. An echo from a more distant target will arrive after one or more subsequent pulses are transmitted. It is called a multiple-time-around echo or a multiple-trip echo. It will be displayed at a range equal to its true range minus a multiple of 150 km.

In general, maximum unambiguous range is

$$r_{max} = \frac{c}{2(\text{p.r.f.})} \tag{1.7}$$

1.2.5 Maximum Unambiguous Velocity

The echo from a moving target will be Doppler-shifted as already seen. The maximum Doppler shift F (called the Nyquist frequency) which can be detected is:

$$F_{max} = \frac{\text{p.r.f.}}{2} \qquad (1.8)$$

Therefore the maximum target velocity, which can be unambiguously determined, is:

$$v_{max} = \frac{(\text{p.r.f.}).\lambda}{4} = \frac{(\text{p.r.f.}).c}{4f} \qquad (1.9)$$

It will be seen from equations (1.7) and (1.9) that the maxima of unambiguous range and unambiguous velocity are related as follows:

$$v_{max} = \frac{c^2}{8fr_{max}} = \frac{\lambda.c}{8r_{max}} \qquad (1.10)$$

Hence to maximise r_{max}, p.r.f. has to be low and to maximise v_{max}, a high p.r.f. is needed. If both are to be measured unambiguously, a compromise is necessary. Also note that a higher unambiguous velocity can be obtained by using a longer wavelength.

1.2.6 Signal Reception and Display

The echo received is detected and amplified by the radar receiver and can be fed as a voltage to the vertical deflection plates of a cathode ray tube (CRT). The horizontal sweep of the CRT beam is started at the instant of transmission of each pulse and ended before the next pulse and is repeated at the p.r.f. Thus the horizontal axis of the display represents time and hence also range while the vertical axis represents signal amplitude. Such a display (Fig. 1.3) is called an "A-scope" and has been widely used in weather radar observation, calibration and maintenance. Echoes from the same target due to successive pulses will be superimposed, augmenting the brightness of the display. The top of the echo signal pulse is shown fluctuating in the figure, as is normally the case with echoes from precipitation particles. The weak fluctuating background represents the "noise" at the output of the receiver and is known as "grass" from its appearance on a green display. Any echo to be detected has to be stronger in the mean than the noise and should be "discernible" amidst the noise. This defines a quantity known as "Minimum Discernible Signal" or "Minimum Detectable Signal" (MDS) which is a measure of the receiver sensitivity.

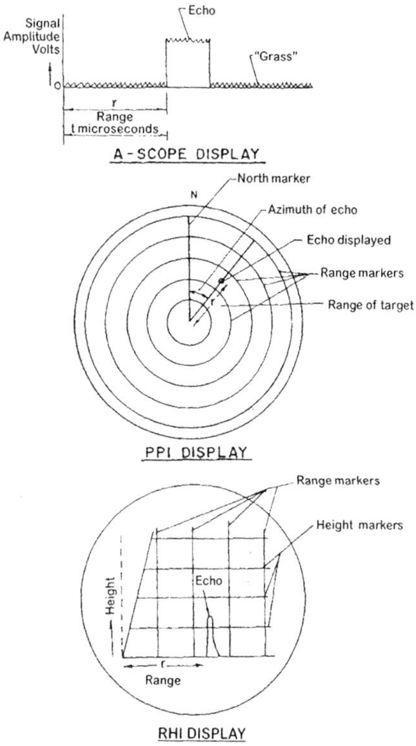

Fig. 1.3. Common weather radar displays.

The above arrangement will give the range of the target but two other co-ordinates are needed to determine its position. The normal way of getting these is to transmit the radiation in a very narrow beam and rotate the antenna around a vertical axis so that the horizon all round is scanned. On a CRT display, the signal is made to modulate the intensity of the beam (Z-modulation). The spot is swept from centre of the tube face to the periphery. Thus range is represented by radial distance from the centre. The trace is rotated in step with the antenna. Therefore the azimuth to which the antenna points, can be read off the scope. This display (Fig. 1.3), called a Plan Position Indicator (PPI) provides a quasi-horizontal map of the area of coverage of the radar with target positions shown correctly. While scanning, the radiation can be directed horizontally or at an angle to the horizontal. Because of the earth's curvature, the beam, even if directed horizontally, will be at increasing heights above the surface as range increases. Hence the display is described as quasi-horizontal. The reference azimuth (0 degree) is usually the true North represented as pointing above the centre. Range markers appear in the form of circles. Instead of a 360 degrees scan, any chosen azimuth sector can be scanned and displayed. Having determined the target position, its movement can be determined by plotting its position at successive instants of time. Similarly the antenna can be rotated around a horizontal axis in a sector, say, from 0 to 90 degrees elevation, keeping the azimuth constant. In this case the horizontal axis of the display depicts range while the elevation is represented by the angle the sweep trace makes with the horizontal axis. This display is called a Range Elevation Indicator (REI). The scope is intensity-modulated as in the case of the PPI. Since the height of the target can be computed knowing the range and elevation, it is convenient to have a Cartesian coordinate display with range on the horizontal axis and height as the vertical axis. The height of the target being small in comparison with the range, the height scale is usually magnified. This display, called Range Height Indicator (RHI) is illustrated in Fig. 1.3. Typical photographs of PPI and RHI scopes exhibiting echoes of objects at the earth's surface are shown in Figs. 1.4 and 1.5.

Other types of displays have also been used in weather radar. The R-scope is a variation of the A-scope in which the sweep starts at some range other than zero enabling magnified representation of a chosen small range interval. If the antenna is pointed vertically up, the A- or R-scope reads the height of the echo

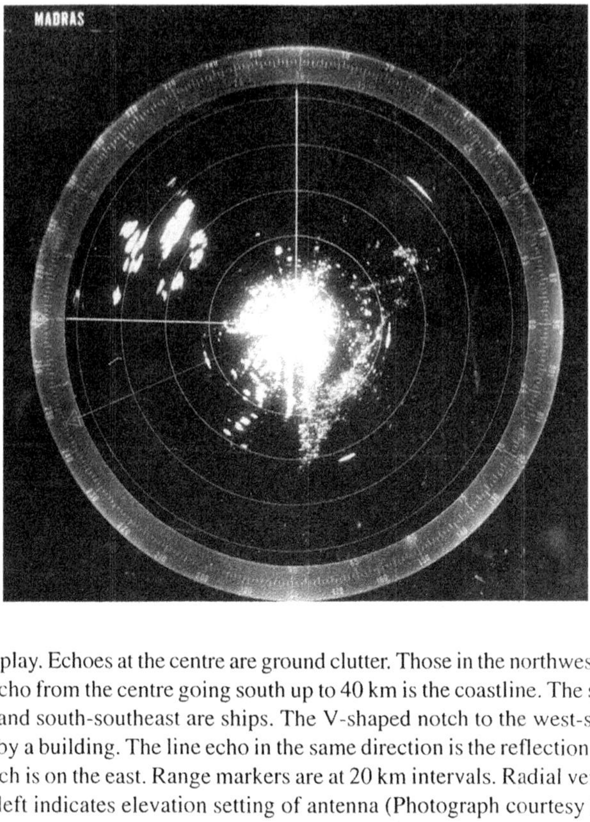

Fig. 1.4. PPI display. Echoes at the centre are ground clutter. Those in the northwest sector are from hills. The line echo from the centre going south up to 40 km is the coastline. The short lines in the northeast, east and south-southeast are ships. The V-shaped notch to the west-southwest is due to shadow cast by a building. The line echo in the same direction is the reflection by the building, of the ship, which is on the east. Range markers are at 20 km intervals. Radial vertical line points north. Line on left indicates elevation setting of antenna (Photograph courtesy IMD).

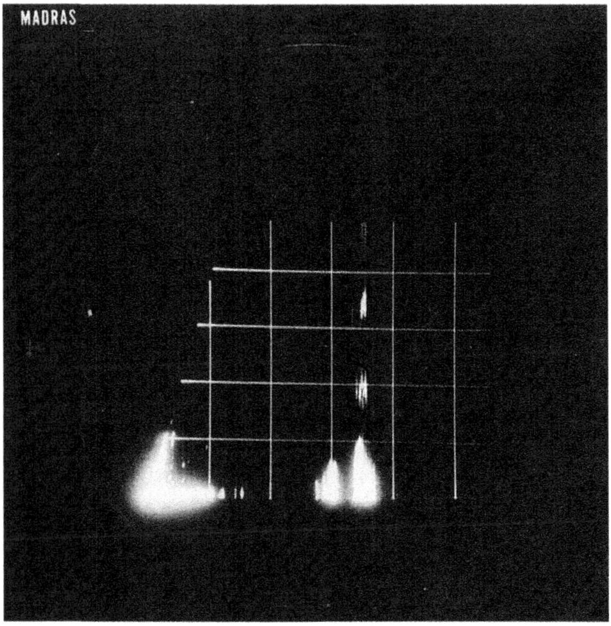

Fig. 1.5. RHI display. Echo of a hill at a range of 70 km, showing effect of finite beam width and side lobes. Range and height markers are at intervals of 20 and 5 km respectively (Photograph courtesy IMD).

directly on the X-axis. The B-scope is an intensity-modulated rectangular coordinate display with azimuth on one axis and range on the other. A helical scan, also called a "volume scan", is often used. The antenna is rotated in the horizontal plane as in a PPI-scan but the elevation is stepped up by a small increment (1 or 2 degrees) at the end of each rotation. A series of PPI's result and an annular area in each corresponds to a particular altitude. By combining annuli from the various PPI's relating to one altitude, a "Constant Altitude PPI" (CAPPI) can be constructed. This was done initially by a photographic process by Langleben and Gaherty (1957), later electronically by East (1958). Now with computerised equipment this is carried out through software. With digitising equipment it is also now routine to display the processed signal on a Cartesian coordinate TV type scope with a map superimposed. Echo intensity, reflectivity factor, velocity or derived products such as rain rates of various magnitudes can be displayed in a grey-scale or false colour representation, a particular grey shade or colour being assigned to a particular range of values of the quantity displayed. Any of the displays can be repeated at a remote location. For this purpose, digital data are transmitted by dedicated communication links or through the Internet.

1.2.7 Power Received by the Radar

Consider the radar antenna to be an isotropic point radiator radiating equally in all directions, a power of P_t watts. At a distance r, the power is spread over the surface area $4\pi r^2$ of a sphere. Power incident on unit area (power density) is

$$\frac{P_t}{4\pi r^2} \tag{1.11}$$

Since a radar antenna is directional it transmits a power in a certain sector, which is G times the power that an isotropic point radiator with the same input would have transmitted in that direction. G is called the gain of the antenna. Being a ratio, it is a dimensionless quantity. Hence, the power density at range r is

$$S_i = \frac{P_t.G}{4\pi r^2} \tag{1.12}$$

Let a target at range r present an effective area σ to the radiation and let this area be small in comparison with the area occupied by the radar beam. It intercepts a power σS_i, which, if radiated isotropically, would produce a power density at the receiving antenna equal to that actually received from the scatterer. The value of σ satisfying this condition is called the radar cross section of the target. Power reradiated towards the radar is

$$\frac{P_t.G.\sigma}{(4\pi r^2)} \tag{1.13}$$

The reradiated power will spread over the surface of a sphere of radius r. Assuming a monostatic radar with the receiver sharing the antenna with the transmitter, power density at the radar receiver is

$$\frac{P_t.G.\sigma}{(4\pi r^2)^2} \tag{1.14}$$

If the effective aperture area of the radar antenna is A, power P_r received by the radar is

$$P_r = \frac{P_t.G.A.\sigma}{(4\pi r^2)^2} \tag{1.15}$$

From antenna theory it can be shown that (e.g. Kraus 1950),

$$A = \frac{G\lambda^2}{4\pi} \tag{1.16}$$

Therefore,

$$P_r = \frac{P_t G^2 \lambda^2 \sigma}{(4\pi)^3 r^4} \tag{1.17}$$

When P_r equals the Minimum Detectable Signal S_{min} of the receiver, it represents the maximum range limit r_{max} of the equipment.

$$r_{max}^4 = \frac{P_t G^2 \sigma \lambda^2}{(4\pi)^3 S_{min}} \tag{1.18}$$

The last two equations show that for a target small in relation to the area of the beam, the received echo intensity is inversely proportional to the fourth power of the range. Hence, to obtain a measurable echo from any appreciable distance, P_t has to be very high.

1.2.8 Back-scattering Cross Section

Since we have assumed a monostatic system, the scattering back to the radar only is relevant. Hence the radar cross section σ may be called the back-scattering cross section. σ is a quantity with the dimensions of an area but is not equal to the physical area of the target. It is the area intercepting that quantity of power, which, if scattered equally in all directions will produce an echo signal at the radar equal to that from the target. σ depends not only on the surface area of the target illuminated by the radiation but also on the shape, the orientation of the target (with reference to the radar beam), the dielectric constant of the material as well as the wavelength of the radiation. A further treatment will follow in Chapter 2, but we may state here that the back-scattered energy is inversely proportional to the fourth power of the wavelength if the size of the target is small in comparison with the wavelength. This is known as Rayleigh scattering (Rayleigh 1871). When the target size becomes comparable to the wavelength, the wavelength dependence becomes more complicated (Mie scattering; Mie 1908; see Chapter 2). The case of a spherical target (which is the simplest to deal with) of radius a, is illustrated in Fig. 2.5, Chapter 2. At the commonly used radar wavelengths of 3 to 10 cm, which are much larger than the dimensions of most, but not all, precipitation particles, the Rayleigh law is usually assumed to be applicable. But this is not true at shorter wavelengths.

1.2.9 Ensemble of Scatterers

The meteorological target, however, is not a single raindrop. Normally the radar beam, at any instant, illuminates a volume containing a large number of water or ice particles. Each particle scatters back some energy and therefore has a certain back-scattering cross section. If we assume that the scattering is incoherent, i.e. the particles are randomly placed and the phases of the echoes from individual scatterers are distributed over an interval 2π, the total back-scattering cross section is the sum of the individual cross sections. A "back-scattering cross section per unit volume" is thus defined. It is also called "radar reflectivity", represented by the symbol η and expressed in $cm^2\ m^{-3}$. This quantity multiplied by the volume illuminated by the radiation will give the radar cross section of the assembly of precipitation particles. It can be shown (see Chapter 2) that, for the Rayleigh case, the back-scattering cross section of a single spherical drop of diameter D is

$$\sigma = \left(\frac{\pi^5}{\lambda^4}\right) \cdot \left|\frac{m^2 - 1}{m^2 + 2}\right|^2 .D^6 \tag{1.19}$$

where m is the complex refractive index of the substance (water or ice in our case) (see Chapter 2 for details). If the scattering volume contains drops of various sizes, the sixth powers of the diameters of the individual drops will have to be summed up over a unit volume to obtain η. For the Rayleigh case,

$$\eta = \left(\frac{\pi^5}{\lambda^4}\right) \cdot \left|\frac{m^2 - 1}{m^2 + 2}\right|^2 \cdot \sum_{\text{unit vol.}} D^6 \tag{1.20}$$

(For a more general expression for η, see Doviak and Zrnić 1993).

The sum of D^6 over unit volume is an important quantity in radar meteorology and is known as "radar reflectivity factor" represented by the symbol Z. It is not to be confused with the radar reflectivity η. Unlike η, Z is exclusively a property of the assembly of precipitation particles and is not dependent on radar wavelength. Z is expressed in the unit $mm^6\ m^{-3}$, admittedly a rather clumsy unit.

1.2.10 Radar Equation for Hydrometeors

The radar antenna may consist of an array of individual antenna dipole elements or more often a paraboloid dish over which radiation is directed from a waveguide horn (see section 1.3 for a brief treatment of antenna principles and definitions). The radiation from the antenna extends over a finite angular width on either side of the direction of maximum radiation. The angle between the two directions on either side of the maximum where the power density is half of that in the direction of the maximum is called the beam width of the antenna. If the beam widths in the horizontal[3] and vertical planes respectively are ϕ and θ the beam at range r spreads over an area equal to,

$$\pi.\left(\frac{r\phi}{2}\right).\left(\frac{r\theta}{2}\right) \tag{1.21}$$

neglecting any radiation transmitted outside the beam width. At any instant, the travelling pulse from the radar will occupy a finite range interval r to $r+h$, h being the pulse width. The back-scattered power from the particles at a range $r+h/2$ from the front end of the outgoing pulse will arrive at the radar at the same time as the power scattered by particles at range r from the rear end of the pulse. Hence the effective depth in the radial direction from which echo is received by the radar at any one instant is $h/2$. Thus the volume V sampled by the radar at any instant is,

$$V=\pi.\left(\frac{r\phi}{2}\right).\left(\frac{r\theta}{2}\right).\left(\frac{h}{2}\right) \tag{1.22}$$

This is illustrated in Fig. 1.6.

The total radar cross section of the precipitation particles in this volume is,

$$\sigma=\eta.V \tag{1.23}$$

Substituting in equation (1.17) we get,

$$P_r=\frac{P_t h G^2\lambda^2\phi\theta\eta}{512\pi^2 r^2} \tag{1.24}$$

Now from equation (1.16),

$$G=\frac{4\pi A}{\lambda^2} \tag{1.25}$$

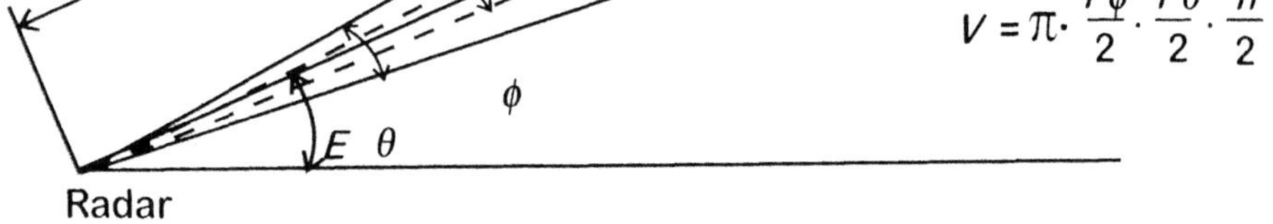

Fig. 1.6. Volume illuminated by a radar beam. E= beam elevation angle, θ= elevation beam width, ϕ = azimuth beam width, P, Q, R, S, are half-power points. PQ = $r\phi$; RS = $r\theta$.

[3]The terms horizontal and vertical are strictly correct only if the direction of propagation is horizontal.

and by definition of G, ϕ and θ, if the power is assumed to be uniformly distributed over the beam width,

$$\phi.\theta = \frac{16}{G} \tag{1.26}$$

Hence (1.24) can also be rewritten as,

$$P_r = \frac{P_t h A \eta}{8\pi r^2} \tag{1.27}$$

the form used by Marshall et al. (1955).

From equations (1.24) and (1.27) it will be seen that, for a target consisting of precipitation particles, the received echo varies inversely as the square and not the fourth power of the range. This is because the target is taken to fill the entire beam volume. If this is not the case, the P_r will be less than the value yielded by these equations. Even if the beam is filled, the above equations were found to give too high an estimate of P_r. Ad hoc corrections to the equation were proposed until Probert-Jones (1962) found that the error was due to the assumptions regarding relationships between G, A, ϕ and θ being in error. He gave a more accurate form of the radar equation for a meteorological target for the case of a paraboloid aerial of circular cross section. His equation is

$$P_r = \frac{P_t h G^2 \lambda^2 \phi \theta \eta}{512(2.\ln 2)\pi^2 r^2} \tag{1.28}$$

(ln represents the natural logarithm).

If now we substitute for η from equation (1.20) we get,

$$P_r = \frac{P_t h G^2 \phi \theta . \pi^3 . \left| \frac{m^2 - 1}{m^2 + 2} \right|^2 . Z}{512(2.\ln 2)\lambda^2 r^2} \tag{1.29}$$

Hence knowing the attributes of the radar and the range of the target, the radar reflectivity factor Z can be uniquely determined from a measurement of P_r, subject to the assumptions made above.

1.2.11 Large or Non-spherical Scatterers

Equation (1.29) is valid only for spherical raindrops small in comparison to the wavelength. Large drops will not scatter according to the Rayleigh law nor will they be spherical. Solid precipitation particles will have complex shapes and their scattering cross sections will depend on their orientation. The determination of Z

from P_r will not give information on the size distribution of drops unless independent measurements of the drop sizes are made. Hence Z cannot be uniquely related to the quantities such as rainfall rate (R mm h^{-1}) or liquid water content (M g m^{-3}) which normally interest the meteorologist. These quantities can be derived through empirical relationships between Z, R and M (see Chapters 2 and 7).

1.2.12 Attenuation

The radar equation also assumes that the propagation of the radiation is unaffected by the intervening medium. In the space between the radar and the target the radiation is attenuated by the atmospheric gases including water vapour and by liquid water or ice. The attenuation by atmospheric gases in the wavelength range 3 to 10 cm usually used for weather radar is small and can be corrected for. Attenuation by aerosols or dust or cloud droplets is neglected at these wavelengths. At shorter wavelengths attenuation by all these agents becomes more significant. The attenuation by liquid or solid precipitation is difficult to correct, as it requires a knowledge of the precipitation rate along the path as well as its spatial distribution. At a wavelength of 10 cm, attenuation by rainfall is small and is usually neglected. At 5 cm it can still be neglected at low rainfall rates. These wave bands are generally used if quantitative measurements with radar are required in tropical and temperate regions respectively. A detailed discussion of attenuation is given in Chapter 2.

1.2.13 Refraction

The other important effect of the intervening medium is refraction. The refractive index of air at microwave wavelengths in the lower atmosphere departs from unity by about 300×10^{-6}. It varies with temperature and moisture content and to some extent with pressure. Hence the radiation from a radar at the surface of the earth is usually refracted towards the earth or sometimes away from it depending on the profile of temperature and humidity. The change in radar range due to the longer path length can be neglected for meteorological purposes. The effect that matters is an error in the elevation angle (and therefore in derived height) of the target, which has to be corrected. Refraction can also cause targets not normally visible, to be picked up. This is known as anomalous propagation. Targets beyond the unambiguous range may also be displayed as multiple-trip echoes at nearer ranges and with distorted shapes or sizes (Fig. 1.7). In case of refraction away from the earth (sub-refraction) the effective range of the radar

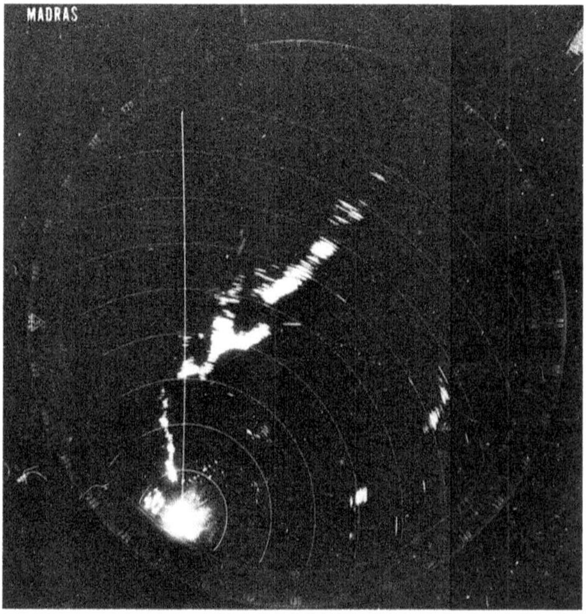

Fig. 1.7. Echoes due to anomalous propagation. This is an off-centred PPI picture seeing east and northeast from Chennai (formerly Madras) (Photograph courtesy IMD). Range rings are at 100 km intervals. The echoes in the north and north-northeast are from the Indian coast (Andhra Pradesh and Orissa). The echoes in the east-northeast at about 600 km range are distorted second-trip echoes of the coast of Myanmar (Burma). To get the correct range of these echoes, add 940 km.

may be reduced. The radiation can also be partially reflected or scattered at refractive index discontinuities and give rise to echoes, which can be used to detect certain meteorological phenomena. Propagation phenomena are discussed in detail in later chapters.

1.2.14 Polarisation

Radiation from a radar can have one of several polarisations. Polarisation in a single linear direction - horizontal or vertical - is normally employed. A spherical drop will return the echo with the same polarisation. Circularly polarised radiation will be returned with opposite circular polarisation and will not be detected. This property is used in non-meteorological radars to minimise clutter echo due to rainfall. Precipitation particles, which are not spherical, will, however, give echoes with a cross-polarised component. Hence if waves with horizontal and vertical polarisation (or opposite circular polarisation) are transmitted alternately and signals arriving with these polarisations are measured, the ratio of the two reflectivity factors will yield information on the departure from spherical shape of the particle. Such dual polarisation radars and their applications are dealt with in later chapters.

1.2.15 The Doppler Spectrum

In paragraph 1.2.9 it was mentioned that the radar cross section of the sampled volume is the sum of the cross sections of the individual precipitation particles. The individual scatterers are, however, moving with respect to each other and at the same time the ensemble may be moving as a whole due to larger scale air motions. The back-scattered signal thus contains information on the velocities of the individual particles in the form of a spectrum of Doppler shift frequencies. The total power P_r received by the radar is the sum of the individual contributions of particles of all velocities taking account of the phase of the scattered radiation from each particle. This is also called the zeroth moment of the Doppler power spectrum. To get a representative value of the back-scattered power it is necessary to average the fluctuating signal over a long enough period of time. The symbol P_r in the previous equations is usually replaced by $\overline{P}_r$ to indicate the averaging.

1.2.16 Doppler Mean Velocity and Spectrum Width

While P_r can be monitored by any radar, it requires additional equipment to monitor power as a function of Doppler shift. The shift is very small and therefore it is necessary to use a radar transmitter with a high degree of frequency stability and to compare the transmitted and received frequencies. With such a "Doppler radar", the mean Doppler velocity (i.e. velocity component of the sampled volume along the line of sight), also called the first moment of the Doppler power spectrum can be determined. The scatter of individual velocities around the mean or the spread of the Doppler power spectrum is a measure of the random motions of the particles or in other words the turbulence in the sampled volume. The spread is expressed in terms of the variance (second moment) or its square root the "spectrum width" of the Doppler spectrum. Thus a Doppler radar can give not only the reflectivity of the precipitation volume but information on the radial velocity and turbulence. The three-dimensional velocity field can be obtained by observing the same target by two or more Doppler radars placed some distance from each other or inferred with a single radar by techniques which will be described in later chapters.

1.2.17 Typical Magnitude of P_r

It will be instructive to look at the typical magnitudes of the various quantities in equation (1.29). P_t will typically be several hundred kilowatts while the minimum detectable signal may be of the order of 10^{-14} watt. Because of the large range of magnitudes involved it is convenient to employ the logarithmic unit, the decibel (dB). Power expressed in decibels is ten times the common logarithm of the ratio between the actual power and a reference value. Taking 1 mW as the reference level, P_t of 1 MW becomes 90 dBm i.e. 90 decibels above 1 mW and 10^{-14}W is equal to −110 dBm. Similarly Z, which may have a range of values from 1 to about 10^7 is expressed in dB above a value of unity. G being a ratio is also expressed in dB.

Let us put $\phi = \theta$, express θ in degrees in equation (1.29), put the equation in logarithmic form and multiply all terms by 10:

$$10\log P_r = 10\left\{\begin{array}{l}\log P_t + \log h + 2\log G - 2\log\lambda + 2\log(\theta_{\deg}\times\frac{\pi}{180}) + \log\left|\dfrac{m^2-1}{m^2+2}\right|^2 \\ +3\log\pi - 10\log 2 - \log(\ln 2) - 2\log r + \log Z\end{array}\right\} \quad (1.30)$$

Let us take radar parameters as, $P_t = 500$ kW = 87 dBm; $h = 1$ μs = 300 m; $10\log G = 38$ dB; $\lambda = 10$ cm = 0.1 m and $\theta = 2$ deg.

$\left|\dfrac{m^2-1}{m^2+2}\right|^2$ for water is 0.93.

Substituting the above values in equation (1.30),

$$\begin{aligned} P_r \text{ in dBm} &= 87 + 24.77 + 76 + 20 - 29.14 - 0.32 + 14.91 - 30.1 + 1.59 \\ &\quad - 20\log r + (\text{dBZ}) - 180 \end{aligned} \quad (1.31)$$

(The last term 180 is inserted because Z should be in $\text{mm}^6\ \text{m}^{-3}$). Therefore,

$$P_r = -15.29 - 20\log r + (\text{dBZ}) \quad (1.32)$$

If a standard relationship, say, $Z = 200R^{1.6}$ (see Chapter 2) is assumed, R being rainfall rate in mm h^{-1}, a rain rate of 1 mm h^{-1} will correspond to 23 dBZ. At, say, a range of $r = 100$ km = 10^5 m this will give $P_r = -92.29$ dBm. In practice it is likely to be lower because of some attenuation due to atmospheric gases in the path and head losses in the radar (between the antenna and the receiver input). These can be taken account of by adding terms in equation (1.32) to represent these losses. At shorter wavelengths, the attenuation due to intervening rainfall may seriously affect the radar performance. The calculations above are for the return from a single pulse. In practice the use of longer dwell times and the averaging of signals will improve the sensitivity of the radar. Doppler signal processing will also effect improvement. Hardware used for signal processing may in some cases weaken performance. These aspects will be discussed in Chapters 2, 3 and 5.

1.2.18 Choice of Pulse Width and p.r.f.

As seen from equation (1.22), the sampling volume at any instant is a function of pulse length and beam width. Two or more targets separated in the radial direction by a distance less than $h/2$ cannot be seen separately. Hence the range resolution of the radar is $h/2$. Decreasing h will improve range resolution but reduce the power received as can be seen from equation (1.29). Considering the dimensions of most meteorological targets, pulse durations of up to 5 μs are used in weather radars. Often a choice of two or more pulse widths with different repetition rates is provided. The shorter pulse provides higher resolution and because of the higher p.r.f. it is possible to average a larger number of samples for measurement of P_r. The use of two p.r.f.'s, one of which is not a multiple of the other, enables identification of multiple-trip echoes as they will appear at different ranges at the two p.r.f.'s. Higher p.r.f. also provides a higher maximum unambiguous Doppler velocity.

1.2.19 Effect of Beam Width

From equation (1.22) it can also be seen that the angular resolution is limited by the beam width. Hence a point target will appear on the PPI as an arc of length $r\phi$ where r is the range and ϕ the azimuth beam width. Thus the resolution in the tangential direction deteriorates with increasing range. A ship at a range of 100 km may be shown as a few km long (Fig. 1.4). If there are two targets separated by less than the resolution limit they will be seen as one. If, now, the targets move closer to the radar, the resolution improves and the one target may appear to split into two. Similarly there could be an apparent merger of two receding targets. The apparent enlargement of the tangential dimension means that in RHI presentations the echo will appear to extend to a greater height than it really does (In Fig. 1.5, the hill at a range of 70 km is not 5 km in height as it seems to be). To correct for this effect, half the beam width is subtracted from the observed elevation angle of the top of the echo before calculating its height. Similarly for evaluation of the height of the base of the echo, half a beam width may be added.

1.2.20 Sidelobes

The presence of sidelobes in the antenna pattern (see section 1.3) means that targets in the sidelobe directions may give echoes, which will be displayed in the main lobe direction. The most common effect is the appearance of a "pillar" extending to a great height on top of the echo of a convective cloud on the RHI. In Fig. 1.5 the echoes at a range of 70 km and heights of 10 and 17 km are the first and second sidelobe echoes of the hill below.

1.2.21 Choice of Wavelength

As mentioned earlier, most ground-based operational weather radars for observation of precipitation, operate at wavelengths between 3 and 10 cm. At longer wavelengths the echo power is much lower. Longer wavelengths also require larger antennas to get a specified beam width. This means greater weight and bulk and higher cost. At shorter wavelengths, attenuation of radiation increases, rendering quantitative measurements difficult. The radar beam cannot penetrate extensive areas of rainfall. Nevertheless, since the echo return is greater and resolution is high, very short wavelengths can be used to detect cloud drops at short ranges. Thus they can be used for cloud physics studies and for cloud height measurements. Alternatively the attenuation experienced by the radar beam in travelling through the rainfall can be used as a measure of the rate of rainfall (see Chapter 7). This method can be used in satellite-borne radar where various limitations require very short wavelengths to be used. For observation of clear air phenomena by detection of radio-refractive index discontinuities, longer wavelengths corresponding to UHF and VHF frequencies are more suitable (see Chapter 10). Table 1.1 lists the frequencies of interest for various meteorological radar applications.

1.2.22 Dual Wavelength Radar

A radar operating at two different wavelengths can be used to observe the same target synchronously (Sulakvelidze and Dadali 1968, Heymsfield and Jameson 1981). For instance, if 10 and 3 cm radiation is used simultaneously, raindrops, which are usually less than 5 mm in diameter will scatter according to the Rayleigh law at both wavelengths. But in the case of hail the diameter is comparable to the 3 cm wavelength and Mie scattering will occur. At 10 cm Rayleigh scattering occurs. Hence the reflectivity factor will be different in the two cases. The ratio of the two Z-values can be regarded as a "hail signal" enabling detection of hail (see Chapter 6). However it is necessary that the radar samples the same volume at both

Table 1.1. Frequencies of interest for various meteorological radar applications.

Band designation	Frequency range	Range of wavelengths	Main applications
VHF	30–300 MHz	10–1 m	Observation of clear air phenomena in the troposphere and stratosphere, wind-profiling,
UHF	300–1000 MHz	1–0.3 m	turbulence, refractive index structure
L	1–2 GHz	30–15 cm	Clear air and precipitation phenomena
S	2–4 GHz	15–7.5 cm	Precipitation measurement, tropical cyclone observation, local severe storms, radio wave propagation
C	5–7 GHz	6–4.5 cm	Local severe storms, precipitation measurement, tropical cyclone observation, radio wave propagation, use on aircraft
X	9–11 GHz	3.3–2.7 cm	Thunderstorm and gust front detection, radio wave propagation, use on aircraft
Ku	12–18 GHz	2.5–1.7 cm	Cloud physics, ceilometers, air- and space-borne radar, Synthetic Aperture Radar (SAR)
Ka	27–40 GHz	1.1–0.75 cm	for sea surface studies, Precipitation measurement from attenuation, tornado observation
millimetre	40–300 GHz	7.5–1 mm	Ceilometers, cloud microphysics and
W	94 GHz	3.2 mm	dynamics, tornado observation
F	140 GHz	2.14 mm	
G	220 GHz	1.30 mm	

wavelengths. The two beam widths and beam patterns of the radar must be identical, which is not easy to achieve. Dual wavelength radar can also be used to measure precipitation. The attenuation at the shorter wavelength and the reflectivity at the longer (non-attenuated) wavelength can be used to determine precipitation intensity without making assumptions about drop-size distribution (see Chapter 7).

1.3 Weather Radar Technology

1.3.1 Essentials of a Radar System

Fig. 1.8 is a block diagram showing the essential constituents of a simple microwave radar system. The block shown as "operating console and display"

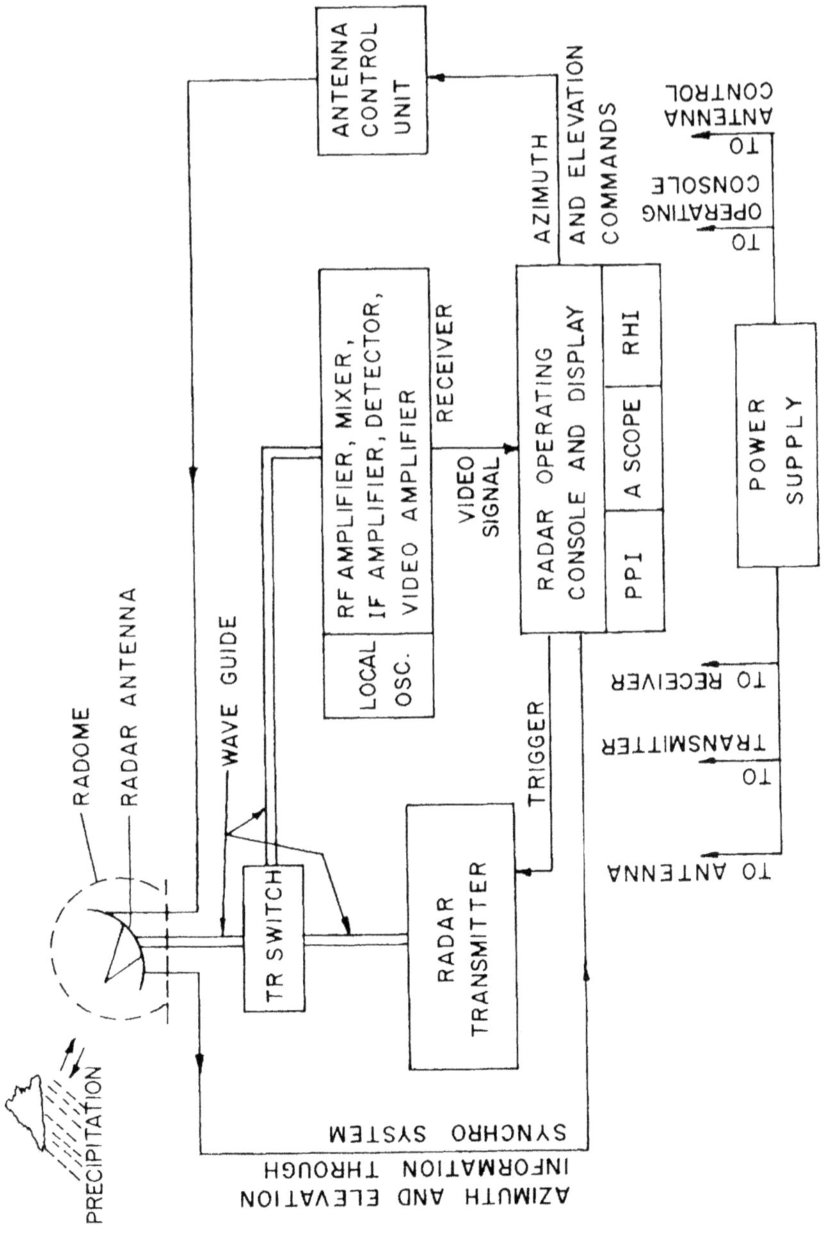

Fig. 1.8. Weather radar block diagram.

contains the controls by which the radar is operated. A series of pulses at the chosen p.r.f., is generated and distributed to all the units to synchronise their operation. These pulses trigger the radiation from the transmitter and, at the same time, start the sweep on each of the displays. In the transmitter, on receipt of the trigger, a high voltage negative pulse is applied for the chosen duration to a magnetron tube which generates radio frequency oscillations. The radiation passes through a waveguide to the antenna. In the waveguide is interposed a T-R switch which is a gas discharge tube fired by the transmitted pulse or it may be a ferrite device. This blocks the path to the radar receiver and thus prevents the high power radiation from damaging the sensitive receiver circuits. The far end of the waveguide is flared into a horn, which is placed at the focus of a parabolic reflector. The reflector may be a paraboloid of revolution or a section of a cylindrical parabola. The radiation from the horn is directed by the reflector into a nearly parallel beam radiated along the axis of the paraboloid. The antenna assembly is mounted such that it can be rotated about a horizontal or vertical axis by drive motors controlled from the operating console through the antenna control unit. Information on the instantaneous position of the antenna is communicated by a synchro system (shaft encoders or fibre-optic systems may be used in modern radars) to the display where the PPI or REI sweep is rotated in synchronism with the antenna. In the case of the RHI, the elevation angle information is used to generate an appropriate vertical sweep.

1.3.2 Signal Reception

The signal, which returns from the target, is received by the same antenna and goes through the same waveguide but by now the T-R switch has recovered from its discharge. The recovery time may typically be a few microseconds, which implies that targets within a few hundred metres of the radar will not be detected. The T-R switch routes the signal to the receiver. The signal is amplified by an RF amplifier and fed to a set of crystal mixers, which mix it with the output of a local oscillator. The latter may be a klystron tube or a solid state device tuned to a frequency slightly different from the magnetron frequency. The difference signal which results, is the Intermediate Frequency (IF), typically 30 or 70 MHz , which is subjected to several stages of amplification. The signal is then detected and the video signal fed to the vertical deflection plates of the A-scope or to the cathode of the PPI or RHI scope. Or the signal is converted into a digital form and fed to a monitor.

1.3.3 Antenna

Any antenna can be considered to be made up of a number of point radiators [see Kraus (1950) for a detailed treatment of antenna theory]. If, for instance, an alternating current is flowing in a short straight wire, the magnitude and phase of the current will vary along the wire. The electric and magnetic fields at a distant point will be composed of the fields due to each radiator. Depending on the path length from each radiator the fields will reinforce in some directions and interfere destructively in others. The ratio of the electric field in any angular direction ϕ in the plane containing the wire, to its maximum value can be plotted as a function of ϕ. The relative field in any direction is given by the radius vector in that direction, of the figure so plotted. Since power is proportional to the square of the field, the relative power pattern can also be plotted. This is also called radiation intensity pattern or radiation pattern, radiation intensity being defined as power incident over unit solid angle. In a plane perpendicular to the plane of the antenna, similar figures can be plotted. (In the case of the wire, the radiation is uniform in all directions in the perpendicular plane and the field pattern will be a circle). The three dimensional pattern can also be visualised.. It should be noted that the field pattern discussed, applies only to the far field i.e. at distances from the antenna large in comparison with the wavelength. The far field is taken to start at the distance $2D^2/\lambda$ from the antenna where D is the dimension of the antenna and λ the wavelength. The same pattern applies when the antenna is used for reception of signals.

Similarly, if there are several such antennas placed along a line or in the form of a square or circular array and the magnitudes and phases of current in each antenna are different, the field at any point will be the resultant of the individual fields. The radiation from a horn or reflector can also be considered to be made up of contributions from various points on the antenna, which may have different magnitudes or phases. A typical radiation pattern of a paraboloid dish in one plane is shown in Fig. 1.9. Fig. 1.9(a) is plotted in polar co-ordinates and Fig. 1.9(b) is in Cartesian coordinates. The relative radiation intensity in any direction is given by the radius vector OA (or OB etc.) in Fig. 1.9(a). A similar figure will result in an orthogonal plane. The three-dimensional pattern, of which the two plane figures are sections can be visualised by rotating Fig. 1.9(a) around the axis ON. As we go away from the direction of maximum radiation (ON) the intensity decreases, reaches a minimum, but again increases exhibiting subsidiary maxima. Each of the oval figures or their three-dimensional counterparts are called lobes. Thus there are several secondary lobes besides the main or primary lobe containing the direction of maximum radiation. The points P and Q in the main lobe are called half power points as the power density at these points is half

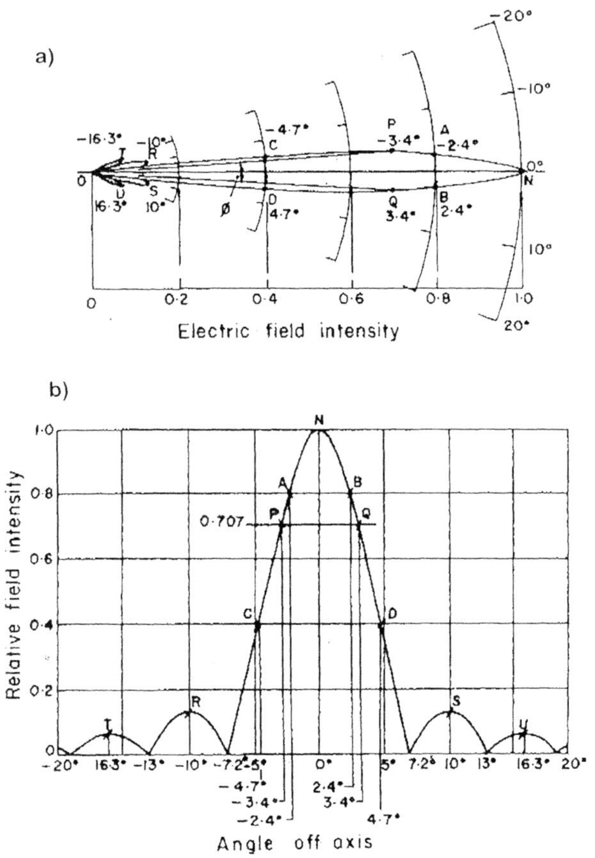

Fig. 1.9. Radiation pattern of a circular dish (a) in polar co-ordinates (b) Cartesian coordinates.

that in the direction ON. The angle POQ is known as the beam width. Most of the power is radiated within this angle.

1.3.4 Beam Width and Sidelobes

It is obviously desirable to keep the beam width as small as possible. The power radiated outside the beam width can also produce echoes from objects in those directions. These echoes will appear to come from the direction of the main lobe and are therefore spurious. The beam width and the relative amplitude of the sidelobes can be reduced by increasing the size of the antenna relative to the wavelength. To some extent these can be varied by changing the relative illumination of various points of the reflector dish but this is likely to improve one of these factors at the expense of the other. The sidelobes cannot be completely eliminated. In a typical ground-based weather radar the sidelobe power may be 25 dB or more below the power in the main lobe and the beam width may be 1 to 2 degrees. The beam width of a circular dish antenna of diameter D is approximately,

$$\phi = \frac{1.22\lambda}{D} \text{ radians} = \frac{70\lambda}{D} \text{ degrees} \tag{1.33}$$

Thus the antenna size needed to get a specified beam width is proportional to the wavelength. The cost and weight of the antenna and of the supporting structures and drive systems needed, go up considerably with size and are a major consideration in choice of wavelength and beam width. In airborne systems, because of space limitations it is usual to choose a wavelength of 3 or 5 cm and a beam width of 3 degrees or more.

1.3.5 Antenna Gain

The gain (G) of an antenna is defined as the ratio of the maximum radiation intensity from the antenna to the radiation intensity from a lossless isotropic point radiator with the same power input. When the antenna is exposed to a radiation field it extracts the power over an effective area known as its aperture A which is not necessarily equal to its geometrical area. As we have seen in section 1.2,

$$G = \frac{4\pi A}{\lambda^2} \tag{1.34}$$

In replacement of equation (1.26) the relationship,

$$G^2 = \frac{\pi^2 k^2}{\theta\phi} \tag{1.35}$$

was substituted by Probert-Jones (1962), where k is a factor to take account of the fraction of the total power from the horn intercepted by the reflector and the degree of uniformity of illumination of the reflector. It was taken as unity in the derivation of Probert-Jones' equation (1.28).

1.3.6 Electronic Scanning

Instead of a rotating dish antenna it is possible to use an array of dipoles producing a narrow beam perpendicular to the plane of the array. The direction of the beam can be altered by changing the relative phase of the current fed to the various dipoles, avoiding the need for physical movement of the antenna. Such arrays are used for VHF and UHF radars for clear air observations and may be considered for weather radars to implement faster scans (see Chapters 10 and 11).

1.3.7 Transmitter

The transmitting device in a conventional (non-Doppler) pulsed radar is usually a magnetron which is a thermionic vacuum diode, to the cathode of which is fed the high voltage pulse. The anode consists of a re-entrant cavity made of a copper block and surrounds the cathode. The cavity consists of holes and slots, which act as resonant circuits. The device is placed in the field of a strong permanent magnet, the magnetic field being perpendicular to the electric field. The tube produces microwave oscillations for the duration of the voltage applied, at a frequency determined by the resonant frequency of the cavities. The output is usually fed by a co-axial line to the waveguide. Because of the high power the tube usually requires air-cooling. Because of the high voltage, arcing can occur inside the tube. Hence a new magnetron is normally subjected to an ageing process by gradually increasing the high voltage applied. The peak power radiated, P_t, increases with the high voltage. Since the oscillation occurs only during the pulse duration, the mean power P_m dissipated is given by,

$$P_m = (\text{pulse duration} \times \text{p.r.f.}) \times P_t = D_u \times P_t \qquad (1.36)$$

D_u being called the duty cycle of the magnetron. P_t here is that measured at the magnetron output. What is radiated by the antenna may be less as there will be some loss between the magnetron and the antenna

Early magnetrons were prone to frequent failure and instability in frequency of operation. They could operate in several modes giving unwanted frequencies. The magnetron usually fails more often than most other components of the radar. It is also one of the most expensive components of the radar. Most magnetrons operate at some random frequency within the assigned band and the frequency is not stable and many tubes cannot be tuned. Hence to keep the IF in the receiver constant, the local oscillator is tuned to the frequency of the magnetron ± the IF Some magnetrons can be tuned but the frequency is not stable. As high frequency stability is essential for Doppler radars, coaxial magnetrons with arrangement to phase-lock the receiver for every pulse have been used in recent years. However the favoured arrangement for Doppler radars is a klystron power amplifier fed by a master oscillator (Master Oscillator Power Amplifier, MOPA). Klystrons capable of delivering high peak power, of the order of 1 MW, have been developed in recent years. These need cooling by circulating water or oil.

Another type of transmitting tube which has been used in radars is the travelling wave tube (TWT) amplifier.

1.3.8 Noise Figure

Because of the very weak signals received, the performance of the receiver is limited not by the amplification that can be achieved but the noise figure of the receiving system. Noise is unwanted electromagnetic radiation, part of which comes from external sources through the antenna and part is produced in the receiver circuits. The latter is mainly due to the random thermal motion of conducting electrons. In order to detect and amplify weak signals, the receiver-produced noise has to be kept a minimum by proper design.

$$\text{The thermal noise power} = kTB \tag{1.37}$$

where T is the temperature of the receiver in degrees Kelvin (usually taken as 290 K), B is its bandwidth in Hz, and ,

$$k = \text{Boltzmann's constant} = 1.38 \times 10^{-23}\ \text{J K}^{-1} \tag{1.38}$$

B is limited by the bandwidth of the IF amplifier chain of the receiver (the RF portion has usually a wide bandwidth) which has to be wide enough to pass all the frequency components of the pulse signal, but the wider the bandwidth the greater the noise power. The shorter the transmitted pulse the greater will be the bandwidth needed. Taking a typical value $B = 1$ MHz the thermal noise will be 4×10^{-15} W. If the signal and noise at the input of the receiver are S_i and N_i and the corresponding quantities at the output are S_o and N_o, the noise figure N_F of the receiver may be defined as,

$$N_F = \frac{S_i / N_i}{S_o / N_o} \tag{1.39}$$

S/N is the signal-to-noise ratio and N_F is a measure of the degradation of this ratio as the signal passes through the receiver. S/N and N_F being ratios are expressed in dB. Typically the noise figure of a modern radar receiver will be less than 2 dB.

1.3.9 RF Amplifier and Mixer

Early radars were not using RF amplifiers as the amplification was offset by the additional noise produced. Modern radars use solid state RF amplifiers which have low noise figures. The RF amplifier is a wideband device to accommodate variations in transmitter frequency. This is followed by a balanced mixer consisting of two crystals. The balanced arrangement ensures a low noise figure. The mixer mixes the signal with the output of a local oscillator which is a klystron tube or a solid state device, the output of which is coupled to the mixer cavity. The difference frequency is the IF. To keep it constant despite fluctuations of the magnetron frequency, an Automatic Frequency Control (AFC) circuit is incorporated.

1.3.10 IF and Video Amplifiers

The IF is amplified by a series of amplifiers before being applied to a detector diode to give a video output. To provide a uniform response over the required bandwidth, the successive IF stages may be stagger-tuned i.e. tuned alternately to slightly different frequencies on either side of the centre of the band. The bandwidth should be wide enough to pass the Fourier components of the signal but should be limited to keep the noise low. This is achieved by a "matched filter"[4]. If the transmitter has a choice of pulse widths, the receiver IF bandwidth will also be appropriately switched according to the pulse width chosen. Since the echo signal may range from about 10^{-14} W to say, 10^{-5} W the receiver has to operate over a range of signals of about 90 dB. This range is called the dynamic range of the receiver. It is difficult to produce truly linear amplification over this range without the stages getting saturated. A linear amplification is carried out only if we can choose some limited range of values of the signal[5]. The output then obtained after detection is called linear video. For handling the entire range of signals, a logarithmic video output is obtained. This can be done by having a detector after every IF stage and combining all the detector outputs. This output is proportional to the logarithm of the input over the entire range except at the top and bottom of the response curve. In Fig. 1.10 the video output is proportional to the logarithm of the input signal between –100 and –35 dBm. Beyond –35 dBm the receiver saturates. The dynamic range of the receiver in this case is 100 – 35 = 65 dB. See Chapters 3 and 5 for further discussion of the input signal vs. video output relationship.

---

[4] The IF amplifier having a frequency response which maximises the peak signal to the mean noise power ratio is termed a matched filter (Skolnik, 1980). The normalised amplitude spectrum of the matched filter determined by the IF bandpass charateristic is the same as the amplitude spectrum of the input signal. The phase spectrum of the matched filter is the negative of the phase spectrum of the input signal plus a phase shift proportional to frequency.

[5] See Chapter 3 for methods of achieving linear amplification over the entire dynamic range.

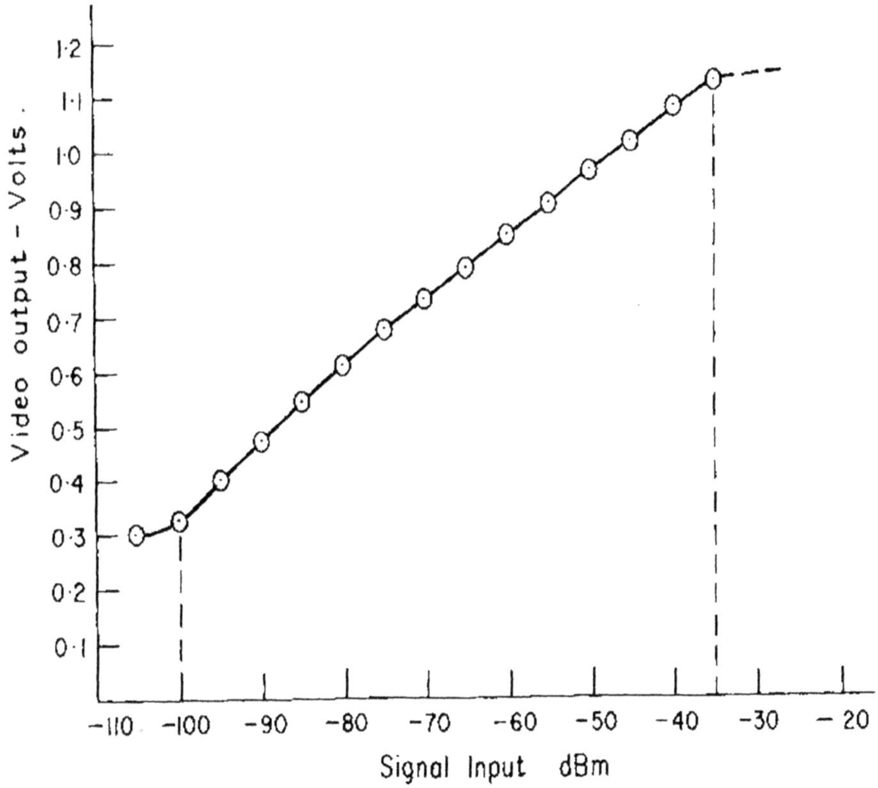

Fig. 1.10. Log video output of a radar receiver as a function of signal input.

1.3.11 Gain Controls and Attenuators

To avoid saturation or to suppress strong echoes or to enable determination of echo intensity or to distinguish weak echoes or again to compensate for attenuation due to range or other causes, it is necessary to vary the gain of the radar receiver. Manual controls for this purpose can be provided. These vary the amplification of the IF or the video stages. These controls can be calibrated in terms of the number of dB of attenuation they produce. It may also be necessary to vary the gain for echoes at different ranges differently. For this purpose a suitable waveform is applied to the IF or video amplifiers. It is also possible to provide variable RF attenuators for these purposes ahead of the mixer. Automatic Gain Control (AGC) is also used for obtaining a wide dynamic range for the linear receiver channel. These are discussed in Chapter 3. In modern radars these functions are all computer-controlled.

1.3.12 Signal Averaging

It was mentioned earlier that a number of successive pulses have to be averaged to get a more representative value of the signal intensity. In a conventional analogue display, the returns from a number of pulses are superposed on the PPI and the CRT screen phosphoresces for sometime after the trace passes any particular point. Hence a certain amount of averaging occurs on the screen. The number of pulses integrated depends on the dwell time on any individual range-azimuth bin, which in turn depends on the antenna scan rate. Intensity estimation was attempted in the early years by photographing the screen with a long time exposure and measuring the density of the image. Averaging of the video signal could also be done by passing it through a Resistance-Capacitance (RC) filter. Depending on the time constant of the filter a certain number of pulses will be integrated. If a logarithmic signal is averaged a correction is necessary because the logarithm of the average is not equal to the average of the logarithms. Signal averaging is further discussed in Chapter 3.

1.3.13 Digitisation

In modern radar systems the video signal is digitised. The raw information along with the antenna position information also converted to digital form can be stored on a computer disk or tape and processed either in real time or off-line in a variety of ways. Digital signal processing is dealt with in Chapter 3. The signal can be averaged as desired and displayed on a Cartesian coordinate display with distances from radar on the two axes and intensity gradations in grey scale or in various colours. Such a display uses a computer monitor (black and white or colour) which unlike the conventional CRT can be viewed in a brightly lit room. It is also cheaper than a dedicated PPI type display. Digitised data lend themselves to easy transmission over long distances by telephone lines or radio links to enable viewing at any other place. The computer used for digital processing can also be used for control of operation of the radar either from the radar site or from a remote position.

1.3.14 Antenna Control

The azimuth and elevation positions of the antenna are sensed by synchros or digital shaft encoders which transmit the information to the operating console. The trace of the PPI or other display is rotated to align with the antenna position. The operator can also set any other position to which it is desired to point the antenna. A servo system in the antenna control unit senses the difference between the desired position and the actual position and generates an error voltage, which is used to drive the motors such that the error voltage is reduced to zero i.e. the antenna reaches the desired position. A continuous rotation, a sector scan or a helical scan can also be set up. The last consists of a stepping up of elevation after each revolution in azimuth.

1.3.15 Power Supply

The other block shown in Fig. 1.8 is the power supply. Mains A.C. power is converted to the required voltages for various units. As the power supply voltage and frequency may fluctuate at some locations or power may completely fail, additional equipment is needed. An automatic voltage regulator is always provided. The provision of a diesel engine generator is necessary for operational radars as power may fail or be unstable in bad weather when the radar is most needed. The generator may be manually started when necessary or it may come on automatically when power fails. In either case there will be a short interruption of power supply which is usually of no consequence, if an uninterrupted power supply is provided for the computers and data storage devices which may lose data due to power interruption.

1.3.16 Cameras and Archival Devices

In conventional radars, cameras (manually operated or linked to the radar operation so that the shutter opens when a scan starts and closes at the end of the scan) were the best means of archival of radar data. Facilities to display the date and time and any other legend to be added were provided. A map of the area of coverage could be displayed electronically on the scope. As the CRT has a period of afterglow, the faint afterglow would appear in the picture if successive pictures are taken quickly after changing display settings. If all PPI pictures in a sequence were taken at the same radar settings the processed film could be run as a "time lapse movie" to get a fast motion view of the development of any weather system. A Polaroid camera could be used to get pictures immediately. A positive film roll or a videotape formed a convenient way of archival of radar data in the absence of digital storage devices. In modern digital radars the raw or processed data can be stored on disk or tape and recalled or reprocessed at any time, the only limitation being the storage space for the immense amount of data generated. Colour printers can yield hard copy and the need for a camera is eliminated.

1.3.17 Overlays

Transparent overlays with a map or a grid printed on them have been used for a long time. These are superposed over the scope for tracing the echoes in pencil. To avoid parallax errors due to the scope face being in a different plane from the glass surface of the CRT a "reflection plotter" could be used. In modern systems the map or grid is electronically produced and can be stored or printed along with the data.

1.3.18 Communication Devices

A variety of communication devices has been used for remote control of the radar and to transmit the radar data. Examples of the latter are a modem to transmit digital data over telephone lines or a radio link, a microwave link to transmit the analogue data to a remote display or a facsimile transmitter to send annotated sketches of the radar scope over telephone lines (or a radio link). Where none of these existed, a teleprinter was used to send a manually coded message. With digital systems, raw or processed data can be transmitted in real time through terrestrial or satellite communication links to any distance. Data are being disseminated on the Internet. But of crucial importance is the need to ensure functioning of the "last mile" communication link from the radar site to the nearest communication terminal in all weather.

1.3.19 House-keeping

Other auxiliary devices relate to house-keeping. Test and measuring equipment for monitoring radar performance, for calibration and for maintenance are provided (see Chapter 5). Some of them may be built into the radar (BITE – Built-In Test Equipment) and may monitor some parameters automatically. Periodical recheck of some of these devices with reference to standard instruments may be necessary. A device is usually provided to pressurise the waveguide with dry air to prevent arcing while high power is transmitted. Since very high voltages are used in the transmitter and display, these units are shielded and provided interlocking switches which disable the unit when it is opened. Shielding against X-Rays may also be needed for the high-power klystrons. Air-conditioning is almost always provided for indoor equipment to optimise performance and reduce failures. Some particular devices may need constant temperature enclosures for correct performance. All units of the radar need to be securely earthed. Lightning protection is also needed. "Obstruction lights" may be necessary above the antenna for warning aircraft.

1.3.20 Radar Configuration

Early radars had the transmitter and receiver (at least the RF part of it) located with the antenna as it was difficult to transfer the RF signal between a stationary transmitter or receiver and a rotating antenna. The video signal and control voltages would be transmitted through slip rings in the supporting pedestal of the antenna. IF signal could be transmitted through a transformer one winding of which would rotate inside the other. These arrangements were unsatisfactory as it was difficult to protect the electronics of the exposed units from extreme temperatures, dust and rainfall. It also rendered maintenance difficult. With the development of low loss rotary waveguide joints it is possible to transmit the RF between the antenna and indoor units situated several metres away. The joint consists of a section of cylindrical waveguide rotating co-axially with its stationary counterpart maintaining the field configuration. Alternatively the antenna and transmitter-receiver (or at least the RF part of the receiver) can be collocated if a radome (see Para 1.3.21) is provided. A Low Noise Amplifier (LNA) forming the first stage of the receiver, alone, is located at the antenna in some radars.

1.3.21 Radome

An antenna dish is made of a continuous metal sheet or a perforated sheet to reduce weight. It is also possible to make the dish of fibreglass and sputter it with a metal coating. The antenna and support structures have necessarily to be at an exposed site like open ground or the top of a building or tower, to get unrestricted exposure. This makes all its components vulnerable to the weather. The drive motors and other electrical and electronic devices can be put in an enclosed space in the support structure. Precautions have to be taken to provide suitable enclosures, gaskets and netting to prevent entry of water, dust, insects or rodents. Maintenance, during bad weather, of units situated outdoors becomes difficult. Above all, in severe weather, the wind can be strong enough to stop the movement of the antenna or even to damage it. Fig.1.11 is a photograph of a PPI showing the stoppage of the antenna by a severe squall.

To overcome these problems, a radome is used to enclose the antenna completely. This is usually a major segment of a sphere made of fibreglass panels

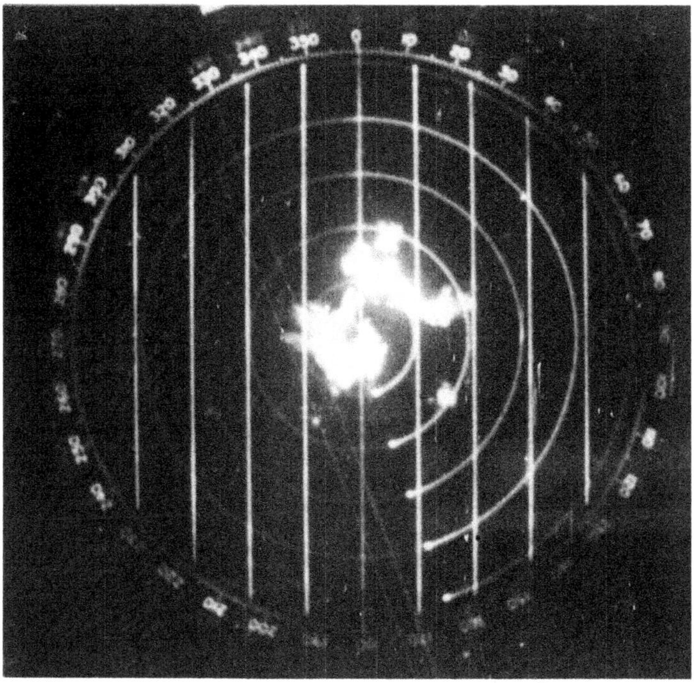

Fig. 1.11. PPI showing stoppage of antenna scan by a squall. (From Soundararajan and Raghavan 1962, courtesy IMD).

and polyester resin. The material of the radome offers very little attenuation to microwaves. When its surface gets coated with rain water there may be some additional attenuation (see Chapter 2). This can be corrected for in the radar equation. Thin sheets of fibreglass may be embedded in metal frames or thick sandwiches of non-metallic material enclosed in fibreglass sheets can be used. Sometimes an inflatable radome made of thin fabric is used and the air pressure inside maintained by a pump. The design and geometry ensure that the boresight i.e. the electromagnetic axis of radiation is not shifted and no diffraction pattern is created. Joints between panels are caulked and surface maintained clean to reduce attenuation. In cold climates, arrangements for preventing accumulation of snow may be necessary. Ventilation may be needed in hot climates. Entry to the radome may be by a door on the side or preferably from below to prevent wind or rain blowing in while opening. The radome is subject to the lifting force of the wind as well as a tilting moment. Hence its base has to be grouted securely to the structure below. The radome can withstand winds up to 250 or 300 km h^{-1} and enable operation and maintenance of the radar in any weather. Even so, a weather radar antenna was toppled by hurricane Andrew (1992) at Miami, USA (Wakimoto and Black 1994).

1.3.22 Health Hazard

Exposure to microwave radiation is injurious to health. Microwave components used indoors are usually well shielded but going in front of the waveguide mouth or the antenna is hazardous. Hence, if people are likely to have access to the surroundings of the antenna, warning signs are to be displayed. It is assumed that no occupied buildings taller than the antenna are in the vicinity.

1.3.23 Pulsed Doppler radar

Fig. 1.12 shows a simplified block diagram of a pulsed Doppler radar. As the small Doppler shift F is to be detected by comparing the received with the transmitted frequency a high stability of transmitting frequency is needed. This is achieved usually by using a klystron amplifier as transmitting tube. A STAble Local Oscillator called a STALO generates a frequency f_L which is mixed with a stable reference frequency equal to the IF f_I, produced by another stable oscillator called a COHO. The resulting frequency ,

$$f = f_L + f_I \tag{1.40}$$

drives the transmitter. The received signal, which has a frequency, $f + F$ (F being the Doppler shift) is mixed with the STALO output to yield the IF signal $f_I + F$.

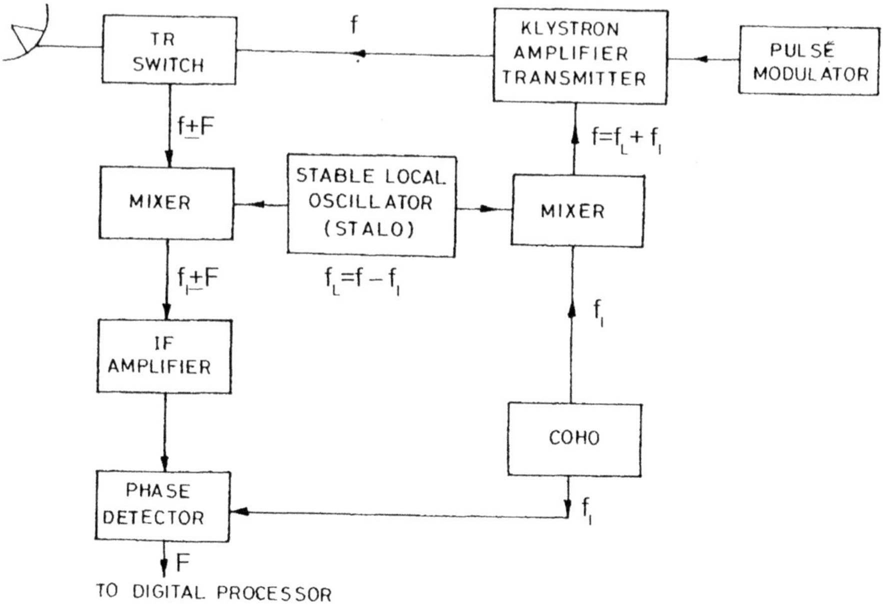

Fig. 1.12. Simplified block diagram of a pulsed Doppler radar.

This is amplified and fed to a phase detector which compares its phase with that of the COHO output. Since the Doppler frequency is represented by the rate of change of phase of the echo signal [see equation (1.3)], it is available at the output of the phase detector. Since the various particles contributing to the signal have different velocities, a spectrum of Doppler frequencies is available. Two outputs in phase quadrature are obtained. These are called the in-phase component I and the quadrature component Q of the signal. They are digitised and passed to the Doppler signal processor (see Chapter 3) for extracting the moments of the Doppler spectrum.

The Doppler radar has usually two displays one to exhibit the reflectivity factor distribution and another to display simultaneously the distribution of the mean radial velocity, both in PPI or RHI formats. Reflectivity factor and velocity in various magnitude ranges are displayed in a colour code (see Chapter 3). Since computer monitors are used for the purpose they can also display the spectrum width as well as various derived parameters.

1.3.24 Typical Doppler Radars

Tables 1.2, 1.3 and 1.4 give the basic specifications of some operational Doppler radar systems. Some of these are discussed in Chapter 3. Several research radars have been designed to incorporate dual wavelength and dual polarisation capabilities also. These are known as multi-parameter radars (see Chapter 3).

1.3.25 Platforms

The platform on which the radar is based may be a fixed site on land, a vehicle which can be moved on land, a ship, aircraft or satellite. Fixed land-based radars can use heavy and elaborate equipment but mobile units have weight and space constraints. Ships have low power radars (usually 3 cm wavelength) for navigation which can also detect meteorological phenomena at short ranges. Research ships may have weather radars. These may need a stabilised platform and also get the input of the ship's heading for determination of absolute direction. Instead of a stabilised platform, a means of electronically correcting the display for the ship movements can be used. Commercial aircraft use weather radar for the limited purpose of detecting hazardous weather systems ahead of the craft. These are usually 3 cm radars with scan in the forward sector (say 120 degrees in azimuth) at a negative angle of elevation. The small antenna is housed in a radome integrated with the profile of the aircraft.

Table 1.2. Characteristics of the WSR-88D Doppler Radar (NEXRAD) of the USA.

Range	Reflectivity 460 km	Velocity 230 km
Transmitter	Klystron	
Frequency	2.7 to 3.0 GHz	
Peak power	750 kW	
Pulse width/p.r.f.	1.57 μs / 318–1304 p.p.s. 4.7 μs / 318–452 p.p.s.	
Radome	fibre glass skin, foam sandwich	
Diameter	39 ft;	
R.F. Loss one way	0.3 dB	
Antenna	centre-feed paraboloid dish	
Diameter	28 ft (8.5 m),	
Beam width	0.95 deg pencil	
Gain	45.5 dB	
Polarisation	Linear Horizontal	
Side lobes (with radome)	25 dB down	
Pointing accuracy	0.2 deg	
Coverage	Azimuth 360 deg; Elevation −1 to + 45 deg	
Rotation rate (max)	36 deg. per sec.	
Receiver channels	Linear output I/Q; Log output	
Dynamic range	95 dB	
Bandwidth (3 dB)	0.79 MHz	
Clutter suppression	30 to 50 dB (Infinite Impulse Response Design)	
Velocity	Complex Covariance Argument	
Calculation algorithm	Pulse Pair Processing	
Pulses averaged	40 to 200	
Spectrum width calculation	Autocorrelation	
Algorithm	Single Lag Correlation	
Pulses averaged	40 to 200	
Reflectivity algorithm	Linear Power Average	
Pulses averaged	6 to 64	
Resolution	Spatial	Magnitude
Reflectivity	1.00 km	0.5 dBZ
Velocity	0.25 km	0.25 m s^{-1}
Spectrum width	0.25 km	0.25 m s^{-1}

The US Terminal Doppler Weather Radar (TDWR) used for aviation meteorological application (Chapter 9), is similar to NEXRAD but operates at 5 cm wavelength and has a lower maximum range.

Table 1.3. Specification of Indian Space Research Organisation (ISRO) Doppler Weather Radar System (courtesy ISRO).

System performance specifications	Reflectivity	Velocity & spectrum width
Range km	500	250
Resolution m	300	300
Accuracy @ 23 dBZ & 2 rpm	±2 dB	±1 m s^{-1}

Operating frequency	2.7 to 2.9 GHz. (at a chosen spot frequency)	Receiver	Coherent
Radome	Foam sandwich rigid, spherical	Dynamic range	80 dB Linear channel with IAGC 95 dB Log. channel
RF loss	0.2 dB one way (dry radome)	Noise Figure	3 dB.
Maximum wind	250 km per hour (kmph) gusting to 300 kmph	I.F	30 MHz
Antenna type	Prime focus parabolic dish	Signal Processor	Floating Point DSP
Diameter	9 m nominal	ADC	14 bits
f/D ratio	0.35	Range bins	Selectable up to 2048
Beam width	0.95 degree @ 2.8 GHz.	Range Resolution	Compatible with ADC sampling rate; 300 m default
Feed	Scalar with OMT (OrthoMode Transducer)	Clutter Suppression	30 dB and 50 dB IIR filter
Polarisation	Linear Horizontal (default) and Linear Vertical	DSP Algorithms	
Steerability	360 deg. Az., -2 to +90 deg. in El.	Reflectivity	Averaging
Rotation rate	3 rpm	Velocity/Spectrum width	Pulse Pair; FFT (optional)
Pointing Accuracy	±0.1 deg.	Computer System Configuration	Distributed Architecture UNIX workstations & PC's (Windows-NT) on LAN
Transmitter	Coherent, Klystron based	Graphics	OpenGL & Motif
Peak Power	1.0 MW	Compilers	C, C++, VC++
Pulse width	1 μs; 2 μs (selectable)	Display	21" Colour Monitor 1280 X 1024 resolution
p.r.f.	250–1200 p.p.s. selectable	Mass storage (data only)	30 GB Hard disk total, expandable to 60 GB
Modulator	Tetrode based hard tube	Data Archive	(i) 8 mm and 4 mm DAT (ii) CD-writer on PC-NT
		Hard copy	Black and White Laserjet and Colour Inkjet Printers.

1.3.26 Airborne Radar

The use of aircraft-borne instrumentation for meteorological studies has been attractive because the aircraft can get into the weather system of interest and move with it. Observations can be made at several altitude levels and from the most suitable vantage points. Sondes can be dropped from the aircraft for vertical sounding. For collection of statistics about any particular phenomenon, it is possible to take the aircraft to wherever it occurs and collect a large population of data. The same considerations apply to airborne weather radar. Radar observation can be complemented by insitu instrumentation on the craft including micro-physical measurements. Aircraft equipped with such instruments and one or more radars have been used extensively to penetrate the core areas of tropical cyclones and to study local severe storms and other precipitation systems in various field experiments.

Table 1.4. Characteristics of Environnement Canada Radar WSR82D; King City Weather radar Research Station, from Crozier et al. 1989, courtesy AMS.

Parameter	Units	Conventional	Doppler
Frequency	MHz	5625	5625
Wavelength	cm	5.33	5.33
Peak power	kW	260	260
Pulse duration	μs	2.0	0.5
p.r.f.	pps	250	892, 1190, 650*
Scan rate	rpm	6.0	0.75, 2*
Reflector diameter	m	6.1 parabolic	
Polarisation		linear horizontal	
Gain	dB	48	
Beam width	degrees	0.65	
Radome		fiberglass laminate	

* long range mode only.

Constraints are the availability and cost of aircraft of the required flying range, flight altitude limitations, aircraft stability in turbulence, safety considerations in flying into severe weather and space, weight and power limitations in the aircraft. Severe limitations in the design of the radar arise. Because of space limitations shorter wavelengths (5 cm or less) are used. The peak power outputs at these wavelengths are limited. Precipitation attenuation is a problem. The antenna and radome have to be fitted into the profile of the aircraft. The choice of antenna size and beam width is therefore limited. The aircraft movement including roll, pitch and drift introduces errors in radar position, velocity and pointing angle. The speed of movement of the aircraft limits the number of samples in each beam volume. These will introduce corresponding errors in the measurements of reflectivity and Doppler parameters and complicate the signal-processing. In the case of Doppler radars two legs of traverse in orthogonal directions can be made and the data combined in what is known as a pseudo-dual-Doppler mode; here the two observations have to be synthesised in space and time. Since the two observations are not simultaneous there can be errors. The antenna usually scans in the vertical plane normal to the aircraft track; as the aircraft moves, a helical scan is achieved. The vertical velocity derived from the vertical incidence data is corrected by assumed values of the fall-speeds of precipitation particles. Airborne Doppler radar has enabled, for the first time, gathering of wind data within severe weather systems at several levels in addition to the flight level of the aircraft. Airborne radar can be used in conjunction with ground radar as a multiple-Doppler system. Specifications of airborne radars used in the US WP-3D aircraft are given in Table 1.5. The tail radar in the table, was subsequently made a Doppler system with characteristics as in Table 1.6. The characteristics of the ELDORA/ASTRAIA Doppler radar developed by the USA and France and used on board US research aircraft Electra are also given in Table 1.6.

Table 1.5. Airborne radar parameters for US NOAA WP-3D aircraft (from Jorgensen 1984, courtesy AMS).

Parameter	Nose radar	Lower fuselage radar	Tail radar*
Transmitter Frequency MHz.	5445 ± 6.6	5370 ± 6.7	9315 ± 11.6
Transmitter wavelength cm	5.51	5.59	3.22
Transmitter pulse duration μs	3.0	6.0	0.5
Transmitter peak power kW (min)	70	70	60
p.r.f. p.p.s.	400	200	1600
Receiver dynamic range dB	80	80	80
Antenna gain main beam dB	34	37.5	40
Antenna sidelobe dB down	23	23	23
Beam width degrees			
horizontal	3.6	1.1	1.35
vertical	3.6	4.1	1.9
Antenna stabilisation degrees	± 20 pitch and roll	± 10 pitch and roll	± 25 pitch and drift
Maximum range km	185	371	93

* The tail radar has subsequently been modified to have Doppler capability. See Table 1.6.

Apart from conventional radar, airborne Synthetic Aperture Radar (SAR) – used primarily for ocean surface remote sensing – can be used for precipitation studies (Atlas and Moore 1986). Airborne radar is dealt with further in Chapter 3.

1.3.27 Satellite-borne Radar

Radars on board satellites have the advantage of large area coverage and can yield observations over the oceans. The constraints of airborne systems like space, weight and power apply to them too. The very large range involved also dictates the use of higher frequency bands. Apart from the methods mentioned in section 1.2, satellite altimeters, scatterometers and real and synthetic aperture radars can be used for precipitation studies (Hildebrand and Moore 1990). Satellite-borne radar is discussed in Chapter 7.

1.3.28 Other Radars

Radars on land meant for other purposes are also often used as weather radars, at least qualitatively. The prominent example is the Airport Surveillance Radar (ASR). This is a relatively short range, high resolution radar designed for locating aircraft. It has a fast scan and a large vertical beam width and therefore not ideal as a weather radar. It will also not have circuitry for quantitative processing of

Table 1.6. Characteristics of US Airborne Doppler radars (from Hildebrand and Moore 1990, courtesy AMS).

Characteristic	NOAA P3 radar	ELDORA/ASTRAIA* radar
Wavelength cm	3.2	3.2
Beam width degrees – horizontal	1.35	1.8
Beam width degrees – vertical	1.9	1.8
Pulse volume depth km	0.075	0.03–0.15
Gate spacing km	0.15	0.03–0.15
p.r.f. pps	1600	2000–5000
Unambiguous velocity m s^{-1}	25.8	25.8
Unambiguous range km	94	94
Elevation step degrees	1	1 (nominal)
Elevation scan rate deg s^{-1}	48	0–144
Along track spacing of beams km	1.05	0.3–1
MDS dBZ at 10 km		
with 10 dB S/N	11	0
with 10 dB S/N	–9	–20
Receiver band width MHz	2	2
Receiver temperature deg K	1600	1600
Antenna gain dB	40.3	40.3

* Slightly different final specifications are given by Hildebrand et al. (1996).

weather data. In the ASR-9 radar, developed in the USA, a weather channel has been built-in for aviation weather requirements (Mahapatra and Zrnić 1991; see also Chapter 9 on use of weather radar in aviation). HF radars known as Over-The-Horizon (OTH) radars which achieve long ranges by using reflection by the ionosphere have also been used for studying weather systems (Chapter 11).

1.3.29 Siting of Weather Radars

Siting considerations are important for fixed radars on land. The site to be chosen will depend on the proposed application. For instance, observation of tropical cyclones will need a coastal site with a good exposure towards the sea. Ground clutter at nearer ranges may not matter but the view on the land side will also be needed as the cyclone moves over land and precipitation needs to be mapped. A good example is the Mt. Fuji radar in Japan located 3776 m above sea level to gain an effective range of 600 km but the elevated site implies ground clutter extending over 200 km. On the other hand, if studies at short range are required, it is important to reduce ground clutter. It is useful to place the radar in a shallow bowl surrounded by ridges up to an elevation of about 0.5 degree to shield ground reflections from greater distances (Smith 1972). Reflection from ground, besides giving clutter, also distorts the antenna beam, introducing errors in measurement of reflectivity. It is difficult to separate ground clutter superposed on precipitation echoes while making reflectivity measurements with conventional radars except by the operator's judgement. Automated systems may print out ground clutter (or anomalous propagation echoes) as rainfall unless special processing methods are used (see Chapter 3). The direction and speed of approach of weather systems is also a consideration in choosing an optimum site. If rainfall mapping over a river catchment is needed, the radar has to be placed to cover as much of the catchment as possible. If the area is mountainous, shadows will be produced, severely limiting the effective range. The effective range for qualitative echo detection, for rainfall mapping and for Doppler parameters may all be different. The user will find it convenient to locate the radar close to their laboratory or forecast centre but a site in a built-up area implies physical shadowing of the radar view in some directions unless the antenna is put over a very tall tower or building. Such elevated installations are expensive and may increase the ground clutter. Nearby buildings will produce spurious echoes by reflection. (In Fig. 1.4 the ship echoes to the east of the radar are real. The "ship" seen at an azimuth of 246 degrees is over land and is the image of one of the real ships, produced by reflection from a nearby building). On the other hand, location of a radar at a well-exposed remote site may involve difficulties in obtaining stable power supply, and living arrangement or transport of personnel for operation and maintenance, as well as accessibility and reliable operation and communication in bad weather. One solution is complete automation (except for occasional maintenance visits) and remote control of operation, if the communication links are reliable. In the case of aviation weather radars there is an obvious advantage in locating the radar at the airport where the forecaster and the pilot have ready access to the display. But location some distance away enables low angle operation to observe low level phenomena, provided remote transmission of data (and possibly remote-controlled operation) can be achieved by reliable communication channels. This avoids the congestion of facilities at the airport.

1.3.30 Electromagnetic Interference (EMI)

One of the most important considerations in siting is electromagnetic interference (EMI). The radar should not get interference from other radars or communication equipment nor should it interfere with them. The high power of the radar should not cause damage to other electronic equipment or hazard to personnel. No physical obstruction should be caused to (or by) other installations or to aircraft movement. These conditions are increasingly difficult to achieve because of proliferation of communication equipment and pressure on the electromagnetic spectrum. A national authority regulates allocation of operating frequencies and siting of radars and communication equipment.

1.3.31 Site Survey

Yet other considerations are availability of a desirable site, expenditure in developing it, its safety from vandals and from natural disasters. Hence a detailed survey is desirable before a radar site is chosen. Table 1.7 reproduced from Leone et al. (1989) illustrates the contents of a typical site report.

Table. 1.7. Contents of a typical radar site report (After Leone et al. 1989, courtesy AMS)

ABSTRACT

PREFACE

I INTRODUCTION
- A. General Description
- B. Site Ownership and Availability
- C. Site Size and Acquisition
- D. Present Use
- E. Climate
- F. Topography and Coordinates
- G. Flood Zone
- H. Soils
- I. Surrounding Terrain
- J. Future Development Plans

II RADAR COVERAGE
- A. Weather Approach Directions
- B. Radar Coverage
- C. Clutter Estimate
- D. Anomalous Propagation

III RADAR TOWER HEIGHT

IV FREQUENCY ASSIGNMENT

V ROADS AND UTILITIES
- A. Roads
- B. Water Supply
- C. Sewage Disposal
- D. Fuel Storage
- E. Electric Power
- F. Communications

VI ENVIRONMENTAL IMPACT
- A. Potential Impact on Environment
- B. Radio Frequency Radiation
- C. Key Environmental Impact Issues

VII FAA REGULATIONS COMPLIANCE

VIII LONG LEAD-TIME ITEMS
- A. Zoning and Other Approvals
- B. Building Permits
- C. Power Service
- D. Telephone Service

APPENDIX A ECAC FREQUENCY ASSIGNMENT ANALYSIS
B CORRESPONDENCE AND BACKGROUND INFORMATION
C LOCAL POINTS OF CONTACT

Chapter 2

PROPAGATION, SCATTERING AND ATTENUATION OF MICROWAVES IN THE LOWER ATMOSPHERE

In this chapter, in section 2.1 we discuss the propagation of electromagnetic radiation in the lower atmosphere. The refractive index and its variations with meteorological conditions and the effects of such variations on radio wave propagation are dealt with. In section 2.2, the scattering of radiation by hydrometeors and the derivation of meteorological parameters from radar-measured parameters is discussed. Scattering in clear air due to refractive index inhomogeneities is also considered. In section 2.3, the attenuation of radiation in the atmosphere by atmospheric gases and hydrometeors is discussed.

2.1 Propagation of Electromagnetic Radiation

2.1.1 The Poynting Vector

When there is a varying electric current in an antenna, an alternating electric field and an alternating magnetic field in mutually perpendicular directions are associated with it. These propagate as an electromagnetic wave. The energy of the radiation is stored alternately in the electric and magnetic fields. If these two fields are represented by vectors **E** and **H,** the flow of energy is represented by the complex Poynting vector **S** directed perpendicular to both the fields. The time average S_a of the magnitude of the Poynting vector is given by (Kerr 1951),

$$S_a = (\tfrac{1}{2})\mathrm{Re}(\mathbf{E} \times \mathbf{H}^*) \tag{2.1}$$

where Re stands for "Real part of" and the * represents the complex conjugate. The electric displacement (representing the flux lines associated with the field **E**) is denoted by the vector **D** where

$$\mathbf{D} = \varepsilon'\mathbf{E} \tag{2.2}$$

ε' is known as the permittivity of the medium through which the wave is propagated. Similarly the magnetic induction or flux density (associated with the field **H**) is denoted by the vector **B** where

$$\mathbf{B} = \mu'\mathbf{H} \tag{2.3}$$

μ' being known as the magnetic inductive capacity of the medium. For detailed study of basic electromagnetic theory, reference may be made to books such as Stratton (1941) or Jordan (1967). A treatment of electromagnetic concepts with particular reference to weather radar is given by Bringi and Chandrasekar (2001).

2.1.2 Refractive Index

If the medium is homogeneous the velocity of propagation is given by

$$v = (\varepsilon'\mu')^{(-1/2)} \tag{2.4}$$

In free space, the velocity (of light) is c and is

$$c = (\varepsilon_0\mu_0)^{(-1/2)} \tag{2.5}$$

ε_0 is the permittivity of free space and is 8.85×10^{-12} farads per metre. μ_0 is the magnetic inductive capacity of free space and is $4\pi \times 10^{-7}$ henries per metre. This yields

$$c = 2.9986 \times 10^8 \text{ m s}^{-1} \tag{2.6}$$

The ratio $\varepsilon'/\varepsilon_0 = \varepsilon$ is the dielectric constant of the medium and the ratio $\mu'/\mu_0 = \mu$ is the magnetic permeability of the medium. The ratio of the velocity in free space c to that in the medium v is

$$\frac{c}{v} = n = (\varepsilon\mu)^{(1/2)} \tag{2.7}$$

n is defined as the refractive index of the medium. For the media of propagation considered by us, we may take μ as unity and write

$$n^2 = \varepsilon \tag{2.8}$$

If the dielectric is not perfect i.e. if there is absorption, the refractive index is a complex quantity m of which n is the real part.

$$m = n - ik \tag{2.9}$$

The imaginary part k represents the absorption.

2.1.3 Refractive Index of Air

The refractive index of air (in a non-ionised atmosphere[1]) is a function of its temperature, pressure and humidity and is usually taken, subject to certain assumptions, as (Bean and Dutton 1968),

$$(n-1)\times 10^6 = N = 77.6PT^{-1} + 3.73\times 10^5 eT^{-2} \tag{2.10}$$

where P = air pressure in hPa (including water vapour pressure), e = water vapour pressure in hPa and T = air temperature in degrees K.

It is convenient to use the quantity N called radio "refractivity" or "refractive modulus" instead of n. N represents the departure of n from unity in parts per million and is expressed in "N-units"[2]. Since n for air at sea level at microwave frequencies is about 1.0003, N has a value of about 300 (at the surface) and its variations can be considered more conveniently. In equation , the second term on the right hand side applies only when the air is moist and is therefore called the wet term. The first term is known as the dry term. The value of N can be computed[3] from measurements of P, T and e.

2.1.4 Measurements of Refractive Index

Except at the surface level, these measurements are generally obtained by radiosondes and are often not accurate enough for refractive index calculations. In the lowest layers which are the most important for terrestrial communications, P, T, e can be measured by captive balloons or other special instrumentation for greater accuracy (see for example Sarkar et al. 1992), or N can be measured directly by a refractometer. N decreases with increasing height because of the pressure fall. The vertical variation of N also depends on the profile of T and e. The profile can differ from this on occasion depending on the actual vertical lapse

---

[1] In the ionosphere there will be an additional term in equation (2.10) due to the electron density.

[2] Note that N is a number and therefore the coefficients on the right hand side of equation 2.10 have the dimensions K hPa^{-1} and K^2 hPa^{-1}.

[3] Atlases are available showing the spatial distribution in various seasons of N and other refraction parameters. See for example Sarkar et al. (1985).

rate of temperature and humidity; there can often be layers in which N decreases less rapidly with height or even increases with height[4].

2.1.5 Geometrical Optics

While considering radar wave propagation in the atmosphere it is convenient to treat the radar beam as a ray and employ geometrical optics. If the atmosphere is homogeneous, the ray will propagate in a straight line. Since the refractive index varies as a continuous function and decreases with height, the ray transmitted at a small elevation angle will usually be refracted towards the earth's surface. Fig. 2.1 illustrates a ray transmitted from a radar at height z_0 above the surface at a small elevation angle θ_0. When it reaches a height z the elevation is reduced to θ_z. If s is the distance along the curved surface of the earth (of radius a) it can be shown that

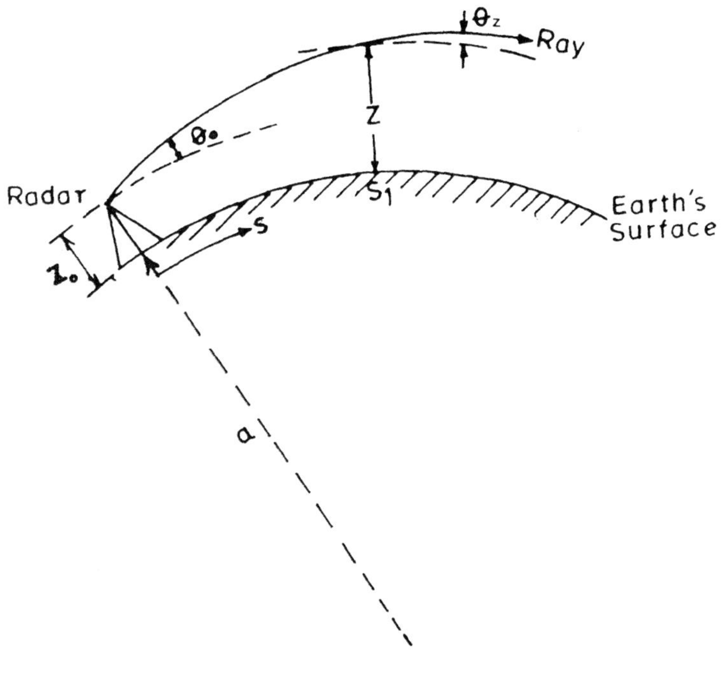

Fig. 2.1. Ray path over a curved earth.

[4] It is of interest to note here that instead of deriving refractivity from temperature and humidity profiles, techniques are now available for doing the converse operation i.e. deriving temperature and humidity profiles from refractivity determined from ray paths of satellite transmissions.

$$\frac{d^2z}{ds^2} = \frac{1}{a} + \frac{dn}{dz} \tag{2.11}$$

By integrating this we get

$$\theta_z^2 - \theta_0^2 = 2\frac{z - z_0}{a + n - n_0} \tag{2.12}$$

or

$$\theta_z^2 - \theta_0^2 = 2\times 10^{-6} \times \left\{ \left[\frac{z}{a}\times 10^6 + (n-1)\times 10^6 \right] - \left[\frac{z_0}{a}\times 10^6 + (n_0 - 1)\times 10^6 \right] \right\} \tag{2.13}$$

n and n_0 being the refractive indices at the two points considered. The quantity on the right hand side can be written as

$$2\times 10^{-6} \times (M - M_0) \tag{2.14}$$

where M is defined as

$$M = N + \frac{z}{a}\times 10^6 \tag{2.15}$$

and is called the "modified refractivity" (expressed in the same units as N). Even if N decreases with height, M increases with height because of the z/a term. We may also define another quantity called "modified refractive index"[5] which is $n+z/a$. A "standard" atmosphere is a hypothetical atmosphere in which N decreases at the rate of 39 units per km (Glickman 2000, Shibuya 1987) and M increases by about 118 units per km. Generally, as large gradients in T and e occur in the lowest layers, M gradients are also likely to be large in these layers (Sarkar et al. 1992).

2.1.6 Effective Earth's Radius

The curvature of the ray relative to the earth is given by

$$\frac{d\theta}{ds} = \frac{1}{a} + \frac{dn}{dz} = \frac{1}{a'} \tag{2.16}$$

If we assume an imaginary earth with radius a', the ray propagating over that

[5] The terms modified refractive index and refractivity modulus have also been used by some authors for M, causing some confusion. The definitions in this book follow Glickman (2000).

earth will be a straight line. a' is known as the "effective earth's radius". This fiction simplifies the analysis of ray propagation. Since dn/dz is normally negative, a' is usually greater than a. The ratio a'/a is called the "effective earth's radius factor" k and is taken as 4/3 for a standard atmosphere. However it can vary widely with meteorological conditions.

2.1.7 Normal Refraction

When $k = 4/3$, the refraction of the ray is said to be normal. Fig. 2.2 shows the heights above the earth's surface, of the centre of a beam at various distances from a radar located at the ground and transmitting at various angles of elevation, if the refraction is normal. The height corresponding to the zero degree beam gives the height of the radar horizon below which the radar cannot see (assuming a beam of zero width). This limits the maximum range at which the radar can see an object at a given height.

2.1.8 Anomalous Propagation

If the M-profile differs from the normal, the beam may be refracted abnormally i.e. more than normal or subnormally i.e. less than normal. If sufficient downward bending occurs, the ray may come down to earth, get reflected again and even make several hops on the ground. In such a case, the ray is trapped in a layer close to the ground called a "duct" and may travel long distances in the duct. The back-scatter from the point of contact with the ground or sea surface will produce an echo on the radar. Thus the radar will see ground objects at much greater distances than can normally be expected. A duct can also occur at some height above the ground (Chatterjee and Mathur 1982, Niranjan et al. 1990). In a less extreme case, an object which is below the normal radar horizon will be illuminated by the radiation and return an echo. This extends the maximum range of the radar. It will also increase the ground clutter echoes around the radar. If the height of a target is determined from a figure such as Fig. 2.2 or from an RHI calibrated therefrom, the radar will overestimate the height. Such refraction more than normal is called superrefraction and the propagation is said to be superstandard or abnormal. In the opposite case of subnormal refraction or subrefraction, the beam will be bent down less than normal or even bent upwards and the propagation is said to be substandard or subnormal. In this case, the height of a target is underestimated. Targets, which are normally detected, may go below the horizon and the maximum range of the radar is diminished. These phenomena are illustrated in Fig. 2.3. The height of a target at S1 in the figure will be estimated by radar, as z_1 while its actual

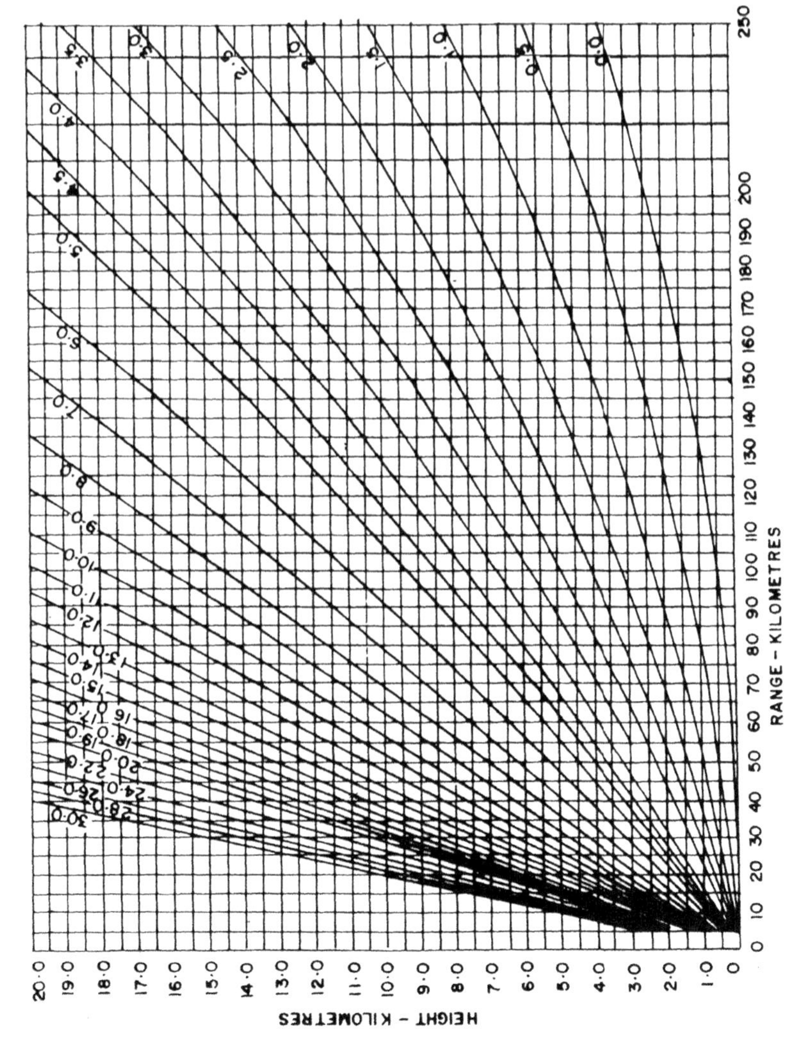

Fig. 2.2. Radar Range-height diagram. Height of centre of radar beam at various antenna elevations (degrees). Corrected for standard refraction taking effective earth radius as (4/3) x 6371 km.

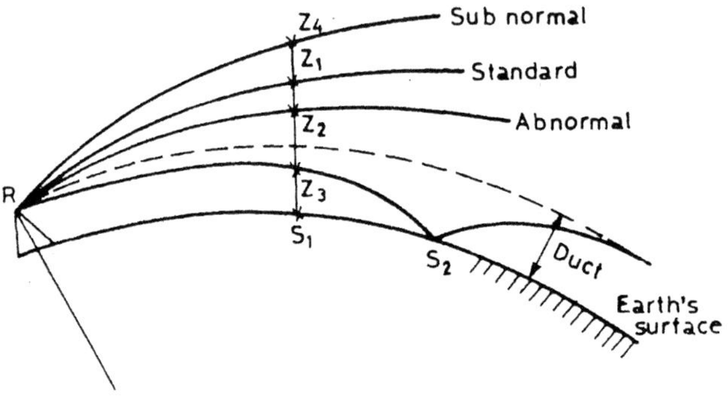

Fig. 2.3. Normal, abnormal, and subnormal propagation.

height may be z_2, z_3, or z_4 depending on the propagation conditions. The profiles may change rapidly with time, causing changes in the echo pattern.

2.1.9 Meteorological Conditions Leading to Anomalous Propagation

Examples of meteorological conditions leading to other than normal propagation are:

(1) On winter mornings, a temperature inversion often occurs close to the ground. This is usually accompanied by a high humidity at the surface. Because of the inversion, the air is stratified and the high humidity does not get mixed with higher layers. Both the inversion and the steep lapse of humidity with height ensure a steep fall of N with height. Hence there is abnormal propagation. Echoes from ground targets may extend to long ranges. Fig. 2.4 is an example of a tephigram[6] associated with such a situation. Such a condition is conducive to formation of mist or fog. Hence the observation of abnormal propagation can be used to predict fog. However when fog has actually formed, propagation may become normal again.

(2) If a warm and dry airmass is advected over a cold surface, a temperature inversion and steep lapse of humidity will occur and abnormal propagation can arise. This is particularly important over sea or coastal locations where the surface humidity will be high. In the case of a sea breeze, the lowest levels have cool and moist air, which undercuts relatively dry land air and hence causes abnormal propagation. A land breeze can produce the effect because of its dryness.

(3) If there is an anticyclone in the vicinity, the subsiding dry air over relatively moist surface air can cause anomalous propagation. Fig. 1.7 is an example.

[6]A tephigram is an aerological diagram with temperature T as abscissae and entropy (ϕ) or potential temperature (θ) as ordinates.

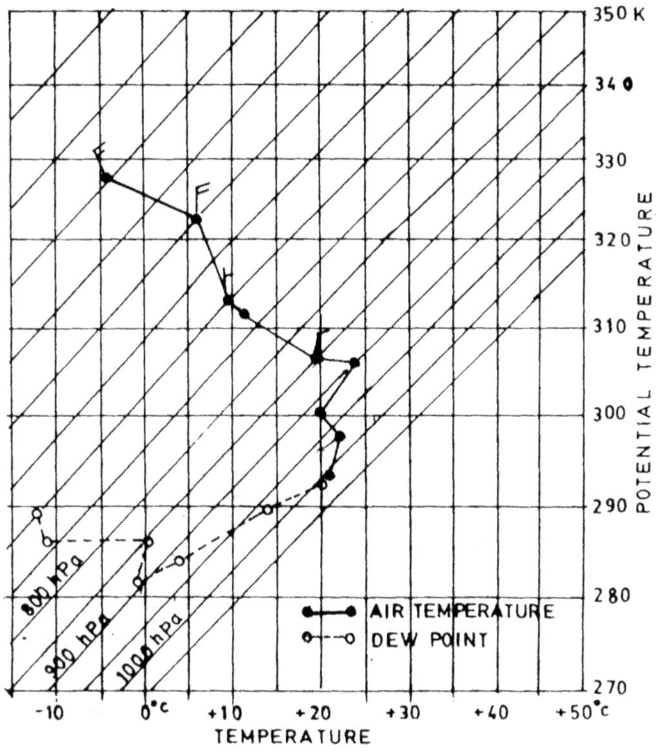

Fig. 2.4. Tephigram of Chennai, India, on a winter morning showing ground inversion and steep lapse of humidity leading to abnormal propagation. Winds are also plotted (from Rajagopalan and Raghavan, 1980; courtesy IMD).

(4) When thunderstorms occur in an area, the strong convective mixing prevents the formation of layers conducive to abnormal propagation. But in the wake of the thunderstorm, cold downdraft air spreads over the surface with relatively warm air above it. The downdraft air may sometimes have a higher humidity mixing ratio and sometimes a lower mixing ratio than the air above it. The former case will produce anomalous propagation echoes. It may sometimes be difficult to distinguish these from the precipitation echo of the thunderstorm as they will also show considerable apparent height due to the anomalous propagation. Further discussion and explanations of this phenomenon may be seen in Coons (1947), De (1959b), Mathur and Kulshrestha (1961) and Raghavan (1965).

(5) During a rain spell, high humidity extends to considerable height. Hence subnormal propagation may occur and reduce the effective range of a radar. A particularly interesting case is that of a tropical cyclone or depression approaching

a stationary radar from a distance. Initially when the weather over most of the line of sight is fair, the precipitation associated with the weather system may be seen clearly over distances longer than the normal range of the radar. When the weather system comes closer to the radar, rainfall occurs over the line of sight reducing the humidity lapse. Hence propagation may become subnormal, reducing the ability of the radar to see the weather system (see Chapter 8 and Rajagopalan and Raghavan, 1980).

Extensive anomalous propagation is common over the Arabian sea and Bay of Bengal (Fig. 1.7). It was first noticed with military radars during world war II (Durst 1946, Hatcher and Sawyer 1947).

2.1.10 Horizontal Variation of *N*

In the above discussion we have considered only the vertical profile of *N* over one point. Since *N* varies horizontally also along the line of sight, the actual ray path depends on these variations. Anomalous propagation echoes may not therefore be seen equally in all directions around the radar. Where air movement is involved, the winds at various levels will influence the ray path.

2.1.11 Superposition of Echoes

Besides producing errors in height measurements of targets, non-standard propagation may also superimpose different echoes. Besides the superimposition or juxtaposition of thunderstorm and anomalous propagation echoes mentioned in Para 2.1.9.(4) above, mutiple-trip echoes from beyond the maximum unambiguous radar range may get superimposed on echoes within range. Multiple-trip echoes can be identified by changing the p.r.f. of the radar (see section 1.2.). Clutter echoes from the sea surface may also get superposed on precipitation echoes.

2.1.12 Refractive Index Discontinuities

Variation of atmospheric refractive index also produces partial reflection or scattering from refractive index discontinuities. This will give rise to a relatively weak radar echo from the surface of discontinuity. Many such echoes were termed "angels" in the early radar meteorology literature for want of a clear understanding of their origin. Echoes of meteorological origin in this category include those from sea breeze fronts, airmass boundaries, gust fronts, mantle echoes from cumulus clouds and those from the turbulent inhomogeneities in the atmosphere. These are further discussed in section 2.2. and in chapters 4 and 10.

2.1.13 Heights of Echoes

In Para 2.1.7 the beam height was considered for a beam of zero width. Since the beam has a finite vertical width it will illuminate layers of the atmosphere below and above the heights shown in Fig. 2.2. The vertical spread of the beam is limited on the lower side by the earth's surface; there is no such limitation in the horizontal plane. Considering this, the horizontal and vertical widths of a 2-degree beam directed horizontally, are given in Table 2.1. For measuring the height of the top of an echo the radar antenna may be raised until the echo just disappears. This will happen when the top of the echo is just below the bottom of the beam. The radar antenna indicates the elevation of the centre of the beam. Hence half the beam width should be subtracted from the indicated value to get the elevation of the bottom of the beam. The corresponding height obtained at the range of the echo from Fig. 2.2 gives the height of top of the echo if propagation is normal. By the same reasoning, to get the height of the bottom of the echo, half the beam width should be added to the indicated elevation at which the echo disappears on the lower side. Fig. 2.2 takes account of the earth's curvature and normal refraction. Usually radar RHI's as well as digital equipment are also calibrated likewise. The measured heights will be in error if propagation is other than normal. As there is no means of determining the *k* value at every instant it is not possible to correct these errors quantitatively in real time.

Table 2.1. Horizontal and vertical extent of a beam of width 2 degrees (radiated horizontally) at various ranges.

Range (km)	Height of base of beam* (km)	Height of top of beam (km)	Vertical extent of beam (km)	Horizontal width of beam (km)
10	Almost zero	0.17	0.17	0.33
50	0.15	0.98	0.83	1.66
100	0.6	2.25	1.65	3.32
150	1.35	3.81	2.46	4.98
200	2.4	5.67	3.27	6.63
300	5.4	10.28	4.86	9.94
400	9.5	16.06	6.56	13.25

* Limited to zero elevation.

2.1.14 Airborne Radar

The above discussion related to ground-based radar. In the case of airborne radar, similar principles, of course, apply but the layers involved in propagation are at higher levels where the vertical gradients of refractivity are much smaller and refraction may be less important.

2.2 Scattering

2.2.1 Scattering by Precipitation Particles

What happens when electromagnetic radiation meets a precipitation particle? The simplest case is that of a water or ice sphere. The diffraction of a plane wave incident on a spherical surface was treated by Mie (1908) and further discussed by Stratton (1941) and Kerr (1951). The incident radiation induces electric and magnetic dipoles, quadrupoles etc., which oscillate and produce the scattered radiation. The power W intercepted from the incident wave by the sphere is given by,

$$W = S \times Q(a,\lambda) \qquad (2.17)$$

where S is the magnitude of the incident Poynting vector expressed in W m^{-2} and $Q(a,\lambda)$ is a quantity with the dimensions of an area and is a function of the radius a of the sphere and of λ the wavelength. $Q(a,\lambda)$ is called the total cross section and represents the depletion of energy by absorption and dissipation as heat on the one hand and by re-radiation at the same frequency in all directions i.e. scattering, on the other hand.

$$Q(a,\lambda) = Q_s + Q_a \qquad (2.18)$$

Q_s is the scattering cross section and Q_a the absorption cross section. According to the Mie theory, these are given by,

$$Q(a,\lambda) = \left(\frac{\lambda^2}{2\pi}\right)(-\text{Re})\sum\{(2n+1)(a_n + b_n)\} \qquad (2.19)$$

$$Q_s = \left(\frac{\lambda^2}{2\pi}\right)\sum\left\{(2n+1)\left(|a_n|^2 + |b_n|^2\right)\right\} \qquad (2.20)$$

Re means "Real part of", the symbol$|\;|$ represents the modulus of the quantity concerned and the summation is from $n = 1$ to ∞. The coefficients a_n and b_n represent respectively the amplitudes of the field distribution on the sphere due to the magnetic and electric multipoles which are induced by the incident wave (n here is not to be confused with the symbol for refractive index; $n = 1$ represents the dipole; $n = 2$ corresponds to the quadrupole and so on). a_n and b_n are functions of the third and higher powers of α where α represents $2\pi a/\lambda$ (Kerr 1951, Gunn and East 1954).

2.2.2 Back-scattering Cross Section

The back-scattering cross section σ defined in Chapter 1 is a part of Q_s. The "normalised" back-scattering cross section i.e. the ratio of σ to πa^2 is given by,

$$\frac{\sigma}{\pi a^2} = \alpha^{-2} . \left|\sum(-1)^n . (2n+1) . (a_n - b_n)\right|^2 \tag{2.21}$$

When a is small in comparison with λ, powers of α higher than 3, and therefore all a_n and b_n except b_1 can be neglected. Kerr (1951) gives,

$$b_1 = -i\frac{2}{3}\frac{m^2-1}{m^2+2}\alpha^3 \tag{2.22}$$

where m is the complex refractive index defined in equation (2.9).

Then (2.21) becomes,

$$\frac{\sigma}{\pi a^2} = \alpha^{-2} \cdot \left|(-1)\cdot 3 \cdot (-1) \cdot (-1) \cdot \frac{2}{3} \cdot \frac{m^2-1}{m^2+2} \cdot \alpha^3\right|^2 \tag{2.23}$$

or

$$\sigma = 64 a^6 \cdot \pi^5 \cdot \lambda^{-4} \cdot |K|^2 \tag{2.24}$$

where

$$|K| = \left|\frac{m^2-1}{m^2+2}\right| \tag{2.25}$$

If $D = 2a$ is the diameter of the sphere, we have,

$$\sigma = D^6 \cdot \pi^5 \cdot \lambda^{-4} \cdot |K|^2 \tag{2.26}$$

Since $m^2 = \varepsilon$, the dielectric constant, the above equation can also be written as,

$$\sigma = D^6 \cdot \pi^5 \cdot \lambda^{-4} \left|\frac{\varepsilon - 1}{\varepsilon + 2}\right|^2 \tag{2.27}$$

These equations represent the Rayleigh approximation to the Mie theory i.e.

the case of a particle small in comparison with λ. A detailed treatment of back-scatter and attenuation by precipitation particles was given by Ryde (1946). Later discussions may be found in Gunn and East (1954), Atlas (1964), Battan (1973) and Probert-Jones (1984). A review of scattering by hydrometeors has been published by Oguchi (1981) from a communications point of view.

2.2.3 σ for Water and Ice Spheres

At higher values of α, higher orders of a_n and b_n have to be considered. Fig. 2.5 after Herman and Battan (1961) gives the variation of $\sigma/\pi a^2$ with α for water and ice spheres. We can designate a σ_{MIE} as a general case and a $\sigma_{RAYLEIGH}$ when α is small. The range of values of α where the ratio of σ_{MIE} to $\sigma_{RAYLEIGH}$ is between 0.5 and 2 is called the Rayleigh region (Gunn and East 1954), where the Rayleigh scattering can be assumed without appreciable error. For water, the value of the dielectric constant ε is about 80; hence, $|K|^2$ is taken as 0.93 in the wavelength range 3 to 10 cm and temperature range 0 to 20°C. For ice of unit density, it is 0.208; but ice exists in various forms (with admixture of air or water) whose

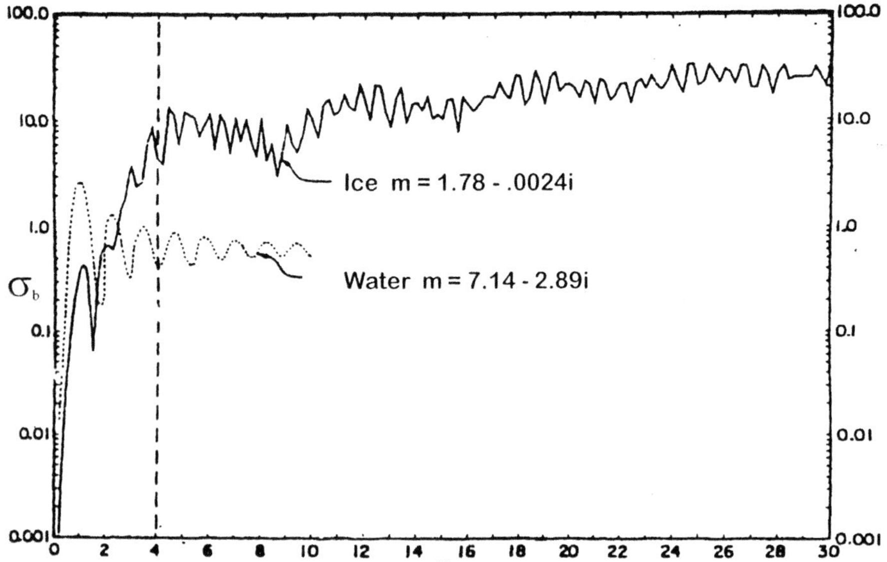

Fig. 2.5. Calculated values of normalised cross section $\sigma_b = \sigma/(\pi a^2)$ for Mie backscattering from water and ice spheres (from Herman and Battan, 1961, courtesy AMS). The water curve is for a temperature of 0°C and a wavelength of 3.21 cm. The ice curve is for the wavelength range 1 to 10 cm. The dashed ordinate at $\alpha = 4$ represents the limit upto which σ_b was calculated by Ryde.

densities vary widely[7]. Snow can be taken as a mixture of ice and air. The ratio of K to the density ρ for a mixture of two dielectrics 1 and 2 is obtained by adding the corresponding ratios for the two components in proportion to the mass (M_1 and M_2) of each, provided neither dielectric is a significant absorber (Ryde 1946, Battan 1973), i.e.,

$$\frac{K}{\rho}\cdot(M_1+M_2)=\frac{K_1}{\rho_1}\cdot M_1+\frac{K_2}{\rho_2}\cdot M_2 \tag{2.28}$$

the suffixes 1 and 2 referring to the two dielectrics. Hence the blowing up of an ice particle by admixture of air does not alter the product $D^6|K|^2$. In the case of Rayleigh scattering we can use the value 0.208 for unit density ice and the melted diameter of the ice particle for calculating σ.

2.2.4 Ice–water Mixture

It will be noted from Fig. 2.5 that for small particles the normalised cross section of an ice sphere is less than that of a water sphere. If the ice melts, the back-scatter should increase. A melting ice sphere can be treated as composed of two concentric spheres, the outer one of water with higher dielectric constant and the inner one of ice with lower dielectric constant (Aden and Kerker 1951, Herman and Battan 1961). If the water shell is sufficiently thick it will act as an all-water sphere (i.e. no radiation interacting with the ice) and the effective dielectric constant is that of water. This may not be true of melting snow which may need to be treated as a mixture of water and ice. Since water is an absorber at some of the microwave frequencies it is not correct to use equation (2.28) as in the case of the ice–air mixture. Further, as particle diameter increases, the normalised cross section of ice exceeds that of water and undergoes wide fluctuations with increase in particle diameter (Fig. 2.5). This is relevant to the case of hailstones which will be discussed later.

2.2.5 Drop Size Distribution (DSD)

We have considered above, the back-scattering cross section of single (spherical) particles. As seen in paragraph 1.2.9, we have to sum the cross sections

---

[7] The value of 0.208 is given by Gunn and East (1954), as corrected vide Corrigendum in Quart. J. Royal Meteor. Soc., 1955, **81**, p653. The value is independent of temperature and of wavelength in the microwave region. For ice of density 0.917 gm cm^{-3}, a value of 0.18 is normally used.

over unit volume to get the radar reflectivity η. For the Rayleigh case, from equations (1.20) and (2.26), this is given by,

$$\eta = \left(\frac{\pi^5}{\lambda^4}\right)|K|^2 \sum_{\text{unit vol.}} D^6 \tag{2.29}$$

ΣD^6 represented by Z, is the radar reflectivity factor expressed in mm^6 m^{-3}. Using the value 0.93 for water for the quantity $|K|^2$ we can also write,

$$Z = 3.52\times 10^{-3}\cdot\lambda^4.\eta \tag{2.30}$$

Note that this equation is correct only if the quantities are expressed in consistent units i.e. λ in cm, η in cm^2 cm^{-3} and Z in cm^6 cm^{-3}. Otherwise the right hand side has to be multiplied by a suitable power of ten.

D can have a large range of values. Let $N_D\delta D$ be the number of drops per unit volume of air (cm^3) with diameter between D (cm) and $D+\delta D$. Based on measurements of drop sizes at the ground (Laws and Parsons 1943, Marshall et al. 1947), in temperate latitude rainfall, Marshall and Palmer (1948) fitted the drop size distribution (DSD) by an exponential relation,

$$N_D = N_0 e^{-\Lambda D} \tag{2.31}$$

where N_0 and Λ are parameters to which they assigned the values,

$$N_0 = 0.08 \text{ cm}^{-4} \tag{2.32}$$

and

$$\Lambda = 41R^{-0.21} \text{ cm}^{-1} \tag{2.33}$$

R representing rate of rainfall in mm h^{-1}. Equation (2.31) obviously implies that small drops are large in number and the number decreases rapidly with increasing *D*. This distribution known as the Marshall-Palmer (M-P) distribution was found to fit observations well, except at diameters less than about 0.1 cm and has been usually taken as valid for raindrops though the values of N_0 and Λ (determined empirically) could differ with place or type of precipitation. Some values for N_0 and Λ obtained for different types of precipitation are given in Table 2.2. Note that in these cases N_0 is also dependent on *R*. Waldvogel (1974) found that N_0 values jump from moment to moment even in the same rainfall.

Table 2.2. Some empirical values for N_0 and Λ.

Precipitation type	Authors	N_0 cm^{-4}	Λ cm^{-1}
Thunderstorm Rainfall	Sekhon and Srivastava (1971)	$0.07R^{0.37}$	$38R^{-0.14}$
Snowflakes	Gunn and Marshall (1958)	$0.038R^{-0.87}$	$25.5R^{-0.48}$
Snowflakes	Sekhon and Srivastava (1970)	$0.025R^{-0.94}$	$22.9R^{-0.45}$

The volume of one water drop is

$$\tfrac{1}{6}\cdot\pi D^3 \text{ cm}^3 \tag{2.34}$$

Volume of water in $N_D\delta D$ drops in one cubic metre of air

$$\tfrac{1}{6}\cdot\pi D^3 N_D \delta \mathrm{D} \text{ cm}^3 \tag{2.35}$$

(N_D is taken as drops per m^3). Hence the liquid water content (LWC), M g m^{-3} of raindrops (excluding non-detectable cloud droplets) is given by,

$$M = \tfrac{1}{6}\cdot\pi\rho\int_0^\infty D^3 N_D dD \tag{2.36}$$

where ρ is the density of water in gm cm^{-3}. Note that the integration is from $D=0$ to $D=\infty$. Where it is desired to truncate the DSD at some chosen maximum diameter D_m the integration is done from 0 to D_m.

Further, the drop population can be divided by a "Median Volume Diameter" D_0 into two parts each contributing $M/2$ to the LWC, i.e.

$$\tfrac{1}{6}\cdot\pi\rho\int_0^{D_m} D^3 N_D dD = \tfrac{1}{6}\cdot\pi\rho\int_{D_m}^\infty D^3 N_D dD = \frac{M}{2} \tag{2.37}$$

For the exponential distribution of equation (2.31), if D_m is not less than $2.5D_0$, it can be shown (Atlas 1964, p357) that ,

$$D_0 = \frac{3.67}{\Lambda} \tag{2.38}$$

Hence (2.31) can be written as,

$$N_D = N_0 \cdot \exp\left[\left(\frac{3.67}{D_0}\right)D\right] \tag{2.39}$$

It has been increasingly recognised that the Marshall-Palmer exponential distribution is not valid for all types of rain. Hence, a more general relationship in the form of a gamma function has been proposed by several authors (e.g. Sulakvelidze and Dadali 1968, Ulbrich 1983).

$$N_D = N_0 D^m \exp\left(\frac{-3.67+m}{D_0}\cdot D\right) \tag{2.40}$$

m being a positive or negative integer and N_0 has the units m^{-3} cm^{-1-m}. This is called a three-parameter drop size distribution. When $m = 0$, this reduces to an exponential distribution. A plot of log N_D as ordinates against D as abscissae is a straight line in the case of an exponential distribution. In the gamma distribution it will be a curve concave upwards if m is negative and convex upwards if m is positive. The former (i.e. the sag in the concave curve) is interpreted as a deficit of medium size drops compared to an exponential distribution; and the convex as an excess.

A log-normal distribution has also been proposed by Feingold and Levin (1987),

$$N_D = \frac{N_T}{(2\pi)^{0.5} \ln \beta D} \exp\left\{ \frac{-\ln^2\left(\frac{D}{D_g}\right)}{2\ln^2 \beta} \right\} \text{ m}^{-3} \text{ mm}^{-1} \qquad (2.41)$$

where N_T is the drop concentration (m^{-3}), D_g the geometric mean diameter (mm) and β a breadth parameter.

Finally it may be noted that, assuming the M-P distribution, N_0 is a constant and,

$$Z = N_0 \int_0^\infty D^6 \left(e^{-\Lambda D}\right) dD = N_0 \cdot 6! \cdot \Lambda^{-7} \qquad (2.42)$$

2.2.6 Particle Terminal Velocity

When a small spherical water drop of diameter D cm falls, the gravitational force on it is soon balanced by the air drag on it and a constant velocity of fall results. This is called the terminal velocity V_D. Following Spilhaus (1948) we equate the two forces as follows:

$$\tfrac{1}{6}\pi D^3 \rho g = \tfrac{1}{4} C_D \pi D^2 \rho_0 V_D^2 \qquad (2.43)$$

where, ρ and ρ_0 are the densities of water and air respectively, g the acceleration due to gravity and C_d is the drag coefficient. Hence

$$V_D^2 = \frac{\frac{2}{3}\rho g D}{C_D \rho_0} \qquad (2.44)$$

i.e. V_D is proportional to $D^{0.5}$. But this is a small spherical drop. Bigger drops and solid particles will not be spherical. In general, a relationship of the form

$$V_D = aD^p \qquad (2.45)$$

is used. According to Spilhaus (1948), a = 1420 cm^{1-p} s^{-1} and p = 0.5 as mentioned. Sekhon and Srivastava (1971) obtained 1690 and 0.6 respectively for a and p.

Experimentally, the relation between terminal velocity of water drops in still air at sea level pressure and temperature 293 K for spherical water drops was determined in the laboratory by Gunn and Kinzer (1949). The equivalent drop diameter was calculated from the mass. The maximum diameter was about 0.6 cm. Their data have been assumed by most subsequent workers in fall speed calculations. However, according to their data, the maximum velocity approaches an asymptote of 9.2 m s^{-1} at $D = 0.6$ cm. As this does not agree with the relation aD^p, Atlas et al. (1973) fitted an equation,

$$V_D = a_1 - a_2 \exp(-600D) \text{ m s}^{-1} \tag{2.46}$$

where $a_1 = 9.65$ m s^{-1}, $a_2 = 10.3$ m s^{-1}, V_D is in m s^{-1} and D is in metres. This will give a negative V_D for very small drops, but it is claimed that this does not matter in practice.

Equations (2.45) and (2.46) are for sea level at which air density was taken as ρ_0. For any higher altitude h at which air density is ρ_h, their right hand sides should be multiplied by a correction factor $(\rho_0/\rho_h)^{0.4}$ (Foote and du Toit 1969). Fall speed corrections for raindrops versus altitude for the standard atmosphere, computed by Atlas et al. (1973) are reproduced in Table 2.3.

A value of $p = 0.67$ is usually taken for rain [$0.5 < D < 5$ mm] following Atlas and Ulbrich (1977). Sulakvelidze and Dadali (1968) give, on the basis of experiments in the former Soviet Union,

$$V_D = 13d^{0.53} \text{ for } d \le 3 \text{ mm} \tag{2.47}$$

$$V_D = 9.33d^{0.22} \text{ for } d > 3 \text{ mm} \tag{2.48}$$

where d is the diameter in mm. Values of p of 0.3 for snow to 0.8 for hail have been found by various workers.

2.2.7 Relation between Z and Precipitation Parameters

There are $N_D \delta D$ drops in one cubic metre, having diameters in the interval D to $D+\delta D$. Volume of water in these drops, is

$$\tfrac{1}{6}\pi D^3 \cdot 10^{-6} \cdot N_D \delta D \text{ m}^3 \tag{2.49}$$

Volume of water in these drops falling in one second over one m^2 of the earth's surface is,

$$\tfrac{1}{6}\pi D^3 \cdot 10^{-6} \cdot N_D \delta D \cdot V_D \text{ m}^3 \tag{2.50}$$

if the vertical motion of the air itself is neglected. The rate of precipitation is given by this volume divided by the area of one m^2 i.e.

$$\tfrac{1}{6}\pi D^3 \cdot 10^{-6} \cdot N_D \delta D \cdot V_D \text{ m s}^{-1} \tag{2.51}$$

Rate of precipitation in mm h^{-1}, due to these drops, is,

$$10^3 \cdot 3600 \cdot \tfrac{1}{6}\pi D^3 \cdot 10^{-6} \cdot N_D \delta D \cdot V_D \text{ mm h}^{-1} \tag{2.52}$$

Table 2.3. Fall speed corrections for rain drops versus altitude for the standard atmosphere (Atlas et al. 1973).

	Height kilometres									
D mm	0.5	1.0	1.5	2.0	2.5	3.0	3.5	4.0	4.5	5.0
$\left(\frac{\rho_0}{\rho_h}\right)^{0.4}$	1.010	1.040	1.060	1.082	1.103	1.127	1.150	1.175	1.200	1.226

The total rate of precipitation R is obtained by integrating this over all D (0 to ∞)

$$R = 0.6\pi \int_0^{\infty} N_D V_D D^3 dD \ \ \text{mm h}^{-1} \tag{2.53}$$

Taking Atlas and Ulbrich (1977) value of $p = 0.67$,

$$R \propto \int_0^{\infty} N_D D^{3.67} dD \tag{2.54}$$

Since, by definition,

$$Z = \int N_D D^6 dD \tag{2.55}$$

it is reasonable to expect a power law relation between R and Z of the form,

$$Z = AR^b \tag{2.56}$$

where the exponent b is between 1 and 2 and A is a constant. Since this is not a linear relation, the values of b and A will vary considerably with drop size distribution. On the basis of the experiments of Marshall and Palmer (1948) in temperate latitude stratiform rain, the values of $A = 200$ and $b = 1.6$ are used in some countries as standard values. It should be noted that the M-P drop size measurements were made at the ground, while the distribution could have been different aloft when the radar was observing the drops. Numerous experiments all over the world have produced widely different values of A and b for various types of rain and snow (Chapter 7). A similar power law,

$$Z = A'M^{b'} \tag{2.57}$$

can be thought of between Z and M. Since M is proportional to D^3 the exponent b' in this case should be about 2.0. Also since V_D, for raindrops, is proportional to $D^{0.67}$, we may expect a power law relation between V and Z of the form,

$$V_D = A''Z^{b''} \tag{2.58}$$

where the exponent b'' will be of the order of 0.1. This will be discussed further in Chapter 3.

Another useful parameter is Vertically Integrated Liquid Water Content (VIL). It is the liquid water content, M, integrated over the entire vertical column of a precipitating cloud and may be defined as (Greene and Clark 1972, Greene 1972),

$$\text{VIL (kg m}^{-2}) = 10^{-3} \cdot \int_b^t M.dh \tag{2.59}$$

where h is the height in metres, M is in g m^{-3} and b and t are heights of base and top of the cloud respectively. Since M can be related to observed Z by equation (2.57) it is possible to compute VIL from a vertical profile of Z. By integration of the Marshall-Palmer size distribution [equation (2.31)], Greene (1972) obtained the relation,

$$M = 3.44 \times 10^{-3} \times Z^{4/7} \tag{2.60}$$

Hence,

$$\text{VIL (kg.m}^{-2}) = 3.44 \times 10^{-6} \cdot \int_b^t Z^{4/7}.dh \tag{2.61}$$

In the case of ice particles, it is possible to relate the "ice water content" I (g m^{-3}) to Z by a power law similar to that for M. However the coefficient and exponent in the I–Z relationship are much more variable than in the case of the M–Z relationship owing to the various forms of ice particles that are possible. An example of calculation of VIL is shown in Chapter 3.

2.2.8 Equivalent Radar Reflectivity Factor Z_e

An expression for the power received by the radar was given in equation (1.29). It can be rewritten as ,

$$P_r = \frac{C \cdot |K|^2 \cdot Z}{r^2} \tag{2.62}$$

where C is a constant depending only on the attributes of the radar. In practice, one cannot be certain that the Rayleigh approximation holds. Hence it is common to replace Z by Z_e in equation (2.62). Z_e is called the "equivalent radar reflectivity factor". It is the reflectivity factor, which would yield the P_r actually observed. From P_r, η can be computed using equation (1.28) and from equation (2.30), Z_e can be obtained. In day-to-day operation the reflectivity factor values indicated by a radar set are all Z_e values. Z_e computed from measured P_r is taken as a measure of the intensity of the radar echo. We should also replace Z by Z_e in equations (2.56) onwards. Putting $Z = Z_e = AR^b$ in equation (2.62),

$$AR^b = \frac{P_r r^2}{C|K|^2} \tag{2.63}$$

This equation forms the basis for the most common method of quantitative estimation of precipitation by radar (Chapter 7).

2.2.9 Scattering by Non-spherical Particles

We have considered above only spherical particles. Solid precipitation particles are far from spherical. Raindrops are also non-spherical when they are large. As the size of a raindrop increases, the drop tends to have an elliptical cross section and the shape of a solid of revolution about one of the axes of the ellipse. The major axis tends to be horizontal in still air. If the solid is formed by rotation around the minor axis, the shape is known as an oblate spheroid. (A prolate spheroid is one that results from revolution around the major axis). The minor axis may depart from the vertical (the angle of departure is called the canting angle) due to air motions and particle collisions. The particle may also oscillate about an axis. The ratio b/a of the minor axis ($2b$) to the major axis ($2a$) is called the axial ratio (some authors define a/b as the axial ratio). If D_{eq} in cm is the diameter of the sphere of the same volume as the drop, Pruppacher and Beard (1970), based on wind tunnel studies, gave the formula,

$$\frac{b}{a} = 1.03 - 0.62 D_{eq} \tag{2.64}$$

for D_{eq} between 0.1 and 0.9 cm (some authors have used an exponent higher than unity for D_{eq} in this equation). The incident electric vector will produce dipole (and multipole) moments along the three axes. If the particle is small, the mutipole moments can be neglected as was done in Para 2.2.2. The magnitude of each dipole moment will be proportional to the component of the electric vector along the particular axis. Atlas et al. (1953) worked out the back-scatter from randomly oriented spheroids on this basis and found that a non-spherical particle can return a much larger power than a spherical one having the same volume. But if the particles are preferentially oriented with reference to the plane of polarisation of the incident wave, the returned power will be higher if the major axis of the particle is aligned with the incident electric vector and lower if the minor axis is so aligned.

Consider a small drop with major axis horizontal. The back-scattering cross section σ_H for electric vector horizontal (i.e. horizontal polarisation) is given (in cm^2) by (Seliga and Bringi 1976),

$$\sigma_H = \frac{16}{9}\frac{\pi^7}{\lambda^4} D_{eq}^6 \left| \frac{m^2 - 1}{4\pi + (m^2 - 1)P'} \right|^2 \tag{2.65}$$

and the cross section for vertical polarisation as,

$$\sigma_V = \frac{16}{9}\frac{\pi^7}{\lambda^4} D_{eq}^6 \left| \frac{m^2 - 1}{4\pi + (m^2 - 1)P} \right|^2 \tag{2.66}$$

P and P' are geometrical factors defined by,

$$P = 4\pi - 2P' = \frac{4\pi}{e^2}\left\{1 - \frac{b/a}{e}\sin^{-1} e\right\} \tag{2.67}$$

e being the eccentricity of the elliptical cross section given by,

$$e^2 = 1 - \frac{b^2}{a^2} \tag{2.68}$$

Note that for a spherical drop $a = b$, $D_{eq} = D$, $P = P' = 4\pi/3$, and equations (2.65) and (2.66) reduce to equation (2.26). The effects of polarisation will be large for water particles and relatively small for ice. Oscillation of the particle as it falls will produce fluctuations of the returned power.

A rigorous treatment of scattering by non-spherical particles is available in Bringi and Chandrasekar (2001).

2.2.10 Dependence of Scattering on Polarisation

Normally a weather radar transmits a plane polarised wave and detects only the signal received with the same plane of polarisation. In the case of an ellipsoid scatterer none of whose axes are aligned with the incident electric vector, the signal received by the radar will have components in two orthogonal planes of polarisation. The sum of the received power in the two components is the total power received. Only the component in the original plane of polarisation will be detected by a conventional single polarisation radar. Since a large raindrop tends to flatten i.e. the longer axis is horizontal, horizontally polarised component of received power is greater (this being one reason for the preferred use of horizontal polarisation[8] in most weather radars). But the orthogonal component, called the cross-polarised component, can also be received and measured if appropriate hardware is provided. Two modes of linear polarisation can be transmitted – Horizontal (H) and Vertical (V) – and for each, two components H and V can be received. The back-scatter properties of a drop or ensemble of drops can be expressed as a back-scatter matrix **S** of complex numbers,

$$\mathbf{S} = \begin{vmatrix} S_{HH} & S_{HV} \\ S_{VH} & S_{VV} \end{vmatrix} \tag{2.69}$$

the two suffixes of each S representing respectively the planes of polarisation of

[8] For a discussion of relative merits of horizontal and vertical polarisation for weather radars see Balakrishnan et al. (1993).

the transmitted and received radiation. The back-scattered electric field vector is the product of **S** and the incident field. Hence, there can be four signals with four associated Z_e values – Z_{HH}, Z_{HV}, Z_{VH} and Z_{VV}. The ratio of the reflectivity factor of the received signal in the cross-polarised plane to that in the original (i.e. transmitted) plane of polarisation is called Linear Depolarisation Ratio (LDR); (it is usually expressed in decibels).

$$\mathrm{LDR_H} \text{ in dB} = 10\log\frac{Z_{HV}}{Z_{HH}} = Z_{HV} \text{ in dB} - Z_{HH} \text{ in dB} \tag{2.70}$$

$$\mathrm{LDR_V} \text{ in dB} = 10\log\frac{Z_{VH}}{Z_{VV}} = Z_{VH} \text{ in dB} - Z_{VV} \text{ in dB} \tag{2.71}$$

LDR is obviously zero (i.e. $-\infty$ dB) in the case of a spherical particle or one oriented with one axis along the direction of the incident electric vector. Note therefore that a flattened drop will also have a zero LDR if it falls with its axes vertical and horizontal. In practice, the drops will be tumbling and have random orientations, and the incident beam will not be horizontal; there will also be some depolarisation in the path to the target. Hence a finite LDR will result. Typically LDR may be between –35 and –15 dB depending on precipitation rate and presence of hail or bright band[9].

2.2.11 Dual Polarisation Measurements

A dual polarisation radar provides for switching the plane of polarisation from horizontal to vertical and back to horizontal in successive pulses or sets of pulses (see Chapter 3). The horizontally and vertically polarised signals are received by separate channels in the receiver. If the Z_e values in the horizontal and vertical channels are denoted by Z_H and Z_V (they are really Z_{HH} and Z_{VV} as defined in the last paragraph), their ratio Z_{DR} is called the differential reflectivity (and is expressed in dB).

$$Z_{DR} \text{ in dB} = 10\log\frac{Z_H}{Z_V} = Z_H \text{ in dB} - Z_V \text{ in dB} \tag{2.72}$$

From equations (2.26) and (2.55) we get,

$$Z = 10^6\lambda^4\pi^{-5}|K|^{-2}\int_0^\infty \sigma_D N_D dD \tag{2.73}$$

where D is the diameter of a spherical drop in cm. Z_H and Z_V for non-spherical

9 Bright band is discussed in Chapter 4.

drops can be obtained from equations (2.65) and (2.66) as

$$Z_H = C\int_0^{\infty} D_{eq}^6 S_H N_{D_{eq}}\, dD_{eq} \tag{2.74}$$

$$Z_V = C\int_0^{\infty} D_{eq}^6 S_V N_{D_{eq}}\, dD_{eq} \tag{2.75}$$

where

$$C = \frac{16}{9}\times 10^6 \pi^2 |K|^{-2} \tag{2.76}$$

$$S_H = \left|\frac{m^2-1}{4\pi + (m^2-1)P'}\right|^2 \tag{2.77}$$

$$S_V = \left|\frac{m^2-1}{4\pi + (m^2-1)P}\right|^2 \tag{2.78}$$

S_H and S_V being called the shape functions for a drop of refractive index m and axial dimensions a and b.

For drops of equivalent diameter D_{eq}, equation (2.39) can be rewritten as,

$$N_{D_{eq}} = N_0 \cdot \exp\left[\left(\frac{3.67}{D_0}\right) D_{eq}\right] \tag{2.79}$$

Hence values of Z_H and Z_V normalised to N_0 are:

$$\frac{Z_H}{N_0} = C\int D_{eq}^6 S_H \exp\left[\frac{-3.67}{D_0} D_{eq}\right] dD_{eq} \tag{2.80}$$

$$\frac{Z_V}{N_0} = C\int D_{eq}^6 S_V \exp\left[\frac{-3.67}{D_0} D_{eq}\right] dD_{eq} \tag{2.81}$$

Therefore Z_{DR} in dB is ten times the logarithm of the ratio of the right hand sides of (2.80) and (2.81). C cancels out. Z_{DR} depends only on D_0 and is independent of radar parameters *provided* the same particles are observed at both polarisations and the radar performance does not change between the two observations.

For drops of D_{eq} less than 0.8 cm, Seliga et al. (1981) evaluated Z_H/N_0 and Z_{DR} from the above equations and plotted them against D_0 (Fig. 2.6). They assumed that drops were oblate spheroids falling with major axis horizontal. Using a radar measurement of Z_{DR}, D_0 can be obtained from the figure. Using this value, Z_H/N_0 is also got from the same figure. Since Z_H is known from measurement, N_0 is determined. Hence both the parameters of the drop size distribution of equation

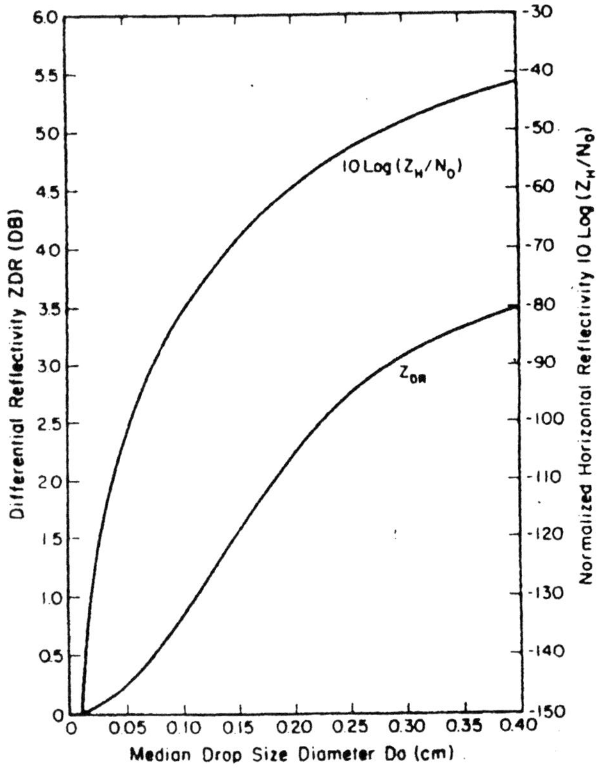

Fig. 2.6. Differential Reflectivity Z_{DR} in dB and normalised horizontal reflectivity factor $10\log(Z_H/N_0)$ plotted as a function of median volume diameter D_0. Z_H is in cm^6 m^{-3}. Maximum diameter considered is 0.8 cm (from Seliga et al. 1981; courtesy AMS).

are known. For this case, equation (2.53) for R can be rewritten as,

$$R = 0.6\pi \int N_0 \exp\left[\frac{-3.67}{D_0} D_{eq}\right] V_{D_{eq}} D_{eq}^3 dD_{eq} \tag{2.82}$$

From this they obtained the rainfall rate. This is likely to be more accurate than the rain rate estimate obtainable from the statistical relation $Z = A.R^b$ of equation because in the latter case the drop size distribution is not known. This will be further discussed in Chapter 7.

From measurement of Z_{DR} in addition to Z_e (Z_H or Z_V) the departure from sphericity can be evaluated and the phase (liquid or solid) of the particles and the presence of hail can be inferred. Large raindrops exhibit high Z_e accompanied by high Z_{DR}, while large hail which tends to fall with longer axis vertical will have a combination of high Z_e and low or negative Z_{DR}. If an attenuating wavelength is used, the two polarisations suffer different attenuation (see section 2.3). This will result in Z_{DR} being underestimated. An overestimate of LDR will also result.

Golestani et al. (1989) defined another parameter Z_{DP} as,

$$Z_{DP} = 10\log(Z_H - Z_V) \text{ where } Z_H > Z_V \tag{2.83}$$

They found that in the case of rain, Z_{DP} had a near-perfect straight-line relationship with Z_H. Where ice was present, there was a departure from this straight-line and this could be used to evaluate the reflectivity-weighted fraction of ice to rain particles in mixed phase precipitation (see also Chandrasekar et al. 1991).

Another useful parameter is the "correlation coefficient at zero lag" [represented by the symbol $\rho_{HV}(0)$ between the H and V signals received at the same time. If the H and V pulses are not transmitted or received simultaneously, $\rho_{HV}(0)$ has to be derived from the observed correlation between the sequential H and V signals. If only one polarisation is transmitted, the cross-correlation between the received copolar and cross-polar signals is also a quantity of interest.

2.2.12 Circular Polarisation

Consider a radar transmitting a circularly polarised wave (say right-handed). Circular polarisation signifies that the electric vector rotates in the plane perpendicular to the direction of propagation. The rotation may be considered as made up of oscillating electric vectors in two orthogonal directions with a phase difference of $\pi/2$. Rotation of the electric vector clockwise as seen by looking in the direction of propagation is termed right-handed circular polarisation. A spherical scatterer will return a signal, which is circularly polarised in the opposite direction (left-handed in this case), which the radar receiver will not detect. This property is used in non-meteorological radars to minimise echo from rainfall, which is considered unwanted clutter. A non-spherical particle will, however, return components in both senses of circular polarisation. If both the components are detected and measured, the ratio of the equivalent radar reflectivity factor Z_{RR} in the "parallel" (i.e. in the transmitted mode of polarisation, say, right circular) component to that (Z_{RL}) in the "orthogonal" (i.e. in the opposite sense of circular polarisation, left circular) component is called the Circular Depolarisation Ratio (CDR) and is expressed in dB.

$$\text{CDR in dB} = 10\log\frac{Z_{RR}}{Z_{RL}} \tag{2.84}$$

CDR can be used to detect hydrometeor shape, distinguish hail and study lightning discharges. Applications of polarisation diversity observations will be discussed further in later chapters.

2.2.13 Differential Phase Shift

We have thus far considered only the amplitude of the scattered signal and not its phase. In free space propagation the only phase shift, which occurs in the electromagnetic wave is $2\pi r/\lambda$ each way, due to range [see equation (1.2)]. $2\pi/\lambda$ is called the free space propagation constant, also represented by the symbol k_0. E.g. at $\lambda = 10$ cm, k_0 is 3.6×10^6 degrees per km one way. However, if the radiation passes through a scattering medium, the radiation incident on a target beyond this medium contains two components: (1) the transmitted wave, and (2) the forward scattering by the scatterers in the medium. Because of the different origin of these two components, their phase angles will be different. The phase of the net electric vector will thus be different from the phase of the transmitted signal. The forward scattered component depends on the polarisation of the radiation. Hence the propagation phase shift will be different for different polarisations. The difference in phase arising between the horizontally polarised and vertically polarised radiation for one way transmission through the medium is called the one way differential propagation phase shift. If the scatterers in the medium are spherical drops, this difference will be zero. In the case of non-spherical particles, if we assume them to be uniformly oriented (say with major axis horizontal i.e. zero canting angle) and quiet (i.e. not involved in collision, coalescence and break-up) the differential phase shift ϕ for travel over unit distance through the medium is (Jameson 1985),

$$\phi = \lambda \int \mathrm{Re}\left[S_{fHD_{eq}} - S_{fVD_{eq}}\right] N_{D_{eq}} dD_{eq} \tag{2.85}$$

where S_{fHDeq} and S_{fVDeq} are the forward scattering complex amplitude functions for the two polarisations for drops of equivalent diameter D_{eq} *in the path traversed*, Re indicates "Real part of" and the other symbols have their usual meanings. The integration is over all the drops along the path traversed. Hence we can define a specific differential phase shift (K_{DP}) in degrees per kilometre as,

$$K_{\mathrm{DP}} = \frac{180}{\pi} \lambda \int \mathrm{Re}\left[S_{fHD_{eq}} - S_{fVD_{eq}}\right] N_{D_{eq}} dD_{eq} \tag{2.86}$$

The suffix DP stands for Differential Phase. The two-way differential propagation phase shift Φ_{DP} over a path from range r_1 to range r_2 is given by,

$$\Phi_{\mathrm{DP}} = 2\int_{r_1}^{r_2} K_{\mathrm{DP}} dr \tag{2.87}$$

If Φ_{DP} is measured (see Chapter 3), K_{DP} can be calculated. K_{DP} has been related to precipitation rate R by power law relations of the form ,

$$R = AK_{\mathrm{DP}}^{b} \tag{2.88}$$

It is claimed that this relationship is less sensitive to DSD changes than the *Z-R* relationship and hence more accurate rain estimation can be made by measuring K_{DP} (Sachidananda and Zrnić 1986). K_{DP} has also been used to distinguish hail. See Balakrishnan and Zrnić 1990, Bringi et al. 1990, Chandrasekar et al. 1990, Bringi and Chandrasekar 2001, for further discussion on K_{DP} and its applications. Use of K_{DP} for precipitation estimation will be discussed in Chapter 7.

Since the forward-scattered radiation will have a cross-polarised component, the differential phase shift will produce a spurious increase in measured LDR. Since the phase shift increases with path length, rain rate and radar fequency, so will the error in LDR. It should be noted that in the case of circular polarisation, which may be regarded as the resultant of horizontal and vertical polarisations with a 90° phase difference, the differential propagation phase shift will alter this phase difference. Hence the circularly polarised wave becomes elliptically polarised and will lead to errors in interpretation based on assuming circular polarisation.

The nomenclature of 'horizontal' and 'vertical' polarisation implies a nearly horizontal direction of propagation. If, however, the radar antenna is pointing vertically, Z_{DR} and K_{DP} lose their significance. However, since cross-polarised signals can be received, LDR is still relevant. Also, the quantity $\rho_{HV}(0)$ and the cross-correlation between co-polar and cross-polar signals are meaningful and can be used for discriminating shape, size, orientation and phase of hydrometeors (see e.g. Zrnić et al.1993). Matrosov et al. (1996), using K-band radar, show that elevation angle dependence of LDR and CDR can be used to distinguish between different types of ice hydrometeors. Reviews of dual polarisation measurements may be seen in Herzegh and Jameson (1992) and Bringi and Chandrasekar (2001).

2.2.14 Scattering by Hail

Hail is formed in thunderstorms by accretion of water drops over a nucleus of ice. The hailstone may be taken up and down several times within the storm by the up- and down-drafts, undergoing partial melting, refreezing and repeated accretion of ice or water. Hence the stone may grow to a size of several cm in diameter and in extreme cases to tens of cm. The stone may have several layers of solid ice, "spongy ice" and water with occlusion of air at times. The large size, the irregular shape and the mixture of dielectrics complicates the scattering properties. As may be seen from Fig. 2.5, the back-scattering cross section σ may oscillate in value as the particle grows. Further, there may be internal reflections at the boundaries of the various layers leading to an increase of σ. Hail cross sections have been determined empirically or by calculation making several assumptions about size distribution (Atlas et al. 1960, Atlas and Ludlam 1961, Atlas and Wexler 1963, Bohren and Battan 1980). Cheng and English (1983) determined size distributions in Alberta (Canada) of hail samples at the ground and proposed an equation,

$$N_D = 115\Lambda^{3.63}\exp(-\Lambda D) \tag{2.89}$$

Aydin et al. (1984) and Longtin et al. (1987) have analysed the polarisation parameters of large spongy ice oblate spheroids. Equations similar to (2.56) relating Z and hail fall rate (liquid water equivalent) have also been proposed. Hail will produce very large values of Z_e (up to 75 dBZ). Unless this is recognised, an overestimation of R from equation (2.63) will result. The presence of hail can be inferred qualitatively from the very high Z_e, the vertical extent of high values of Z_e or the physical configuration of the echo. On Doppler radar, the fall velocities of hail as well as its size distribution can be measured. Because of the oscillation of the Mie scattering cross section, the Z at 10 cm wavelength is higher than at 3 cm when hail diameter exceeds about 1 cm (Atlas and Ludlam, 1961). Hence dual wavelength radar may be able to identify hail. A more positive identification and quantification of hail is possible (see Chapter 6), by measurement of polarisation parameters (Z_{DR} , LDR, CDR, K_{DP}).

2.2.15 Non-precipitating Clouds

In the case of non-precipitating clouds and fog where the drop diameter is less than 10^{-4} m, the reflectivity will be too small for detection at wavelengths of 3 cm or longer. However a "mantle" echo can be obtained from a cloud boundary because of the refractivity discontinuity. At higher frequencies e.g. the K band, clouds can be detected at short ranges. Hence radars at such frequencies are used for determination of cloud-base heights and for cloud microphysics studies.

2.2.16 Scattering from Clear Air

Radar return as a result of Rayleigh scattering from air molecules is negligible at microwave wavelengths because of the extremely low value of σ but detectable echo is obtainable from scattering or reflection from inhomogeneities in the refractive index. Three mechanisms for the production of radar returns can be identified viz. Bragg scattering, Fresnel reflection and Fresnel scattering. The first one named by analogy with X-ray scattering in crystals, is due to random fluctuations of the refractive index caused by turbulence on a scale approximating to half the radar wavelength. Fresnel reflection occurs if a vertically pointing radar beam is incident on a thin horizontal layer with a sharp discontinuity in refractive index across it. Fresnel scattering occurs if several refractive index discontinuities are present along the pointing direction of the radar beam. Fresnel reflection and Fresnel scattering are important at VHF and UHF frequencies particularly at vertical incidence.

2.2.17 Turbulent Eddies

Turbulent eddy motions in the atmosphere have a wide spectrum of sizes ranging from kilometre to millimetre scales. Eddies are unstable and break up into eddies of smaller sizes. The energy of the smallest eddies is ultimately dissipated by viscous heating. Three spectral regions are recognised.

(1) The energy-containing range which has the bulk of the turbulent energy and where energy is produced by buoyancy and shear,

(2) The inertial subrange where energy is neither produced nor dissipated but handed down to smaller and smaller scales. This contains the intermediate scales of eddies. (Turbulence in this range is assumed to be isotropic implying that the velocity field is independent of translation, rotation and reflection of the spatial axes.)

(3) The dissipation range where kinetic energy is converted into internal energy.

According to the "Kolmogorov hypothesis", the spectral densities within the inertial subrange are proportional to $K^{-5/3}$ where K is the magnitude of the three-dimensional wave number[10]. These spectral densities can be used to find the rate of energy transfer within the inertial subrange, which is called the energy dissipation rate ε (units m^2 s^{-3}). This quantity will be discussed in Chapters 6 and 10.

The theory of turbulent scattering is dealt with in detail in Tatarskii (1961, 1971), and Gossard and Strauch (1983). Only a brief introduction follows. If the refractive index is $n(r_0)$ at position r_0 and $n(r_0+r)$ at position r_0+r, a quantity C_n^2 known as refractive index structure constant is defined by,

$$\overline{|n(r_0+r)-n(r_0)|^2} = C_n^2 r^{2/3} \tag{2.90}$$

The overbar indicates averaging. C_n^2, expressed in units of m$^{-2/3}$, is a measure of the refractive index fluctuations due to turbulence. Tatarskii (1971) gives the relationship,

$$C_n^2 = a^2 \alpha' L_0^{4/3} M^2 \tag{2.91}$$

---

[10] $K = \left[K_x^2 + K_y^2 + K_z^2\right]^{1/2}$ where K_x is the one-dimensional wavenumber along the longitudinal (i.e. say the radar beam direction) direction x and K_y, K_z are the wavenumbers in the transverse directions.

where a^2 is a constant taken as 2.8 (VanZandt et al. 1978), α' is the ratio of eddy thermal diffusivity to eddy viscosity, usually taken as unity, L_0 is the outer scale length of the turbulence spectrum (i.e. the transition scale between the energy input range and the inertial subrange[11]), and M is the mean generalised vertical gradient of the potential refractive index. Potential refractive index ϕ is defined as,

$$\phi = \frac{77.6}{\theta}\left[1000 + \frac{4810e_p}{\theta}\right] \tag{2.92}$$

where θ is the potential temperature(K) and e_p the potential water vapour pressure (hPa). According to Tatarskii (1971), M is given (for metre and centimetre wavelengths) by[12],

$$M = \frac{-79\times10^{-6}P}{T^2}\left\{1+\frac{15{,}500q}{T}\right\}\left\{\frac{dT}{dz}+\Gamma_a - \frac{7800}{1+\dfrac{15{,}500}{T}}\cdot\frac{dq}{dz}\right\} \tag{2.93}$$

q being the specific humidity, Γ_a the dry adiabatic lapse rate and z the vertical coordinate. VanZandt et al. (1978) computed M from radiosonde measurements and C_n^2 from equation (2.91) assuming the values of constants as mentioned above. Sarkar et al. (1985) give values of C_n^2 over India of the order of $10^{-14}\,m^{-2/3}$ at the surface to about $10^{-16}\,m^{-2/3}$ at an altitude of 5 km. Radar, however, will detect turbulence only if the radar wavelength lies in the inertial subrange. If the turbulence fills only a fraction F of the radar-sampled volume, C_n^2 derived from radar measurement will be less than the value computed from radiosonde measurements and may be designated as $C_n^2(r) = F.\ C_n^2$.

The volume reflectivity η ($m^2\,m^{-3}$) due to turbulent Bragg scatter is given by (Ottersten, 1969),

$$\eta = 0.38C_n^2(r)\lambda^{-1/3} \tag{2.94}$$

Note that the wavelength dependence is on the 1/3 power (as distinct from the 4th power in the case of particles). This equation has been verified by comparison of radar observations with airborne microwave refractometer measurements in the volume sensed by radar (Kropfli et al. 1968). The exponent –1/3 has also been verified by Cohn (1994) using radars at wavelengths of 68, 23 and 3 cm. The received power P_r can be got by putting this value in equation (1.28). It is, however, considered more convenient to use the antenna aperture A instead of G, ϕ and θ, in this equation. Using the relations given in equations

---

[11] VanZandt et al. (1978) took L_0 to be 10 m.

[12] This M is not to be confused with M the modified refractivity defined in equation (2.15). Unfortunately the same symbol is commonly used in the literature for both.

(1.25) and (1.35), we get,

$$P_r = \frac{\pi k^2 P_t h A}{128(2\ln 2) r^2} \cdot \eta \tag{2.95}$$

Substituting for η from equation (2.94) we get a relation between P_r and C_n^2, which can be used to compute the latter from radar observations (Nastrom et al. 1981). Chadwick and Moran (1980) determined C_n^2 at a height of 805 m over a period of one year (at Boulder, Colorado, USA) with a 10 cm FM-CW radar. They obtained monthly averages ranging from 3×10^{-17} to 2×10^{-15} m$^{-2/3}$. C_n^2 was maximum at midnight and early afternoon (see Chapter 10 for a further discussion of C_n^2 measurements).

The reflectivity per unit volume for an ensemble of Rayleigh-scattering particles is given by equation (1.20). We can use that equation along with equation (2.94) to define an equivalent radar reflectivity factor Z_e, for the turbulent scatter case. If we put Z_e in place of ΣD^6 in equation (1.20), equate the two expressions for η, and put $|K|^2$ equal to 0.93, the value for water[13], we get,

$$\frac{\pi^5}{\lambda^4}|K|^2 Z_e = 0.38 C_n^2 \lambda^{-1/3} \tag{2.96}$$

or

$$Z_e = \frac{0.38 C_n^2 \lambda^{11/3}}{0.93\pi^5} \tag{2.97}$$

Z_e is the reflectivity factor of a volume of particle scatterers, which will produce the same echo power as the Bragg scatter echo. However this Z_e (unlike its particle counterpart) is dependent on the wavelength (Pratte and Keeler 1986). This definition makes it convenient to evaluate the likely echo return at various ranges from Bragg scatter. Typical values of scale of turbulence in the free atmosphere may be several tens of centimetres and hence the echoes are best observed on longer wavelength radars. At an airmass boundary, a typical value of 10^{-14} m$^{-2/3}$ can be expected for C_n^2, yielding dBZ equal to –25 at a wavelength of 10 cm. C_n^2 calculated from radar observations varies considerably with time (see Chapter 10).

2.2.18 Fresnel Reflection and Scattering

P_r from Fresnel reflection at a discontinuity layer at normal incidence is given by (Gage et al. 1981),

---

[13] By convention, the value for water is used.

$$P_r = \frac{k^2 P_t A^2}{4\lambda^2 r^2} |\rho|^2 \tag{2.98}$$

where k^2 is an antenna efficiency factor and ρ is the amplitude reflection coefficient of the discontinuity which depends on the wavelength of the radar and the vertical gradient of refractive index across the layer. Fresnel scattering is the coherent scattering process that is pertinent when the refractive index irregularities that comprise the scattering medium are coherent in the plane transverse to the probing wave and random in the dimension parallel to the wave direction. Gage et al. (1981, 1985) and Röttger and Larsen (1990) may be seen for a detailed treatment of these phenomena.

2.3 Attenuation

2.3.1 Absorption

The dielectric constant ε of any substance is, in general, a complex quantity. We put,

$$\varepsilon = \varepsilon_1 - i\varepsilon_2 \tag{2.99}$$

The real part ε_1 represents the refraction while the imaginary part represents the absorption in the medium, and the two are not independent of each other. Absorption represents the energy abstracted from the radiation and includes scattering. In the case of air, the absorption may be due to molecules of the constituent gases including water vapour or by liquid or solid particles (dust, or condensed particles in fog, cloud, rain or snow). The reduction dP in the incident power P transmitted over a path length dr is given by,

$$dP = k_L P dr \tag{2.100}$$

where k_L is the fraction of power attenuated per unit path length. Integrating over a range 0 to r,

$$\ln \frac{P_r}{P_0} = -\int_0^r k_L dr \tag{2.101}$$

at frequencies not close to an absorption band. Or,

$$10 \log \frac{P_r}{P_0} = -\int_0^r k_a dr \tag{2.102}$$

where k_a is the attenuation in decibels per unit length, given by,

$$k_a = k_L \cdot 10 \log e \tag{2.103}$$

If a number of constituents are involved, the attenuation due to each in decibels can be added and,

$$10 \log \frac{P_r}{P_0} = -\int_0^r (k_a + k_b + k_c \cdots) dr \tag{2.104}$$

2.3.2 Attenuation by Oxygen and Water Vapour

The main absorbing gases are oxygen and water vapour because of their magnetic and electric dipole moments respectively. Oxygen has an absorption band around 60 GHz. and water vapour at 22 GHz. Data on attenuation by atmospheric oxygen and water vapour are based on the original work of Van Vleck (1947a,b) as reviewed by Gunn and East (1954) and Bean and Dutton (1968). At sea level pressure and temperature of 20°C, oxygen attenuation is approximately 0.007 dB km^{-1} at 3 GHz. At the same temperature, pressure and frequency, the water vapour absorption is 0.07×10^{-3} dB km^{-1} per g m^{-3}. The one way attenuation at sea level in the tropics due to both these constituents can be taken as about 0.01 dB km^{-1} and the two-way attenuation as 0.02 dB km^{-1}. However, since the radar beam does not travel at sea level, lower temperature increases the absorption slightly at higher levels while lower pressure decreases it substantially. Lower water vapour content at higher levels will decrease the absorption, but in humid environments the beam may be refracted down thus passing through higher humidity levels (Bean and Dutton 1968), thus countering the decrease. Skolnik (1970) has given curves of total two-way attenuation corresponding to various ranges, beam elevations and radar frequencies, assuming an ICAO standard atmosphere and surface water vapour content of 7.75 g m^{-3}. In the humid tropics the curves may need modification. In measurements of received power at frequencies well away from absorption bands, a correction may be made in equation (1.30) by adding a term $-2k_a r$ on the right hand side where k_a is the one way attenuation in dB per unit distance.

2.3.3 Attenuation by Cloud (or Fog) Particles

Attenuation by cloud (or fog) particles can be considered as follows. The total attenuation cross section $Q(a, \lambda)$ of a sphere is made up of Q_s the scattering cross section and Q_a the absorption cross section (see Para 2.2.1). In the case of very small particles ($\alpha \ll 1$) higher orders of a_n and b_n in equations (2.19) and (2.20) can be neglected and we get the Rayleigh approximations

$$Q_s = \frac{\lambda^2}{2\pi} \cdot \frac{4}{3} \alpha^6 |K|^2 \tag{2.105}$$

$$Q_a = \frac{\lambda^2}{2\pi} 2\alpha^3 \, \mathrm{Im}|K| \tag{2.106}$$

Im means "Imaginary part of", α is $2\pi a/\lambda$, a being particle radius. Particles in non-precipitating cloud and fog being less than 10^{-4} m in diameter, Q_a, which is proportional to α^3 is far larger than Q_s which is proportional to α^6. Hence the latter can be neglected and the total attenuation Q due to the particle put equal to Q_a. If we consider the attenuation due to all the particles in a path length r of unit cross section, the reduction of power over the distance r is given by [by analogy with equation (2.101)],

$$\ln \frac{P_r}{P_0} = -\int_0^r \left(\sum Q\right) dr \tag{2.107}$$

where ΣQ is the sum of the attenuation cross sections of the particles in unit cross section and unit length i.e. in unit volume. Hence,

$$10 \log \frac{P_r}{P_0} = \log e \frac{\lambda^2}{2\pi} 2 \sum \alpha^3 \, \mathrm{Im}|K|$$

$$= 0.4343 \times \frac{8\pi^2}{\lambda} \sum a^3 \, \mathrm{Im}|K| \tag{2.108}$$

Equation (2.108) will give the attenuation in dB km^{-1} also called specific attenuation if we put $r = 1$ km and take the summation of Q in cm^2 over a volume of 1 m^3, λ being in metres. The liquid water content M in unit volume (g m^{-3}) of the cloud is

$$M = \frac{4}{3} \pi \rho \sum a^3 \tag{2.109}$$

ρ being the density of water. Hence attenuation in dB km^{-1} is,

$$= 0.4343 \cdot 6\pi \cdot M \lambda^{-1} \rho^{-1} \, \mathrm{Im}|K| \tag{2.110}$$

This is independent of the drop size distribution in the cloud. The attenuation increases with frequency. In the case of ice clouds (density of ice taken as 1 g cm^{-3}) the one way attenuation ranges from 8 to 80 in units of 10^{-4} dB km^{-1} per unit M, for the frequency range 3 to 30 GHz and can be neglected. In the case of water clouds the attenuation is 0.005 dB km^{-1} per unit M at 3 GHz and 0.02 dB km^{-1} per unit M at 6 GHz and can again be neglected. At higher frequencies the attenuation increases sharply and cannot be neglected.

2.3.4 Attenuation by Rain

In the case of raindrops (1 mm or more in diameter), Q_s cannot be neglected in comparison with Q_a, and higher orders of a_n and b_n have to be considered. Since higher powers of D (drop diameter) than 3 are involved, the attenuation is highly dependent on drop size distribution – the large drops contributing most – and cannot be simply related to M as in the case of cloud drops. The specific attenuation k_R in dB km^{-1} can be expressed as,

$$k_R = 0.4343 \int N_D Q_D dD \tag{2.111}$$

the integration being over all D (Kodaira 1963). Here Q has been replaced by Q_D to show its dependence on D. It also depends on wavelength and temperature. Atlas and Ulbrich (1974) expressed Q_D by a power law relationship,

$$Q_D = CD^n \tag{2.112}$$

where C and n are constants. Then, assuming an exponential drop size distribution as in equation (2.31), equation (2.111) can be rewritten as,

$$\begin{aligned} k_R &= 0.4343 N_0 C \int_0^\infty D^n e^{-\lambda D} dD \\ &= 0.4343 \Gamma(n+1) C N_0 \Lambda^{-(n+1)} \end{aligned} \tag{2.113}$$

and from equation (2.38)

$$k_R = 0.4343 \Gamma(n+1) C N_0 \left(\frac{D_0}{3.67} \right)^{(n+1)} \tag{2.114}$$

They plotted C and n for various wavelengths and temperatures and showed that at a wavelength of 0.88 cm, $n = 3.67$. Putting this value in equation (2.114) and comparing with equation (2.54) shows that at this wavelength R and k_R are linearly related. At other wavelengths k_R can be related to R by a power law relation,

$$k_R = A'' R^{b''} \tag{2.115}$$

Table 2.4 based on Gunn and East (1954), gives the relationships at various wavelengths at a temperature of 18°C. Similarly power law relations of the form $k_R = cZ^d$ between k_R and Z have been derived from the Z–R and k_R–R relations or from assumed drop size distributions. Some of them may be seen in Battan (1973). Expressions relating Z, k_R and R to the drop size distribution parameters are given in Chapter 7.

Note that the definition of k_R in dB km^{-1} implies rainfall at a rate of R mm h^{-1} extending over a distance of 1 km along the path of the radiation. Hence in the equation (1.30) for the received power P_r from a target at range r, a quantity k_{12} given by,

$$k_{12} = 2\int_{r_1}^{r_2} A''R_i^{b''}dr \tag{2.116}$$

Table 2.4. Rainfall attenuation, one way at various wavelengths.

Wavelength (cm)	k_R dB km^{-1} one way $A''R^{b''}$
10	$0.0003R^{1.00}$
5.7	$0.0022R^{1.17}$
3.2	$0.0074R^{1.31}$
1.8	$0.045R^{1.14}$
0.9	$0.22R^{1.00}$

must be subtracted from the right hand side where R_i is the rate of rainfall in the range interval r_1 km to r_2 km in the path between the radar and the desired target at range r. This requires a knowledge of the rainfall rate throughout the intervening path if one wishes to obtain a quantitative estimate of the precipitation rate at range r. This problem can be avoided by using a wavelength at which the rainfall attenuation is very small. For example, at 10 cm wavelength, the 2-way attenuation due to a rain rate of 100 mm h^{-1} extending[14] over 1 km is only 0.06 dB and can be ignored in comparison with other measurement errors. At a wavelength of 5.7 cm the attenuation is still not too high if the rain rate in the intervening path is low, as is usually the case at temperate latitudes. At this or shorter wavelengths the quantitative measurement may be corrected as above – which is possible but the result may not be satisfactory (see Chapter 7 for a further discussion).

Moreover the attenuation may take the P_r from echoes from long ranges below the MDS of the radar (especially at very short wavelengths) and thus reduce the effective maximum range of the radar. Further, the shape of an echo from a distant area of heavy rain will be distorted even if there is no intervening precipitation. While its nearer edge will be unaffected by attenuation the farther part of the echo may not be seen. Thus the centroid of the echo will be shifted towards the radar. For these reasons a 10 cm wavelength is preferred for precipitation measurement where heavy rainfall over long distances is expected (as e.g. in tropical cyclone detection) unless there are special reasons to choose a shorter wavelength.

In the case of dual polarisation measurements, we can define specific attenuations k_H or k_V as,

$$k_{(H,V)} = 0.4343\int Q_{D(H,V)} N_{D_{eq}} dD_{eq} \tag{2.117}$$

where $Q_{D(H,V)}$ are the attenuation cross sections for horizontally and vertically

[14] Such high rain rates are unlikely to persist over long path lengths.

polarised radiation. The difference $k_H - k_V$ in dB km^{-1} is called the specific differential attenuation k_{Dif} which can also be written as (Jameson 1994),

$$k_{\text{Dif}} = 0.4343\lambda \int \text{Im}\left(S_{fHD_{eq}} - S_{fVD_{eq}}\right) N_{D_{eq}} dD_{eq} \qquad (2.118)$$

The one way differential attenuation for a path length r_1 to r_2 can be obtained by integrating k_{Dif} over the path. Since k_H is greater than k_V, it follows that Z_{DR} may be underestimated (and LDR overestimated) if the measurements are at an attenuating wavelength. The error will increase with path length and rain rate. A correction can be made if the differential phase shift can be measured accurately (Bringi et al. 1990).

2.3.5 Estimation of Rain Rate from Attenuation

From Table 2.4 it is seen that at wavelengths close to 1 cm the attenuation is nearly proportional to R. Hence, if the received power from a fixed target (say, a passive reflector) at range r during rain is compared with that received when there is no rain, the difference will yield the rate of rainfall in the range 0 to r. This is an alternative to the method of precipitation estimation from reflectivity factor measurements (Collis et al. 1963, Atlas 1964, Atlas and Ulbrich 1977; see Chapter 7 for a further discussion). This usually gives a path-integrated value which is useful in radio wave propagation studies. This method has also been proposed for precipitation measurements from space (Chapter 7).

2.3.6 Attenuation by Snow

In the case of snow, Gunn and East (1954)[15] give k_R at 0°C as,

$$k_R = 0.00368 R^{1.6} \lambda^{-4} + 0.00224 R \lambda^{-1} \qquad (2.119)$$

the two terms being due to scattering and absorption respectively. R is in mm h^{-1} of melted water. Battan (1973) gives a slightly different relationship:

$$k_R = 0.035 R^2 \lambda^{-4} + 0.0022 R \lambda^{-1} \qquad (2.120)$$

This is true if the snowflakes can be assumed to be small spheres. Since snow can be in the form of flat plates or needle shapes their attenuation can be greater depending on their orientation, for the reasons discussed in Para 2.2.9. The R values involved being small, the attenuation due to snow at wavelengths of 3 cm or more is usually neglected except at large ranges. In radio wave propagation

[15] as corrected vide Corrigendum Quart. J. Royal Meteor. Soc., 1955, **81**, p653.

calculations also it is convenient to neglect attenuation by snow. When there is mixed phase precipitation, communication engineers define a "rain height" up to which rain exists as liquid drops and neglect the solid precipitation above. However when snow is in the melting stage the outer layer is of liquid water and will cause attenuation as though it was a water particle. Since the snow particle can be considered to be an ellipsoid usually with its long axis horizontal, the attenuation, especially for a horizontally polarised beam, will increase considerably at the melting stage (see Chapter 4 for a discussion of the melting band).

2.3.7 Attenuation by Hail

Calculations of attenuation cross sections of large spheres of water or ice or a mixture of the two have been made by several workers (e.g. Battan and Herman 1962, Battan 1971). The normalised attenuation cross section i.e. $Q/(\pi a^2)$ for water spheres was found to increase gradually with α to a value of about 3. For dry ice spheres this quantity oscillates with increasing α and reaches a value of about 2 at a large value of α. For mixtures, if ice predominates, the attenuation is mostly due to scattering. If water predominates, the absorption is also appreciable. In the case of hail taken as water-coated ice spheres, if the size distribution [an exponential distribution similar to equation (2.31)] is assumed, the attenuation can be computed as a function of maximum stone size and water shell thickness (Battan 1971). Since the stones of large size are small in number, the smaller hail stones up to 2 cm diameter contribute most to the attenuation. Sauvageot (1991) lists values as in Table 2.5 for hail with maximum diameter of 2.89 cm.

2.3.8 Attenuation due to a Radome

If the radar antenna is enclosed in a radome as is often the case, the material of the radome produces a small attenuation which can be taken account of in equation (1.30) by adding a loss term (say–0.3 dB) on the right hand side. However, when there is rain, a thin film of water forms on the radome surface and

Table 2.5 Attenuation due to hail.

λ (cm)	Attenuation dB km^{-1}	
	Dry Hail	Water-coated hail
3.2	1.66	3.79
5.5	0.33	2.60
10	0.034	1.18

causes additional attenuation. The thickness d of the film is given by Gibble (1964) in the case of a spherical radome by the relation,

$$d^3 = \frac{3}{2}\mu r_r \frac{R}{W} \tag{2.121}$$

where R is the rain rate at the radar site, r_r is the radius of the radome, μ the viscosity of water and W equal to ρg the weight density of water. (All the quantities should be in SI units). Based on this Kodaira (1969, personal communication) put,

$$d \text{ in mm } = 0.0075\left(Rr_r\right)^{1/3} \tag{2.122}$$

where R is in mm h^{-1} and r_r in cm. He obtained the two-way attenuation for various rain rates for typical radomes as in Table 2.6.

At high rainfall rates these losses are far higher than the transmission loss due to the radome material itself. Radar measurement of P_r will therefore require a correction during heavy rain at radar site. By comparison, the reflection loss due to an uncovered antenna reflector collecting a thin film of water on its surface is negligible (Blevis 1965). Modern radomes have water-repellant coatings and are claimed to have very low attenuation in rain.

2.3.9 Range Attenuation

Equation (1.29) for the power received from an ensemble of precipitation particles contains the factor r^{-2}. Thus the radar signal gets attenuated in proportion to the square of the range and this is termed range attenuation. It is standard practice to correct for this while comparing echo intensities from precipitation at different ranges (see Chapter 3).

Table 2.6. Two-way attenuation due to water film on radome.

R (mm h^{-1})	Attenuation dB two-way	
	$\lambda = 5.7$ cm; $r_r = 275$ cm	$\lambda = 10$ cm; $r_r = 450$ cm
1	0	0
10	1.4	1.4
20	2.0	2.0
40	3.0	2.4
100	4.6	3.0
200	6.0	4.0
400	8.0	5.6

Chapter 3

WEATHER RADAR SIGNAL PROCESSING AND DISPLAY

In the first section of this chapter we deal with the nature of the received signal, the need for averaging signals, application of various corrections and display of reflectivity. This is followed in section 3.2 with the digital processing of reflectivity and parameters derived from it. Section 3.3 deals with methods of processing of Doppler parameters viz. mean velocity and spectrum width and resolution of ambiguities. The use of multiple-Doppler, dual-Doppler and single Doppler radars and characteristic signatures of wind patterns are also discussed in this section. In section 3.4 we discuss methods of recognition and suppression of ground clutter and anomalous propagation echoes. Section 3.5 deals with techniques used on airborne weather radar. In section 3.6 we describe some modern ground-based operational Doppler radars. Section 3.7 deals with the implementation of polarimetric capability and describes some polarimetric radars.

3.1 Signal Averaging and Display of Reflectivity

3.1.1 Scattering by an Ensemble of Particles

As seen in Chapter 1, the back-scattered power received by the radar is the sum of the back-scattered power from the various particles. The particles are moving with respect to each other and hence the back-scattered fields from the particles are not in the same phase. Therefore the instantaneous signal amplitude at the radar receiver will not simply be the sum of the individual amplitudes. The field *A(t)* from a single particle *j* at instant *t* is given by (Atlas 1964, Battan 1973),

$$A(t) = \mathrm{Re}\left[\delta_j(t) \cdot e^{-i \cdot 2\pi f_0 t}\right] \tag{3.1}$$

where Re means "real part of", $\delta_j(t)$ is the complex amplitude and f_0 the transmitting frequency of the radar. If the particle moves with (radial) velocity V_j corresponding to Doppler shift f_j, the complex amplitude $\delta_j(t)$ will be,

$$\delta_j(t) = \delta_j \cdot e^{-i \cdot 2\pi f_j t} \tag{3.2}$$

where δ_j is a constant amplitude. *A(t)* will then be,

$$A(t) = \mathrm{Re}\left[\delta_j \cdot e^{-i \cdot 2\pi f_j t} \cdot e^{-i \cdot 2\pi f_0 t}\right] = \delta_j \cos 2\pi \left(f_0 + f_j\right) t \tag{3.3}$$

By measuring *A(t)* and its variation with time, the Doppler shift f_j can be determined. The average signal power P_r over one cycle is proportional to $A^2(t)$ and can therefore be written as the product of $\delta_j(t)$ and its complex conjugate. Omitting the constant of proportionality, we have,

$$P_r = \left[\delta_j \cdot e^{-i \cdot 2\pi f_j t}\right]\left[\delta_j \cdot e^{i \cdot 2\pi f_j t}\right] = \delta_j^2 \tag{3.4}$$

which is independent of its velocity. If there are two particles with different velocities we can put $j = 1$ and $j = 2$ in succession. The total instantaneous complex amplitude is,

$$\delta_{1+2}(t) = \delta_1 e^{-i \cdot 2\pi f_1 t} + \delta_2 e^{-i \cdot 2\pi f_2 t} \tag{3.5}$$

$$A(t) = \delta_1 \cos 2\pi \left(f_0 + f_1\right) t + \delta_2 \cos 2\pi \left(f_0 + f_2\right) t \tag{3.6}$$

Multiplying the quantity on the r.h.s. of equation (3.5) by its complex conjugate we get,

$$A^2(t) = P_r = \delta_1^2 + \delta_2^2 + 2\delta_1\delta_2 \cos 2\pi \left(f_2 - f_1\right) t \tag{3.7}$$

i.e. the signal contains a constant term which is the sum of the scattered power from the two particles as well as a fluctuating component. The frequency of fluctuation F_{12} is the difference between the two Doppler frequencies,

$$F_{12} = f_2 - f_1 \tag{3.8}$$

i.e. the fluctuation depends on the relative velocity of the two particles. Lhermitte (1960) extended this to a large number of particles to get,

$$P_r = \sum_j \delta_j^2 + 2\sum_{jk}\left[\delta_j \delta_k \cos 2\pi F_{jk} t\right] \tag{3.9}$$

In the first term on the r.h.s., the sum is over all *j*. The second term represents the interaction between the fields of particle $j = j$ and particle $j = k$ and the summation is over all combinations of j and k. The radar will therefore receive a signal with fluctuations of all frequencies F_{jk} representing the relative velocities of all combinations of particles. If P_r is averaged over a time large in comparison with these frequencies, the fluctuating term becomes zero and we are left with the first term on the r.h.s. which is related to the reflectivity η or reflectivity factor Z.

3.1.2 The Doppler Spectrum

P_r in equation (3.9) may be considered to be made up of contributions of particles having various radial velocities v (considered positive if directed away from the radar and negative if towards the radar). Each v corresponds to a unique value of F the Doppler frequency shift. If there are $N(v)$ particles having velocity in the interval v to $v + \delta v$ corresponding to Doppler shift interval F to $F + \delta F$ and each contributes power $P(v)$ the contribution from all such particles in the velocity interval δv is $p(v)dv$, given by,

$$p(v)dv = N(v)P(v)dv \tag{3.10}$$

The total average power $\overline{P_r}$ is obtained by integrating $p(v)dv$ over all v or all F ($-\infty$ to $+\infty$).

$$\overline{P_r} = \int p(v)dv = \int N(v)P(v)dv \tag{3.11}$$

The relative power in each interval δv as a fraction of the total power is given by the expression,

$$\frac{N(v)P(v)dv}{\int N(v)P(v)dv} \tag{3.12}$$

Since there is a very large number of scatterers in each pulse volume, scattering with all phases 0 to 2π, the signal $S(v)$ as a function of v has a Gaussian distribution as shown in Fig. 3.1. This is known as the Doppler spectrum. The area under the curve gives P_r. The peak of the curve in Fig. 3.1 is at a velocity v_m (shown by the dashed line), which, because of the symmetry of the curve, represents the mean motion of the particles. The mean velocity is given by,

$$v_m = \frac{\int vS(v)dv}{\int S(v)dv} \tag{3.13}$$

which is called the first moment of the Doppler spectrum. Here each v is weighted according to the signal $S(v)$ received from particles having that

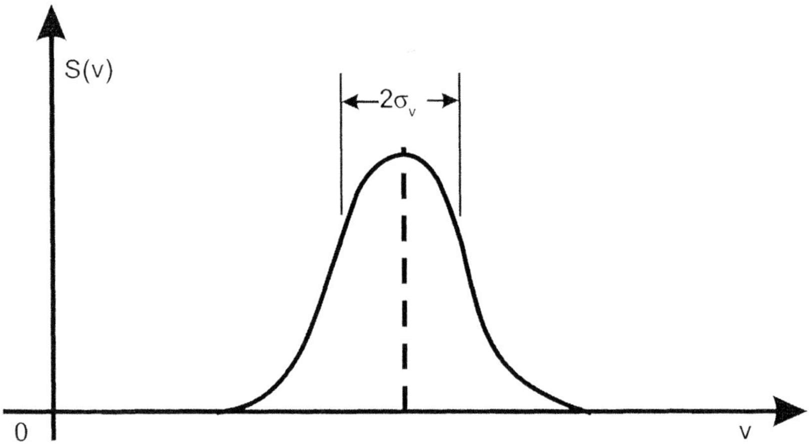

Fig. 3.1. Doppler Spectrum

velocity. The spread of the curve represents the variation of the velocities of individual particles from the mean i.e. the fluctuations. The variance of the spectrum is given by,

$$\sigma_v^2 = \frac{\int (v - v_m)^2 S(v)dv}{\int S(v)dv} \tag{3.14}$$

This is also termed the second moment of the Doppler spectrum. The square root of this quantity i.e. σ_v is the Doppler velocity spectrum width.

The complete Doppler spectrum can be displayed in various ways. One possible presentation has the Doppler velocity on the X-axis and range on the Y-axis for a given azimuth. The variation of the signal as a function of Doppler velocity is plotted for each "range gate". An example is shown in Fig. 3.2. The amplitude at any point of the curves in the figure indicates the power corresponding to the particular velocity. Other types of presentation, for example, an azimuth vs range format can be used. But the raw spectrum constitutes an enormous volume of data requiring vast storage capacity. It is usual therefore to process the signal (as discussed later) and display only the reflectivity factor, mean velocity and spectrum width which are called "base products".

3.1.3 Averaging Samples

The fluctuations imply that an instantaneous signal measured on the

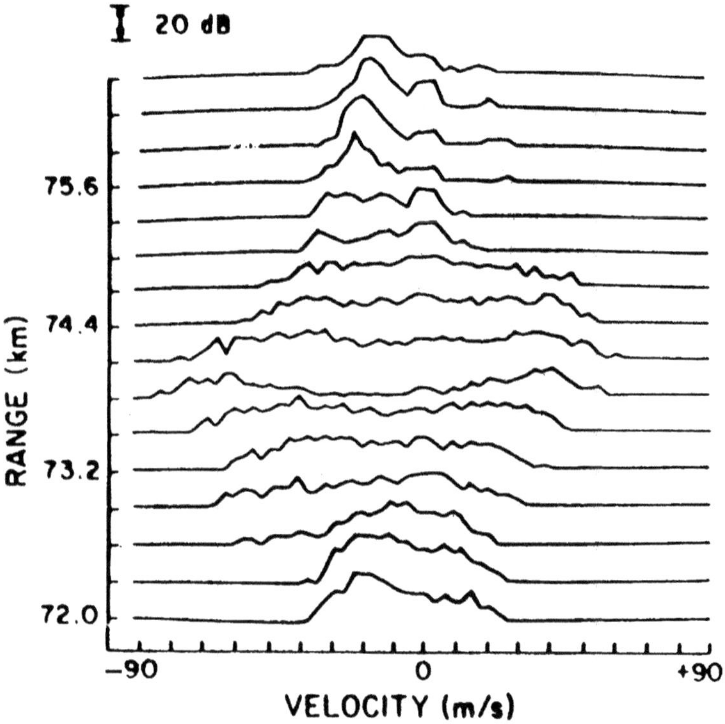

Fig. 3.2. Averaged spectra displayed in 16 consecutive range gates. (from Zrnić et al., 1985; courtesy AMS). The axes are Doppler velocity and range. The amplitudes of the curves are signal power corresponding to each velocity. The scale in dB is shown at top left. The range gates are at intervals of 300 metres. The picture corresponds to a central azimuth of 281.5 degrees but azimuths on either side have been averaged with the centre azimuth to produce the display.

radar display (say an A-scope) will not represent the true reflectivity of the echo. If we collect a number of independent samples and average them we can reduce the effect of the fluctuations. If any one volume V (see Chapter 1) is sampled by the radar receiver in successive pulses, each sample will be independent of the previous one, only if sufficient separation in time is allowed between samples for the particles in the volume to reshuffle to an independent configuration. Echoes from the same transmitted pulse from a few successive pulse volumes in range (Fig.3.3) are independent, if the range over which the average is taken is large compared to the pulse length.

The radar receiver may be,

1. linear i.e. output amplitude proportional to the input amplitude,
2. square law i.e. output amplitude proportional to the input power, or,
3. logarithmic i.e. output amplitude proportional to the logarithm of the input power.

For these three cases, the probability $P(X)d(X)$ that the quantity X will be in

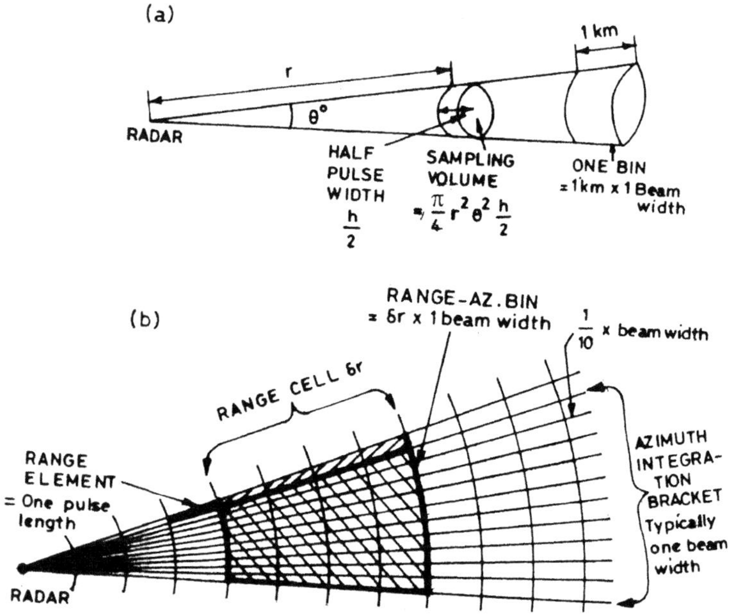

Fig. 3.3. Sampling and Averaging

the interval X to $X + dX$ has been worked out by Marshall and Hitschfeld (see Atlas 1964)). If a sufficient number of samples are averaged, the probability distribution of the average is Gaussian in shape. The variance σ_p^2 of the averaged received power is reduced by a factor k and the standard deviation σ_p by a factor $k^{0.5}$ by averaging k independent samples, if k is greater than[1] about 10. This implies a 95% probability that the true value of the power is within $\pm$ 1.96σ_p of the averaged value. In Fig. 3.3 if the p.r.f. is 900 s^{-1}, the beam width is 2°, the antenna rotation speed is 3 revolutions per minute (rpm), the antenna dwells over one sampling volume for (2/360).(60/3) seconds. In this period (2/360).(60/3). 900 = 100 pulses are transmitted and therefore 100 samples are received. (The effect of continuous motion of the beam over the pulse volume is neglected for the present). However the 100 samples are not independent. The correlation between successive samples (autocorrelation) will depend on the Doppler spectrum width of the particle motion besides the radar parameters. Autocorrelation denotes the degree to which the values of a time function $f(t)$ at two instants of time t and $t + \tau$ are correlated with each other. The autocorrelation function is defined as,

[1] The reflectivity value deduced by averaging the signals will correspond to the actual mean reflectivity only in the case of a square law receiver (Rogers, 1971). A logarithmic receiver will underestimate the mean reflectivity by an amount depending on the fluctuation range in the signal. A linear receiver gives a smaller error.

$$H(\tau) = \lim_{T \Rightarrow \infty} \left(\frac{1}{2T}\right) \int_{-T}^{+T} f(t) f(t+\tau) dt \tag{3.15}$$

The finite limits $-T$ to T should be taken long enough for a sufficient degree of accuracy to be obtained. The autocorrelation function is the Fourier transform of the power spectral density function (Nathanson 1969). A detailed treatment may be seen in Zrnić (1975), Walker et al. (1980) or Doviak and Zrnić (1984). Walker et al. (1980) have given curves showing the equivalent number of independent samples k_e for square law, linear and logarithmic receivers as a function of k, the Doppler spectrum width σ_v, the p.r.f. and the wavelength. k_e will be less than k. In the example above, k is 100, and if we take σ_v as 1 m s^{-1} and wavelength to be 0.1 m, k_e will be 10. By proper choice of the radar parameters the ratio of k_e to k can be improved.

To increase the number of samples we can also integrate a few pulse volumes in "range time" i.e. over a range interval δr along the radial direction as shown in Fig. 3.3. If the pulse duration is 1 μs and δr is chosen as 1 km, there are $1000/(hc/2)$ i.e. 6.7 pulse volumes in the volume consisting of one beam width multiplied by 1 km. This volume is called a range-azimuth bin. If the samples along the range direction are taken to be uncorrelated, the bin contains a total number of $6.7 \times k_e$ i.e. 67 independent samples in the above case. But the reflectivity derived from this averaging is integrated over the range and thus the range resolution becomes poorer. For the case where range samples are also not independent of each other, Walker et al.(1980) have given curves for determining the equivalent number of independent samples.

3.1.4 The Doppler Spectrum Width

The Doppler spectrum width was taken arbitrarily as 1 m s^{-1} in the example of Para 3.1.3. The width of the spectrum arises because of the relative velocities of the particles in the radial direction. Relative motion may be due to (1) turbulence or (2) a shear in the wind within the volume or (3) differences in the fall velocities of the particles because of their different sizes or (4) antenna movement. The last arises in the case of a scanning antenna as the beam shifts to slightly different azimuths (or elevations) resulting in a change in the pulse volume and consequent decorrelation of echo samples. The variance (square of the spectrum width) of the Doppler spectrum is therefore the sum of the contributions, $\sigma_t^2, \sigma_s^2, \sigma_d^2, \sigma_a^2$, due to these four causes (Doviak and Zrnić 1984).

$$\sigma_v^2 = \sigma_t^2 + \sigma_s^2 + \sigma_d^2 + \sigma_a^2 \tag{3.16}$$

Hence the number of independent samples will often be more than indicated in the calculations in the last paragraph. Typical values of σ_t may range from 0.5 m s^{-1} for light turbulence to 10 m s^{-1} for severe turbulence. The variance σ_d^2, due to fall velocity differences may range from less than 1 m^2 s^{-2} for snow and light rain to 20 m^2 s^{-2} for hail. The shear term can also be large but the term σ_a^2 can usually be neglected.

3.1.5 Display of Echo Intensity

In a conventional analogue radar, the video output of the receiver which may be proportional to the input amplitude or power or logarithm of the power is applied to an A-scope, PPI or RHI. If the averaging as discussed above is not carried out, the instantaneous signal is presented with a certain unquantified amount of integration on the PPI or RHI. To obtain a measure of the echo intensity the signal can be attenuated by a continuous or step attenuator until the signal on the A-scope is barely discernible in the midst of the noise. This is a subjective process as the signal and noise are fluctuating all the time. The signal level ($S_{\min}$ dBm) at this stage is the Minimum Detectable Signal (MDS). If the attenuation needed to reduce the echo to the MDS level is A dB the received power is,

$$P_r = S_{\min} + A \text{ dBm} \tag{3.17}$$

Knowing the range of the echo and the radar parameters, Z_e can be computed from equation (2.62). Since P_r values for the same Z_e will be different at different ranges, a "range normalisation" involving corrections for (1) attenuation due to atmospheric gases[2], and (2) attenuation proportional to r^{-2}, needs to be provided. This can be done through hardware by feeding suitable bias waveforms to the RF, IF or video amplifiers such that the amplifier gain at nearer ranges is reduced to compensate for the range-dependent attenuation[3]. An example is shown in Fig. 3.4 where all signals are normalised to a range of 200 km. Curve (1) in this figure is a straight line waveform producing 3.8 dB attenuation at 10 km range (66.7 μs delay with reference to the transmitted pulse) and zero attenuation at 200 km (1333 μs). Taking k_a, the one way attenuation due to atmospheric gases in dB km^{-1} to be 0.1, the two-way attenuation over a range interval of 200–10 i.e. 190 km is 190 × 0.2 = 3.8 dB. This waveform compensates for the attenuation due to atmospheric gases[4]. In curve (2) the attenuation produced at range r km ($10 < r < 200$) is,

---

[2] Correction for attenuation due to intervening precipitation is also necessary if a short wavelength is used.

[3] A simple form of attenuation of short range echoes, to avoid receiver saturation, by applying a bias waveform to the IF or video amplifier is called Sensitivity Time Control (STC).

[4] The first 10 km near the radar is excluded as it is a region of strong ground clutter.

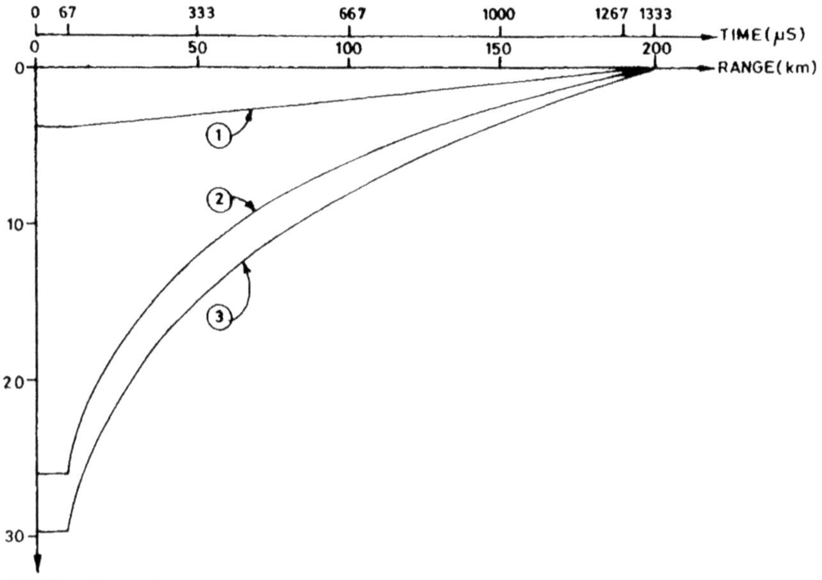

Fig. 3.4. Range normalisation waveform.
Curve (1) is a straight line waveform $2k_a(200\text{-}r)$.
Curve (2) 20 log (200/*r*).
Curve (3) is the sum of curves (1) and (2).

$$A_r = 20\log\left(\frac{200}{r}\right)\text{ dB} \tag{3.18}$$

This compensates for range squared attenuation. Curve (3) is the sum of curves (1) and (2). When this waveform is applied, the signal is brought down to what it would be if it were at 200 km and is said to be range normalised to 200 km. Echoes at any range from 10 to 200 km are comparable if the beam is filled by hydrometeors at all ranges.

As the beam volume increases with range the beam will not be filled with hydrometeors at long ranges. The echo from a fraction of the beam volume filled will be displayed as though it is the echo from the entire beam. Also, the beam may be seeing only the upper portions of a precipitating cloud which will not be representative of conditions below. These effects are illustrated in figure (Fig. 3.5). Hence the display should be limited to a range where these effects are not serious. For a one degree beam this limit is often taken as 200 to 250 km in level terrain but much less in hilly terrain.

A simple analogue method of averaging the signal is to pass the linear video output through a resistance-capacitance (R-C) circuit with a time constant of several milliseconds. This results in averaging several successive pulses in

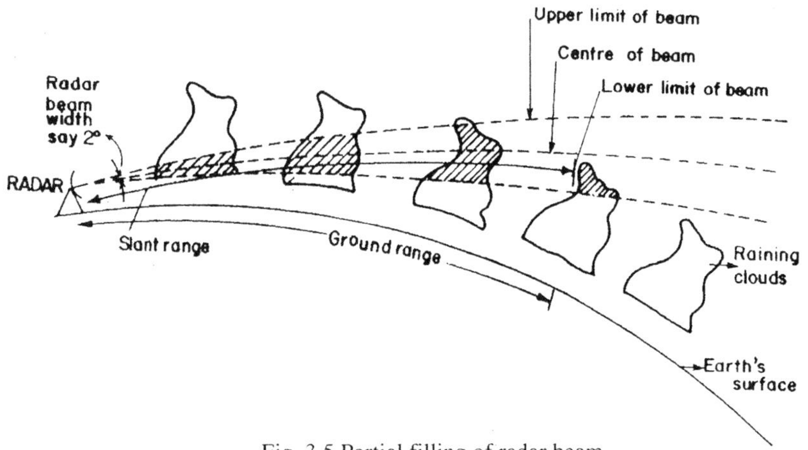

Fig. 3.5 Partial filling of radar beam

range time. By using fixed attenuators in the receiver, only the signal exceeding particular threshold levels designated in dBm or dBZ_e can be displayed, eliminating the weaker echoes. The lowest threshold has to be well above the noise level. The signal exceeding the specified threshold can be amplified to saturation. Such a display is called an iso-echo display[5]. Figs. 3.6, 3.7, and 3.8 show PPI iso-echo

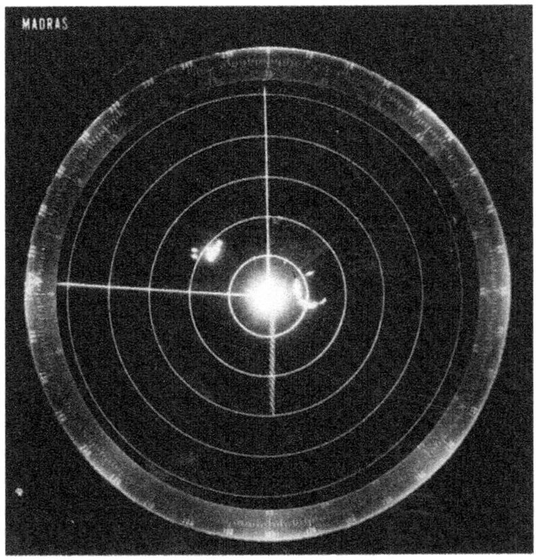

Fig. 3.6. Range normalised iso-echo PPI picture of a tropical cyclone, iso-echo level 4, reflectivity threshold 42 dBZ, showing only a part of the eyewall of the cyclone (courtesy IMD).

[5] In simple early radars, especially in aircraft, a two-level "video inversion" iso-echo system was used. Echo below a specified intensity threshold would be exhibited normally and those above the threshold would be inverted to show a black area indicating stronger echoes.

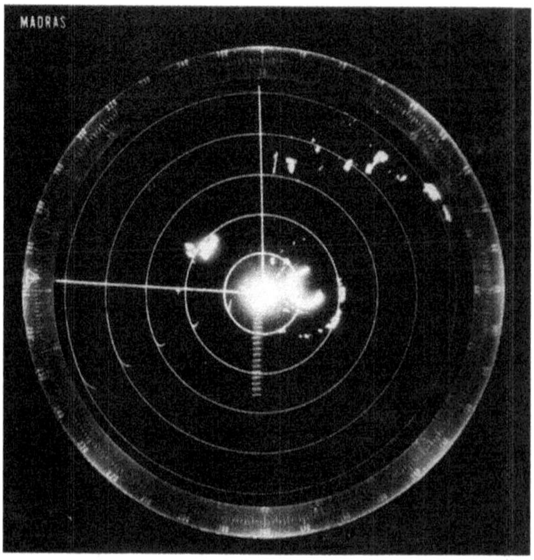

Fig. 3.7. Same as at Fig. 3.6, but iso-echo level 3, reflectivity threshold 37 dBZ. Picture shows more of the eyewall and parts of spiral bands (courtesy IMD).

pictures corresponding to thresholds of 42, 37 and 32 dBZ respectively in the case of a tropical cyclone in the Bay of Bengal. Fig. 3.9 shows a reconstruction of RHI pictures of a thunderstorm cell at five iso-echo levels. Knowing the vertical extent of echo at each signal level the Vertically Integrated Liquid water

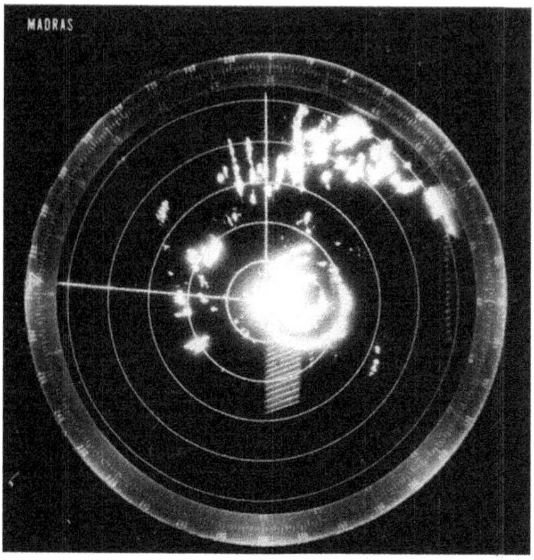

Fig. 3.8. Same as at 3.6. but iso-echo level 2, reflectivity threshold 32 dBZ . More echoes are seen now (courtesy IMD)

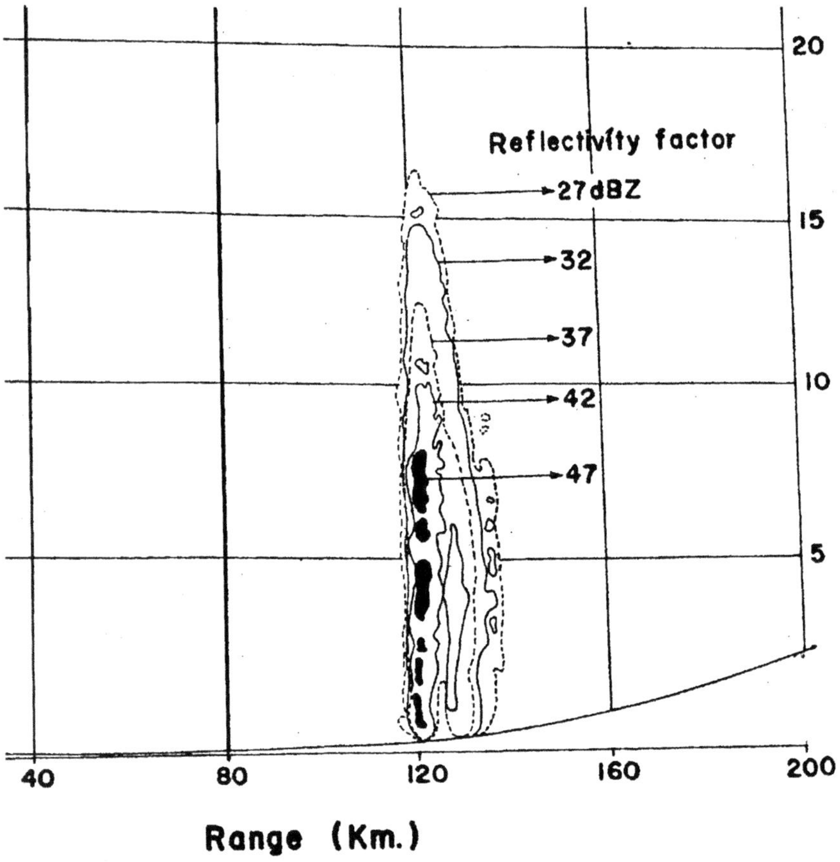

Fig. 3.9. Analog iso-echo picture of a thunderstorm cell.

content (VIL) can be computed from this figure using an M-Z relationship of the form,

$$M \text{ g m}^{-3} = 3.329 \times 10^{-3} Z^{0.57} \tag{3.19}$$

In this case we have,

$$\text{VIL} = 12.1 \text{ kg m}^{-2} \tag{3.20}$$

Circuitry can also be provided to exhibit different ranges of signal intensities in shades of gray (or in false colours) at the same time or alternatively, contours of dBZ_e can be displayed. A gridded overlay can be placed on such a display and each grid square marked with the number of the highest dBZ_e level at which echo appears in it. Knowing the grid area and the Z-R relation, an areally integrated rain rate map can be obtained. This forms a simple way of near real time portrayal of echo intensity and rain rate distribution in the absence of elaborate computer processing. Such a technique with a 40 km × 40 km grid was in operational use in the USA, until superseded by digital techniques, and is called Manually Digitised Radar (MDR). The MDR values were sent in the form of a coded message and were incorporated in a radar summary map of all radars in the network and were used in hydrological forecasting (e.g. Hudlow, 1985).

3.1.6 Range Ambiguity

It was mentioned in Chapter 1 that echoes of targets beyond the maximum unambiguous range r_{max} will be displayed at a range equal to the actual range minus an integral multiple of r_{max}. For example if the p.r.f. is 1000 p.p.s., r_{max} is 150 km. An echo extending from 100 to 120 km may be due to any of three targets, T_1 extending from 100 to 120 km or T_2 from 250 to 270 km or T_3 from 400 to 420 km. The confusion can be resolved if the p.r.f. is changed to, say, 400 p.p.s. T_1 and T_2 will then be displayed at their correct positions and T_3 will appear from 25 to 45 km. Similarly in Fig. 1.7, the p.r.f. of the radar used was 160 pulses per second which corresponds to r_{max} = 937.5 km. The true range of the echoes in the east north east in the picture is the displayed range (about 600 km) plus 937.5 km. i.e. 1537.5 km. If the p.r.f. is changed to, say, 590 p.p.s. r_{max} becomes 254 km. Hence the same target will now be displayed at 1537.5 – 6 X 254 = 13.5 km. Thus a change of p.r.f. shifts the location of a mutiple-trip echo and helps to identify it. Use of a low p.r.f. helps to minimise multiple-trip echoes as echoes from very large ranges are rare because of the height of the beam and the range attenuation. It should also be noted that the dimension of the multiple-trip echo in the radial direction remains correct but the azimuthal dimension is reduced. This distorts the shape of the echo. Rinehart (1989) has devised an ingenious overlay map projection which will correctly identify the multiple echo positions from terrain features.

3.2 Digital Signal Processing of Reflectivity

3.2.1 Digital Video Integrator and Processor

In general analogue methods of signal processing are rather crude compared to modern digital methods.. For example, one of the effects of the R-C filter mentioned in section 3.1, for integration is to convert a square pulse input to an exponential waveform. While amplifying signals to saturation, the tail of the exponential waveform is also amplified and the apparent area of the echo is increased. Also if the threshold is set too low the noise may be amplified to saturation and presented as signal.

Several types of Digital Video Integrator and Processor (DVIP) have therefore been developed in various countries. The DVIP quantises the echo signal in the form of a digital value, averages the fluctuating echo, stores the data relating to various bins on tape or disk, and converts the data into analogue form for display on the CRT. It can also apply various corrections such as for range attenuation, clutter suppression etc. The Z_e values can be computed for each bin and converted to rain rate values applying any chosen *Z-R* relationship. The *R* values can be integrated further over cartesian squares of any desired size and the rainfall distribution displayed in cartesian co-ordinates on a monitor without specially designed (and therefore expensive) displays such as PPI and RHI. The digital data from a suitably designed scan programme (which itself is under computer control) can be used to compute other derived products such as echo top heights and Vertically Integrated Liquid water content (VIL) in near real time. Numerous algorithms have been devised to compute various other derived products and will be referred to in the appropriate contexts. Digital data also lend themselves to transmission on narrow-band telephone or radio links or on the Internet for display or recording elsewhere and for off-line computation of derived products or for inclusion in numerical models.

Digital control of the radar enables programming of a "volume scan" involving scanning in the azimuthal plane successively at various elevations. This facilitates displays involving data from various elevation scans e.g. a CAPPI or an X-Z section. The latter is similar to an RHI but can be a vertical section through any two points (x_1, y_1) and (x_2, y_2) on the earth's surface. The need in an analogue radar to make RHI scans at various azimuths is eliminated[6].

The logarithmic video voltage output from the radar receiver can be fed to an Analogue-to-Digital (A/D) converter whose output is in the form of a digital number. Video samples are collected and averaged as described in section 3.1. The DVIP output voltage values are converted to equivalent P_r values knowing the "transfer function" of the receiver i.e. the relationship between the signal input (dBm) and the video output (volts). The relationship is determined by calibration with a signal generator and a curve as shown in Fig. 3.10 is obtained.

The use of the logarithmic output is resorted to in order to get a large dynamic range over which a plot of DVIP output vs signal input in dBm is a straight line. With weak signal the relationship will not be linear and a knee is

---

[6] In an analogue mode of operation it is not feasible to study each echo in this way. The operator would choose some echoes which appear important. Hence any statistics gathered from manually operated radar may contain a bias. However the X-Z section has the disadvantage that one has to wait for the completion of a volume scan before the section can be generated.

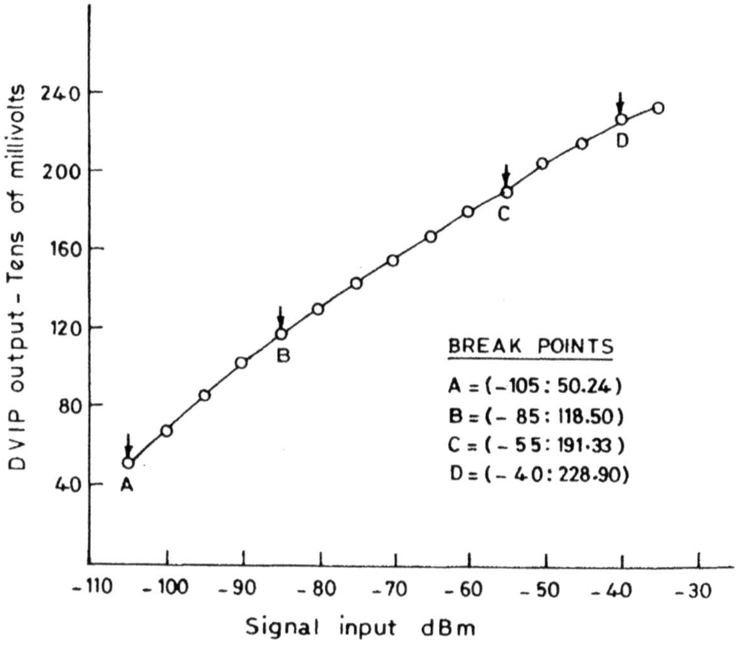

Fig. 3.10. DVIP output vs signal input from signal generator.

formed as in Fig. 1.10. At the high power end there will be saturation. It is possible that the curve may not be a straight line even within these limits In Fig. 3.10 the calibration curve is treated as made up of three straight lines. The break points A,B,C,D, are fed to the computer which uses them for the conversion of DVIP output into P_r. Alternatively a polynomial regression of the form (Schroeder and Klazura, 1978)

$$y = a + bx + cx^2 + dx^3 \tag{3.21}$$

is fitted, where x is the DVIP output corresponding to a signal of y dBm. The coefficients a, b, c and d are stored in the computer and updated by repeating the calibration at frequent intervals. The calibration data along with "house-keeping information" must be archived with each file of observational data.

All P_r values have to be corrected for the radar head losses which include loss in the waveguide from antenna to receiver, loss due to radome, if any, and insertion loss in any other hardware component en route e.g. T-R cell or RF attenuators or any connecting cables. These values, measured at the time of installation, may usually stay constant unless some component or configuration is changed. In a logarithmic receiver, logarithms of the power are being averaged. This average is lower than the logarithm of the average power which is the quantity we need to know. Following Lhermitte and Kessler (1966) the average difference between the two is taken as 2.5 dB. A correction of +2.5 dB is added to the average obtained from the digitiser[7].

A range normalisation and gas attenuation correction (see Para 3.1.5) can be carried out by software operation on the digitised video signal. However, some consider it advantageous to do it at the R.F. stage using a PIN[8] diode modulator, which provides a fast response and constant noise level besides accuracy and stability of the output.

3.2.2 Linear Video with Large Dynamic Range

The log video has traditionally been used because of the need for a large dynamic range. A linear video channel will saturate at relatively low signal levels. But with the availability now of fast response solid-state components, it is possible to use the linear video as input to the DVIP over small segments of the total dynamic range. For this purpose the log video signal is used to select the segment depending on the echo intensity by applying a suitable attenuation to the linear receiver (Mueller and Silha 1978). Thus the log video is used for preliminary selection while the actual quantitative measurement is made with the linear video which by itself may have a much smaller dynamic range. In the French RONSARD[9] radar, dynamic range is obtained by using a fast AGC (Automatic Gain Control). A 1 μs response time ensures full gain control between successive range gates. The AGC maintains the received echo intensity within an interval 9 dB wide. The AGC circuitry sets a 6-dB quantised PIN diode attenuator allowing 48 dB gain control. When the echo intensity approaches the saturation level, the parametric amplifiers are automatically switched off and their 30 dB gain is replaced by a 2 dB insertion loss. This prevents receiver saturation. The total AGC dynamic range is increased to 80 dB and the total linear receiver dynamic range is increased to over 110 dB. A rather similar circuit is used in the ISRO Doppler weather radar in India and in the CSU-CHILL radar in USA, both described later in this Chapter.

3.2.3 Computation of Z_e

For conversion of P_r to Z_e, the correct values of the parameters of the

[7] As the difference is smaller at signal levels below about −80 dBm, a two-step correction is sometimes incorporated e.g. + 1.25 dB for signals below −80 dBm and + 2.5 dB for stronger signals. See Chapter 5 for further discussion of receiver calibration.

[8] Positive-Intrinsic-Negative.

[9] Recherche sur les Orages et les Nuages par un Système A deux Radars Doppler. See section 3.7 for other details.

radar must be available to the computer. Nominal values of P_t, h, and p.r.f. are fed as default values to the computer. They are also measured at regular intervals or monitored continuously and adjusted to conform to the nominal values. In case of a departure the new values are to be fed to the computer. The antenna parameters (G, ϕ, θ) are specified by the manufacturer and can be checked at the time of installation of the radar (see Chapter 5). The corrected dBZ_e can be processed through a digital to analogue (D/A) converter and displayed on a conventional PPI. Since however the range-azimuth bins are unequal in size it is convenient to integrate the data further in squares of, say 2 km × 2 km or 5 km × 5 km as shown in Fig. 3.11 and display the data in Cartesian coordinates on a monitor. The advantage of this presentation over a conventional CRT display is that it uses an addressable memory which is repetitively scanned, Hence the picture can be viewed in a normally lit room. It is also possible to enlarge or zoom any chosen area of the image to the size of the whole screen. The dBZ_e values may be divided into a number of class intervals and each category is assigned a gray shade or a false colour for display on the picture. In the latter case it is the convention to use colours at the blue end of the spectrum to indicate weak echoes and the red end to represent strong echoes. While it is difficult for a human observer to distinguish more than 5 or 6 gray shades it is easier to distinguish a larger number of colours. The house-keeping parameters may also be displayed and map overlays can be superposed. The data can also be printed out in alpha-

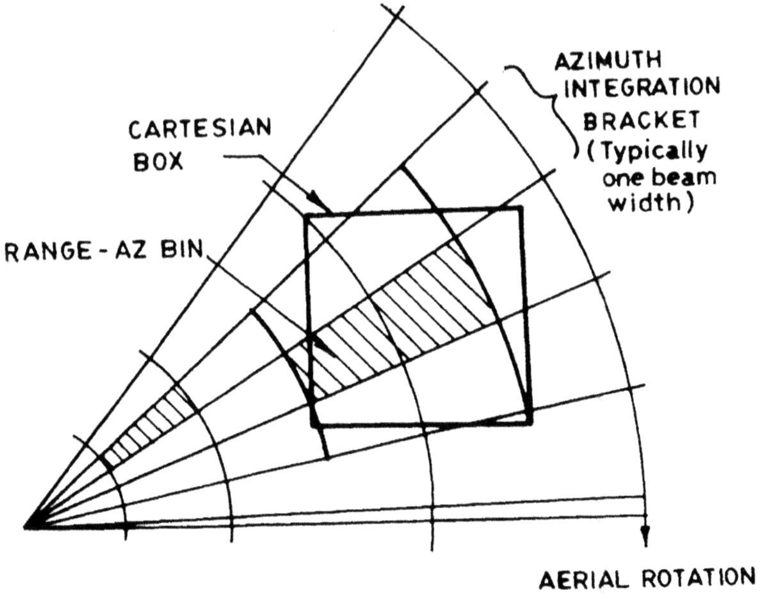

Fig. 3.11. Integration into Cartesian squares.

numeric form by assigning a number to each category of dBZ_e or the corresponding rain rate.

3.2.4 Computation of *R*

Since the *Z-R* relationship is not a linear one it is appropriate to convert the Z_e values in each bin to *R* values before integration into Cartesian squares. For conversion, default values of *A* and *b* in the *Z-R* equation e.g. the Marshall-Palmer values $A = 200$, $b = 1.6$ can be stored in the computer which uses them to convert to *R* values. These are then integrated into Cartesian squares and displayed in categories of rainfall rate. If any other values of the constants are to be used, the operator can insert them in place of the default values. Similar relationships can be used if the operator wishes to display or print out other quantities e.g. liquid water content *M* of rain drops.

3.3 Doppler Signal Processing

3.3.1 I and Q Outputs

To extract the Doppler frequency shift and determine its sign, the phase detector in the Doppler radar receiver (Fig. 1.12) is given two inputs of the reference signal f_I one of which is phase shifted by $\pi/2$ with reference to the other. As each of these is mixed with the signal from the receiver I.F. channel, two outputs result. As seen in section 3.1, the complex signal from a particle *j* moving with radial velocity V_j corresponding to Doppler shift f_j, is,

$$S(t) = \delta_j \cdot e^{-i \cdot 2\pi f_j t} \cdot e^{-i \cdot 2\pi f_0 t} \tag{3.22}$$

where δj is a constant amplitude and f_0 the transmitting frequency of the radar. This can be rewritten as,

$$S(t) = \delta_j \left[\cos 2\pi (f_j + f_0)t - i\sin 2\pi (f_j + f_0)t \right] \tag{3.23}$$

f_0 will be substituted by the intermediate frequency f_I after conversion. The video outputs from the phase detector will be proportional to the real and imaginary terms in equation (3.23) and are of the form,

$$I(t) \text{ proportional to } \cos 2\pi f_j t \tag{3.24}$$

$$Q(t) \text{ proportional to } \sin 2\pi f_j t \tag{3.25}$$

The sum of I^2 and Q^2 is equal to the signal power averaged over a cycle of the carrier signal. The two outputs I and Q are known as in-phase and quadrature outputs and are necessary to process the complex signal and derive the sign as well as magnitude of Doppler velocity. Q lags or leads I by $\pi/2$ according as Doppler frequency shift is positive or negative. A positive shift in frequency represents a component of radial motion towards the radar which by convention is considered a negative velocity, and a negative shift in frequency corresponds to motion away from the radar which is a positive velocity.

3.3.2 Fast Fourier Transform (FFT) and Pulse Pair Processing

The outputs contain pulse samples taken in each range element corresponding to successive transmitted pulses at intervals of T_s where T_s is the pulse interval i.e. the reciprocal of the p.r.f. Each sample contains returns from a large number of scatterers with different velocities. To derive the Doppler frequencies a spectral analysis of a tremendous volume of data is needed. This can be carried out by digital processing through a computer.

Any complex time function can be expressed as the sum of a large number of terms corresponding to the component frequencies known as a Fourier series. The coefficients of the terms in the series give the magnitudes of the spectral components. The Discrete Fourier Transform (DFT) of a signal sampled M times at a spacing T_s is defined as,

$$Z(kf_f) = \sum V(mT_s)e^{-2\pi f_f T_s mk} \tag{3.26}$$

where $V(mT_s)$ is the complex voltage of the mth sample, $Z(kf_f)$ is, the complex amplitude of the kth spectral coefficient, f_f is the fundamental i.e. the lowest frequency (the other frequencies in the Fourier series being harmonics of f_f),

$$f_f = \frac{1}{MT_s} \tag{3.27}$$

and the summation is from $m = 0$ to $m = M-1$. Fast computations of the coefficients are done by algorithms known as Fast Fourier Transforms (FFT). The FFT is a computational tool for efficient computation of the DFT. It avoids the need to analyse the entire spectrum which will require large computer capacity and slow down data acquisition. A good introduction to the FFT may be found in Cochran et al. (1967).

As an alternative to this "frequency domain" processing, "time domain" processing in the form of autocovariance processing or pulse pair processing can be used. Here the autocorrelation [see equation (3.15)] between two successive pulses is used to determine the Doppler frequency. If a sequence of M pulses are transmitted at intervals T_s the autocorrelation is estimated from the sum,

$$\hat{R}(\tau) = \left(\frac{1}{M}\right)\sum_{m=1}^{m=M-1} V^*(m)V(m+1) \qquad (3.28)$$

where the caret (^) is used to denote that it is an estimate, $V(m)$ is the complex voltage of the mth sample, the asterisk indicates the complex conjugate and the lag between the two pulses is τ, in this case equal to T_s. The mean Doppler velocity will be,

$$\hat{v} = -\frac{\lambda}{4\pi\tau}\arg\hat{R}(\tau) \qquad (3.29)$$

where arg stands for the argument (in radians) of the complex quantity R (Doviak and Zrnić 1984, 1993).

The time interval between the pair of pulses can be different from the time interval from one pair to the next. This variation in the p.r.f. is advantageous to get optimum sampling rates for Doppler and reflectivity processing. The relative merits of FFT and pulse pair processing are discussed by Doviak and Sirmans (1979).

The complete theoretical treatment of signal processing is beyond the scope of this book. It may be found in Zrnić (1979), Doviak and Zrnić (1984, 1993), Bringi and Chandrasekar (2001) and in books on digital signal processing e.g. Rabiner and Gold (1978). A very good review of signal processing for atmospheric radars may be seen in Keeler and Passarelli (1990).

3.3.3 Mean Velocity, Variance and Spectrum Width

The two basic Doppler products are the mean velocity v_m in the pulse volume and the variance of v, i.e. σ_v^2 or its square root the spectrum width σ_v. Both the quantities can be displayed on a conventional PPI (or RHI or a time-height) display or in a Cartesian coordinate display in false colours. By convention the blue end of the spectrum is used for negative (towards the radar) velocities and the red end for positive velocities (directed away from the radar). For spectrum width the colour code is similar to that for reflectivity i.e. the red end of the colour spectrum denotes higher values.

3.3.4 Range and Velocity Ambiguities

In the case of a Doppler radar, ambiguities in range as well as velocity can arise as explained in Section 1.2. For example, for a 10 cm wavelength radar, if r_{max} of 300 km is desired, the p.r.f. will be 500 and v_{max} will be 12.5 m.s.

A velocity v higher than v_{max} will be "folded over" i.e. brought within the interval $+v_{max}$ to $-v_{max}$ and displayed as,

$$v' = v - 2nv_{max} \tag{3.30}$$

where n is a positive or negative integer called the Nyquist folding number. For example, in this case, a velocity of $+20$ m s^{-1} will be displayed as $+20-2\times1\times12.5$ i.e. -5 m s^{-1}. This is known as aliasing. In a colour display this will create confusion by juxtaposing colours representing widely different velocities e.g. a negative velocity next to a positive velocity[10]. As radial velocities much greater than 12.5 m s^{-1} can be encountered in severe weather systems, velocities may even be folded over more than once. For shorter wavelengths, v_{max} is further reduced. There can be a multiple-trip echo overlying or close to a correctly positioned echo and the two may have widely different velocities. If the overlying multiple-trip echo is of comparable intensity to the first trip echo, the velocity of the latter cannot be determined correctly. Or again, an echo produced at a wrong azimuth by a side lobe or by reflection from a building will have a velocity different from that of other echoes displayed near it.

3.3.5 Resolving Ambiguities

How do we resolve this confusion? If the Doppler shift within one pulse width can be measured there will be no ambiguity but this is not possible for the low velocities met with in meteorology. Depending on the application, the p.r.f. has to be chosen to obtain optimum v_{max} and r_{max}. One can carefully look for incongruous discontinuities and reinterpret them. This is difficult to do in the case of isolated echoes. Algorithms for "unfolding" or "dealiasing" the display have been developed using the continuity principle. In automated systems, an assumption can be made of the environmental wind using radiosonde data. The difference between its radial component and the displayed radial wind is used to estimate the integer n in equation (3.30). This succeeds in many cases but not in all situations. Besides this, special methods of transmission in conjunction with software algorithms have been attempted. These involve complication of hardware as well as higher needs of computing power (and also computing time which is important in operational use). These are listed below.

(1) Dual (or staggered) p.r.f.: Two different p.r.f.'s may be used in alternate scans resulting in two estimates of velocity each having a different Nyquist velocity. The effective v_{max} will be the least common multiple of the two Nyquist velocities.

---

[10] Note, however, that real velocities of opposite signs can be juxtaposed e.g. in the case of a tornado vortex or a downburst.

(2) Dual (or staggered) Pulse Repetition Times (PRT): If two PRT's T_1 and T_2 are used, (usually in the ratio 3/2 or 4/3), v_{max} will be decided by $T_2 - T_1$ and r_{max} by $T_2 + T_1$. This method has the advantage that the second trip echoes will be rendered incoherent and will not affect the velocity estimates of the first trip echo. However, ground clutter filter is difficult to implement in this method.
(2) Interlaced sampling: Two p.r.f.'s are used; a higher one to sample velocities and after a sufficient "clearing" time till the last range-ambiguous echoes, a lower p.r.f. to sample reflectivity. Thus getting higher v_{max} and r_{max}.
(3) Transmission of pulses in pairs: The velocity is determined from the pair of pulses. The spacing between one pair and the next is long enough to get high r_{max}.
(4) Phase diversity: A large v_{max} and a corresponding small r_{max} are chosen. A fraction of the transmitted pulse is monitored in the receiver, its phase information is stored and used to correct the phase of multi-trip echoes in successive pulse intervals. Phase-coded pulse sequences can also be used.
(5) Dual wavelength transmission: one with high p.r.f. for velocity determination and the other with low p.r.f. for reflectivity processing.,
(6) Polarisation coding: Alternate pulses are transmitted in different polarisations i.e. vertical and horizontal. This was first proposed by Doviak and Sirmans (1973) but appears to have been implemented only recently (see Chapter 11). The pair of H and V pulses have a short time difference and hence give a high v_{max}. Two successive H-H or V-V pulses have a long time interval between them and provide high r_{max}. The H and V pulses are derived from the same source and therefore there is coherence. The two signals have to be processed by separate receiver channels. Coupling betwen the channels can occur in the radar system. The depolarisation in precipitation has to be corrected for.

Albers (1989); Bargen and Brown (1980); Bergen and Albers (1988); Boren et al. (1986); Carbone et al. (1980); Dazhang et al. (1984); Desrochers (1989); Doviak et al. (1976); Durden and Werner (1996); Eilts and Smith (1990); Ekengren and Dahlberg (1984); Frush and Daughenbaugh (1999); Golestani et al. (1995); Hennington (1981); Jing and Wiener (1993); Keeler and Carbone (1986); Merritt (1984); Sachidananda and Zrnić (1999, 2000); Sirmans et al. (1976); Yamada and Chong (1999); Zrnić and Mahapatra (1985), may be seen for descriptions of some methods of making unambiguous range/velocity measurements (see also Paragraph 3.3.10 regarding the method of Tabary et al. 2001). A review of various techniques is available in Keeler et.al. (1999). Besides reviewing many of the earlier techniques cited above, James and Houze (2001) have formulated a new dealiasing scheme called four-dimensional Doppler dealiasing (4DD) which is said to work in real time even in hilly terrain where dealiasing is most difficult because of the variations in the wind caused by orography. It examines one elevation scan at a time, beginning at the highest elevation. It attempts to adjust the value of each gate by Nyquist intervals such that it agrees with both the nearest gate in the next higher tilt and the nearest gate in the previous volume.

None of these methods is perfect and some of them may render ground clutter suppression difficult. On the display, doubtful values of velocity are flagged or shown in a different colour.

3.3.6 Effect of Antenna Rotation

A relatively minor source of error in velocity estimation is the rotation of the antenna[11]. Since the source of radiation viz. namely the feedhorn is located (typically 3 or 4 metres in a 10-cm radar) away from the axis of rotation of the antenna, a component of its velocity is added to the radial velocity of the target. This may need to be corrected in the case of large antennas. One effect of this is that ground targets will have an apparent radial velocity and may escape being rejected by clutter suppression techniques (Rinehart 1991).

3.3.7 Vertical Incidence Doppler Radar

The particle velocities determined above are only the radial components. This was not a problem in the early application of the pulsed Doppler radar for measurement of vertical air motions in precipitating systems. The antenna was pointed upwards and only the vertical motions were studied. This has been termed Vertical Incidence Doppler (VID). If the vertical velocity of air is w, taken as positive downwards, and the mean terminal velocity of fall of precipitation particles in the pulse volume in still air is V_t, the measured Doppler velocity will be $w + V_t$. Since the terminal velocity V_D of a rain drop of diameter D is proportional to $D^{0.67}$ (see Paragraph 2.2.6) and Z is given by equation (2.55) the reflectivity-weighted mean V_t of all V_D's is given by

$$V_t = \frac{\int V_D N_D D^6 dD}{\int N_D D^6 dD} \tag{3.31}$$

V_t may be defined as the mean Doppler fall speed. Using the Marshall-Palmer values for the drop size distribution [equations (2.32) and 2.33)], Rogers (1964) obtained,

---

[11] Incidentally, it should be noted that the effective antenna beam width also increases when the antenna is scanning (see e.g. Wood and Brown 1997).

$$V_t = A''Z^{b''} = 3.8Z^{1/14} \tag{3.32}$$

where V_t is in m s^{-1} and Z is in mm^6 m^{-3}. Note that the exponent b'' is not far different from that estimated in Paragraph 2.2.7. Hence, knowing Z, V_t can be computed (subject to the assumption of small rain drops) and the actual air velocity w determined from the Doppler measurement. Aoyagi (1969) compared V_t values thus obtained with measurements near the ground and found good agreement. Several early studies [Probert-Jones and Harper (1961); Battan (1964); Battan and Theiss (1966)] assumed the minimum fall speed of ice particles or small rain drops to be 1 m s^{-1}. This meant that in the absence of vertical air motion, the smallest Doppler velocity of particles approaching the radar would be 1 m s^{-1} i.e. the "lower bound" of the Doppler spectrum would be 1 m s^{-1}. The departure of the observed lower bound from 1 m s^{-1} can be taken as due to up- or downdrafts and thus w can be calculated. This method called the lower bound method was prone to large error as it was difficult to recognise the signal due to the small particles which correspond to 1 m s^{-1} fall velocity amidst the stronger signal from larger particles. Battan and Theiss (1970) have compared this method with Rogers' method.

If actual measurements of drop sizes are made, better measurements of vertical velocities can be obtained. In light rain with no vertical motion, the measured distribution of Doppler velocities can be attributed entirely to fall velocities and can be used to compute the drop size distribution (DSD). DSD is further discussed in Chapter 7 in the context of rain rate determination. In the case of variance of the Doppler velocity, if the contribution due to spread of fall velocities can be assumed, it can be subtracted from the total variance to obtain that due to turbulence. However measured variance can be in error if there are extraneous echoes due to ground clutter, side lobes, aircraft, birds or insects.

3.3.8 Multiple-Doppler Radar

To obtain the three-dimensional velocity field it will be ideal to have Doppler radars at three different sites observing the same pulse volume. Following Armijo (1969) and Ray and Sangren (1983), if there are three non-collinear radars located at (x_i, y_i, z_i), $i = 1,2,3$, the radial velocity V_i with respect to the ith radar at a point (x,y,z) is given by the set of three equations ($i = 1,2,3$),

$$V_i = \left(\frac{1}{R}\right)\left[u(x - x_i) + v(y - y_i) + w(z - z_i)\right] \tag{3.33}$$

where,

$$R^2 = (x - x_i)^2 + (y - y_i)^2 + (z - z_i)^2 \tag{3.34}$$

u,v, are the mean horizontal Cartesian components of velocity of the precipitation particles which are taken as tracers of the air motion and $w + V_t$, as explained above, the sum of the vertical component of air velocity w and the mean terminal velocity V_t of the precipitation particles. Since we need a fourth equation to determine four variables, we take the equation of continuity in meteorology,,

$$\frac{\partial u}{\partial x} + \frac{\partial v}{\partial y} + \frac{\partial w}{\partial z} = -\frac{1}{\rho_a}\frac{d\rho_a}{dt} \tag{3.35}$$

where ρ_a is air density and t represents time. The r.h.s. of equation (3.35) can be written as,

$$\frac{d(\ln \rho_a)}{dz}\frac{dz}{dt} = -w\frac{d(\ln \rho_a)}{dz} = kw \tag{3.36}$$

Here it is assumed that the vertical rate of change of air density is logarithmic and that there is no horizontal advection of density. These equations are sufficient to determine u,v,w and V_t in the area around the radar. Since the three-dimensional scan will take some time, the velocity is assumed to be constant during that time. The variances of the estimates of the velocity components will depend on the positions with respect to the radar sites of the points at which the estimate is made. Hence the placement of the radars has to be such that the errors are minimised (Ray and Sangren 1983). Another source of error is the lack of matching of the volumes observed. If there are more than three radars, the wind field is overdetermined and the final wind field is a least-squares fit of the winds from all combinations of the radars, leading to greater accuracy.

3.3.9 Dual-Doppler Radar

Considerable simplification is achieved if there are only two radars, provided certain further assumptions are made and constraints accepted. This is called the dual-Doppler technique (Lhermitte 1970, 1975). If the same pulse volume is observed by the two radars, the mean radial velocities V_1 and V_2 observed by them are given by,

$$V_1 = u\sin\phi_1\cos\theta_1 + v\cos\phi_1\cos\theta_1 + (w + V_t)\sin\theta_1 \tag{3.37}$$

$$V_2 = u\sin\phi_2\cos\theta_2 + v\cos\phi_2\cos\theta_2 + (w + V_t)\sin\theta_2 \tag{3.38}$$

where ϕ and θ are the azimuth and elevation angles respectively, of the radar beam and the suffixes 1 and 2 refer to the two radars. From these equations we get,

$$u = \frac{1}{\sin(\phi_1 - \phi_2)}\left\{V_1\frac{\cos\phi_2}{\cos\theta_1} - V_2\frac{\cos\phi_1}{\cos\theta_2} - (w+V_t)\left[\cos\phi_2\tan\theta_1 - \cos\phi_1\tan\theta_2\right]\right\} \quad (3.39)$$

$$v = \frac{1}{\sin(\phi_1 - \phi_2)}\left\{V_2\frac{\sin\phi_1}{\cos\theta_2} - V_1\frac{\sin\phi_2}{\cos\theta_1} - (w+V_t)\left[\sin\phi_1\tan\theta_2 - \sin\phi_2\tan\theta_1\right]\right\} \quad (3.40)$$

If the elevation angles are small, the last terms in equations (3.39) and (3.40) can be neglected and u and v computed, provided $\sin(\phi_1 - \phi_2)$ is not zero. At higher elevation angles, for small rain drops, V_t can be computed from Z as in Paragraph 3.3.7 and w can be obtained from the equation of continuity (3.35). An iterative process using the above equations will produce mean values of u,v,w. A boundary value of zero for w at the storm top (or the ground surface; the former produces less error) is assumed and a downward (upward) integration carried out. Instead of coordinated observation of the same pulse volume by the two radars, independent three dimensional scans can be used. Data for all range gates from each radar can be stored and processed together by a computer.

Alternatively, a "Coplane" scanning[12] (Lhermitte and Miller 1970) can be used, i.e.the azimuth and elevation angles of the two radars can be controlled such that they both scan in a plane inclined at an angle α to the horizontal plane containing the baseline between the two radars (Fig. 3.12). If u is taken as parallel to the baseline between the radars (and v perpendicular to it) and ϕ is measured with reference to the perpendicular to the baseline it can be seen that,

$$\frac{\tan\theta_1}{\cos\phi_1} = \frac{\tan\theta_2}{\cos\phi_2} = \tan\alpha \quad (3.41)$$

Equations (3.39) and (3.40) then reduce to,

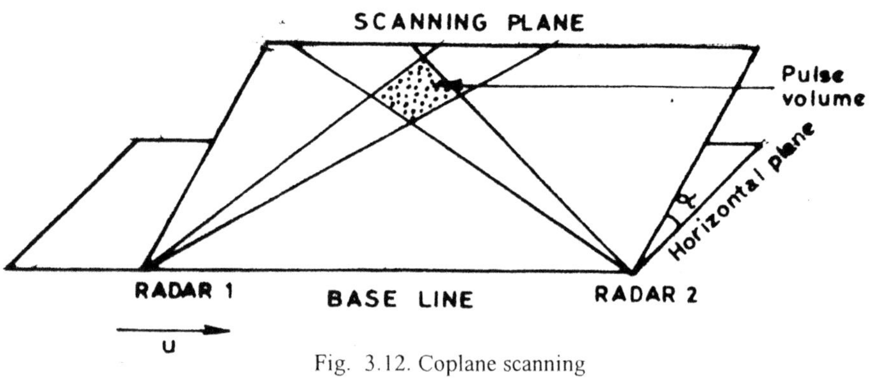

Fig. 3.12. Coplane scanning

[12] Also called "Coplan" scanning.

$$u = \frac{1}{\sin(\phi_1 - \phi_2)}\left\{V_1\frac{\cos\phi_2}{\cos\theta_1} - V_2\frac{\cos\phi_1}{\cos\theta_2}\right\} \tag{3.42}$$

$$v = \frac{1}{\sin(\phi_1 - \phi_2)}\left\{V_2\frac{\sin\phi_1}{\cos\theta_2} - V_1\frac{\sin\phi_2}{\cos\theta_1}\right\} - (w + V_t)\tan\alpha \tag{3.43}$$

i.e. u and v can be determined without the constraint of elevation angle. Another way of using the dual-Doppler system is to keep (Browning et al. 1968) one antenna pointing upwards and the other pointing to the first radar and scanning in the vertical. In this case V_1 is equal to the vertical velocity plus fall speed (over the first radar) and u is given by,

$$u = \frac{V_2 - V_1\sin\theta_2}{\cos\theta_2} \tag{3.44}$$

Besides the assumptions mentioned above, storm motion and earth curvature among others may produce errors and have to be corrected for. Despite the complexities and several sources of error, dual and multiple-Doppler configurations have been used very successfully in field experiments. Fig. 3.13 from Wilson et al.(1984) gives an example of dual-Doppler radar analysis. In this figure, winds are plotted as arrows on a rectangular grid, the axes being distances in two orthogonal directions from one of the radars. The length of the wind arrow is proportional to the magnitude of the wind. Reflectivities are shown as contours on the same diagram. A comprehensive treatment of three-dimensional wind field analysis from dual-Doppler radar may be seen in a set of three papers (1983) by Testud and Chong, Chong et al. and Chong and Testud. Given (1989) has given a good review of dual-Doppler processing algorithms. New formulations of dual-Doppler wind analysis have been made by Shapiro and Mewes (1999) and Mewes and Shapiro (2002).

3.3.10 Single Doppler Radar – the Velocity-Azimuth Display (VAD) Technique

Though a single Doppler radar yields only the radial component of the wind, Peace et al., (1969) demonstrated that in principle the horizontal wind field can be derived. A technique to obtain the wind field with single Doppler radar is the Velocity-Azimuth Display (VAD) first proposed by Lhermitte and Atlas (1961). The radar scans in azimuth at an elevation θ. Corresponding to any particular range r the height of the beam is fixed, let us say, h. If the horizontal

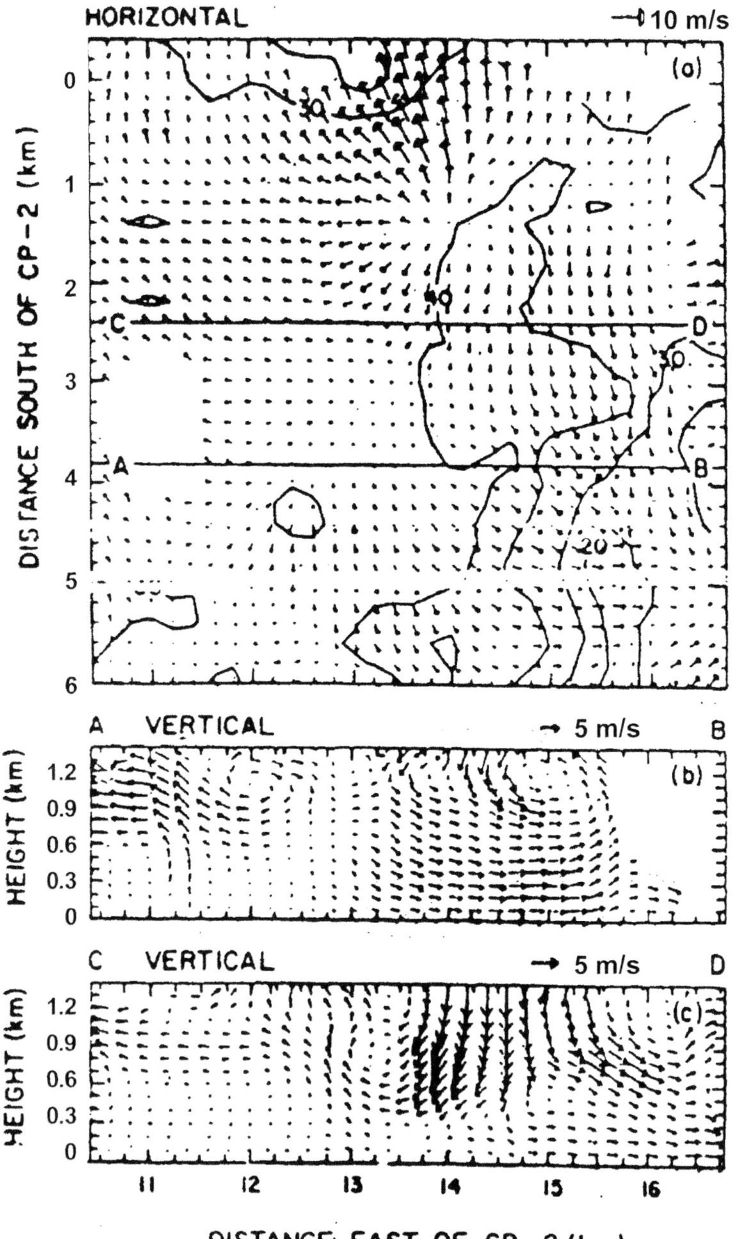

Fig. 3.13. Dual Doppler analysis of horizontal and vertical velocity field (from Wilson et al. 1984; courtesy AMS). (a) Horizontal eddy wind (mean wind subtracted) and reflectivity factor contours in dBZ, at 50-100 m above ground. (b) Vertical cross section of winds along line AB (c) Vertical cross section of winds along line CD. Wind speed scale is given at the top of each figure.

particles are uniform around the radar as may be the case in widespread stratiform precipitation, the mean radial velocity V (taken as positive when directed away from the radar) derived from the Doppler velocity spectrum for a point (r, ϕ, θ) is, (r, ϕ, θ defined as in Fig. 3.14a.),

$$V = V_h \cos\theta \cos(\phi - \phi_0) - V_f \sin\theta \tag{3.45}$$

where ϕ_0 is the direction of V_h. If the elevation angle is small, the second term on the r.h.s. can be neglected especially in stratiform precipitation. If V is plotted as a function of azimuth (Fig. 3.14b and c) it will have a maximum positive or negative value when $\phi - \phi_0 = 0$ or π (the upwind and downwind directions) and zero value when it is $\pi/2$ or $3\pi/2$ i.e. when the radar is looking at right angles to the wind. The maximum value gives the magnitude of V_h. The direction of V_h is given by the azimuth corresponding to the negative maximum of V. In case V_h is outside the range $+V_{max}$ to $-V_{max}$ the velocity curve will be aliased as shown in Fig. 3.14c. The dotted portion of the curve shows the correct value. In practice the V-ϕ curve will not be the smooth sine curve shown in Fig.3.14b and c. The curve will be broadened by the fluctuations of particle velocities[14] and will also be affected by weak signals amidst noise. The positive and negative maxima may not be equal because of local variations in V_h. If the elevation angle is not low, the $V_f \sin\theta$ term will become important. In this case equation (3.45) becomes,

$$\text{when } \phi - \phi_0 = 0, \quad V_0 = V_h \cos\theta - V_f \sin\theta \tag{3.46}$$

$$\text{when } \phi - \phi_0 = \pi, \quad V_\pi = -V_h \cos\theta - V_f \sin\theta \tag{3.47}$$

$$\text{Hence, } V_h = \frac{V_0 - V_\pi}{2\cos\theta} \text{ and } V_f = \frac{-V_0 + V_\pi}{2\sin\theta} \tag{3.48}$$

If a scan is taken at a fixed elevation, each successive range gate corresponds to a different height. Thus a vertical profile of the horizontal wind may be obtained in the vicinity of the radar provided wind does not vary in the horizontal and detectable precipitation is available at ranges of interest. An example of VAD wind profiles obtained from scans at regular time intervals is shown in Fig. 3.15.

Browning and Wexler (1968) considered the case when V_h varies in the horizontal. If V_x and V_y are the two components of V_h,

$$V_x = V_{x0} + x\frac{\partial V_x}{\partial x} + y\frac{\partial V_x}{\partial y} \tag{3.49}$$

---

[13] V_f is the sum of w and V_t and is positive downwards.

[14] The fluctuations in a snowstorm were used by Lhermitte (1968) to deduce the turbulent air motion. See Chapters 6 and 10 for a discussion of turbulence measurements.

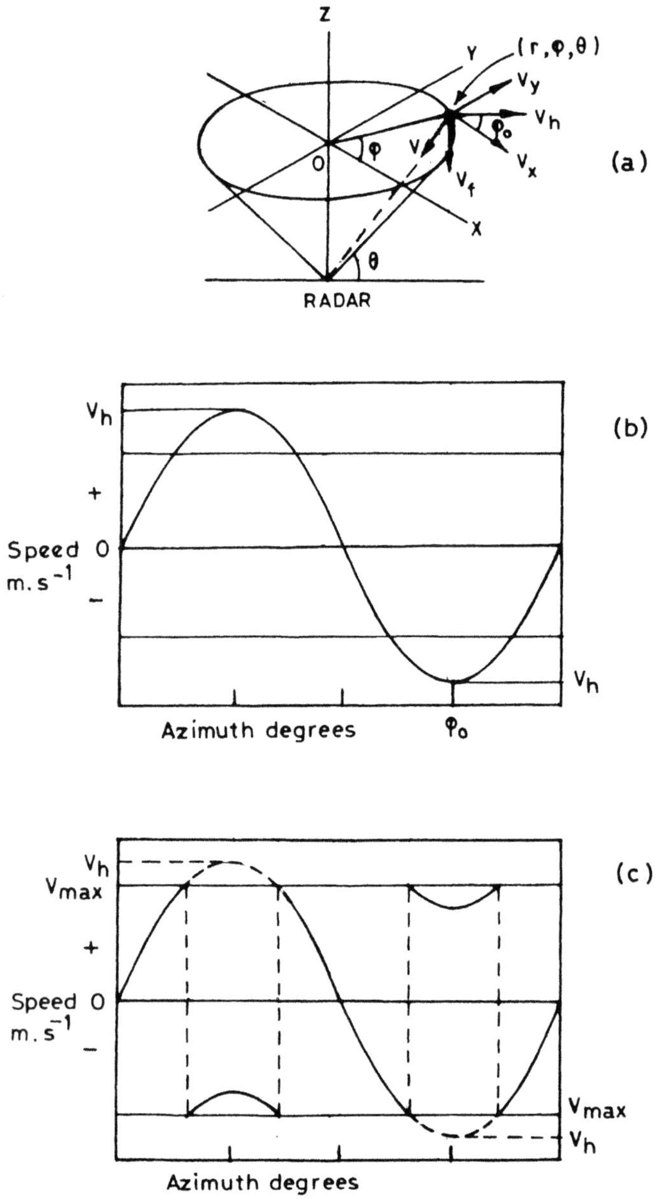

Fig. 3.14. Principle of Velocity Azimuth Display.
a) Geometry. The radar beam scans at elevation θ over a horizontal circle centre O. OX, OY are the horizontal axes through O and OZ is the vertical axis. ϕ is the azimuth angle with reference to X-axis. At any point (r, ϕ, θ) on the circle, the horizontal velocity is V_h; its azimuth is ϕ_0; its X and Y components are V_x, V_y. V_f is the precipitation fall velocity and V is the radial Doppler velocity.
b) Velocity Azimuth plot.
c) Velocity Azimuth plot when V_h exceeds V_{max}.

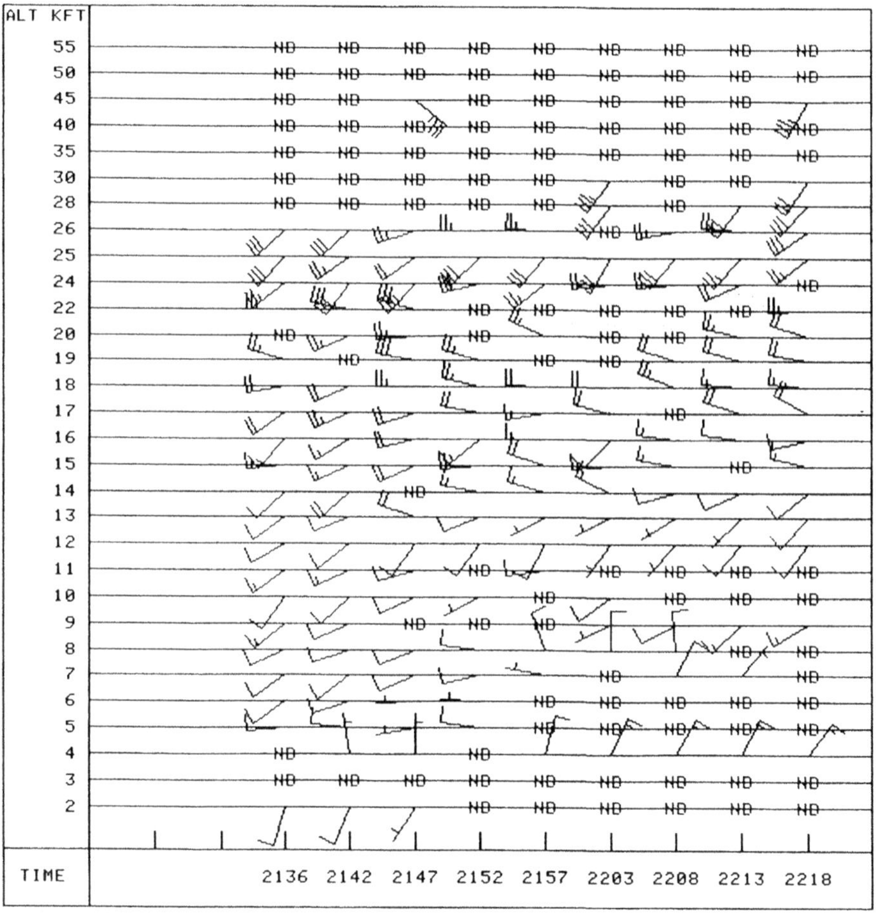

Fig. 3.15. VAD wind profile from US WSR-88D radar. ND indicates "No Data". Wind barbs follow standard meteorological practice. (From Klazura and Imy, 1993, courtesy Dr. G.E.Klazura, NOAA, and AMS).

$$V_y = V_{y0} + x\frac{\partial V_y}{\partial x} + y\frac{\partial V_y}{\partial y} \tag{3.50}$$

where the additional suffix 0 refers to the values of the wind components at the centre of the scanned circle (point O in Fig. 3.14a). Equation (3.45) can then be rewritten in terms of these quantities and expressed as a Fourier series,

$$V_h = a_0 + a_1 \cos\phi + b_1 \sin\phi + a_2 \cos 2\phi + b_2 \sin 2\phi \tag{3.51}$$

The Fourier coefficients on the r.h.s. are given by,

$$a_0 = \frac{r\cos\theta}{2}\left[\frac{\partial V_x}{\partial x} + \frac{\partial V_y}{\partial y}\right] - V_f \sin\theta \tag{3.52}$$

$$a_1 = V_{x0}\cos\theta \tag{3.53}$$

$$b_1 = V_{y0}\cos\theta \tag{3.54}$$

$$a_2 = \frac{r\cos\theta}{2}\left[\frac{\partial V_x}{\partial x} - \frac{\partial V_y}{\partial y}\right] \tag{3.55}$$

$$b_2 = \frac{r\cos\theta}{2}\left[\frac{\partial V_x}{\partial y} + \frac{\partial V_y}{\partial x}\right] \tag{3.56}$$

It will be recognised that the quantities in square brackets in equations (3.52), (3.55) and (3.56) are respectively, the two-dimensional horizontal divergence(Div), the stretching deformation (St.D) and the shearing deformation (Sh.D). These can be computed besides V_h, if V_f is assumed constant. It should be noted that Div., St.D, Sh.D and V_f are averages over the VAD circle.

To obtain reasonably accurate estimates of these quantities, Browning and Wexler (1968) suggested that instead of scanning at a single elevation, VAD's may be obtained at several elevations. They called this Stepped Elevation Velocity-Azimuth Display (SEVAD). V_f(overhead) may also be determined by having a vertically pointing observation by the same radar (see Paragraph 3.3.7). It should be noted however that vorticity defined as,

$$\left[\frac{\partial V_y}{\partial x} - \frac{\partial V_x}{\partial y}\right] \tag{3.57}$$

cannot be determined directly by the VAD method.

Browning and Wexler's (1968) work effectively extended the VAD concept to observation of convective echoes where constancy of the horizontal wind cannot be assumed. Srivastava et al.(1986) extended the VAD (Extended VAD or EVAD) to a number of elevation scans to compute both the divergence and fall speed. This assumes steady conditions over several scans and can be used in stratiform precipitation. Equation (3.52) can be rewritten as,

$$\frac{2a_0}{r\cos\theta} = \text{Div} - 2V_f h r^{-2} \tag{3.58}$$

where h is the height corresponding to range r and elevation θ neglecting earth curvature and refraction. Based on scans at several elevations, a plot of $(2a_0)/(r\cos\theta)$ vs $h.r^{-2}$ will yield a straight line, the slope and intercept of which will give V_f and Div respectively. Horizontal uniformity of these quantities is assumed. Figs. 3.16(a) and (b) show the vertical profiles of these two quantities obtained by them in the stratiform anvil of a squall line. Knowing the divergence, Srivastava et al. (1986) also computed the vertical profile of vertical air velocity, w, [Fig. 3.16(c)] from the continuity equation (3.35). These figures will be referred to in appropriate contexts in later Chapters.

Scialom (1984) used the VAD with a dual-Doppler radar system (called Double VAD or DVAD) to analyse the wind field over the overlapping area of the two VAD circles. Roux and Sun (1990) developed a method called ABCD[15] of combining successive scans of a single Doppler radar to determine the three-dimensional wind field in an approaching squall line.

Among extensions of the VAD technique proposed for use over limited areas for determining wind field parameters, which may be mentioned are the method of Easterbrook (1975) called Velocity ARea Display (VARD), Volume-Velocity Processing (VVP) of Waldteufel and Corbin (1979) and Modified Velocity-Volume Processing (MVVP) of Koscielny et al. (1982). All the techniques assume the constraints of linear variation of wind velocity in the horizontal and constancy of velocity during the period of observation. Subject to these constraints, divergence and deformation can be estimated and used to predict severe storm development. The VVP on which algorithms of most modern radars are based, makes use of a volume scan of several successive PPI scans at various elevations. If the wind components (u, v, w) vary linearly around a value of (u_0, v_0, w_0) at a point (x_0, y_0, z_0),

$$u = u_0 + u'_x(x - x_0) + u'_y(y - y_0) + u'_z(z - z_0) \tag{3.59}$$

with similar equations for v and w. Converting to polar coordinates (r, ϕ, θ) and considering only situations where variation of the vertical velocity in the horizontal plane can be neglected, Waldteufel and Corbin (1979) get the expression for radial velocity V as,

$$\begin{aligned} V = {} & u_0 \cos\phi \cos\theta + v_0 \sin\phi \cos\theta + w_0 \sin\theta + u'_x r \cos^2\phi \cos^2\theta \\ & + v'_y r \sin^2\phi \cos^2\theta + \left(u'_y + v'_x\right) r \sin\phi \cos\phi \cos^2\theta \\ & + w'_z \sin\theta \left(r \sin\theta - z_0\right) + u'_z \cos\phi \cos\theta \left(r \sin\theta - z_0\right) \\ & + v'_z \sin\phi \cos\theta \left(r \sin\theta - z_0\right) \end{aligned} \tag{3.60}$$

It is also assumed that conditions do not change during the volume scan. Hence there is a need to keep the total duration of the volume scan as small as possible consistent with the need for coverage of a number of elevations and a low scan speed to maximise the sampling. Koscielny et al. (1982) point out that divergence

[15] ABCD = Analyse de Balayages Coniques Doppler.

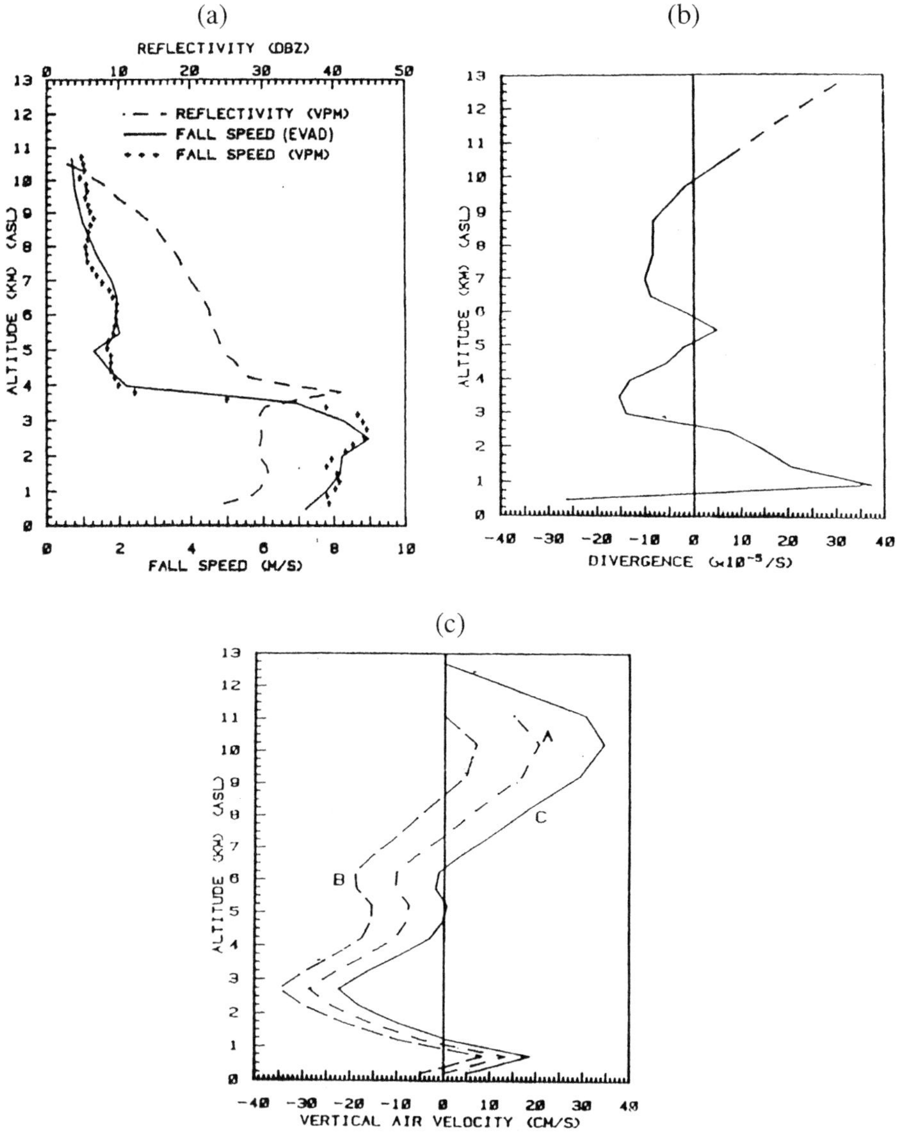

Fig. 3.16. (a) Vertical profile of hydrometeor fall speed from the EVAD technique (solid line) and from measurements in a Vertically Pointing Mode (VPM) in the trailing stratiform anvil of a mid-latitude squall line. The dashed curve shows Z profile from the VPM measurements. (b) Vertical profile of divergence of the horizontal wind from the EVAD technique. The dashed line is an extrapolation of the profile to the satellite-estimated cloud top. (c)Vertical profile of vertical air velocity w from integration of the divergence in (b). Curve A assumes $w = 0$ at the ground; curves B, C take $w = 0$ at radar and satellite cloud tops respectively. Curve B uses the solid divergence curve in (b) while curve C uses the dashed extrapolation also. (From Srivastava et al., 1986; courtesy AMS).

can be estimated with an accuracy of about 4×10^{-5} s^{-1} and a resolution of about 30 km which is sufficient to sense pre-thunderstorm convergence (negative divergence). Johnston et al. (1990) proposed a Band-Velocity-Processing (BVP) analysis for cases where the precipitation system has a small width compared to length and can be treated as a two-dimensional system. This can extract a X-Z cross section from a single volume scan. Other recent literature on single Doppler measurements of wind field and errors involved, which may be referred to are, Donaldson and Harris (1984, 1989), Harris et al. (1993), Lilly and Moeng (1984), May et al. (1989), Smith and Rabin (1989, 1989b) and Wilson et al. (1980). The case of non-linear windfields has been discussed by Caya and Zawadski (1992). A QVAD (Quadratic Velocity Azimuth Display) analysis has been presented by Scialom and Lemaître (1994) to cover the case of horizontal wind varying quadratically and the vertical velocity varying linearly in the horizontal plane.

A Synthetic Dual-Doppler (SDD) analysis technique has been described by Klimowski and Marwitz (1992). This uses observations from a single Doppler radar at position R_1, at two positions P_1 and P_2 of a moving precipitation cell at times t_1 and t_2. Radial velocity measurements V_1 and V_2 are obtained at the two positions of the cell as it moves. If now the radar is assumed to be shifted to a position R_2 by a distance equal to P_1P_2 in a direction parallel to but opposite to the cell motion, V_2 can be treated as a measurement taken from a fictitious radar at R_2 at time t_1. V_1 and V_2 are then subjected to dual-Doppler analysis. This analysis assumes that the cell is in a steady state in the time interval t_1t_2 and there is only translatory motion. It is also necessary that the azimuth angle subtended by P_1P_2 at the radar is large.

Tabary et al. (2001) have described a method of velocity ambiguity resolution involving a modified VAD analysis on the azimuthal derivative ($\partial V/\partial \phi$) of the radial velocity rather than on the radial velocity itself. They claim that this method can provide an estimate of the real wind directly from aliased wind measurements without prior dealiasing, allowing the method to be used in real time by operational radars, saving computing time.

3.3.11 Doppler Velocity Displays – Signatures of Weather Phenomena

A Doppler velocity display shows the values of the radial component of velocity in suitable false colours. The pattern seen on the display depends on the wind distribution as well as the relative positions of the radar and the weather system and the viewing direction. It is therefore necessary to recognise weather systems by their "signatures" i.e. particular patterns of Doppler velocities corresponding to particular spatial distributions of wind velocity. Wood and

Brown (1986) have produced computer-simulated pictures for various assumed non-divergent wind velocity distributions in the horizontal and vertical. Some of these are reproduced in Fig.3.17. An isopleth of constant Doppler velocity is called an isodop. In the figure the isodop patterns seen on a radar located at the centre of the circle, corresponding to the assumed vertical profiles of wind speed and direction shown on the top and the left, are portrayed. Wind is assumed not to vary in the horizontal. Solid lines represent flow away from the radar; short dashes represent flow towards the radar. Heavy long dashes represent zero Doppler velocity i.e. where wind direction is perpendicular to the radar beam. Heavier solid and dotted lines indicate locations of maximum and minimum Doppler velocity values within the display. *S* in the speed diagrams is an unspecified constant wind speed. Contour interval is 0.2*S*. Thus in Fig. 3.17a the wind being westerly, the radar sees maximum inward velocity to the west, maximum outward velocity on the east and zero velocities on the north and south. In other directions the radial component of the westerly wind is seen. In Fig. 3.17c two oppositely directed Doppler wind maxima are seen at the range at which the radar beam sees the wind maximum. Fig. 3.18a shows a case of wind varying in the horizontal plane in association with a cold front approaching the radar and Fig. 3.18b the corresponding isodop pattern. On the radar display the Doppler velocity distribution is depicted in various colours. The so-called "cool" colours – shades of blue and green – are used for inward velocities and warm colours – yellow and red for outward ones. Brown and Wood (1987) have presented simulated colour pictures of Doppler velocity patterns for various wind distributions. The isodop pattern for a divergent flow (e.g. a microburst) is shown in Fig. 3.19a and that for an axisymmetric vortex (e.g. a mesocyclone or the eyewall of a tropical cyclone) in Fig. 3.19b. In each case the radar is assumed to be 200 km south of the picture.

The Doppler velocity pattern of a vortex is particularly important and is discussed in detail. Armstrong and Donaldson (1969) devised a Plan Shear Indicator (PSI) display which produces a series of concentric arcs. Radial shear is indicated by gaps or bunching of the arcs while tangential shear (which may imply the presence of a vortex) is indicated by wrinkles in the arcs. Donaldson (1970) showed how a small scale vortex (such as a mesocyclone or tornado) can be identified from the observed tangential shear of the radial velocity. An axisymmetric vortex can ideally be considered to be a Rankine Vortex (Rankine 1901). The circle EFNGPHE in Fig. 3.20 is the base of a solid cylinder rotating with its axis vertical. It has a radius r_c and centre at C_T. The tangential[16], $v_t(r)$, and radial, $v_r(r)$, wind speeds at any point at radius r from the centre are given by (see e.g. Brown and Wood 1991, Donaldson 1970, Wood and Brown 1992, Zrnić and Doviak 1975, for a full discussion),

[16] The terms "tangential" and "radial" are with reference to the vortex.

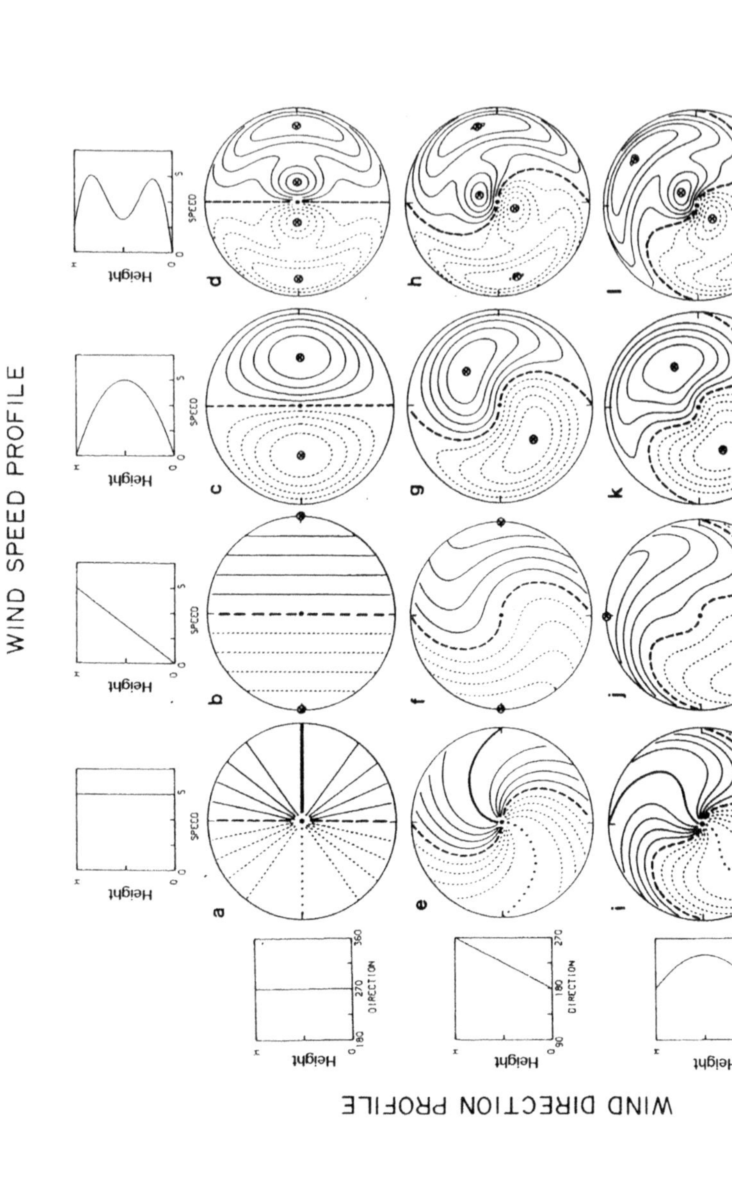

Fig. 3.17. Doppler velocity patterns (PPI scans at constant elevation angle) for various vertical profiles of wind speed and direction. See paragraph 3.3.11 for details. (from Wood and Brown, 1986, courtesy Drs. V.T. Wood and R.A. Brown, NSSL, USA).

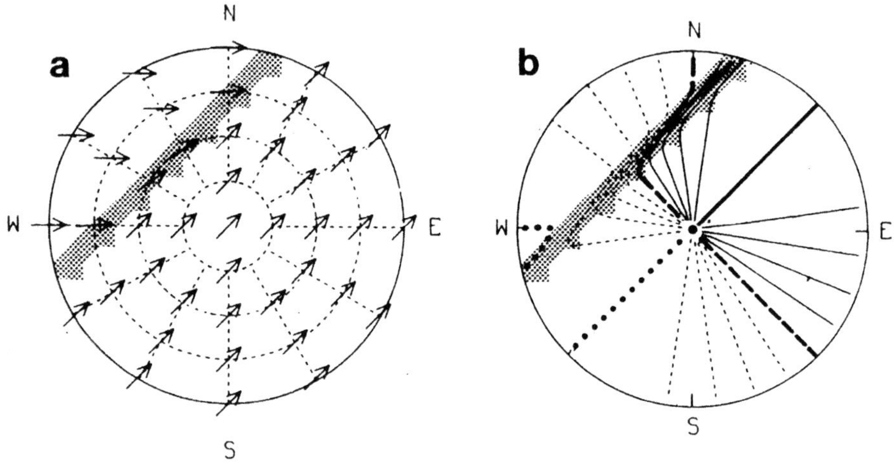

Fig. 3.18. (a) Assumed wind pattern around the radar in association with an approaching cold front. The front is shown by stippling.
(b) Corresponding single Doppler velocity pattern at zero elevation angle.
(from Wood and Brown, 1986, courtesy Drs. V.T. Wood and R.A. Brown, NSSL, USA).

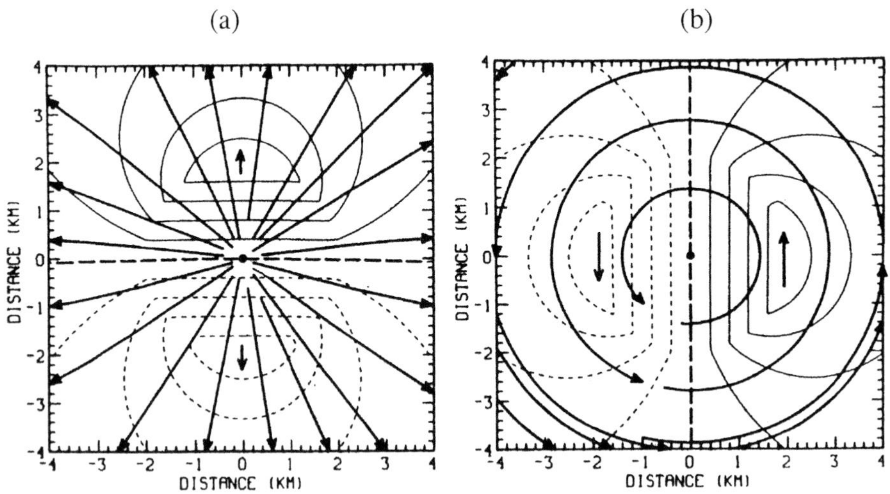

Fig. 3.19. Doppler velocity pattern corresponding to (a) axisymmetric divergent flow and (b) an axisymmetric vortex, seen from a radar located 200 km to the south of the picture (from Brown and Wood, 1991, courtesy Drs. V.T. Wood and R.A. Brown, NSSL, USA). Thick lines are streamlines. Thin solid lines are contours of Doppler velocity away from the radar and dashed lines, velocities towards the radar. Arrows on either side of the centre indicate the location of peak velocity.

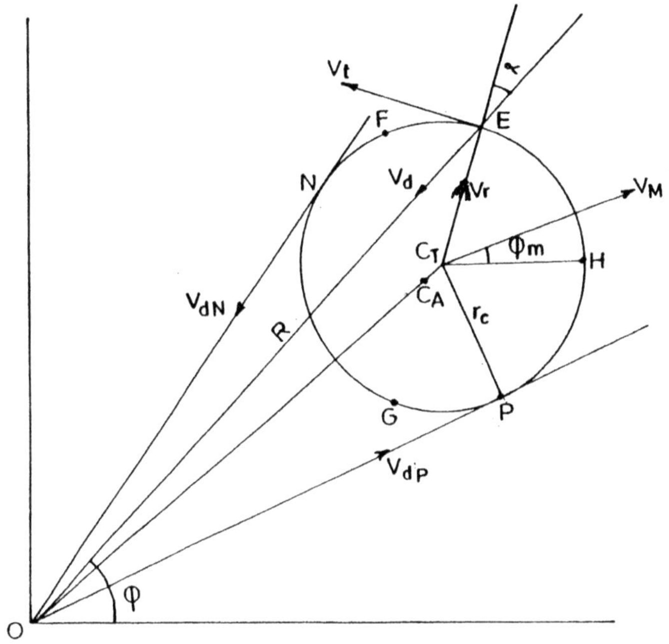

Fig. 3.20. Geometry of Doppler radar detection of an axisymmetric Rankine vortex. (adapted from Lee et al., 1993, courtesy AMS and Dr. Lee).

$$v_t(r) = V_t\left(\frac{r}{r_c}\right)^p \text{ and } v_r(r) = V_r\left(\frac{r}{r_c}\right)^p \tag{3.61}$$

where V_t, V_r, are the peak tangential and radial velocities at vortex radius r_c, $p = 1$ for $r <$ or $= r_c$ and $p = -1$ for $r > r_c$. The Doppler velocity V_d, measured by the radar located at O in Fig.3.20, at any point E (coordinates R, ϕ, with respect to the radar) on the vortex circumference will have contributions from V_t and V_r and the mean motion V_M, in the direction ϕ_m. The contributions from vertical motions and fall speeds of hydrometeors may be neglected if the radar is operated at a low antenna elevation angle θ. V is given by (Lee et al., 1993),

$$V_d = \cos\theta\left[V_r\cos\alpha - V_t\sin\alpha + V_M\cos(\phi - \phi_m)\right] \tag{3.62}$$

where α is the angle betwen $C_T E$ and OE in Fig. 3.20. If the vortex is stationary and there is no other motion superimposed, $V_M = 0$. The isodop pattern as seen by a radar due south of the vortex will then be as in Fig. 3.19b. There is a couplet of two oppositely directed Doppler velocity peaks. In the general case, the mean motion is known from tracking of the reflectivity pattern and can be subtracted. At the points N and P in Fig. 3.20, $\alpha = 90$ degrees and $V_r\cos\alpha$ vanishes.

Hence ,

(1) The couplet of opposite Doppler velocities V_{dN} and V_{dP} at the points N and P can be taken to a first approximation to be equal to the peak tangential velocity in the vortex.,

(2) The apparent centre of the vortex as seen from this pattern is C_A whose range is the mean of ON and OP and azimuth is the mean of the azimuths of N and P.

If the two maxima are not at the same range it indicates the presence of convergence or divergence in addition to the axisymmetric circulation. Brown and Wood (1991) showed that even when the wind field specific to the phenomenon is superposed on a larger scale flow, the signature can still be recognised although the pattern will change.

Wood and Brown (1992) studied the cases of an axisymmetric circulation as a function the ratio of core diameter of the circulation to range of the vortex centre. They showed that at low values of this ratio (i.e. when the circulation is far away or small in size) the two maximum radial velocity vectors are parallel to each other (as in Fig. 3.19b). The line joining the two velocity maxima is a diameter of the core and the mid-point of the line is the centre of the circulation. When the storm approaches the radar, the ratio increases and the isodop pattern distorts. The line joining the points N and P of maximum velocity towards and away from the radar forms a chord and not the diameter of the core. Hence the apparent centre position C_A of the core determined from the Doppler display will be nearer the radar, as in Fig. 3.20, than the true circulation centre C_T. The true centre C_T can be obtained using a correction factor which depends on the ratio of core diameter to range. The apparent diameter of the core will be smaller than the true diameter. Wood and Brown (1992) also showed that similar distortion will occur in the case of the divergent flow field of Fig. 3.19a, as the ratio of the size of the feature to the range increases.

Lee et al., (1993, 1999) have expressed V_t and V_r as truncated Fourier series and devised a means of their calculation from harmonic analysis of V_d collected from a VAD scan. They call this GBVTD (Ground Based-Velocity Track Display)[17].

The simulations are based on certain assumptions such as narrow beam width (in comparison to the scale of the phenomenon), low elevation scanning to avoid contamination by the component of the vertical motion of air and precipitation, idealised air flow fields in the weather system, and presence of scatterers in the entire region of interest. There can be distortions in the patterns due to all these assumptions not being true. The beam width at long range may be larger than the size of the feature. Also as discussed above, the signature

[17] See Chapter 8 and also the companion papers Lee and Marks (2000) and Lee et al. (2000).

distorts when the weather system approaches the radar. Distortion can arise from other causes too. Wood and Brown (2000) simulated a mesocyclone changing range relative to a Doppler radar and found that the deduced core diameter and mean rotational velocity of the Doppler velocity mesocyclone signature oscillate back and forth, even though the physical width of the radar beam changes uniformly with range. May (2001) has pointed out that the dual-PRT technique for resolving ambiguities can introduce errors in velocity measurements when horizontal gradients of velocity are large, even in cases where there is no velocity folding. The evaluation of the actual flow pattern from the observed Doppler pattern therefore needs careful interpretation. In this sense the signature recognition technique may be considered qualitative. An experienced observer may be able to detect many signatures from a colour display. However because of the vast mass of data involved, automated systems are necessary for recognition of Doppler signatures (see e.g. Moninger, 1986). Signatures of mesocyclones and tornadoes and algorithms for their detection are discussed in Chapter 6 and those of tropical cyclone cores in Chapter 8.

3.3.12 Other Methods of Deriving the Wind Field

Evidently, the horizontal wind (at the altitude level of the echo) can be inferred by tracing the motion of the reflectivity pattern over successive observations without the facility of Doppler measurements. The pattern can be from clear air or insects or from a group of cells, if these can be assumed to be tracers of air motion. Such methods are in operational use and are discussed in Chapter 5. Implicit in using such a method is the assumption of conservation of the reflectivity pattern as it moves. Techniques have been developed by combining reflectivity data with Doppler radial velocity data in successive volume scans to derive the three-dimensional wind field and even thermodynamic parameters. These are not in operational use at present and will be briefly dealt with in Chapters 5 and 6.

3.4 Suppression of Ground Clutter and Anomalous Propagation (AP) Echoes

3.4.1 Clutter and AP Echoes

Besides meteorological echoes, a weather radar picks up echoes from many other phenomena - from the land or sea surface, from ground features such as hills, trees and buildings and from aircraft, ships, birds or insect swarms. For example in Fig. 1.4, the echoes at the north-west (60-80 km) and southwest (about 40 km) are hills, the line going south from the radar site is the coastline, the other echoes around the radar are ground features and the short arcs are ships. In the reflectivity display, the ground clutter can be recognised by its correspondence to known features around the radar such as hills, buildings, and towers or the sea coast. The operator can have a photograph or printout of the echoes taken (at various elevation settings) in fair weather conditions for reference to distinguish these echoes from weather echoes. The ground clutter close to the radar is usually very strong and cannot be suppressed by gain reduction procedures. On the contrary, the precipitation echoes are more likely to be eliminated by such techniques. Clutter echoes from vertical sidelobes can also be quite strong. Most clutter can usually be eliminated by raising the antenna elevation, but many precipitation echoes will also disappear. As can be seen on an A-scope, the ground clutter being from solid targets will be much more steady i.e. have a much smaller variance than precipitation echoes which fluctuate considerably. Some fluctuation of clutter echoes will also be there because of moving or vibrating objects, effect of antenna rotation and radar system instability. Moreover ground clutter may be augmented in area as well as intensity by occurrence of anomalous propagation (AP). Such conditions may also enable clutter echoes to appear at greater elevations than usual. Sea clutter echo shows a greater variance than other ground clutter. Even though the operator can distinguish and ignore clutter on most occasions, it is not possible to make quantitative observations on the precipitation echoes superposed on the ground clutter unless the latter is removed. Hence the ground clutter is minimised on ground-based radars by careful siting (see Chapter 1). Moores and Harrold (1975) described a method of estimating the extent and intensity of ground clutter from a digitised contour map. Areas covered by clutter may have to be excluded from precipitation measurements. This is a major problem in mountainous terrain.

AP echoes constitute a more serious problem as they vary considerably with propagation conditions and are often mixed up with precipitation echoes. It has been estimated that they can create an error of about 15% in long-period rain accumulation estimates, unless recognised and eliminated. In instantaneous or short-period rain rate distribution they can cause larger errors if they are not distinguished from precipitation echoes.

3.4.2 Elimination of Clutter and AP

One method of tackling the problem in digitised systems is to store the clutter video signal exceeding a specified threshold, acquired in fair weather in respect of each range-azimuth bin. In operation, the current data from these bins are ignored leaving the clutter areas blank. Values for the blank bins can be interpolated from observed values for adjacent bins (Ball et al. 1979). In case of widespread precipitation where horizontal gradient of reflectivity is small, the interpolation does not cause large error. This, will not be valid in convective precipitation where the horizontal reflectivity gradients are large. In such a case, interpolation from bins vertically above can be done (Sánchez-Diezma et al. 2001). This procedure does not take into account clutter variation with time. Also, this method will not work for AP echoes. If there is a dense network of radars, anomalous propagation can be recognised by criteria to correlate its occurrence at the same points in different radars of the network (Atkinson and James, 1991). It is also useful to prepare a climatology of usual anomalous propagation patterns observed at various times and seasons for each radar site (e.g. Rajagopalan and Raghavan, 1980). Pamment and Conway (1998) devised an automated scheme to estimate the probability that an echo is caused by anomalous propagation, using a climatology of AP echoes, hourly surface observations, geostationary satellite infrared images, and detected lightning.

3.4.3 Variance of Clutter and Precipitation Signals

The fact that the variance of a signal from a fixed ground target is much smaller than that from precipitation echo can be used for rejecting ground clutter. This is done by passing the linear output signal from the radar receiver (the sum of the signals from the ground clutter and precipitation) through a "delay line canceller". A delay line is a device which delays the signal by a specified time. The delay time chosen is the PRT, T_s or a multiple thereof. The signal passing through the delay line and the direct signal are combined at the output. The former represents the return from the previous pulse and the latter from the latest transmitted pulse. The signal can be considered to consist of a DC component and an AC component. At the output, the delayed and undelayed signals will cancel each other if the return from the target is constant with time. When there is a variance it can be shown that the output is proportional to the variance. In the case of a logarithmic receiver the signal is passed through an anti-log circuit before the delay line and the canceller output made logarithmic again. Hence the ground clutter which has low variance is reduced in comparison with the precipitation echo. By proper design of the canceller circuit a clutter reduction of as much as 40 dB can be achieved (see Johnson et al. 1975, Tatehira and Shimizu 1978, Aoyagi 1983). This technique will also reject anomalous propagation echoes besides normal ground clutter.

3.4.4 Clutter Suppression using Doppler Parameters

In Doppler radars, the ground clutter can be identified from its zero or near zero velocity. By using a "notch" filter to reject the DC and one or two adjacent spectral coefficients (i.e those corresponding to low velocity, say, less than 1 m s^{-1}) the clutter is rejected[18]. A canceller design has been given by Groginsky and Glover (1980). The drawback is that precipitation echoes with near-zero velocities may also get eliminated. Spectrum width also can be used as a criterion. Upto 50 dB clutter suppression can be achieved. A parametric estimation method for retrieval of Doppler moments from clutter-contaminated measurements of the autocorrelation function has also been proposed (Passarelli, 1983). But complication is created if multiple-trip echoes are present. These have to be eliminated by changing p.r.f. or other methods (see section 3.3.). In Doppler polarimetric radar the alternate transmission of two polarisations reduces effective maximum unambiguous velocity. The velocity-aliased precipitation echoes may appear to have near-zero velocities and get rejected by the clutter filter. As mentioned earlier, range-velocity ambiguity resolution techniques may affect the clutter-filtering. Sachidananda and Zrnić (2000) have developed a method to suppress clutter with the staggered PRT waveform with little distortion using sophisticated matrix methods.

3.4.5 Use of Polarisation

In radars for aircraft detection, rain which is clutter, is sought to be suppressed by using circular polarisation. As discussed in Chapter 2, the return in the cross-polarised component from spherical rain drops is zero though non-spherical drops and snow will give cross-polarised return also. In polarimetric weather radar, the polarisation properties of ground echoes which are different from those of rain can be used to distinguish them. In the case of dual linear polarisation radars Z_{DR} and Z_H clutter vary widely because the shapes of ground objects are highly irregular. The Z_{DR} to Z_H correlation is poor in clutter and produces a pattern on the Z_{DR} display which can be used to identify clutter (Giuli et al., 1991). In general, at radar wavelengths shorter than 10 cm the ground and sea clutter is stronger for vertical polarisation than at horizontal polarisation. Even at 10 cm, sea clutter is stronger at vertical polarisation (Balakrishnan et al., 1993).

---

[18] Duplicate cancellers are needed for the I- and Q- channels.

3.4.6 Statistical Techniques

Moszkowicz et al., (1994) formulated a statistical technique of recognition of anomalous propagation echoes based on observed vertical structure , horizontal gradient of reflectivity etc., which can be used to distinguish them from precipitation echoes. Their algorithm is based on examination by human experts of a large volume of past data, visually estimating the features and classifying them as precipitation or AP echoes. This laborious process can be automated by "training" a neural network (NN) with past data of radar features[19] so that it can then automatically recognise AP (Grecu and Krajewski 1999, 2000; Krajewski and Vignal 2001). This technique has the advantage that it can be used to remove AP from past data also, but requires considerable computer power and its performance in the operational time frame is yet to be evaluated. Training is also needed for each radar site. The NN can recognise only situations, which it has been trained to, and may misclassify on some occasions. However its training can be updated with new data. It is not likely to be able to eliminate AP which is mixed up with precipitation echo, as in the case of AP following a thunderstorm.

The combined use of several Doppler and reflectivity parameters is considered to give better results in clutter and AP elimination rather than the use of any one feature. It should also be noted that Doppler parameters will usually be available up to a shorter range than reflectivity parameters. For example Pratte et al. (1997) have considered the "texture" of signal-to-noise ratio, mean and standard deviation of the velocity moments and the spectrum width estimates in a cell. Lee et al. (1995) used criteria relating to a number of parameters viz., MDS, wide-band noise, Doppler velocity and vertical gradient (of reflectivity) in conjunction with an adaptive clutter map. More elaborate algorithms involving parameters such as computed vertical profile of reflectivity (VPR), horizontal or vertical gradient of reflectivity, vertical integration of Doppler velocity and comparison with satellite imagery (e.g. Bellon and Kilambi 1999, Fiore et al. 1986, Keeler et al. 1999b, Kessinger et al. 1999, VanAndel et al. 1999) have been attempted with some degree of success. A method of correcting for the bias in reflectivity estimates caused by the clutter filtering has also noisebeen developed (Ellis et al., 1999). Steiner and Smith (2002), have developed an algorithm using a 3-dimensional structure (without using Doppler parameters) and claim that it rejects even sea clutter and AP embedded in precipitation echo[20]. Even so, operator familiarity with normal

---

[19] The features considered by Grecu and Krajewski (2000) in each pixel were, absolute velocity and its coefficient of variation, maximum height of echo, reflectivity and its coefficient of variation around the pixel, coefficient of fluctuation and maximum horizontal gradient in the neighborhood. The velocity parameters were derived from successive reflectivity scans and not from Doppler measurements.

patterns of clutter at any given location, still remains an important means of clutter and AP recognition.

3.4.7 Occultation and Screening

Besides giving echoes, ground objects may partially or fully obscure areas behind them and cast a radar shadow. This is a particularly serious problem in hilly terrain. The shadow can often be got rid of by raising elevation but this will mean looking at the precipitation echo at greater altitude. The echo may disappear or not be representative of the precipitation at the ground. Where the shadow is not extensive, the signal from affected bins may be rejected and interpolated values from adjacent bins may be substituted. Harrold et al. (1972, 1974) in the UK, applied corrections for partially filled (as a result of screening) beam volumes and found that such correction is valid if less than half the beam is screened.

3.5 Airborne Radar

3.5.1 Early Airborne Radars

Some of the advantages as well as constraints in the use of weather radar on board aircraft were discussed in section 1.3. Airborne radar has been used qualitatively even in the forties of the last century for observation of weather systems (e.g. Maynard, 1945). Simple radars with a 120 degrees sector PPI scan in the forward sector have been in use in passenger aircraft for a long time for the pilot to see severe weather ahead. These had a video inversion iso-echo or an iso-echo display with a few reflectivity levels in false colour to judge the intensity of a thunderstorm (Helfrick 1984).

3.5.2 Types of Scan

For serious meteorological studies more elaborate equipment is needed. The radar may be in the nose of the aircraft and scan in a PPI mode in the forward sector or it may be in the belly of the aircraft and scanning in the PPI mode. Or, the radar may be vertically pointing up or down. Or, it can be a tail radar scanning in a vertical plane perpendicular to the aircraft motion (see typical specifications of some airborne radars in section 1.3.). The type of scan desired may depend

---

[20] A negative aspect is that their algorithm tends to remove boundary layer features, such as gust fronts.

on the flight level and the purpose of the observation. Observation may be made from below a storm or from above or a penetration can be made into the storm. In case of convective storms of limited area which are unsafe to penetrate, observation can be made by flying on the sides. Consider, for instance, a radar on board an aircraft flying at 10 km altitude (Merritt 1969). The PPI sector scan will be pointing at a negative elevation to see most weather systems. If the beam width in the vertical is, say, 3°, the lower edge of the beam will make a tangential contact with the earth's surface at a slant range of 412 km in normal propagating conditions, if the beam is directed at −1.3° elevation. At short range the beam illuminates the upper part of the precipitating cloud but as range increases the altitude of the beam decreases and it will intercept the lower and therefore higher reflectivity portions of the precipitation. Hence the power received from uniform precipitation will increase with range for some distance and after that the power decreases with range but not so fast as the r^{-2} term in the radar equation would indicate. This behaviour is opposite to that encountered with ground-based radar. At still greater ranges the beam will not be filled and the received power will decrease faster than r^{-2}. This is similar to what happens with a ground radar. If the aircraft is flying lower, the maximum range will be reduced. Since the aircraft can, in most cases, go close to the weather system it is usually sufficient to observe at short ranges where high resolution can be obtained and the wide beam width is not a limitation. Attenuation at long range can be serious because of the use of shorter wavelengths for airborne radar.

3.5.3 Data Processing

The position and orientation of the aircraft must be accurately known for the proper processing of the data. The antenna scanning needs to be stabilised for the pitch and roll of the aircraft (see footnote[21] and Fig. 3.21 for a glossary of aviation terms). Information for this is obtained from the inertial navigation

---

[21] Some aviation terms (see also Fig. 3.21).
Roll: Oscillation of the craft about a horizontal axis parallel to the direction of motion. Roll angle R is the deviation from the horizontal in the pitch-yaw axes plane. Right wing down is positive.
Pitch: Oscillation of the craft about a horizontal axis perpendicular to the direction of motion. Pitch angle P is the angle of deviation from horizontal in the roll-yaw axes plane. Nose up is positive.
Yaw : Oscillation of the craft about a vertical axis. Nose to the right is positive yaw angle (Y).
Heading: Angular direction of the centre-line of the craft measured clockwise from North.. It represents the motion with reference to air at flight level.
Track: Direction of aircraft motion with reference to the ground, measured clockwise from North.
Drift: Angular difference D (clockwise) between the heading of the aircraft and its track. This is caused by the wind.

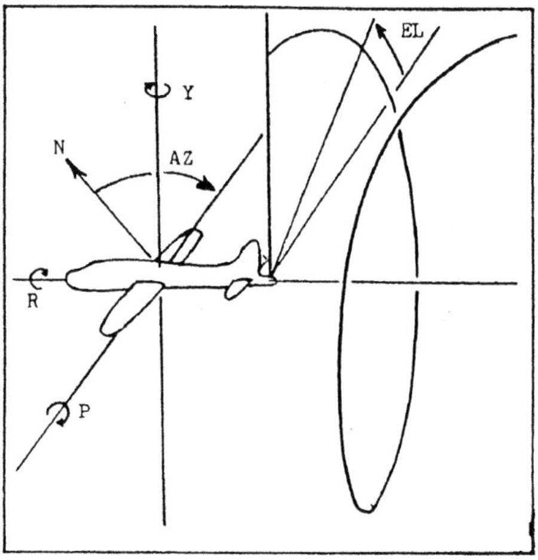

Fig. 3.21. Geometry of helical-scan airborne radar. (From Hildebrand et al. 1983, courtesy AMS).

system of the aircraft. Because of the motion of the aircraft the pulse volume is continuously changing and this can be a constraint on the number of samples that can be integrated. For example, if the aircraft speed is 125 m s^{-1} and radar p.r.f. is 400, the translatory motion is 20 m in the time taken to collect 64 samples. The number of samples can be increased by increasing p.r.f. at the cost of decreasing unambiguous range. If the scan rate is say 8 rpm the aircraft travels nearly 1 km during the period of one 360° scan. If the aircraft speed is further increased, the spatial resolution will suffer. If the speed is decreased, the time taken to cover a storm is increased. The digital processing of reflectivity is similar to that in the case of ground radar. The picture can be displayed on board as a PPI with aircraft as centre or as an RHI in the case of scanning in a vertical plane.

3.5.4 Airborne Doppler Radar

In the case of a Doppler system, the component of the aircraft velocity in the viewing direction gets added to the Doppler velocity. This is minimised (ideally made zero) if the radar is scanning in a plane perpendicular to aircraft motion as in Fig. 3.21. Even so, in practice there will be a small component added due to the drift of the aircraft. Also, since the location of each data point is measured relative to the aircraft, the absolute location of each data point with reference to the ground has to be computed, to get the ground-relative winds. Another effect is the broadening of the spectrum width due to the motion of the aircraft perpendicular to the radar viewing direction. Since the beam has a finite width the aircraft velocity is not perpendicular to the viewing direction at the edges of the beam; hence it will have a component in that direction. This is similar to the effect due to the rotating feed horn of a ground-based radar antenna. Following Hildebrand et al. (1983) and Jorgensen et al. (1983) the scan geometry is shown in Fig. 3.21. In the figure, the scan is in a plane perpendicular to the paper. The AZimuth ϕ is measured clockwise from North (N). The ELevation θ is measured from the horizontal plane containing the aircraft. As the aircraft moves, the radar effectively scans a helical pattern. Because of the pitch and drift the aircraft velocity is not perpendicular to the scan. Hence there will be a component of the aircraft motion in the measured Doppler velocity. To prevent this, the azimuth instead of remaining at right angles to the heading, is varied by the antenna servomechanism according to the relation,

$$\sin\phi = \cos(\theta - R)\sin P - \sin(\theta - R)\sin D\cos P \qquad (3.63)$$

(see footnote 21 for explanation of symbols R, P and D).
Thus the scan is not in a plane perpendicular to the heading of the aircraft and is a conical one.

The unambiguous velocity is rather low at the 3-cm wavelength radars (see specifications of NOAA P3 radar in Table 1.6, Chapter 1). To extend this, Jorgensen et al. (2000) have developed a batch-mode dual pulse repetition frequency technique. This utilizes two batches of pulses that are each averaged to produce two velocity estimates at each range bin, with each batch at a different sampling rate. Comparison of the two velocity estimates produces a difference velocity that is used to dealias the second estimate within an extended velocity range that is larger than the Nyquist velocities of either of the two original samples.

3.5.5 Pseudo-dual-Doppler Technique

The I and Q outputs are fed to a pulse pair processor as in a ground-based radar and the Doppler velocities are recorded on magnetic tape. The reflectivity data are separately recorded. In order to get the horizontal velocity, the aircraft is flown in the pattern shown in Fig. 3.22 in orthogonal legs. At each of the two positions shown, the radar monitors the same pulse volume where the horizontal velocity is represented by V_H. From the two positions of the aircraft, radial velocities Vr_1 and Vr_2 which are in planes at right angles to each other, are observed. These are given by,

$$V_{r1} = u\sin\beta_1\cos\theta_1 + v\cos\beta_1\cos\theta_1 + (w + V_t + V_a)\sin\theta_1 \qquad (3.64)$$

$$V_{r2} = u\sin\beta_2\cos\theta_2 + v\cos\beta_2\cos\theta_2 + (w + V_t + V_a)\sin\theta_2 \qquad (3.65)$$

β and θ being the azimuth and elevation angles respectively. These equations are the same as equations (3.37) and (3.38) for a ground-based dual-Doppler system

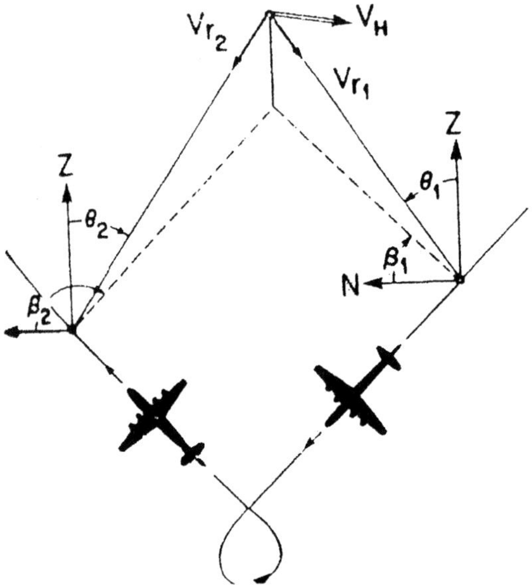

Fig. 3.22. Geometry of pseudo-dual-Doppler radar horizontal wind calculation. (From Jorgensen et al. 1983, courtesy AMS).

except for the addition of the term V_a which is the vertical velocity of the aircraft. As in the ground-based case the third terms on the right hand side can be neglected especially in stratiform rain situations if θ is kept less than 20°. Hence we get,

$$u = \left(\frac{1}{\sin(\beta_1 - \beta_2)}\right)\left(\frac{V_{r1}\cos\beta_2}{\cos\theta_1} - \frac{V_{r2}\cos\beta_1}{\cos\theta_2}\right) \tag{3.66}$$

$$v = \left(\frac{1}{\sin(\beta_1 - \beta_2)}\right)\left(\frac{V_{r2}\sin\beta_1}{\cos\theta_2} - \frac{V_{r1}\sin\beta_2}{\cos\theta_1}\right) \tag{3.67}$$

Thus, the horizontal velocity components can be computed assuming that the wind field at the pulse volume does not vary within the 10 or 15 minutes that it takes for the aircraft to go from position 1 to position 2. The vertical velocity w is to be obtained from the equation of continuity (3.35). Preliminary estimates of u and v are obtained from equations (3.66) and (3.67) and used in equation (3.35). V_t is computed from a knowledge of Z [see equation (3.32)]. Different constants in the V_t-Z relation may have to be used for the rain and snow altitude regions. Refined estimates of u and v are then obtained from equations (3.64) and (3.65) by iteration. This method has been used in convective weather systems where the vertical motion cannot be neglected (Jorgensen and Marks 1984). Alternatively the vertical motion can be obtained from a vertically pointing Doppler radar on board (see below).

The pseudo-dual-Doppler technique is similar to the Synthetic Dual-Doppler (SDD) technique described in section 3.3, except that in this case the radar moves while in the SDD case the target moves.

Area-average divergence can be computed from the known Doppler radial velocities in the two flight legs. For this purpose the area chosen is a parallelogram whose adjacent sides are parallel to the two flight legs in Fig. 3.22. Hence, the horizontal component of the Doppler radial velocity is the component of wind normal to the boundary of the parallelogram (Mapes and Houze 1993).

As mentioned in section 1.3, the tail radar of the NOAA P-3 aircraft of the USA was converted into a Doppler system. Details of this system are given by Trotter et al. (1980) and Parrish (1989). The design of the ELDORA/ASTRAIA radar of the US NCAR and the French CNET is discussed by Hildebrand et al. (1986, 1996), Frush et al. (1986) and Walther et al. (1986). An analysis of the errors involved in airborne Doppler radar observations may be seen in Ray et al., (1985). The mapping of airborne Doppler radar data is discussed in detail in Lee et al., (1994), corrections for coordinate transformation errors in Testud and Hildebrand (1991) and corrections for navigation errors using the echo returned from the earth's surface in Testud et al., (1995). An improved procedure for the latter may be seen in Bosart et al. (2001).

Heymsfield et al. (1989) have described the development of an airborne system with a forward- and a nadir-looking beam. The latter will enable a direct measurement of the vertical hydrometeor motions. By subtracting the fall speeds, as discussed earlier, the vertical air motion can be obtained. This may be particularly useful for studies of severe thunderstorms. A high degree of aircraft stability will be required to achieve accurate measurements.

3.5.6 Velocity Track Display (VTD)

An algorithm known as Velocity Track Display (VTD) has been developed (Lee et al., 1994b) to deduce the circulation in vortices, mainly tropical cyclones. When the airborne Doppler radar penetrates a circulation, VTD decomposes Doppler velocities on cylindrical rings into tangential, radial and mean cross-track components of velocity. Data from only one leg of the flight are needed. This is further discussed in Chapter 8 relating to tropical cyclones.

3.5.7 Dual Helical Scan Radar

One of the principal sources of error in the above technique is the assumption of constant wind field between the two legs of observation. Advection and storm evolution will vitiate this assumption.To minimise this problem an "ELDORA-ASTRAIA"[22] radar has been designed by the National Center for Atmospheric Research, USA and the Centre d'études des Environnements Terrestre et Planetaire (CNET), France. This consists of two helical scan radars on the same aircraft. They scan in cones respectively about 20° forward and about 20° aft of the plane normal to the longitudinal axis of the aircraft (Walther et al.1986; Hildebrand et al.1996). As the aircraft flies, the two observe the region around the radar and can be treated as a dual-Doppler pair. The difference between the two viewing angles (about 40°) has been selected on the basis of the needed accuracy of the velocity measurements and the maximum allowable time lag between the fore and aft observations. The geometry of this Fore/Aft Scanning Technique or FAST technique is shown in Fig. 3.23. It eliminates the time difference and also enables a safe straight path parallel to the storm without penetrating it. Hildebrand et al., (1996) also point out that the higher resolution of their radars gives higher peak updraft velocities in storms which are probably smoothed out by lower resolution radars.

Jorgensen and DuGranrut (1991) have described a method of doing this with a single radar on board the NOAA P3 aircraft. The single vertically scanning antenna is made to point alternately 25° forward and aft. This requires the antenna to be mechanically slewed to the positions alternately and reduces the resolution by a factor of two. An Alternative FAST (AFAST) technique is to do the forward scan as the aircraft approaches the region of interest and do the aft scan when the craft passes the region. This restores the resolution but halves the area of coverage. Jorgensen and Smull (1993) used this successfully in a mesoscale convective system.

Another possibility is to fly two separate aircraft with radars, in orthogonal legs. Each radar scans in a plane perpendicular to the aircraft heading. The disadvantages of this are the high cost and the problem of co-ordination of the observation from the two aircraft. Gamache et al. (1995) have compared the use of all the three methods in a hurricane and found that the dual aircraft and FAST techniques enable considerable improvement in accuracy of wind determination over the original pseudo-Doppler technique. Two P3 aircraft flying on parallel tracks in the same storm (Hildebrand et al., 1996) have been used to collect what are known as quad-Doppler data. Quad-Doppler analysis makes use of fore and aft data from the two airborne radars to provide four radial velocity measurements at each grid point. They have also compared this

---

[22] ELectra DOppler RAdar/ Analyse STéréoscopique par RAdar à Impulsions Aéroportes. The use of the "stereoradar" technique on this radar is discussed in Chapter 7.

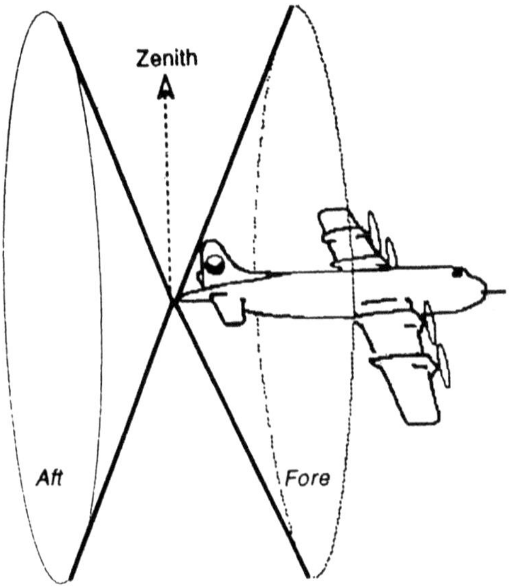

Fig. 3.23. Geometry of FAST scan. (From Jorgensen and Smull, 1993, courtesy AMS).

analysis with the dual-Doppler analysis from the Eldora radar and found good agreement. Chong and Campos (1996) have proposed an Extended Over-determined Dual-Doppler (EODD) technique to overcome certain geometrical limitations in the analysis techniques and to mitigate errors in estimating vertical velocity.It is possible to use more positions of an aircraft or these positions in conjunction with ground-based Doppler radars as a multiple-Doppler radar system. Combining airborne and ground-based Doppler data is discussed by Ray and Jorgensen (1988).

3.6 Descriptions of Some Modern Radars

3.6.1 The WSR-88D Radar

The US NEXRAD WSR-88D radar (see specifications in Chapter 1) is the operational Doppler Weather radar currently used in the largest numbers. Details of the hardware, the output products and the data recording and archival relating to this radar, are given by Heiss et al. (1990), Klazura and Imy (1993), Crum et al. (1993), and Crum and Alberty (1993). NEXt generation RADar intended to replace the conventional Weather Surveillance Radar (WSR-57) network of the USA was conceived in the seventies and over several years its specifications were worked out, evaluation tests made on prototypes and algorithms developed. This 10 cm wavelength radar renamed WSR-88D (Weather Surveillance Radar-1988-Doppler) has been installed in large numbers all over the USA. and in some places elsewhere. A 5 cm radar of somewhat similar specifications known as the TDWR (Terminal Doppler Weather Radar) has been installed at US airports exclusively for aviation weather service (see Chapter 9).

The WSR-88D consists of three major functional components: Radar Data Acquisition (RDA), Radar Product Generator (RPG), and Principal User Processor (PUP) along with the communication lines that link them (Fig. 3.24). The RDA consists of the antenna, pedestal and radome mounted on a tower, transmitter, receiver, minicomputer and signal processor. The transmitter is a coherent one using a very stable oscillator and a klystron power amplifier. This ensures high frequency stability, negligible signal distortion and spurious signal emission, good ground clutter suppression and high reliability. The receiver down-converts the signal to IF at which, amplification, matched filtering and automatic gain control are carried out. The signal is divided into I and Q components.

The RDA has a status control processor controlling antenna scanning patterns, signal processing, ground clutter suppression, status monitoring, error detection, automatic calibration and the capability to record the base data. Mean radial velocity and spectrum width data are provided with 0.25 km spatial resolution and 0.5 $m.s^{-1}$ velocity resolution. Reflectivity data have a spatial resolution of 1 km and Z resolution of 0.5 dBZ_e. These, known as base data or Level II data are obtained in digital form in polar format.

The RPG is either located near the RDA and directly connected to it or it is at a remote location and linked to the RDA by a duplex wideband communication link. The RPG executes algorithms for converting the base data to meteorological and hydrological products (listed in Table 3.1). It also provides data storage, velocity dealiasing and control and status monitoring of the RDA and RPG. The products are recorded on optical disk and also distributed as "Level III data". The RPG has an application terminal from which the operator controls the RDA and RPG.

The PUP is a user terminal. Several PUP's can be connected to a single RPG by narrow band communication links. The PUP consists of a minicomputer, system console, colour printer, graphics processor, work station and communication system. It displays, manipulates and records products known as Level IV data. Colour graphics with a 16-colour scale are used for depicting reflectivity, velocity, spectrum width, echo top heights and other

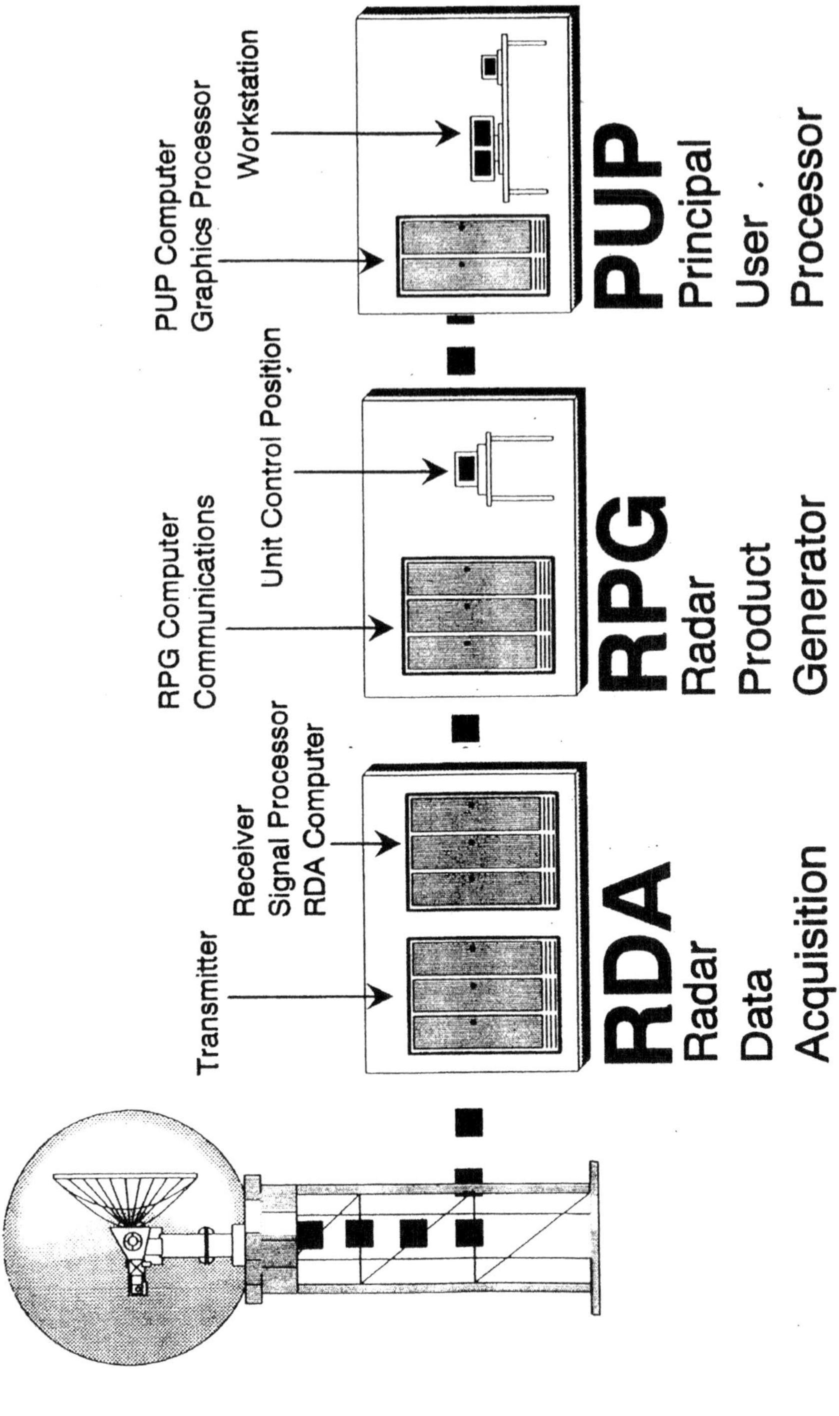

Fig. 3.24. The WSR-88D system (after Crum and Alberty 1993, courtesy AMS)

Table 3.1. Output products of WSR-88D radar; adapted from Crum and Alberty (1993) and Klazura and Imy (1993); courtesy AMS.

Base data	Reflectivity	Velocity	Spectrum width
Single elevation Products		Combined moment	
	Base reflectivity Severe weather analysis - reflectivity	Mean radial velocity Storm relative mean radial velocity (map) Storm relative mean radial velocity (regional) Combined shear Combined shear contour Severe weather analysis - radial shear Severe weather analysis - velocity	Spectrum width Severe weather analysis - spectrum width
Volume Scan Products	Cross section - reflectivity Composite reflectivity Composite reflectivity contour Echo tops Echo tops - contour Weak Echo Region (WER) Layer Composite reflectivity - average Layer Composite reflectivity - maximum One-hour digital precipitation array One-hour precipitation accumulation 3-hour precipitation accumulation Supplemental precipitation data Vertically Integrated Liquid (VIL) Severe weather probability Hail Index Storm structure Storm track information	Cross section - velocity Velocity Azimuth Display (VAD) Velocity Azimuth Display wind profile Mesocyclone Tornadic Vortex Signature (TVS)	Cross section - spectrum width Layer Composite turbulence - average Layer Composite turbulence - maximum
		Radar coded message	

products. The PUP operator can manipulate the display in various ways, magnify it or view four displays simultaneously. The four displays can, for example, be the same product viewed at four elevations. Map or other overlays can be included. Signatures of severe weather can be identified and alarms sounded.

3.6.2 Output products, Capabilities and Performance

A description of the output products of the WSR-88D (Table 3.1) may be seen in Klazura and Imy (1993). Combinations of these are obtainable. "Level_II" data (i.e. high resolution spectral moments data of reflectivity, mean radial velocity and spectrum width) at the full spatial and temporal resolution of the radar are recorded on a 8-mm tape with a jukebox capable of automatic sequential loading of new tapes to avoid data interruptions. Copies of these tapes are made available to researchers (Crum, 1995).

In operations, the WSR-88D is designed to provide reflectivity to a range of 460 km and velocity to a range of 230 km for clear air as well as precipitation phenomena. The scanning is done over 360° at successive elevation angles resulting in a volume scan. The elevations are chosen depending on the phenomena of interest. The main modes of operation are (1) clear air strategy using azimuth scans at 5 elevations every 10 minutes (2) precipitation strategy (9 elevations in 6 minutes) and (3) severe weather strategy (14 elevation angles from 0°.5 to 19°.5 every 5 minutes). There will be a cone of silence close to the radar as the highest elevation is not sufficient to see the tallest echoes. Also, there are gaps between azimuthal sweeps at elevation angles higher than 6.2°. This could lead to errors in echo tops displayed. Other strategies can be chosen depending on requirements. Azimuth rotations can be made successively with 2 p.r.f.'s — low p.r.f. for unambiguous range, followed by a higher p.r.f. scan for Doppler data acquisition. The acquisition time is 5 to 10 minutes depending on the scan mode.

A preliminary review of the performance of six weather service offices in the USA after installation of a WSR-88D radar in severe local storm and flash flood warning shows a dramatic improvement (Polger et al. 1994) in Probability of Detection (POD) and False Alarm Ratio (FAR). (POD and FAR are defined in Chapter 6). Bieringer and Ray (1996) compared tornado detection and warning performance before and after replacement of the conventional non-Doppler radars by WSR-88D's at some sites in the USA. They found significant improvements in tornado warning lead times and detection after installation of the WSR-88D. The improvement may be partly attributable to other observational improvements and forecaster training but is mainly credited to the WSR-88D. Performance in a tropical cyclone has been documented by Stewart and Lyons (1996). A review of the overall performance of the WSR-88D network and planned improvements in hardware and products, taking advantage of the rapid advances in computer technology, is given by Crum et al. (1998)

3.6.3 The ISRO-IMD Doppler Weather Radar

The Indian Space Research Organisation (ISRO) has developed a Doppler Weather Radar (DWR) for the India Meteorological Department (IMD). Its specifications are given in Chapter 1. The radar installed on a tall building is shown in Fig. 3.25 and its block diagram in Fig. 3.26. This 10 cm wavelength radar is designed, with the observation of tropical cyclones as the prime objective. Though, for the present, only horizontal polarisation is used, the radar has a provision for future implementation of dual polarisation. The antenna feed is a "scalar feed" capable of transmitting both vertical and horizontal polarisations. It uses an "orthomode transducer" (OMT) which

Fig. 3.25. Doppler Weather Radar made by Indian Space Research Organisation (ISRO), installed on tall building; courtesy ISRO.

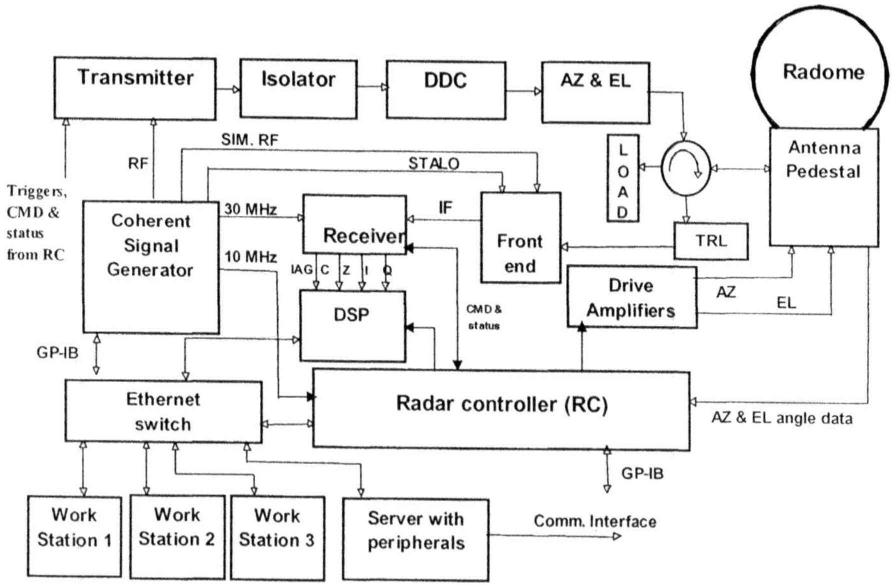

Fig. 3.26. Block diagram of ISRO radar; courtesy ISRO. DDC – Dual Directional Coupler, for sampling forward & reflected power; CMD – command; DSP – Digital Signal Processor; TRL – Transmit/Receive Limiter (following a T/R Switch; limits the residual power leakage into receiver); BITE – Built–In Test Equipment; GP–IB General Purpose Interface Bus for interconnecting standard test instruments; Comm I/F – Communications Interface for data exchange.

consists of a common waveguide of square cross section and two branch waveguides corresponding to the two modes. A few other interesting features are given below.

The received signal is amplified by a Low Noise Amplifier (LNA) and converted to IF. The RF portion of the receiver is located at the antenna, only the IF being taken down to the receiver room. The Signal-to-Noise ratio (SNR) variation with range and the corresponding lowest detectable dBZ values are shown in Fig. 3.27. The receiver has log and linear video channels. While the linear channel is used for reflectivity and rainfall computation, the log channel is used to select linear channel range through an Instantaneous Automatic Gain Control (IAGC) (as discussed in section 3.2) to achieve an adequate dynamic range.

The linear output is demodulated by a synchronous detector providing I and Q signals. The signal processor digitises the range-gated radar analogue data and generates the base products viz. reflectivity, mean radial velocity and spectrum width. A Doppler filter for clutter rejection is used before moments estimation. This is a 30 or 50 dB (selectable) notch filter of variable width, nominally 20 Hz corresponding to 1 m s^{-1} velocity.

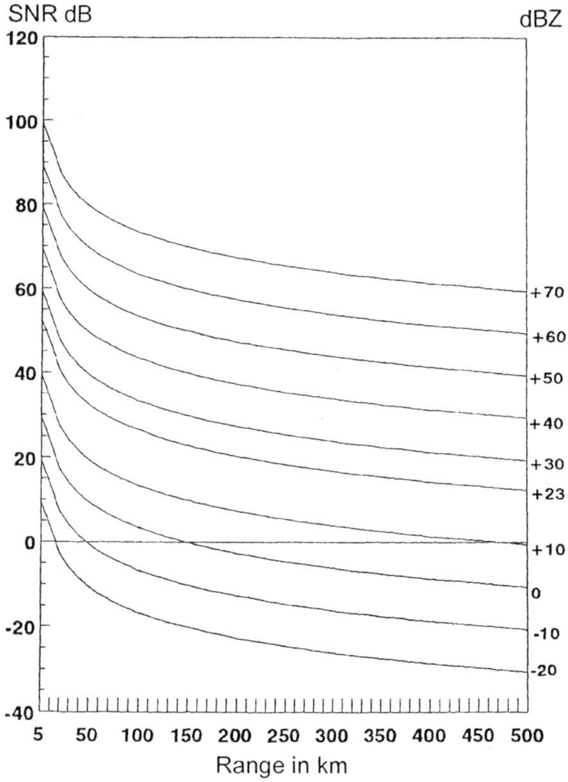

Fig. 3.27. SNR-Range curves for ISRO radar with corresponding dBZ values. (Courtesy ISRO)

After every volume scan, a calibration of linear and log output for each range bin at 1 dB power level steps is carried out. This gives a look-up table which is used for subsequent real time computation of Z.

3.7 Polarimetric Radar

3.7.1 Transmission and Reception of Polarised Radiation

A polarimetric radar may transmit and receive linear polarisations H and V or circular polarisations R and L. For transmiting H and V polarisations alternately, a mechanical switch may be used. For example in the "Chilbolton" radar in the UK, a rapidly rotating chopper vane has windows opening in synchronism with the transmission of H and V pulses. These direct the pulses arriving at a T-junction in the waveguide, alternately to one or other of two arms. The waveguide arms are twisted appropriately to polarise the radiation in the required plane. The energy from these is recombined and directed into a circular waveguide and through it to a "scalar feed" which distributes the energy over the antenna reflector. In this manner the beam formed by the reflector is identical in shape for both polarisations. The mechanical vane enables switching of the high power (500 kW) at the p.r.f. of 610. The received signal follows the same path in reverse. Alternatively a ferrite circulator is used to switch the polarisations. A circulator is a magnetic ceramic device incorporated in the waveguide which can be switched to allow radiation to pass in the desired direction only. Figs. 3.28 (a) and (b) show a schematic diagram of the US National Severe Storms Laboratory (NSSL) "CIM" dual polarisation Doppler radar[23] (after Carter et al., 1986). When H polarisation is to be transmitted, 16 microseconds prior to the transmitted pulse the circulator is switched to State A by the control logic. The transmitted pulse passes clockwise from port 1 to port 3 with a transmission loss of 0.75 dB only. It cannot pass to port 2 as there is a loss of 24 dB. The path of the signal is reversed by switching to State B, 16 microseconds after the transmitted pulse, so that the received H polarised signal only passes to the receiver. Suitable sequence of switching enables H and V polarisations to be transmitted or received alternately. It is possible to transmit a sequence of H pulses followed by V pulses but pulse to pulse switching gives the best accuracy for Z_{DR} computation. The 16 microseconds delay is to allow for the switching time of the circulator. The feed assembly in Fig. 3.28 is a three-port waveguide device. Two separate rectangular ports couple polarised radiation into a dual mode rectangular to circular waveguide transition. The third port radiates from a circular waveguide aperture a H or V wave depending on which port is excited. An important consideration in any polarimetric radar is the need for the beam patterns to be the same in both polarisations (Herzegh and Carbone 1984). Even if they are the same in the main lobe it is difficult to achieve this in the sidelobes. A mismatch which enhances Z_H relative to Z_V will obviously increase Z_{DR} (and vice-versa). In the case of LDR measurement, since the cross-polarised signal is some 30 dB below the main signal, a high degree of polarisation purity is needed.

Separate transmitters and receivers can be used for the two polarisations. The use of separate transmitters for H and V polarisations is expensive but dispenses with the high power waveguide switch and improves the isolation between the two signals. It also leads to greater flexibility in varying the p.r.f. and other parameters. A radar in which the tramsmitted polarisation can be changed from

---

[23] This radar has been upgraded. Originally it could record only the time series data; the polarimetric parameters had to be computed offline. Now it is a remotely operated facility capable of giving real time outputs of all Doppler and polarimetric parameters (Zahrai and Zrni 1993).

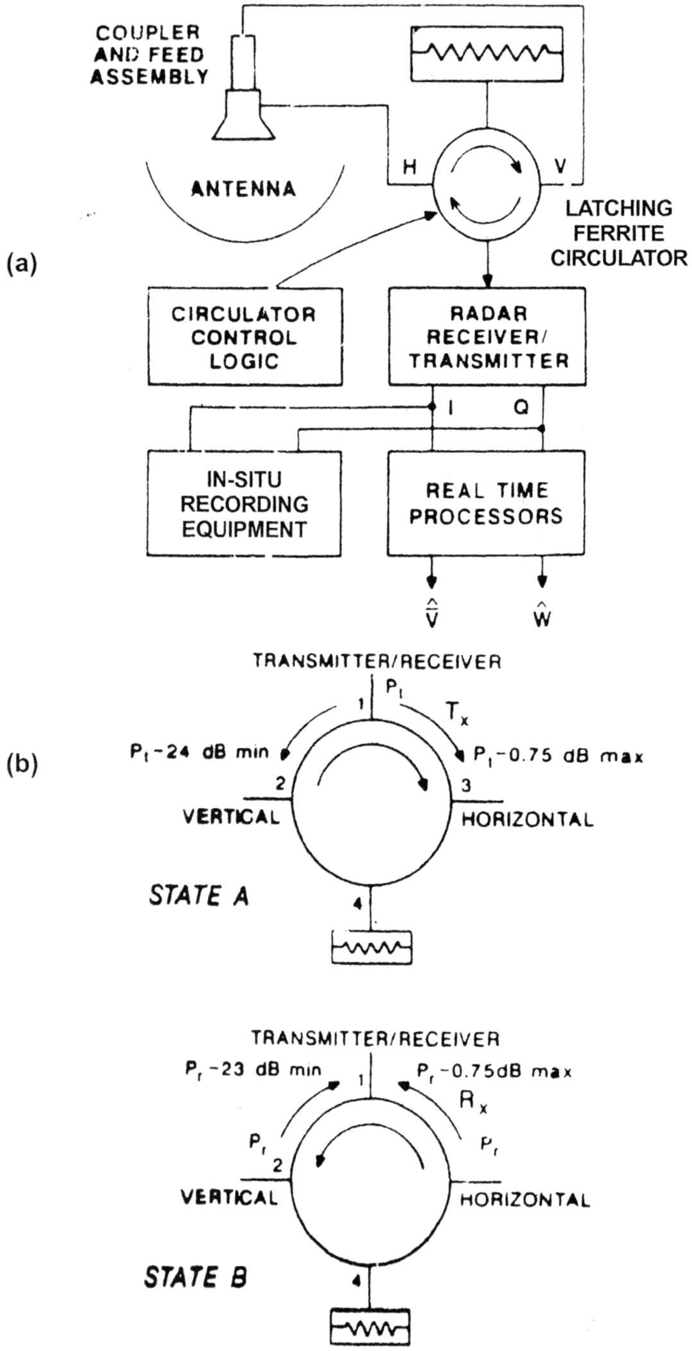

Fig. 3.28 a and b. Schematic diagram of the US NSSL CIM polarimetric radar (after Carter et al., 1986, courtesy AMS).

pulse to pulse is called a polarisation agile radar. There are also radars which transmit only one polarisation but can be changed after a period of time to another. It is also possible to have only single polarisation transmission with the capability to measure LDR with two channels in the receiver. Such a radar is said to have polarisation diversity. Another configuration is the use of two receivers facilitating simultaneous[24] transmission of H and V waves by a single transmitter and their reception by the separate receivers. Circular polarisation can be transmitted and received instead of linear polarisation, or in a hybrid approach H and V signals can be transmitted simultaneously, in the form of circular polarisation, and received in parallel H and V channels (Scott et al. 2001). Bringi and Chandrasekar (2001) may be referred to for a complete treatment of polarimetric radars.

The received signal goes to the appropriate path depending on the state of the switch and can be processed to measure Z_H or Z_V in the normal way by averaging required number of samples in a logarithmic or linear receiver. If alternate pulses are in different polarisations the effective number of samples for each polarisation is halved. Z_{DR} can be computed as the ratio of sample mean powers of H and V polarisations. To measure Z_H, Z_V, Z_{DR} or LDR it is not necessary to have a coherent Doppler radar. As Z_{DR} is small (less than 5 dB) in comparison with Z_H or Z_V it has to be measured to an accuracy of 0.1 dB.

One advantage of LDR measurement is that it does not need a dual-polarisation capability in the transmitter. LDR measurements are, however, very sensitive to ground clutter as the signal is very weak.

3.7.2 Polarimetric Doppler Radar

A dual-polarisation Doppler radar enables measurement of the differential phase in addition to the above polarisation parameters and the mean velocity and spectral width. If pulse pair processing is used H-H and V-V pulses are to be used and the effective p.r.f. for velocity calculation is halved i.e. the maximum unambiguous velocity is halved. Sachidananda and Zrnić (1989) have demonstrated a signal-processing scheme to overcome this. Zahrai and Zrnić (1993) have described a technique using autocovariances from the alternately polarised echoes to compute differential phase. The phase difference introduced in the radar system between the different polarised signals has to be allowed for. The differential phase is also more affected by sidelobe

---

[24] The advantages of simultaneous transmission of H and V are discussed by Doviak et al. (2000) and Scott et al. (2001).

contamination than the signal amplitude is. Sachidananda and Zrnić (1987) have discussed the errors involved. Sachidananda and Zrnić (1985) proposed transmission of + 45°and −45° polarisation alternately as well as simultaneous reception of H and V signals and range averaging. This avoids reduction of V_{max}, reduces data acquisition time and avoids bias error due to propagation. Reduction of the total acquisition time for a volume scan at various elevations is important in the case of rapidly changing meteorological phenomena (e.g. downbursts). However for accurate measurement of Z_{DR}, a longer acquisition time is desirable.

Mueller (1984) has presented methods for calculating the differential phase shift. Hubbert et al.(1993) have presented a method of separating the differential phase shift due to propagation from the differential phase on back-scatter (the latter is not negligible at C-band and shorter wavelengths).

3.7.3 The French RONSARD System

An interesting research radar system with polarimetric capability is the French RONSARD C-band dual Doppler facility developed in the 1970's (Nutten et al., 1979). This consists of two transportable sub-systems which can be assembled in a short time. The full azimuth motion is restricted to 720 degrees to eliminate the need for RF/IF rotary joints. The transmitter normally uses only linear horizontal polarisation which is not switchable. But the polarisation can be changed in about 30 minutes by dismantling and changing the RF circuitry. The receiver, however, has an orthomode coupler and a computer-controlled polarisation switch for selecting V or H polarisation. Hence the cross-polarised signal can be received. A linear receiver with large dynamic range is used as explained in section 3.2. A self-calibrating facility is available (see Chapter 5).

3.7.4 The CSU-CHILL Radar System

Another example is the CSU-CHILL radar at Colorado State University (CSU) USA which was mentioned in section 3.2. The CHIcago-ILLinois radar originally located at Chicago (Mueller and Silha, 1978; Mueller, 1989) has been moved to CSU and upgraded into a dual-transmitter, dual-receiver, dual polarisation radar. It is now a national standard radar facility which is moved once in a few years to other locations. A complete description may be seen in Brunkow et al. (2000).

A dual-channel waveguide rotating joint is used to maintain signal separation all the way to the orthomode transducer on the feedhorn. Power received at the antenna returns along the same waveguide system and passes through the transmit/receive circulators, through ionizing power limiters to low noise amplifiers (LNA's). A high frequency solid state transfer switch after the LNA's allows power to be routed to either of two matched receivers which handle the co- and cross-polarised signal returns. A pre-selector filter limits bandwidth to 5 MHz. The signal is mixed with the STALO to down-convert to 60 MHz and then split to drive log and linear receivers. The log receiver is digitised by the signal processor and used to set the gain of the linear channel on a gate by gate, pulse by pulse basis. The linear signal passes through a 700 kHz bandwidth matching filter and is delayed slightly to allow the step attenuator to be set for each gate based on the sample from the log receiver. This prevents the saturation of the following IF stages and allows for approximately 90 dB dynamic range in the linear receiver. The quadrature detector provides I and Q base-band video signals that are digitised and combined with the step attenuator setting to produce floating point samples that are used in the signal processor to generate the various moments. The signals are digitally processed in real time, and are presented on an interactively controlled colour display. All data fields are archived at full resolution on magnetic tape for post analysis. The radar's operational control system is highly flexible, readily allowing real time modification of the data acquisition parameters. Available data fields are, returned power (dBm or dBZ), normalized coherent power (NCP), radial velocity (pulse pair method), spectral width (second lag method), unprocessed complex voltages (I and Q), differential reflectivity Z_{DR}, differential propagation phase ϕ_{DP}, zero lag HV correlation $\rho_{HV}(0)$, linear depolarisation ratio LDR and complex correlations.

The radar has an antenna of diameter 8.5 metres, which can be dismantled and moved. Beam width is 1.1 degree, side lobe level –25 dB and cross-polarisation level –35 dB. By the use of suitable geometry of the struts the beam patterns for the two polarisations are matched well up to the first sidelobe. An inflatable radome (kept pressurised) withstanding winds up to 120 miles per hour is used.

Data are available on the Internet.

Figs. 3.29 and 3.30 are RHI pictures from this radar showing reflectivity and differential reflectivity respectively in a hailstorm. Fig. 3.31 depicts reflectivity, differential reflectivity, cross correlation signal between two polarisations and linear depolarisation ratio in stratiform precipitation. A bright band signature can also be seen (see Chapter 4 for a discussion on bright band).

3.7.5 Other Polarimetric Radars

A research radar having dual polarisation and Doppler capability is the

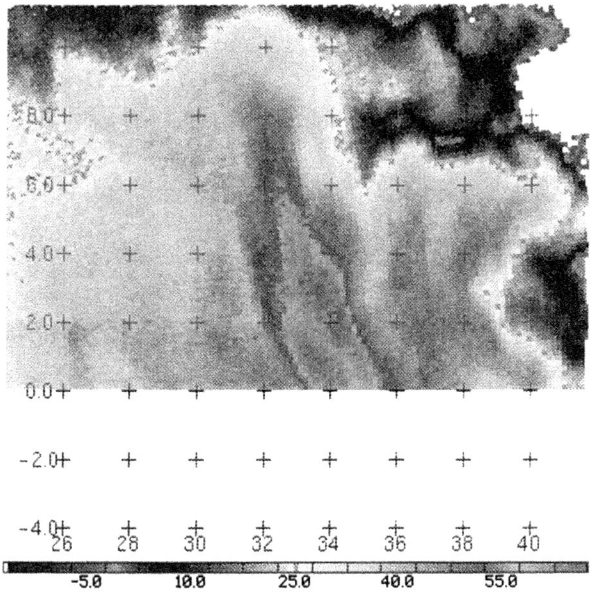

Fig. 3.29. RHI scan of reflectivity Z_H in a hailstorm observed by the CSU-CHILL radar at (courtesy Dr. V. Chandrasekar, Colorado State University). The + marks indicate distances and heights in km. The most intense part of the storm is at 34 km range.

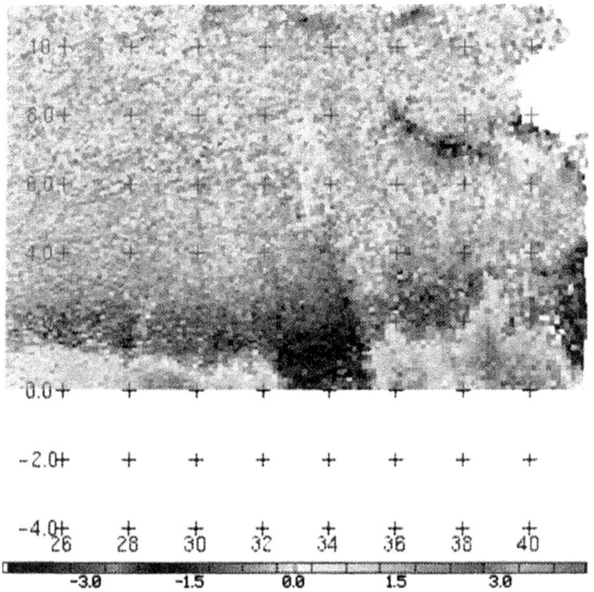

Fig. 3.30. Same storm as in Fig. 3.29, but RHI of Z_{DR} (courtesy Dr. V. Chandrasekar, Colorado State University). The + marks indicate distances and heights in km. Z_{DR} values below 1 km are positive, indicating ice to rain transition. At the most intense part of the storm (34 km range) Z_{DR} values are slightly negative or near zero indicating hail reaching the surface.

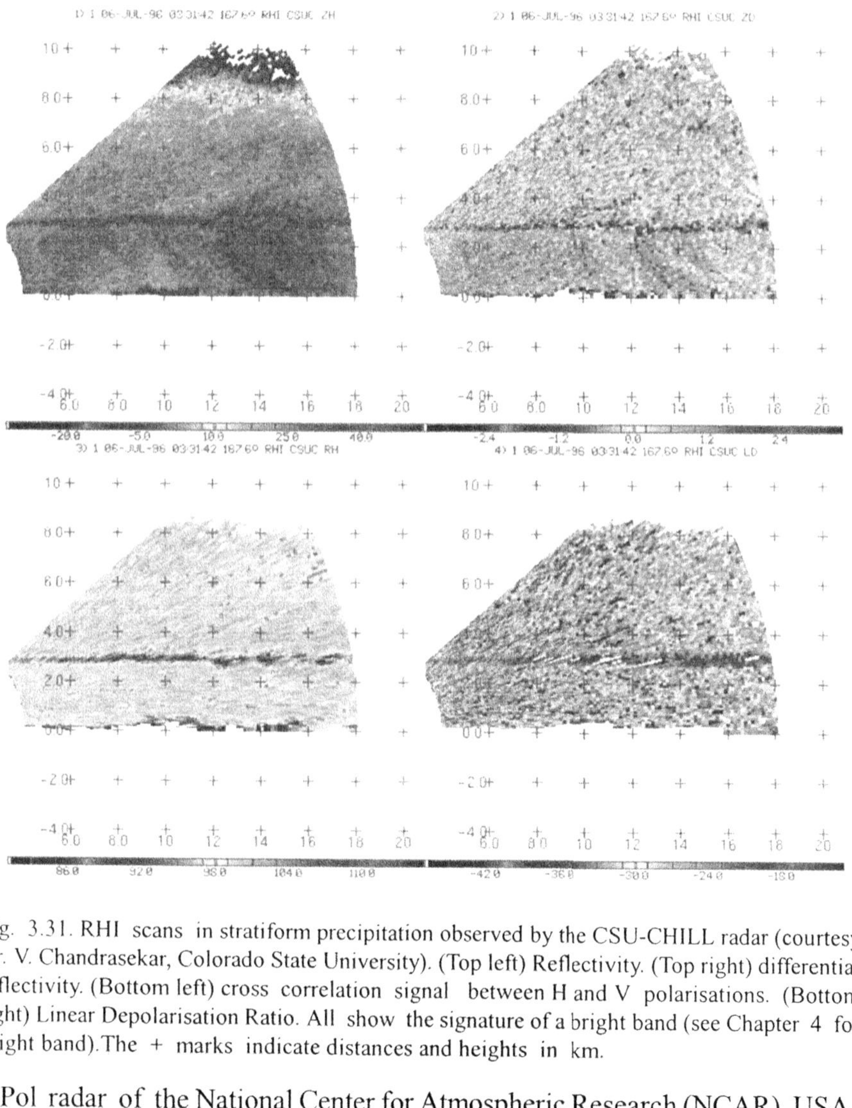

Fig. 3.31. RHI scans in stratiform precipitation observed by the CSU-CHILL radar (courtesy Dr. V. Chandrasekar, Colorado State University). (Top left) Reflectivity. (Top right) differential reflectivity. (Bottom left) cross correlation signal between H and V polarisations. (Bottom right) Linear Depolarisation Ratio. All show the signature of a bright band (see Chapter 4 for bright band). The + marks indicate distances and heights in km.

S-Pol radar of the National Center for Atmospheric Research (NCAR), USA. This is a transportable radar, packaged in shipping containers, intended to be moved to field experimental sites. It uses a high speed mechanical rotary switch which gives transmit isolation of 47 dB. It has a provision for adding up to three bistatic receivers i.e. receivers located at points away from the transmitter (Lutz et. al 1997, Randall et al. 1997). In Australia, a portable C-band polarimetric radar system, called C-POL, has been developed (Keenan et al. 1998) by upgrading an existing operational C-band radar. It transmits linear horizontal and vertical polarizations on a pulse-to-pulse basis providing simultaneous Doppler and standard polarimetric parameters. Among the most versatile polarimetric Doppler radars may be mentioned the DFVLR radar in Germany (Schroth et al., 1988) and the U.S. Air Force Geophysics Directorate Radar (Metcalf et al., 1993). These can transmit and receive linear and circular polarisations. Bringi and Hendry (1990) who reviewed the technology of polarimetric radars have also given a list of polarimetric radars around the world (at that time). All the equipment so far described, function in the research mode. In view of the advantages such as particle phase recognition and more accurate rainfall estimation, expected from operational use of polarimetric parameters, retrofitting of polarisation capability on existing operational radars is an option to be seriously considered (see Zrnić and Ryzhkov 1999). As mentioned above the ISRO-IMD radar has a provision for this in the hardware. The upgradation of the WSR-88D network is under way. The options available and the configuration that is being adopted are discussed by Doviak et al. (2000).

3.7.6 Airborne Polarimetric Radar

Antipovsky et al., (1993) have described an Aircraft Meteorological Polarisation Radar (AMPR) of the Russian Central Aerological Observatory, mounted on a research aircraft for study of cloud and precipitation microphysics. The antenna is mounted in the forward portion of the belly as in civil aircraft and can be switched to transmit H and V polarisation. The received signal is switched by PIN diodes.

Chapter 4

PHENOMENA OBSERVED BY WEATHER RADAR

In this chapter we discuss the various types of echoes observed by weather radar. These are divided broadly into (1) echoes from hydrometeors, (2) those not directly arising from hydrometeors or precipitation and (3) echoes of non-meteorological origin. The first category is discussed in the first section and the rest in section 4.2. Discussions of the melting or "bright" band and of lightning detection by radar are also included.

4.1 Echoes from hydrometeors

4.1.1 Classification of Echoes

Echoes observed on a weather radar display may be broadly classified into three types:
(a) Echoes from precipitation or hydrometeors
(b) Echoes due to meteorological phenomena not directly arising from hydrometeors in precipitation. These are loosely called "non-precipitation" echoes.
(c) Echoes of non-meteorological origin.
Category (a) is dealt with in this section and in further detail in Chapter 6. Categories (b) and (c) will be discussed in the next section. Although category (b) echoes are termed "non-precipitation" echoes some of them originate from conditions which may give rise to precipitation. They are also of interest to the meteorologist by themselves as they yield information on various atmospheric processes. Category (c) echoes are also of interest to the meteorologist because their sources do interact with meteorological conditions and may be useful as

tracers of air motion. It is also necessary for the meteorologist to distinguish between and identify the different types of echoes. A good guide to interpretation of satellite and radar imagery in extra-tropical latitudes in relation to analysed air flow fields may be seen in Bader et al., (1995).

4.1.2 Convective and Stratiform Echoes from Precipitation

Radars in the 3 to 10 GHz bands are normally used to detect only particles of precipitation. Cloud particles in the micrometre size range are also detectable but because of the low backscatter at these frequencies they may be difficult to separate from clear-air or other types of echoes (Miller et al., 1998). Cloud particles are detectable by higher frequency radars and these will be discussed in later Chapters. Precipitation echoes can be broadly divided into two types – convective and stratiform. These two categories are not mutually exclusive in the sense that the two may not only exist together but convective echoes may exhibit stratiform characteristics at some stage.

Convective clouds (large cumulus and cumulonimbus) have a large vertical extent, and occur as discrete cells or groups of cells with sharp boundaries. They are characterised by large up- or downdrafts with vertical velocities up to several m s^{-1} and large sizes of precipitation particles (1 mm to 5 mm for raindrops and larger sizes for hail). Instantaneous rate of rainfall is usually high (often several tens of mm h^{-1}) and there are large spatial gradients of rainfall rate. The duration of precipitation from any one convective cell is also small (typically several minutes). Because of the above characteristics, radar echoes from convective precipitation (showers or thundershowers) are seen in the form of intense and discrete cells with sharp edges. They may be in the form of isolated cells or occur in groups or oriented along a line (Fig. 4.1). On the RHI they show up as tall columns. The maximum height may go beyond 20 km as some cumulonimbi overshoot the tropopause. The echo may form aloft initially and grow downwards as well as upwards. It may exhibit rapid changes in height (growth or decay). Because of large drop sizes, reflectivity is high, [30 to 50 dBZ and still higher if hail is present]. Large reflectivity gradients occur in the horizontal plane and high reflectivities extend well beyond freezing level. On the RHI or vertical cross sections the reflectivity contours appear vertically oriented (Figs. 4.2 and 4.3). This is so because up- and downdrafts produce vertical mixing of the precipitation particles. In cumulonimbi the particles may remain in liquid (supercooled water) phase even at heights much above the zero degree isotherm. If there is a vertical shear of the horizontal wind the upper part of the echo may be seen sheared off from the lower part, on the RHI. This will be visible on the same

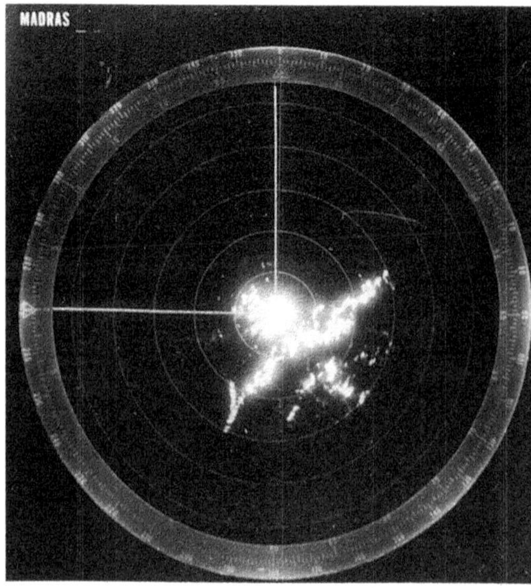

Fig. 4.1. PPI picture of squall line, 400 km long, off Chennai (Madras), consisting of convective cells. Maximum range displayed is 500 km. (courtesy IMD).

RHI if the shear is in the plane of the RHI. If it is in another plane the sheared part will be seen as an echo aloft on the PPI and on the RHI in another azimuth.

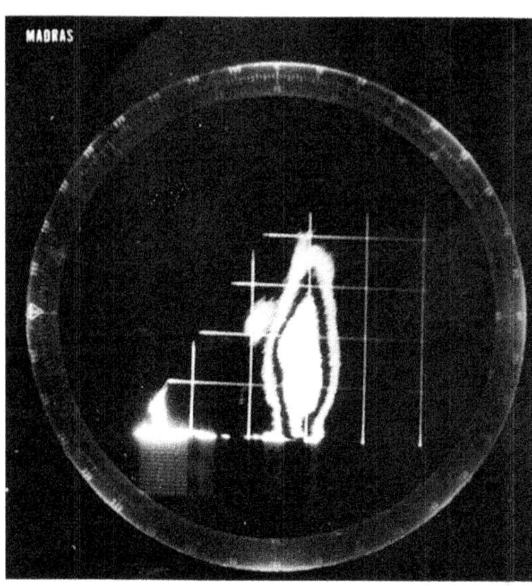

Fig. 4.2. RHI gray scale presentation of echo from a convective cell. The different shades of gray including black in the middle denote reflectivities in steps of 10 dBZ Range and height markers are at intervals of 10 km and 5 km respectively. (courtesy IMD).

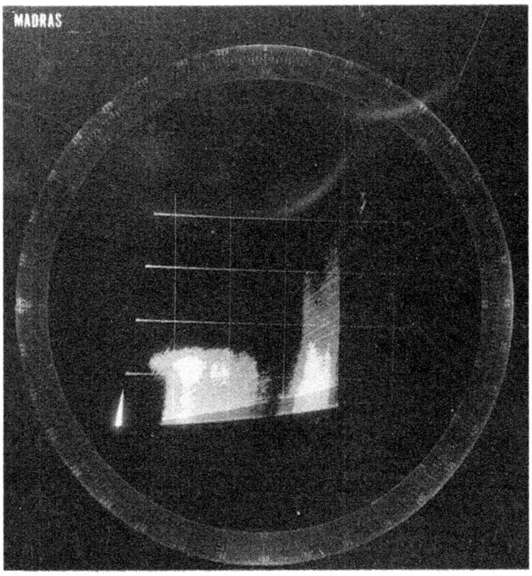

Fig. 4.3. RHI gray scale presentation of echoes from convective and stratiform precipitation. The echo at right is from convective precipitation and exhibits vertically oriented reflectivity contours. Echo at left is stratiform and has horizontally oriented contours with a maximum at freezing level (bright band). Height markers are at intervals of 5 km. (from Raghavan 1977; courtesy IMD).

Reflectivity and horizontal and vertical velocity distribution in a thunderstorm determined by dual-Doppler radar are shown in Fig. 3.13.

Stratiform precipitation, on the other hand, extends usually over a large area without discrete cellular structure and is associated with relatively stable conditions. Vertical air motions may be only a few cm s^{-1}. The rainfall rate is small (a few mm h^{-1}) but the rain is spatially more uniform and persists for a longer time. The vertical depth of the precipitation is also small. Echoes from stratiform precipitation (Figs. 4.3 and 4.4) are usually extensive and have reflectivity factor less than 30 dBZ, as particle sizes are smaller. They are also almost uniform in intensity in the horizontal plane. Echo top heights are usually within 7 or 8 kilometres. Stratiform echo may be seen sometimes at close ranges at higher antenna elevations without any precipitation reaching the ground.

Convective and stratiform precipitation will be further discussed in Chapter 6.

4.1.3 The Bright Band

As the vertical currents in stratiform clouds are small, the particles tend

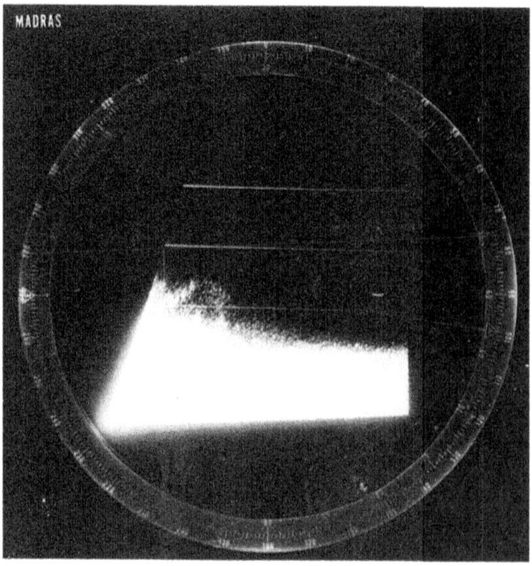

Fig. 4.4. RHI presentation of echo from stratiform precipitation. Range 100 km; height marker intervals 5 km. (courtesy IMD).

to stratify as liquid drops below freezing level and solid particles above. Echo from the solid particles is usually weaker than that from liquid precipitation because of the lower dielectric constant of ice (see section 2.2). The snowflakes vary in orientation and therefore the scattering cross section varies resulting in large fluctuations of the signal. In the case of liquid drops, the drops are spheres or spheroids and do not exhibit such large changes in effective cross section. The particles of solid precipitation which fall through the freezing level melt gradually and get coated with water but retain their large surface area until they melt completely. Also as seen in section 2.2, non-spherical particles can have higher reflectivity than spheres of the same volume. Hence, there is usually a layer, a few hundred metres thick, just below the zero degree isotherm level[1] in which the particles have a large dielectric constant and large cross section[2]. Moreover the fall velocities of the solid particles are less than those of the liquid particles below. Hence, there is an aggregation of particles near this level. There may also be particle growth by coalescence leading to increase in Z_e. For these reasons the reflectivity factor from this layer is higher than that above or below.

---

[1] The zero degree isotherm is usually at an altitude of about 4.5 to 5 km in tropical areas such as India, but much lower in temperate latitudes.

[2] There is usually a preponderance of large drops (formed by melting of ice particles) compared to the Marshall-Palmer (M-P) size distribution. As the drops fall, the bigger ones break up and exhibit at the ground a distribution closer to M-P.

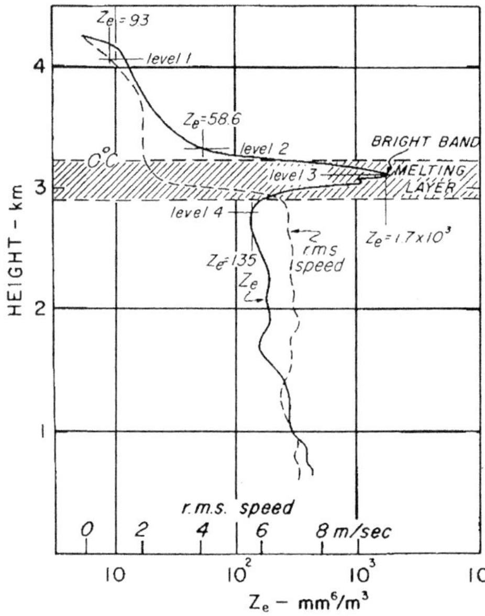

Fig. 4.5. Vertical profile of Z_e and r.m.s. particle fall speed in light stratiform precipitation with a bright band, observed by Doppler radar. (from Lhermitte and Atlas 1963; courtesy AMS).

This layer is called the "bright band" or "melting band". Fig. 4.5 from Lhermitte and Atlas (1963) shows the vertical profile of Z_e and r.m.s. particle fall speed in light stratiform precipitation as observed by Doppler radar[3]. In Fig. 4.6 the weaker echoes of Fig. 4.4 have been eliminated by inserting attenuation in the receiver and the stronger echo of the bright band is alone displayed. The increase in Z_e in the bright band compared to the rain below is typically 10 to 15 dB at 10 cm wavelength. At shorter wavelengths the difference may be smaller because of attenuation of signal. At very short wavelengths, e.g. 1 cm, the bright band may not be so prominent because Mie scattering occurs. If the antenna is directed vertically up when the cloud is overhead the maximum amplitude echo can be seen on an A-scope at the range nearly equal to the height of the freezing level. If the antenna scans in azimuth at high elevation, the bright band will be seen as a ring or arc on the PPI.

Early theoretical treatments of the bright band may be seen in Ryde (1946) and Mason (1955). Fabry and Zawadzki (1995) identified the following effects as contributing to the bright band.

(1) the change in the dielectric constant through melting

(2) the change in fall velocity throughout melting

(3) precipitation growth

---

[3] Similar profiles can be seen in Fig. 3.16a.

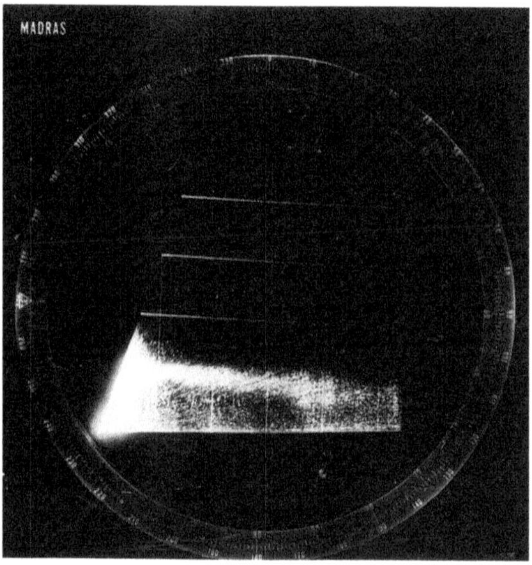

Fig. 4.6. As in Fig. 4.4 but with an attenuation of 15 dB. Only the bright band is seen (courtesy IMD).

(4) the change in particle size distribution (aggregation and breakup)
(5) the combined effect on echo power, of the shape and orientation of melting hydrometeors
(6) the effect of distribution of water within the melting snowflake (termed as density effect).

4.1.4 Structure of the Bright Band.

The bright band is a phenomenon of considerable interest in relation to the physics and dynamics of clouds. In Fig. 4.5, Z_e gradually increases from higher levels up to the peak at the bright band, falls sharply below it and there is no appreciable change further down. The fall speed of particles is typically 1 to 2 m s^{-1} above the band and 5 to 6 m s^{-1} below. It peaks near the base of the bright band but does not decrease thereafter. The melting layer can be identified by the sharp increase in the Doppler velocity of falling precipitation particles at the top of the bright band reaching a maximum at the bottom. From the vertical profile of Z_e, the top can be taken to be where the curvature of the profile changes sign and the base is located where the curvature changes abruptly. The Doppler spectral width also increases sharply at the bright band because of the spread of particle velocities.

Austin and Bemis (1950), Battan (1973), Byers and Coons (1947), Cunningham (1947), Ekpenyong and Srivastava (1970), Gunn and East (1954), Leary and Houze (1979), Ramanamurty et al.(1965), Stewart et al. (1984), Fabry and Zawadzki (1995) and Bellon et al. (1997) among others have discussed the bright band in detail[4]. Observations of the bright band in various parts of India with non-Doppler radar may be seen discussed in Biswas et al.(1962), De (1962), Gupta et al.(1955, 1961), Mani and Venkiteshwaran (1961), Mani et al. (1962), Srivastava et al. (1966), Raghavan (1977) and Raghavan et al., (1983). The thickness of the bright band and its separation from the freezing level should depend on the lapse rate and the particle size distribution. The band is narrow in stable conditions and relatively wide and fuzzy when the lapse rate is steep (Austin and Bemis 1950). Greater thickness (average 0.5 to 0.7 km) and distance of separation (average 1.4 km) have been reported in some early work in India[5] [Biswas et al.(1962), Gupta et al.(1961), Mani and Venkiteshwaran (1961)] than those found in middle latitudes in the USA [less than 300 m; Austin and Bemis (1950), Byers and Coons (1947)]. This has been explained as due to a wider particle size spectrum and differences in lapse rate. A thickness of 600-850 m has been reported from China in association with the "Mei-yu" front (Tang, 1993). Due to the melting, cooling should occur at the melting level, leading to a thin isothermal layer. Using the VAD technique, Atlas et al., (1969) observed, in widespread frontal precipitation in England, significant perturbations of the wind at the melting layer; these are probably produced by pressure perturbations due to horizontal variations in cooling by melting snow.

Some recent studies of the scattering in the melting band may be seen in Klaassen (1988), Willis and Heymsfield (1989) and Hardaker et al., (1991).

The polarisation parameters in and near the bright band give more information on the bright band structure. Humphries and Barge (1979) using a 10 cm radar in Canada found that CDR was about –15 dB above the bright band; below the bright band it was about 5 dB lower. The CDR peak was 300 m below the Z_e peak. Frost et al. (1991) found with the Chilbolton 10 cm radar in the UK, a similar result for LDR. Illingworth and Caylor (1989), with the same radar found the Z_e and LDR maxima in the same height region but the Z_{DR} up to 500 m lower. Steiner et al. (1991) using the "Bright Band Project" data from the same radar found a Z_{DR} peak near the bottom of the bright band. Moninger et. al., (1984) using collocated dual linear

[4] Observation of bright band using airborne radar at X- and Ka-bands is discussed in Chapter 7 and that with clear air radars in Chapter 10.

[5] Narayana Rao et al (1999) using the Indian MST radar found a thickness of 900 metres.

polarisation radar and circular polarisation radar found that the peaks in the bright band from highest to lowest altitude level are in the order Z_e, CDR and Z_{DR}. The cross-correlation between the main and cross-polarised signals had a minimum at the same level as the CDR peak. Hagen et al., (1993) found in Germany that the LDR maximum and the minimum of the cross-correlation occurred at the same level while the Z_{DR} maximum was at a lower height. Balakrishnan et al., (1991) found a sharp drop in the correlation coefficient between H and V signals over a 500 m layer and 200 m further down was the peak of Z_{DR}. They also found K_{DP} to be zero above the freezing level but attaining significant values in the rain below. Zrnić et al., (1993) studied the correlation coefficient between H and V signals, as well as Z_e and Doppler velocity in the melting layer with a vertically pointing radar. Meischner et al. (1991) made, simultaneously, observations with a ground-based polarimetric radar and in situ observations with aircraft. They concluded that above the melting layer horizontally oriented ice needles with positive Z_{DR} dominate, whereas below that, the particles are more spherical; and further down horizontally oriented water-coated snowflakes occur with high Z_{DR}. Fig. 3.31 shows the bright band signature of Z_e, Z_{DR}, cross-correlation and LDR as seen by the CSU-CHILL radar. The fall in the correlation coefficient shows that a mixture of different types and sizes of particles is present in the bright band.

On the basis of a recent long-term experimental study in Canada using a vertically pointing X-band radar and a UHF wind profiler[6], Fabry and Zawadzki (1995), concluded that the bright band is much more intense than predicted by the conventional explanations. They are of the view that shape effects (i.e. non-sphericity) and density effects are important causes of the bright band.

Humphries and Barge (1979) also compared reflectivities Z_{10} and Z_5 at 10 and 5 cm wavelengths. The ratio expressed in dB is a function of particle phase and shape. The ratio was negative just above the freezing level but became positive in the bright band. The attenuation due to melting snow is quite high as explained in Chapter 2. The attenuation will also be higher for horizontal polarisation.

The attenuation in the melting layer at the 3 cm wavelength was studied by Bellon et al., (1997) in Canada by comparing long-term average reflectivity profiles from collocated X-band and UHF vertically pointing radars. They found that in stratiform rain the attenuation in the melting layer usually exceeds that from the rain below it.

Durden et al., (1997) studied the melting layer using an Airborne Rain-

---

[6] See Chapter 10 for wind profilers.

MApping Radar (ARMAR) operating at 13.8 GHz (see also Chapter 7). The radar had Doppler and polarimetric capabilities. They concluded that,

1. greater bright band reflectivity is correlated with lower altitude of the reflectivity maximum, larger bright band thickness and larger cooling rates.
2. ice reflectivity above the bright band is less than rain reflectivity below the bright band; the difference becoming more negative with increasing rain reflectivity.
3. maximum velocity and maximum LDR increase with maximum reflectivity; minimum HH-VV correlation coefficient decreases with increasing maximum reflectivity.
4. maximum reflectivity occurs at higher altitudes than the maximum velocity, maximum LDR and minimum HH-VV correlation coefficient.
5. the offset in altitudes between the maximum reflectivity and the extreme values of the other parameters increases with increasing maximum reflectivity.

4.1.5 Bright band — a Signature of Stratiform Precipitation.

The bright band normally occurs in stratiform precipitation because of the stratification. In convective clouds, because of the vertical mixing the bright band is not expected except in the dissipating stage or in association with the stratiform anvil associated with a convective cloud (see Chapter 6). Mull et al. (1963) reported an observation of a bright band in convective clouds at Delhi said to be not at the dissipating stage. Battan et al. (1964) found bright band occurring in what was probably a decaying stage of convective precipitation but accompanied by lightning activity in the cloud. In the light of later studies (e.g. Leary and Houze 1979, Raghavan et al. 1983) it seems likely that the above cases too may have been from a stratiform anvil. Hence the bright band is taken as a signature of stratiform precipitation and is used as one of the means of distinguishing between stratiform and convective precipitation.

4.1.6 Bright Band Identification

In a digital radar the bright band can be identified by algorithms based on the reflectivity and Doppler features mentioned above. Identification is ofpractical importance in rainfall estimation as a radar beam passing through a bright band experiences high attenuation while the Z_e value in the bright band is high. Radar estimates of precipitation need to be corrected if the beam cuts the band (see Chapter 7). Also, as different Z-R relationships are often used for convective and stratiform precipitation, identification is important. The identification of a bright band with a pronounced ρ_{HV} minimum and Z_{DR} maximum helps to discriminate between snow and rain (Ryzhkov and Zrnić, 1998).

As the vertical resolution of the radar becomes poorer with increasing range the reflectivity variation in the vertical is smoothed out and it is difficult to identify the bright band. Also as the radar sees the band at different elevations as the range changes there will be an oscillation of detected Z_e as range increases. If the volume scan has gaps in elevation coverage it can miss or underestimate the bright band at some ranges (Sanchez-Diezma et al. 1999). The bright band will be discussed further in Chapters 6 and 7.

4.1.7 The Flare Echo.

A special type of echo can occur by scattering of the radar beam incident on a precipitation volume, to the ground, scattering from the ground back to the precipitation volume and thence to the radar. The geometry is shown in Fig. 4.7 (after Wilson and Reum, 1988). Zrnić (1987) derived an equation for Z_e for this three-body scattering and has shown that it varies as r^{-3} (r being the distance from the parent precipitation cell to the ground). Because of

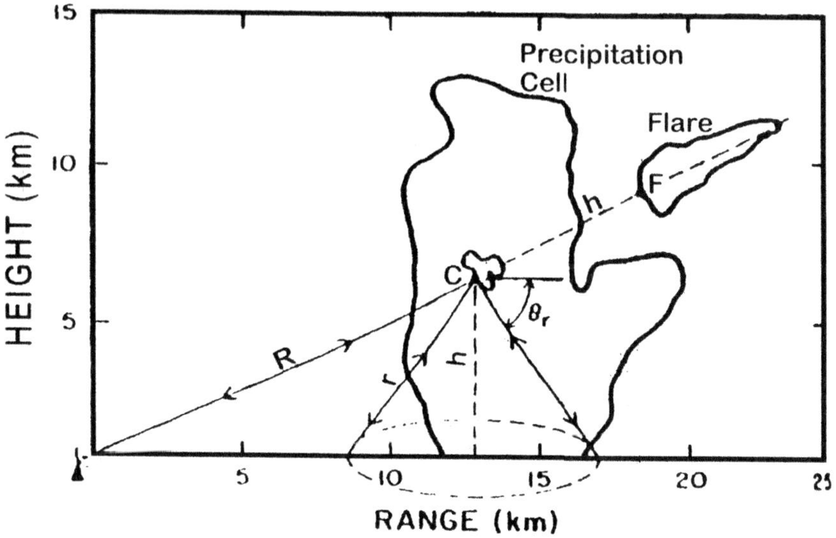

Fig. 4.7. Geometry of flare echo. C = region of reflectivity > 60 dBZ in precipitation cell; R = range to C; h = height of C above ground; r = slant distance from C to periphery of ground region causing scatter. (from Wilson and Reum 1988; courtesy AMS).

multiple scattering, Z_e is low (usually less than 20 dBZ) and arises only if the Z_e of the parent echo is very high (the area shown near point C in the figure). Hail and large rain drops are the prime candidates. The echo, called the flare echo, extends downrange of the parent echo as in the figure and its range is determined by the sum of the range of the parent echo and its height. Flare echo is more common with 3 and 5 cm radars (than with 10 cm radars) and with horizontal polarisation. Its Doppler velocity depends on the radial velocities and particle speeds in the parent echo. Since fall speed is downwards the Doppler velocity is usually towards the radar. The Doppler velocity V is given by,

$$V = V_r + (w + w_t)\sin\theta_r \qquad (4.1)$$

where V_r is the true radial velocity, w the vertical air velocity, w_t the terminal velocity of the hydrometeors and θ_r the angle shown in the figure.

Caylor and Illingworth (1989) using the Chilbolton dual-polarisation radar in the UK, observed a Z_{DR} of +9 dB in association with a flare echo. According to them this extremely high value occurs because the triple scattering is confined to the Z_H channel. It has been suggested that the flare echo can be used for nowcasting of hail and strong downdrafts including microbursts (Zrnić 1987, Roberts and Wilson 1989, Nielsen-Gammon and Read 1995) but it can also cause false indications of converging winds.

4.1.8 Echoes from Non-precipitating Clouds

Non-precipitating clouds have low reflectivity because of the small sizes of hydrometeors and therefore are difficult to detect at the wavelengths of 3, 5 or 10 cm, used in operational radar networks. At these wavelengths the Rayleigh scattering may often be weaker than the Bragg return from clear air due to the refractive index discontinuities at the cloud boundaries[7]. As the cloud grows to a precipitating state the return from hydrometeors will increase. Clouds can be better studied at higher frequencies (e.g. 35 or 95 GHz)[8] at which cloud size hydrometeors give large return. The radars emit low power (which is adequate because of the high reflectivity and short range), are light and small and are portable. They can have quite narrow beams and high resolution with a small antenna. These frequencies are subject to high attenuation but that is not a problem for observation at short range. One disadvantage for Doppler measurements is the small value of the $r_{max} \times v_{max}$ product. Cloud studies will be discussed in Chapter 11.

---

[7] Insect swarms may also give comparable return.

[8] The frequencies have to be chosen to avoid absorption bands of atmospheric gases and water vapour.

4.2 Non-precipitation Echoes

4.2.1 Anomalous Propagation (AP)

Any of the meteorological conditions discussed in Paragraph 2.1.9 may produce downward refraction of the radar beam and lead to anomalous propagation (AP) echoes. Methods of suppression of AP echoes were discussed in Chapter 3. AP in the wake of a thunderstorm [see Paragraph 2.1.9.(4)] is of particular interest because it may be superposed on the thunderstorm echoes. In a study at Delhi, Mathur and Kulshrestha (1961) noted that this occurred only if the surface wind was calm after the thunderstorm. In a study at Chennai (formerly Madras), Raghavan (1964, 1965) found that the AP occurs only on occasions when the downdraft air from the thunderstorm has higher water vapour pressure than the environmental air. He suggested that the observed intensities of these echoes can be explained if these result from partial reflection at a sharp refractive index discontinuity in the boundary layer caused by the downdraft air by a mechanism proposed by Jones (1958) rather than as a result of refraction.

4.2.2 "Angels"- Birds, Insects or Refractive Index Discontinuities?

Echoes in the form of very thin lines or dots or rings have been observed on microwave radars since the early days. Layer echoes aloft, not associated with precipitation were also seen. Some scientists attributed the dot and ring echoes to bird or insect swarms and others to partial reflection or scatter caused by atmospheric refractive index discontinuities or inhomogeneities (see sections 2.1. and 2.2). Because of the uncertainty of their origin they went by the name "angels". That dot or ring angels were in many cases due to birds or insects, was verified by observing them optically and also from their speeds of motion. Radar cross sections of birds have been computed and experiments by dropping assemblies of insects from aircraft and tracking them were carried out (Eastwood 1967). More recently, Sauvageot and Despaux (1996) and Hobbs and Wolf (1996) have studied insect flights using airborne or ground-based polarimetric or Doppler radars. Fig. 4.8 shows a roughly doughnut-shaped Doppler velocity pattern with a maximum radial velocity of about 12 m s^{-1} (away from the radar) seen at around sunset in February about 70 km to the northwest of a 10 cm radar at Chennai (India). The antenna elevation was 0.2 degrees. The echo is believed to be from birds moving in large numbers and getting into the

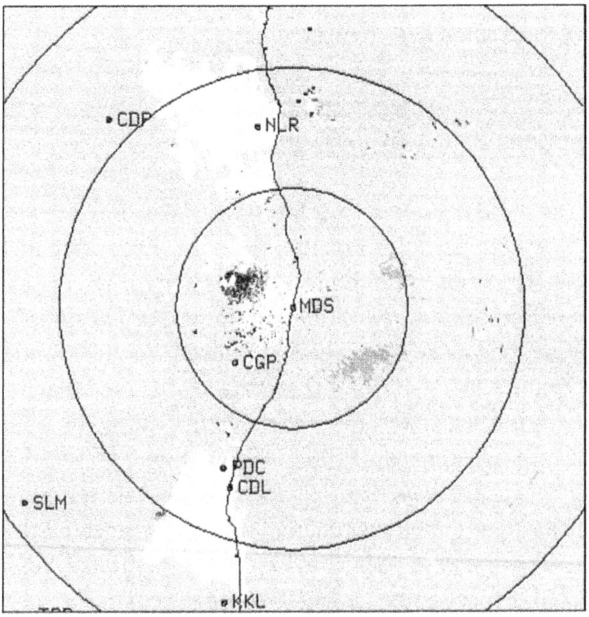

Fig. 4.8. A doughnut-shaped Doppler velocity pattern seen with a 10 cm radar at Chennai (formerly Madras, marked as MDS in the picture). range rings 100 km apart. See text for details. (courtesy IMD).

beam at that range. The view is probably assisted by anomalous propagation, which is common in the evenings in that month.

Swarms of small insects may be carried by the wind and therefore may act as tracers of the wind[9]. Locusts swarms, for instance, are carried over long distances and considerable heights by the wind. Their movement can therefore be forecast by the meteorologist and radar can be used to track the swarms. Further, birds may ride thermals and insects also congregate at meteorological discontinuities and move with them or rise with rising air currents, leading to the speculation whether the radar echo is due to these creatures or due to direct meteorological causes. Since insects can be treated as Rayleigh scatterers at the longer radar wavelengths, their backscatter should be proportional to the inverse fourth power of the wavelength, while echoes due to Bragg scatter

[9] It may not always be true that insects are tracers of air motion. Achtemeier (1991) who studied an "insect cloud" with dual-polarisation Doppler radar found that the insects were confined to a lower layer despite the presence of updrafts which should have taken them up. He thought it likely that they fold their wings and fall down to escape going to colder levels. This hypothesis is supported by observations of the variation of Z_{DR} of the insects which depends on their orientation with respect to the radar beam.

should be proportional to wavelength to the power of $-1/3$. Observation at two wavelengths however indicated that the wavelength exponent was between 1 and 2 in the case of many point angels. To explain the observed echo intensities on the basis of refractivity gradients, large gradients were necessary and it was felt that they were unlikely to occur. However, the line type and layer type angels were found to be mostly from refractive index discontinuities although insects or even birds could be contributing some reflectivity (Russell and Wilson 1997). Early case studies of "angel" observations in India may be seen in Datar and Sikdar (1964), Kulshrestha (1962b) and Rai (1959, 1960, 1961, 1961b) among others. Early reviews of angels may be seen in Atlas (1959, 1964) and Battan (1973). Atlas (1959) classified these echoes into four types:

(1) echoes from thermal columns below cumulus clouds;

(2) mantle-shaped echoes associated with the boundaries of cumulus clouds (caused by Bragg scatter as mentioned earlier);

(3) layer echoes associated with sharp vertical gradients or minima in moisture aloft; and

(4) anomalous low-level echoes believed to be produced by the boundary surfaces between differentially moistened surface air over adjacent cold and warm water.

Lhermitte and Dooley (1966) studied the horizontal and vertical motion of clear air echoes using a VAD scan. Lhermitte (1966) tracked these echoes with Doppler radar and interpreted their movement as air motion. It is now standard practice to determine winds with "clear air" radars in the VHF and UHF bands using the refractive index inhomogeneities as tracers (see Chapter 10).

In a study in Florida and Colorado, USA, Wilson et al. (1994) compared reflectivities of clear air echoes observed with microwave Doppler radars (3, 5 and 10 cm wavelengths) and found that both particulate and Bragg scattering occur in the well-mixed boundary layer but the former dominates. In this layer the echo signal is highly horizontally polarised with Z_{DR} of 5 to 10 dB. These echoes are largely attributed to insects. Above the boundary layer, Bragg scattering dominates and is observed on the S-band. When insect or bird migration was absent, they could derive valid horizontal winds in the boundary layer using the Bragg scatter echoes as tracers.

4.2.3 Types of Discontinuity Echoes

The following are some of the echoes in this category, of operational interest.

(a) Sea breeze front

The sea breeze front is a notable example of discontinuity[10] between the moist and cool maritime air at the surface and dry, warm land air above. The discontinuity produces an echo in the form of a thin line on the PPI with little or no vertical extension on the RHI (see Atlas 1960, Anderson and Lindgren 1992). As the sea breeze advances the line moves and can often be traced some tens of kilometres over land until the mixing of the air destroys the discontinuity. Doppler radar can yield details of the sea breeze circulation.

(b) Thin lines

Discontinuities caused by low level turbulence, gust fronts[11] ahead of thunderstorms or elevated layers of discontinuity can all give rise to echoes in the form of thin lines. Fig. 4.9 shows the line echo of a gust front 10 miles to the northeast of Delhi with a 3 cm wavelength radar. Bhattacharyya et al. (1965) reported similar echoes at Calcutta. Kundu and De (1971) at Agartala (India) have presented the case of a thin line 7 n. miles ahead of a squall line and moving at 56 km h^{-1} while the squall line itself moved at 40 km h^{-1}. An elevated layer echo observed in clear weather on a 10 cm wavelength radar at Chennai is shown in Fig. 4.10. Such echoes have a fine texture, lack vertical extension and may have a Z_e not more than 10 dBZ (at 10 cm wavelength). Thin lines often form boundaries between air masses and may indicate areas where thunderstorms will be initiated (Wilson and Schreiber 1986). Russell and Wilson (1997) estimated the contributions of "aerial plankton" (i.e. weakly flying insects) and birds to thin lines by observing the insects and birds optically. Goodrich et al. (1991), Rogers et al. (1991) and Hagelberg and Helland (1995) have developed algorithms for detection of thin lines from reflectivity and Doppler velocity fields.

(c) Mantle echoes

Edges of cumulus clouds also return weak echoes due to refractive index discontinuity. These are known as mantle echoes (Atlas 1959, Knight and Miller 1993). The reflectivities are usually well below 0 dBZ and may be seen only on very sensitive radars or higher frequency radars.

(d) Sea clutter

When microwave radiation is incident on the sea surface an echo is received which is dependent on the wind and sea surface conditions. At oblique

---

[10] The coming together of the sea air and land air also often leads to convergence and formation of convective clouds which, of course, can be detected by radar. The most prominent example is the Florida peninsula in the USA which has two sea breezes one from the west (Gulf of Mexico) and the other from the east (Atlantic Ocean). (See e.g. Byers and Rodebush 1948, Frank et al., 1967, Bader et al. 1995).These converge and produce convective precipitation. In India, the sea breeze at Chennai has been shown to have an influence on the preferred times of occurrence of convection (see Chapter 6).

[11] The gust front is discussed in Chapter 6.

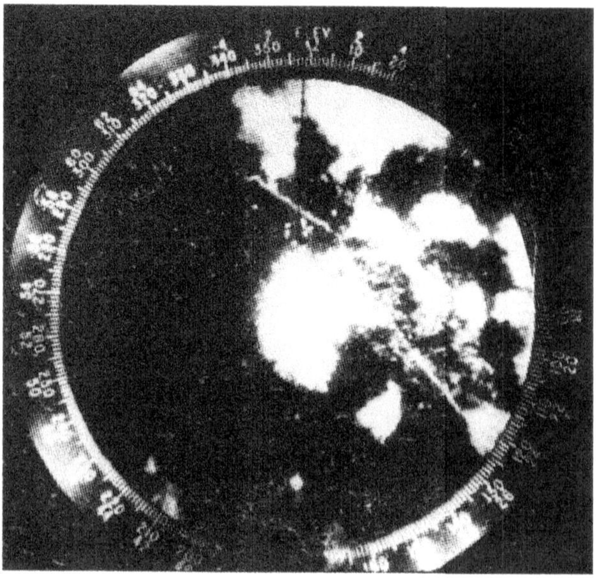

Fig. 4.9. Thin line echo from a thunderstorm gust front 10 miles northeast of a 3 cm radar at Delhi (range setting 25 miles, elevation 1 deg). (after Kulshrestha 1961; courtesy IMD).

incidence the return is due to Bragg scatter from a rough sea surface. Waves on the sea surface which have a wave number equal to twice the horizontal wave number of the incident radio wave, contribute to the signal return. Sea clutter is often observed by coastal radars in anticyclonic situations accompanied by low level windflow from the sea surface towards the radar. The echoes can occur in various interesting patterns (Fig. 4.11), can be quite strong and extend to well over 100 km from the radar. In anticyclonic conditions sea clutter may be accompanied by anomalous propagation and therefore may be seen at antenna elevations of several degrees. It has to be recognised and eliminated so that it is not printed out as rainfall. Zero Doppler filter may not remove it as the echoes have some velocities, say 2 or 3 m s^{-1}. Sea clutter is stronger for vertically polarised radiation than for horizontal. From airborne radar, sea clutter may appear in patterns that tend to move with the aircraft unlike precipitation echoes which will move much more slowly (Houze and Churchill, 1987).

A radar device (usually microwave) known as a scatterometer is used on aircraft or satellite to receive echoes from the sea surface (see Robinson, 1985). From the amplitude of the signal the surface roughness can be estimated. As the roughness is mainly due to wind, algorithms have been devised to derive the wind. There is however an ambiguity of 180 degrees in the wind direction which can be resolved from other meteorological information. The first satellite-borne scatterometer was used on the US SEASAT satellite in 1978.

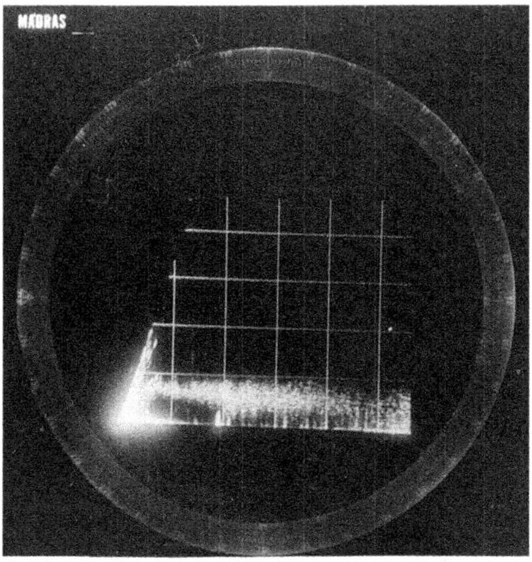

Fig. 4.10. Echo from an elevated layer in clear air on a summer afternoon. Chennai 10 cm radar, RHI range/height markers at 10/2.5 km intervals, (after Rajagopalan and Raghavan 1980; courtesy IMD).

4.2.4 Echo from Lightning

Lightning discharges in thunderstorms generate electromagnetic radiation known as atmospherics or "sferics" at a wide range of frequencies. Very Low Frequency (less than 30 kHz) radiation can be detected at long

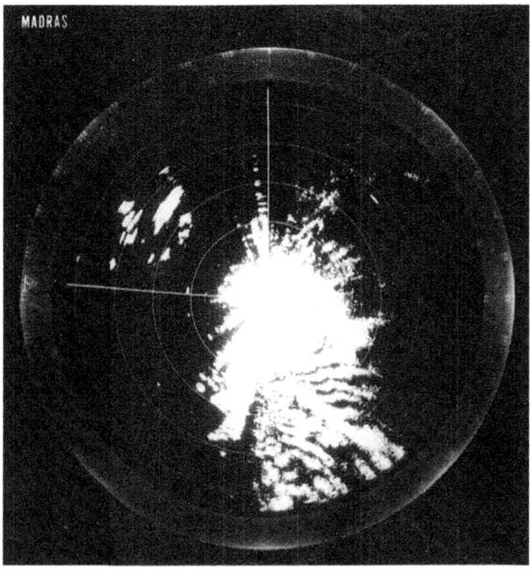

Fig.4.11. Sea clutter to the southeast of the radar, in the presence of southeasterly winds over the sea surface; 10 cm radar PPI picture. Range displayed 100 km, (courtesy IMD).

distances by direction-finding equipment. Using three or more such devices at different places, the location of the thunderstorm can be determined by triangulation. Measurement of the difference in time of arrival at three (or more) locations will also enable location of the source by drawing loci of points of equal time of arrival (hyperbolae). The intra-cloud and cloud-to-ground (C-G) flash rates can be counted. Networks of such lightning detection equipment are available in many countries.

Sferics also occur at microwave frequencies and can be detected on the radar receiver when the antenna is pointed in the direction of the source of radiation. Thus a transient signal can be observed on the radarscope. The direction of this signal is observed correctly by the radar; but the range at which it appears has no significance because the time of origin of the radiation is in no way correlated with the radar-transmitted pulse. Since operational radars are usually in a scanning mode, this transient echo is rarely noticed.

Lightning can also give a true radar echo by other mechanisms. The first such observation was reported by Ligda (1950) who used a 3 cm radar with antenna pointed towards the thunderstorm area. The signal suddenly increased close to the precipitation echo. On the R-scope (see section1.2) the lightning echo appeared with steep edges. The lightning discharge path is a hot air plasma. The electron density in the plasma depends on the temperature which may be a few thousand degrees K. If the electron density is high and the radar frequency is low, the plasma reflects the incident energy as a conductor. The plasma is then said to be "overdense". When the electron density is low (underdense) and radar frequency is high the plasma behaves as a dielectric scatterer . A "volume reflectivity" can be defined for this echo by analogy with that for precipitation particles. At wavelengths less than about 5 cm the echo return is very small in comparison with the precipitation echo from the same storm and will be swamped by the latter. At longer wavelengths the antenna can be pointed in the appropriate direction and the echo observed. The duration of the echo is also very small (typically a few hundred milliseconds).

Another mechanism by which an echo can be obtained is the refractive index discontinuity arising from the shock wave created by the discharge.

The lightning discharge gives a coherent peak in the Doppler velocity spectrum of reflectivities. It is also believed to change the orientation of precipitation particles and to initiate a gush of rain. UHF circular polarisation Doppler radar may be a very good tool for study of lightning and its relation to precipitation formation. For detailed treatment of this topic see Bhattacharya (1992, 1992b, 1994), Bhattacharya et al. (1994), Dawson (1972), Doviak and Zrnić (1984, 1993), Ligda (1950, 1956), Mazur et al. (1984,1986), Rust and Doviak (1982),Williams and Geotis (1986) and Williams et al. (1989).

The occurrence of lightning as observed by independent lightning detection systems has also been correlated with radar echoes from hydrometeors in thunderstorms (e.g. Sarkar and Bhattacharya, 1980). Not all precipitating clouds produce lightning and not all clouds producing lightning produce precipitation. Even so, as lightning is often associated with the initiation of precipitation from convective clouds a correlation can be expected between radar-observed features and lightning activity. The presence of an ice-phase, large updraft velocities and large numbers of precipitation particles is associated with lightning. Hence radar-observed echo area, echo-top heights, reflectivities at various levels, vertical velocities and precipitation rates have been correlated with lightning activity represented by intra-cloud and C-G flash rates. [e.g. Battan (1965), Black et al. (1986), Holle et al. (1984), Kinzer (1974), López et al. (1986, 1991), Marshall and Radhakant (1978), Reap (1986), Richard (1991), Shackford (1960), Shafer et al. (2000) and Watson et al. (1995)].

Based on radar observations of thunderstorms and visual observation of associated C-G flashes, Battan (1965) found in Arizona, USA, a good positive correlation between frequency of C-G lightning and rainfall amount. Shackford (1960) found a positive but not very good correlation between surface-observed lightning stroke rate and reflectivity/echo top height. Kinzer(1974) found in Oklahoma (USA) thunderstorms that areas of high Z_e are apt to be areas of high rates of C-G lightning. The threshold Z_e at which lightning started was about 27.5 dBZ but this was not sharp. There was 1.6×10^{10} grams of rainfall per C-G flash. Marshall and Radhakant (1978) related the sferics count per minute (L) with area (A km^2) of radar echo above 30 dBZ at 6 km altitude at range r by an equation

$$L = 2.7A^{1.64}r^{-1.62} \qquad (4.2)$$

They also suggested that L may be related to Ah rather than A, where h would be height above some elevated base height.

Holle et al., (1984) correlated radar reflectivity from hydrometeors in Florida thunderstorms up to a range of 150 km with frequency of C-G lightning flashes from a lightning direction finder. There was a sharp increase in frequency when the dBZ level reached 37 and there was a peak at 46 dBZ. Above 52 dBZ the frequency of lightning fell rapidly. According to them, when high Z_e occurs at low levels the clouds are in a dissipating stage, the updrafts are not present; hence lightning is absent. In the western USA, Reap (1986) correlated lightning with the Manually Digitised Radar (MDR; see section 3.1) intensity level. He found that most of the lightning strikes occurred in association with reflectivity less than 41 dBZ.

Crozier and Scott (1981) have described a radar incorporating, on the display, the sferics data from a lightning detection system. The azimuth of the sferic and the flashing rate can be lined up with the most promising echo. Watson et al. (1995) found that the number of lightning flashes detected correlates well with echo top heights, but the relationship with Vertically Integrated Liquid water content (VIL) is ambiguous. Precipitation rates were well-correlated with C-G flash rates with a time lag of a few minutes (Richard, 1991). Polarimetric radar may reveal particle characteristics in the mixed phase region of convection where electrical charge may be generated (e.g. Goodman et al. 1989, Krehbiel et al. 1991, Bringi and Chandrasekar 2001). For example, high Z_{DR} above the 0°C level may be indicative of supercooled water while LDR may be from frozen particles. Carey and Rutledge (2000) related the Z_H, Z_{DR}, and K_{DP} measurements from C-band radar to lightning and surface electric field measurements and inferred that lofting of supercooled drops to temperatures beyond −10°C and their freezing led to the thunderstorm electrification.

Eilts et al., (1991) found in Florida that the first radar echo aloft greater than 10 dBZ close to a boundary layer convergence zone precedes the first lightning strike by 8 to 21 minutes, a finding of obvious forecasting value. At the Kennedy Space Center in the USA, Gremillion and Orville (1999) found that for summer storms, persistence for two consecutive volume scans of 40 dBZ at the −10°C temperature height was a good predictor of lightning initiation. More studies of these relationships using radar and lightning detection systems (at each radar site) may be useful in practice for short period forecasting, forest fire detection, electric power distribution, aviation forecasting and spacecraft launches[12].

4.2.5 Non-meteorological Echoes

Sources of common non-meteorological echoes are ground and sea clutter, ships, birds and insects all of which have been discussed earlier. Aircraft will also give echoes but because of their high speeds in relation to the scan rate of most weather radars they will be transient and will not significantly affect observation of meteorological phenomena. Interference from other radars can produce echoes usually in the form of dots in spiral patterns on the PPI. These can be avoided by staggering frequencies and by regulation of operational timings.

---

[12] The launching rocket may itself trigger lightning (see Chapter 11).

4.2.6 Echoes due to Radiation outside the Beam Width

As discussed in section 1.3. the power radiated outside the nominal beam width and in sidelobes will produce detectable echoes from nearby ground targets or strong convective clouds. These will be displayed at the azimuth/ elevation of the main lobe. Particularly on the RHI these may appear as additional images above the principal echo or an upward extension of the principal echo (Fig. 1.5). Their Doppler velocities will also appear in the Doppler spectrum of the relevant range gate.

4.2.7 Other Spurious Echoes

Image echoes due to reflection from other obstructions has been mentioned earlier (Fig. 1.4). These will have a spurious velocity depending on the geometry of the reflection (Raghavan et al., 1976). The image may be that of a precipitation echo or a non-precipitation echo such as the ship in Fig. 1.4. In the former case it may get superposed on other echoes and create confusion in the reflectivity and velocity fields. On a ship-borne radar this problem can be acute due to reflection from other structures on the ship. Buildings giving strong reflection can also give echoes from sidelobes. A sidelobe echo may exhibit a radial velocity in the Doppler spectrum as the feedhorn of a rotating antenna has a radial component of motion with respect to a target detected by a sidelobe. This will not be eliminated by a clutter filter which usually rejects echoes with near-zero velocities (see section 3.4). Another type of spurious echo was reported by Kulshrestha and Sharma (1961) at Delhi. They observed dot angels apparently moving at about 200 miles per hour. They explained them as due to the radar beam getting an echo from a different blob in each successive scan, the blobs being due to turbulence at low levels on a hot afternoon.

Chapter 5

OPERATIONAL WEATHER RADAR

In this chapter, we discuss in section 5.1, with particular reference to operational ground-based radars, the observational procedures, scan strategies and common sources of observational error. This is followed by a discussion of the tracking of echoes and various techniques for derivation of echo motion and wind. We then deal with the remote transmission and archival of radar data. A brief treatment of radar networking and integration with other sources of data is also given. Section 5.2. deals with calibration and maintenance. Methods of monitoring radar parameters and evaluation of system gain are discussed.

5.1 Radar Operational Techniques

5.1.1 Observational Procedures

Radar observational procedures depend on whether the radar is used for research or routine operation. In the former case the personnel using the radar are likely to have specific research objectives and have a plan of collecting data in desired modes and schedules of operation, often in conjunction with other instrumentation. They have, nevertheless, the flexibility to vary radar parameters, time schedules or scanning strategies. In routine operation, however, the radar is part of an observational network to feed raw or processed or interpreted data at intervals to various centres dealing with e.g. synoptic analysis, aviation forecasting, hydrometeorology and flood forecasting, tropical cyclone disaster warning or local severe storm warning, the requirement of each being somewhat different from the others. The data may be directly accessed by several customers, or archived data may be accessed by yet others for climatological purposes or research. Modern radars are usually operated continuously, in some cases totally automatically. The radar may be part of a network of radars, the products of which have to be compatible with each other. For these reasons they have to have an observational programme with predetermined radar parameters, schedules, scanning strategies and product generation and archival procedures.

Manually operated radars were operated at regular intervals of time (interval depending on phenomena observed), and time allotted for photography of the scope and tracing of the echo patterns on superposed overlays. This enabled mapping the reflectivity distribution and evaluating the echo motion. The Manually Digitised Radar (MDR) (see Chapter 3) sketches were transmitted by facsimile with interpretative remarks by the operator who would be familiar with clutter and other local peculiarities. In digital radars the base products are displayed and various algorithms for derived products are applied off-line but in near real time. Even so, it is important for an operator or a meteorologist to use his judgement in interpretation of the observed patterns.

5.1.2 Digital Processors

The development of digital radar processors has progressed along with general developments in computer technology. One of the earliest digital systems to be developed in the USA was the D/RADEX (Digital Radar Experiment) in 1971 (Bigler et al.1970, McGrew 1972). This led to an augmented operational network with updated equipment called RADAP-II (Radar Data Processor) (Greene et al. 1983, NOAA 1988, Saffle 1976). An improved product display system called ICRAD (Interactive Color RAdar Display) was later added (NOAA, 1988). The principles of data processing for the RADAP-II have been dealt with by Ahnert et al. (1983) and Hudlow et al. (1983). The RADAP-II led to the development of the WSR-88D system which has replaced it now in the USA.

In Britain the "Dee Weather Radar Project" was conducted in the early seventies and led to the development of an operational digital weather radar system (Ball et al. 1979, Collier and James 1986). This led to the development of FRONTIERS (Forecasting Rain Optimised using New Techniques of Interactively Enhanced Radar and Satellite data) (Browning 1979, 1986) which enabled compositing radar and satellite data. FRONTIERS has recently been replaced by "NIMROD" (Golding 1998) which produces very short range forecasts of a variety of meteorological elements including precipitation, every 15 minutes.

A wide variety of workstations and software have been developed for processing the signals to provide displays of reflectivity, velocity and other outputs. One of the important functions is to convert data from the radar polar coordinate system to a Cartesian format. Other software is necessary for producing vertical sections, computing particular parameters, inserting maps and so on. Software is used for controlling and programming the radar operation either at site or through remote control using a communication link. Other algorithms have been devised to classify echoes or to detect specific features such as cells, thin lines, vortex signatures etc. Some of these can be run on a suitable PC in near real time and some can be used in off-line processing. Constant altitude planes for various levels and reconstructed RHI's can be displayed in real time by a workstation using volume scan data. Time lapse movie loops can also be generated (e.g. Lutz et al. 1989). The operational use of such algorithms is still evolving; the algorithms themselves will not be discussed in this book. Principles of some of these are discussed in later chapters while dealing with phenomena such as mesocyclones, tornadoes and tropical cyclones. Some relevant references regarding the workstations and algorithms are: Albers (1991), Anderson and Green (1991), Austin et al. (1986), Corbet et al. (1994), Forsyth et al. (1989), Goodrich et al. (1991), Hagelberg and Helland (1995), Heymsfield et al., (1983), James et al. (2000), Joe (1991), Mohr and Miller (1983), Mohr et al., (1981, 1986), Oye and Carbone (1981), Rison et al. (1991), Rogers et al. (1991), Walther (1991), Wiener and Yuter (1991) and Witt and Johnson, (1993).

5.1.3 Operational Sequence

In an operational radar there is a need to look for echoes at the longest ranges as well as operate at various elevations in order to get complete coverage. If the radar is located in hilly terrain (e.g. a hill top) scanning at a negative elevation angle is also desirable (Brown et al. 2002). Scanning at more than one p.r.f. is also necessary. The scan rate has to be low to obtain more samples per beam width. At the same time a complete set of scans has to be completed quickly, especially in fast changing weather systems. Hence the number of elevations has to be limited. The conventional method of stopping the antenna at various azimuths and scanning in RHI mode is time consuming and can be done only for a limited number of azimuths. X-Z sections constructed from the volume scan data are a good substitute for RHI scans. To meet conflicting requirements some compromises are necessary. Two or three alternative scan strategies are usually provided. For example in the US WSR-88D radar (Klazura and Imy 1993) there is a "severe weather strategy" in which a scan at 0.5° elevation is made and the antenna is stepped up by one beam width at a time up to 6.2° and by larger angle intervals thereafter up to 19.5°, the volume scan being completed in five minutes at a scan rate of 3 rpm.. This means that at close ranges there will be a "cone of silence" i.e. the radar does not see anything above 19.5°. Also there are gaps where the steps are more than one beam width. There are also a "precipitation" strategy of nine elevations in six minutes and two "clear air" strategies of five elevations in ten minutes.

The severe weather strategy mentioned above will cause errors in computing rainfall, VIL and other products because of the gaps and the cone of silence, particularly in convective storms (Brown et al. 2000). To overcome this, Brown et al. (2000b) have proposed a scheme, which provides more low elevation angles to improve sampling of distant storms and scanning up to higher elevations to improve sampling of storms near the radar. To increase temporal resolution they provide for increasing the antenna rotation rate to the maximum that is consistent with the desired accuracy of the data. Also, a volume scan can be terminated and recycled when the antenna tilts above all echo returns or reaches a specified elevation angle.

5.1.4 Observational Errors

A number of errors arise in radar observation. Some are listed below.

(1) Distortion of the size or shape of an echo due to resolution problems.
Due to finite beam width or range resolution, the echo may appear larger than it is; in the azimuthal (or elevation) direction the echo is elongated. At long range, the range resolution remains the same but azimuthal resolution is poorer; hence the echo appears distorted in shape. For the same reason a multiple-trip echo may be distorted and difficult to identify with any known feature, such as a coastline. Heights of echoes may be exaggerated. Sidelobe echoes may form a continuation of the main echo or overlap another echo. If two echoes are close together they may appear as one because of inadequate azimuthal resolution. If they are moving towards the radar the single echo may appear to split into two at some range where the azimuthal resolution becomes sufficient to resolve them. Similarly, there may be an apparent merger of receding echoes.

(2) Resolution affected by averaging.
When a number of samples are spatially averaged and integrated (e.g. displayed over Cartesian squares of 5 km × 5 km), the resolution is rendered poorer. A single value of P_r or Z_e for the entire square replaces the individual

values in the pulse volumes. Peaks of Z_e will be averaged over the entire square. If the square is not filled by precipitation the radar underestimates the echo intensity. Also, the shape of small echoes may be distorted. This may be critical in cases where the shape matters e.g. hook echoes in thunderstorms or spiral bands and eyewall in tropical cyclones.

(3) Distortion due to rain attenuation.

Distant echoes may be obscured by attenuation due to intervening precipitation especially on short wavelength (5 cm or less) radars. Even if there is only one area of precipitation being observed, with no intervening echoes, the echo from the farther part of the precipitation may be attenuated by the nearer segment of precipitation; this means that the area of the echo observed is less than what it actually is. Also, the estimated centroid[1] of the echo will be nearer the radar than the true position. If the centroid is used for tracking the echo motion, the estimated velocity of the echo will be in error.

(4) Beam height effects.

As an echo moves away from the radar, the latter sees the higher parts of the precipitating cloud. Hence the echo will appear to weaken and eventually disappear. The reverse happens when the echo approaches the radar. The range normalisation does not correct for this effect. The intensity of the echo will also appear to diminish with increasing range as the radar shows the power received from the portion intercepted by the beam as though it is coming from the entire precipitation. Due to the same effect the radar may underestimate the height of a precipitating cloud.

(5) Propagation effects.

With changes in low level temperature and humidity profiles, the refraction conditions will change. For example, with the onset of rainfall along the line of sight an echo which was seen clearly, may be lost or appear weaker due to subnormal propagation. Heights of echoes may be exaggerated or underestimated, depending on the propagation conditions. The correction for normal refraction (ratio k of effective earth's radius to actual earth radius assumed to be 4/3; see Chapter 2) may cause appreciable error in anomalous propagation situations where the k value will bc different.

[1] The centroid may be defined as the reflectivity-weighted centre of a contiguous echo; while in a digital radar it can be determined objectively, in a manual radar it is subjectively estimated by the operator.

(6) Shear.

As seen in Chapter 4, the top of the echo may be seen on an RHI as sheared away from the lower part if there is a wind shear in the plane depicted by the RHI. But if the shear is in a perpendicular plane it cannot be detected on the RHI at the same azimuth but will show up as an echo aloft in the PPI and in the RHI at another azimuth. This should be kept in mind while evaluating echo top heights. In digital systems it is possible to construct and display an X-Z section through any two points on the ground. Hence, it is easier to identify the shear.

5.1.5 Tracking of Echoes and Derivation of Wind

In the case of an echo with clearly identifiable boundaries (such as a convective cell), the successive positions of its centroid can be plotted to determine its speed and direction of motion. The estimation of its centroid may be subjective. The centroid may change due to growth or decay of the echo. Moreover it would depend on the intensity threshold of the radar. If the threshold is changed, the apparent area as well as the position may change. The effective minimum Z_e detected by the radar decreases with increasing range. The radar also sees the echo at different altitude levels at different ranges. Hence, instead of using the centroid, identifiable points on the periphery of the echo (preferably on the side nearer to the radar) can be used for determining the movement. It may however happen that the point(s) chosen may not be identifiable after some time owing to growth or decay of the echo.

Movement of stratiform precipitation is difficult to track as the echo is large and of uniform nature without easily identifiable points.

It was recognised quite early (Ligda 1953) that the wind vectors in the upper air could be derived by observing the motion of radar echoes much as visual surface observation of the movement of clouds is traditionally used to infer winds[2]. Winds estimated from the movement of Small Precipitation Areas are termed SPAwinds. Convective echoes with distinct boundaries or definable centroids lend themselves well to this technique. The wind has to be assigned to some altitude depending on the height of the observed echo. As range from radar increases, the beam height increases. Hence, instead of tracking the base of the echo at one elevation it is useful to track a feature on a CAPPI.

The SPAwind gives the echo motion which is not necessarily the same as

[2] The same principle governs the derivation of cloud motion vector winds from satellite observations.

air motion. The motion of a convective echo or a group of such echoes depends on two factors. One is the wind in its environment. The effective wind steering it will depend on the altitudes over which the cell (or cells) giving the echo is situated. For convective cells, the wind at 700 hPa level is often taken as the steering wind. In the case of tall clouds, the wind at a higher level or the wind averaged over a large vertical layer from bottom to top of the echo may be the effective steering wind. The other factor is "propagation", of the cells. When a convective cell develops to a mature stage, a downdraft from it spreads around it and triggers the formation of new cells. These are not necessarily downwind of the parent cell. Thus a cluster of cells may persist for several hours although each individual cell may have a short life. This process of regeneration is termed propagation. The effective motion of the cluster as seen by radar will be the resultant of the environmental wind and the apparent motion created by propagation. Hence, the cells may seem to move to the left or right of the environmental wind. Also the movement of individual cells may be different from that of the group (cluster or line) as a whole.

Digital methods enable automated tracking of not only motion but growth and decay. If the position of an echo at time t is expressed as (Zittel 1976),

$$X(t) = A_x + B_x t \tag{5.1}$$

$$Y(t) = A_y + B_y t \tag{5.2}$$

$$\text{The speed is } \left(B_x^2 + B_y^2\right)^{1/2} \tag{5.3}$$

$$\text{The direction is } \tan^{-1}\left(\frac{B_y}{B_x}\right) + 180^\circ \tag{5.4}$$

For an echo having a finite size and irregular shape, the reflectivity pattern above a specified Z_e threshold may be considered. Its perimeter can be defined by points $[x(s), y(s)]$ on the arc measured from any initial point. x and y can be treated as periodic functions repeating when the arc length equals the perimeter and can, therefore, be expressed as a Fourier series (Blackmer and Duda 1972). By taking account of only the first few terms of the Fourier series a fairly close approximation of the shape of the echo can be obtained. The Fourier coefficients can be obtained by integration over the arc length. The zeroth harmonic gives the centroid of the echo and the first harmonic yields an approximating ellipse which is a sufficiently good representation of the echo for many purposes. If the Z_e pattern at successively lower thresholds is dealt with similarly, a contour map of Z_e can be obtained. The patterns at successive times can then be compared to establish motion, growth and decay. A cross-correlation technique can be used for this purpose. Cross-correlation represents a measure of the relationship between the members in two sets of quantities (Panofsky and Brier 1965). A successful cross-correlation technique used in Canada for radar has been described by Bellon and Austin (1978). It has been used in analysis of cloud motions from geosynchronous satellite imagery (Endlich and Wolf 1981, Endlich et al. 1971, Leese et al. 1971, Wolf et al., 1977).

Crane (1979) has described an algorithm for cell tracking. A cell is defined by him as a small region of a Z_e maximum surrounded by a contour a fixed number of dB below the maximum value. Such an algorithm would identify individual cells and track their motion.

In the SHort-term Automatic Radar Prediction (SHARP) procedure developed in Canada [Austin and Bellon (1974, 1982), Bellon and Austin (1978)] the best match between two precipitation patterns is obtained by computing a cross-correlation coefficient Γ for all possible displacements until the maximum value Γ_{max} is identified. The magnitude of Γ_{max} provides a measure of the modification with time of the echo shape and structure . The displacement of the centroids of the two patterns can be taken as an initial guess. One pattern is translated over the other by that amount and then a cross-correlation is calculated over this initial displacement. In subsequent forecasts the computed velocity from the previous hour can be used. The initial guess can be based on the local climatologically expected motion or the steering wind (usually 700 hPa wind).

The echo pattern may be a single cell or a cluster or a squall line. It is advantageous to start with a CAPPI at 2 or 3 km altitude so that the input pattern is free of effects of clutter, anomalous propagation and beam occultation. A three-dimensional centroid can be determined from a volume scan and tracked. An additional pattern matching routine can be employed to recognise anomalous propagation echoes and reject them (though not entirely; see Chapter 3).

Rinehart and Garvey (1978) correlated an initial array of radar reflectivity with later arrays. The area of search for the second array was chosen, taking into account the likely motion as indicated by upper wind data or known storm velocity from past history. By locating the second array with the best correlation, a motion vector for the initial array is determined. They called this TREC (Tracking of Radar Echo with Correlations). Further development of similar techniques may be seen in Rosenfeld (1987), Dixon and Wiener (1993) and Witt and Johnson (1993). In these , a cell or a "storm" is considered with Z_e above a certain threshold (typically 30 dBZ) and size (e.g. volume less than 50 km^3). A contiguous region satisfying these criteria may be taken as a storm. This can be approximated by an ellipse of suitable shape and orientation. Such storms from a volume scan at time t_1 are compared with those from the next volume scan (time t_2). The storms at t_2 are matched with those at t_1 to consider which represents the subsequent position of each t_1 storm. By setting certain constraints the most probable motion trajectories can be obtained.

Tracking of clear air phenomena e.g. thin line echoes should be able to give the wind in the boundary layer. The wind obtained from clear air echoes is more likely to represent the true air velocity than the wind from hydrometeor echoes.

Using Doppler radar, the pattern of Doppler radial velocity can be tracked by similar algorithms instead of (or besides) the reflectivity pattern. Since radial velocity is one component of a vector quantity, tracking will indicate advection of these components if they are carried by flow on scales larger than the region being correlated; or it will show propagation if these components have wave-like behaviour. Smythe and Zrnić (1983) who tracked both reflectivity and Doppler velocity in clear air echoes, found that the correlation coefficient was higher in the case of Doppler velocity than for reflectivity. Winds derived from Doppler velocity compared better with dual-Doppler winds than those derived from reflectivity. This therefore provides a means of determining boundary layer winds with single Doppler radar (Tuttle and Foote 1990).

The movement of the reflectivity pattern represents the advection velocity. The time rate of change of the radial velocity is known from successive radar observations. This contains information about the cross-wind components. Sun et al. (1991) used a variational assimilation method to derive the wind field from this information. In a method due to Zhang and Gal-Chen (1996) (see also Gal-Chen 1982), this is used to choose a moving frame of reference. The reflectivity and radial velocities are then interpolated to Cartesian grids in this moving frame. At each grid point, a set of reflectivity equations and radial velocity equations are established from single Doppler observations at frequent time intervals. These equations are solved for the three Cartesian components of the wind in the moving frame of reference. The complete wind field is obtained by adding the mean advection velocity to the retrieved wind. Liou (1999) proposed a modification of this method from which, according to him, the winds can be retrieved (with certain assumptions) even from non-Doppler radar. The capabilities and limitations of these methods in various situations and comparison of the various methods have been discussed in Caya et al. (2002), Gal-Chen and Zhang (1993), Gao et al. (2001), Laroche and Zawadzki (1994, 1995), Lazarus et al. (1999), Liou (1999), Liou and Luo (2001), Liou et al. (1991), Nissen et al. (2001), Qiu and Xu (1992), Shapiro et al. (1995), Sun (1994), Weygandt et al. (1995), and Xu et al. (1994, 1995). Derivation of thermodynamic parameters by related methods is discussed in Chapter 6.

5.1.6 Computer Vision Techniques

A powerful method is the detection of edges of features (lines or areas) (e.g. Hamann 1991, Harris et al. 1993). The data are first converted to Cartesian co-ordinates. Through an iterative process, lines joining maxima in the magnitude fields are constructed. These lines define the edge or boundary of a feature. A series of templates are used to compute the gradient (of reflectivity or velocity). Contours of specified levels of Z_e can be extracted. By identifying a reflectivity contour which envelopes most of a precipitation area, delineating a leading or trailing edge of precipitation or defining a frontal boundary through velocity and reflectivity gradients, the onset or termination of precipitation can be monitored. A cell can be recognised as a volume of contiguous grid points in which Z_e exceeds a specified threshold. These methods are more effective than merely tracking the centroids. Specific contour structures (e.g. a hook in a thunderstorm echo) can be tracked by a least-square method of correlation. Vertical profiles of Z_e (VPR) can be used to detect the bright band and to infer the phase of the hydrometeors (see Chapter 7).

5.1.7 Remote Transmission of Data

In the pre-computer era, the problem of transmitting the outputs from a remotely located radar to the forecast centres was a serious one. The conventional method of keeping a record of the radar data was to make tracings or photographs of the radar scope. Data of echo top heights, intensity levels and qualitative annotations could be added. Successive tracings could be superposed to estimate movement, growth and decay. As already mentioned, a Manually Digitised Radar (MDR) map could be prepared. Although this method is subjective, a skilled operator could prepare such a product in a few minutes. In Japan around 1969, a system (known as telefax) of facsimile transmission of such sketches to the forecast centres was introduced. These sketches were of the same scale as local weather charts and could be superposed on them. The transmission used ordinary telephone lines but could use radiofacsimile where telephone lines were not reliable. Aside from such a transmission, a semi-plain language coded message can be sent on teletype in the meteorological telecommunication network. The message incorporates the points defining the boundary of any radar-observed feature (such as an area of echoes, a line, a rain band, or the eye of a tropical cyclone), the density of echoes in the feature, heights of tops, movement etc. The WMO code FM 20-VIII, RADOB (WMO 1988), which consists of standard five digit groups, provides for international data exchange while some countries have their own codes. A semi-plain language code is used in India. These relatively crude systems were used because transmission of an analogue PPI picture needed an expensive dedicated wide bandwidth microwave communication link. Slow scan systems for scanning the PPI and transmitting at low speed over narrow band width telephone lines were also in use in some countries.

The advent of digitisation has radically changed the picture. Digital data are transmitted over narrow band links and the recipient can view all the displays. However, a modern radar produces an immense amount of data. It is neither economical nor necessary to transmit or archive the raw radar data in most cases. Hence, a certain amount of pre-processing is done at the radar site and base products (Z, v, σ) are usually transmitted along with headers indicating radar settings. The derived products can be computed at the receiving end if compatible software is available. For input to numerical models, digital data are necessary.

Data compression techniques can be used to reduce data volume. Data resolution in space, time and intensity can be reduced by integration over larger grid squares (or polar bins) or a longer time interval or transmitting more widely spaced intensity levels. Zero values (which are usually preponderant in many grid squares) can be omitted if an addressing system for the grid squares is used. Also depending on the needs of the user, selective data sets can be transmitted. The conventional codes for transmission of radar data are also being replaced in many countries by a computer to computer binary transmission code named FM 94 IX Ext. BUFR (WMO 1988, Suppl. 3, 1991). BUFR stands for Binary Universal Form for the Representation of meteorological data. The BUFR message consists of a continuous bit stream made of a sequence of octets (1 octet = 8 bits) and can be transmitted direct from computer to computer on the Global Telecommunication System (GTS). Table 5.1 gives the type of radar data provided for in the code.

5.1.8 Archival of data

Photography of the radar scope has been widely used to keep an accurate record of radar imagery, Auxiliary information such as date, time, radar settings and any relevant notes can be provided by the side of the scope. An automated camera using continuous roll film can be used with the shutter operation linked to the start of the scan. A positive copy of the film can be prepared later for archival. If the radar settings are constant, the positive film can be run as a "time lapse" movie to visualise the development of weather systems. Videocameras have also been used. In this case the videotape is available immediately for playing back and comparing with subsequent pictures.

Table 5.1. WMO code FM 94 IX Ext. BUFR

Table Reference F	X	Y	Element Name	Unit	Scale	Reference value	Data Width Bits
0	21	001	Horizontal Reflectivity	dB	0	−64	7
0	21	002	Vertical Reflectivity	dB	0	−64	7
0	21	003	Differential Reflectivity	dB	1	−5	7
0	21	005	LDR	dB	0	−65	6
0	21	006	CDR	dB	0	−65	6
0	21	011	Doppler Mean Velocity - x direction	m s^{-1}	0	−128	6
0	21	012	Doppler Mean Velocity - y direction	m s^{-1}	0	−128	6
0	21	013	Doppler Mean Velocity - z direction	m s^{-1}	0	−128	6
0	21	014	Doppler Mean Velocity - radial	m s^{-1}	1	−4096	13
0	21	017	Doppler Velocity - spectral width	m s^{-1}	0	−4096	8
0	21	021	Echo tops	m	−3	0	4
0	21	030	Signal-to-Noise Ratio	dB	0	−32	8
0	21	031	VIL	kg m^{-2}	0	0	7
0	21	036	Radar Rainfall Intensity	m s^{-1}	7	0	12
0	21	041	Bright Band Height	m	−2	0	8
0	21	051	Signal power above 1 mW	dB	0	−256	8

These methods have been largely superseded by digital archival on computer-compatible tapes or cartridges which can be erased (and reused) after a prescribed period of time. The storage format must provide for auxiliary information on the radar calibration parameters, the modes of operation and the meteorological situation in which the data were collected besides, of course, details such as date and time. A limited amount of processed data can be stored and recalled on the screen for comparison with subsequent imagery, e.g., on four-in-one displays. They are required also for working out cross-correlations for pattern recognition and short-period forecasting. Facilities for taking colour prints of the displays are also available. From digital outputs, time lapse movies can be made using animation software to visualise the development of weather systems (see e.g. Lutz et al. 1989, Black and Dodge 1993). The problem now is not the transmission or recording but the selection of the type of data for archival without overloading the data storage facilities. The archival can be done at the radar site or at an analysis centre or in a central documentation unit. Copies of selected data can be supplied on electronic media to anyone. This enables radar data to be used extensively in climatology.

5.1.9 Radar Data on the Internet

The Internet is being increasingly used for transmission of radar imagery and data in near real time. Radar summaries and radar-estimated rainfall data are available on the internet in many countries.

5.1.10 Radar Composites and Networking

Radar information in the early days was used only at the nearby forecast centres for local weather monitoring and forecasting. Examples are thunderstorm forecasting at a local level, pilot briefing and airfield warnings. But as multiple radars were deployed for monitoring weather systems over larger areas, it was found necessary to get the data of all the radars in real time at one point and construct composite imagery. The features associated with larger weather systems may be seen only partly by any one radar; compositing pictures is essential to get a better appreciation of the system. With the availability of digital telecommunications, data from several radars can be collected in real time at any analysis centre and presented on a composite display in appropriate geographical coordinates. The radars are said to be networked.

The UK was perhaps the first country to introduce networking (Browning and Collier 1982). Subsequently most of the radars in west Europe (and some in the east too) were networked under a programme known as COST-73 (CO-operation in Science and Technology) of the European Community countries [Collier (1989, 1989b, 1992, 1992b), Collier and Chapuis (1990), Collier et al. (1988)][3]. The US NEXRAD WSR-88D network (see section 3.6) is a homogeneous network of Doppler radars covering most of that country. Canada is establishing a network of 29 Doppler radars (Lapczak et al. 1999). Another major network is the Japanese network which consists of 20 radars on land and two on ships.

Preparation of a composite display in real time involves considerable standardisation. A network (especially if it covers several countries, as in Europe) may consist of radars of different capabilities, and the data formats used by the countries may vary. The coverage area of each radar is not simply a circle with the radar as centre, but may be complicated by uneven terrain. This will mean different coverage areas at various altitude levels (Maddox et al.

---

[3] Maps of the European and US networks may be seen in Meischner et al. (1997) and Maddox et al. (2002) respectively.

2002). The radars may have overlapping coverage in some areas and data gaps in others. This may lead to conflicting data from different radars for the same grid point. It is also necessary to position the data correctly on a map. A polar stereographic projection is usually used. In the UK network, for example, data from each radar in 5 km × 5 km grid is received by a computer at a Radar Network Centre (RADARNET) during a predetermined time window every fifteen minutes. If more than one radar gives data for a grid square, the one with the lowest beam height at that point is only taken and the rest rejected. If data from the lowest beam is not available, the next best radar is chosen. Clutter area near a radar can be filled by data from another radar wherever such overlap is available. The composite images as well as derived products (e.g. severe weather probability assessments, echo tops, echo movement) can be transmitted to other users. Terrestrial or satellite communication links can be used for the network. The composite imagery is utilised for tracking of mesoscale as well as synoptic scale systems, short-period forecasting, and mapping of precipitation.

Cavalli (1984) and Galli (1984) have described the networking of two radars in hilly terrain in Switzerland. A composite picture is produced every five minutes in real time and transmitted with a delay of 4 minutes. Descriptions of other networks may be seen in Collier (1989b) and Meischner et al. (1997). Mohr et al. (1986) among others have discussed data processing for merging of data into common Cartesian formats.

5.1.11 Composites with other data sources

The superposition of satellite imagery (especially geostationary satellite imagery which is available at frequent intervals) with networked radar composites gives an additional dimension to the utilisation of radar data. The earliest was the FRONTIERS (Forecasting Rain Optimised using New Techniques of Interactively Enhanced Radar and Satellite data) system of the UK (Browning 1979, 1986). Information from other sources such as a lightning detection network can also be combined. as part of an integrated system of observations. These are mapped in a common coordinate system. The interconnection of the graphic workstation for wide-area rainfall analysis and other computer systems at Bracknell (UK) is shown in Fig. 5.1. An interactive system for combining digital radar and satellite data has been described by Heymsfield et al. (1983). A software for combining diverse environmental data sets has been described by Corbet et al. (1994). Another system which combines half-hourly geostationary satellite imagery with radar CAPPI product is the

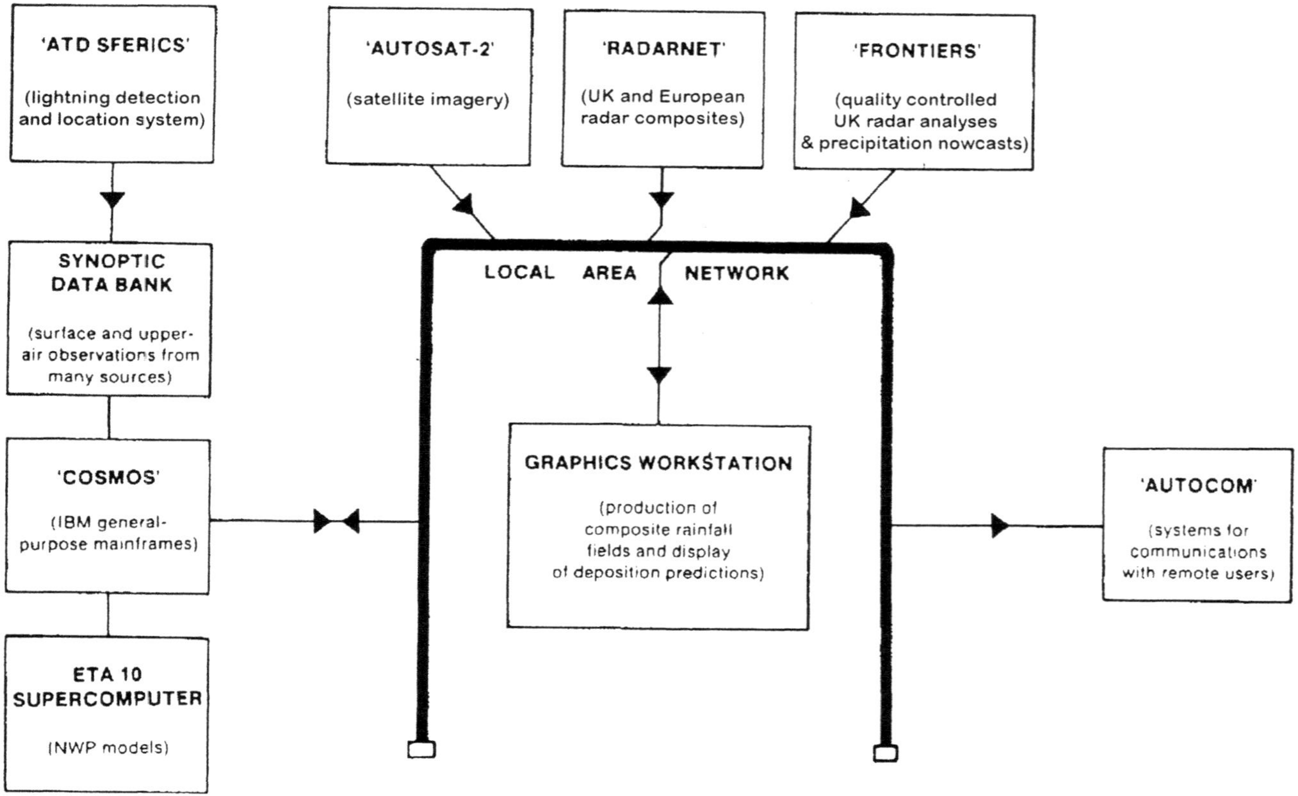

Fig. 5.1. Interconnection of the graphic workstation for wide-area rainfall analysis and other computer systems at Bracknell, UK (from Collier et al. 1989; courtesy HMSO, London).

Canadian RAINSAT (King et al. 1989). Geographical Information System (GIS) is being increasingly used for combining the data in a common format.

The combined use of radar and satellite information is sometimes helpful for a crosscheck of data from either. For example, radar echo top heights can be in error at long ranges and radar may fail to detect echoes at long range due to beam height and shielding problems; these can be crosschecked from satellite imagery and infra-red cloud top temperatures. Anomalous propagation echoes interpreted as precipitation by radar can be checked by identifying cloudfree regions in the satellite imagery (Fiore et al. 1986, Pamment and Conway 1998). Similarly, sea clutter echoes on radar can be identified and rejected if the satellite does not show clouding in the area. Use of radar as ground truth for satellite estimates of rainfall is dealt with in Chapter 7.

5.1.12 Automated radars

Weather radars have often to be located at remote sites having logistic difficulties for personnel to stay and work in any weather. Hence, completely unattended operation is desirable. The first remote-controlled operational weather radar was installed in Japan on top of Mt. Fuji in 1965. The radar was remote-controlled by a microwave link from Tokyo. The same link collected other meteorological and seismic data. However a set of personnel (by rotation) were located on the mountain. A more sophisticated arrangement with a digital radar controlled by a computer at site was made in the UK (Collier et al., 1980) at Hameldon Hill for hydrological purposes. This was a completely unattended station except for maintenance visits. Arrangements for power failure protection and remote monitoring were provided. The radar was networked with other radars in the UK and composite data were available at a central station. All the WSR-88D stations in the USA are automated.

5.1.13 Terrain Problems

Hilly terrain not only produces clutter and beam blockage but modifies the wind flow and precipitation processes. Spatial gradient of rain rate is high. Intersection of the bright band may also be a problem in precipitation estimation. Variations of the refractivity are also large. Higher elevation angles have to be used and maximum range is reduced. If more than one radar is available, one may be able to complement the other. It is advantageous to be able to map the terrain with high resolution and superpose the radar data on it for visualising the terrain effects.

This is done by using a three-dimensional terrain data base and combining with radar PPI or CAPPI products converted to Cartesian coordinates (Doick and Holt 1995, James et al. 2000). Using a known (or climatological) vertical profile of reflectivity (VPR), the observed reflectivity aloft may be corrected to obtain the reflectivity at ground level in occulted areas (see Chapter 7).

5.2 Calibration and Maintenance of Radars

5.2.1 Need for Calibration and Maintenance

To get the best from any equipment it is obviously necessary to ensure that its performance is always according to its specifications. Performance of sub-systems of a radar can vary over a period of time because of degradation or change of characteristics of electronic components, replacement of components, wear and tear of mechanical parts, power fluctuations and environmental effects. The system will also be affected by accumulation of dust, moisture or snow, entry of insects or changes in levelling or alignment. Gross deterioration in performance is detectable by an operator by observing the intensity and extent of permanent echoes, but smaller deficiencies may go unnoticed unless frequent measurements are made. Schedules for periodical maintenance of equipment are usually recommended by manufacturers and will not be part of this book. What are of concern here are calibration checks needed to ensure (1) that the radar parameters which go into quantitative measurements with the radar are known accurately, and (2) that the radar gives as accurate a measure of meteorological parameters as the available techniques will permit. It is important that not only the engineer or technician but the meteorologist or researcher who is using the radar should be aware of the performance of the radar and its limitations, so that erroneous interpretations are not made. Another aspect of calibration is the validation of algorithms to ensure that they represent reality. An algorithm valid at one radar site may not necessarily be valid at another.

5.2.2 Radar Attributes to be Monitored

The factors going explicitly into the radar equation, are transmitted peak power, pulse width, antenna gain and operating frequency. Others which are not explicit in the equation are the p.r.f., pulse shape, losses in the waveguide, T-R switch and the like, receiver sensitivity (MDS), noise figure and receiver bandwidth. The accuracy of pointing of the beam and the settings of the antenna azimuth and elevation angles relative to north and the horizontal are also relevant. Some of the parameters which may fluctuate often need to be measured frequently. With older types of radars use of separate test equipment is necessary. Modern radars usually have Built-In Test Equipment (BITE) for monitoring several radar parameters. But these need, in turn, to be calibrated periodically against standard instruments.

5.2.3 Antenna Parameters

The antenna parameters (gain, beam width and axis alignment, radiation pattern, sidelobe level) are measured by the manufacturer at the factory's antenna range; all these measurements cannot be repeated at field sites. Manufacturers tend to quote the maximum gain without mentioning the system losses. It is, therefore, desirable to measure the system gain, at least initially, at the permanent site, if facilities are available. A standard horn placed at some height on a tower or building as shown in Fig. 5.2 can be used to measure gain and beam width. With the horn in the far field of the radar and the axes of the radar antenna and the horn aligned to maximise the power received at the horn, the average power P_m transmitted by the radar and the average power P_h received at the horn are measured by the same power meter. Knowing the gain G_h of the horn, the gain G_R of the system taking into account head losses (Hudlow et al. 1979), is obtained from the equation,

$$P_h = \frac{P_m G_R G_h \lambda^2}{16\pi^2 r^2} \tag{5.5}$$

However, the signal received from an alternate path involving ground reflection may interfere with the direct signal. An alternative method to measure gain is to observe the backscattered power at the radar receiver from a metallic sphere (large in comparison with λ)[4] suspended from a tethered balloon or a free balloon. To minimise oscillation of the sphere in the wind the measurement should be made in calm conditions[5]. Rinehart and Eccles (1976) have proposed a dihedral target for gain measurements. Another method of system gain measurement is the solar observation technique discussed in Paragraph 5.2.7.

---

[4] The radar cross section of such a sphere is equal to its geometric cross section.

[5] The sphere can be tracked by a tracking radar or a theodolite, whose azimuth and elevation data can be used to control the weather radar antenna during the experiment.

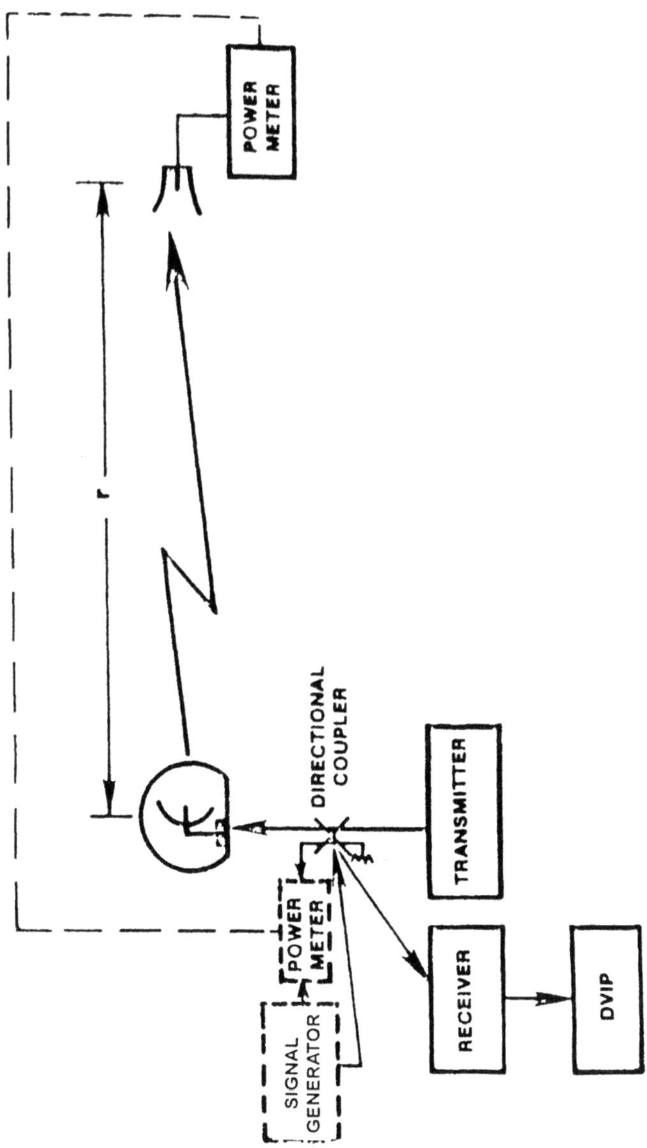

Fig. 5.2. Test equipment set-up for measurement of antenna gain.

To determine the beam width, the horn may be fed a signal from a signal generator and the same received on the radar receiver. The radar antenna may be rotated to locate the azimuths (or elevations) at which the signal power falls to half its maximum value. The positions and levels of the sidelobes can also be determined to a reasonable degree of accuracy.

Rinehart and Tuttle (1981) have given a method of determining the complete beam pattern from observation of a ground target. This is important in dual wavelength radars, as dual wavelength techniques assume matched beams at the two wavelengths. Rinehart and Tuttle (1982, 1984) have shown that if this is not the case there may be large errors, because the sidelobe although weaker, covers a larger echoing volume which contributes to the returned power.

In the case of polarimetric radars the antenna gain, beam pattern and waveguide losses may be different in the different polarisation modes (Bringi and Hendry 1990).

5.2.4 Transmitter Parameters

In the set-up in Fig. 5.3, when radar is radiating and the frequency meter is off-tune, the power meter indicates an average power P_p in dBm. The peak transmitted power P_t is given by,

$$P_t \text{ (dBm)} = L + P_p - 10\log\left(\text{pulse duration} \times \text{p.r.f.}\right) \qquad (5.6)$$

where L is the total loss in dB due to the duplexer, directional coupler, frequency meter and connecting cables, measured independently. If now the frequency meter is tuned to a position at which there is a dip in the power meter reading, it is in resonance and the transmitting frequency can be read off on the frequency meter. If the transmitter is tunable its frequency may be adjusted to the standard value. Otherwise, the receiver local oscillator is tuned to a value equal to transmitter frequency plus (or minus) the IF. Some radars have directional couplers at which the reflected and incident power can be separately measured by the same set-up as in Fig. 5.3. The square root of the ratio of reflected to incident power gives the reflection coefficient α. The Voltage Standing Wave Ratio (VSWR) is given by $(1+\alpha)/(1-\alpha)$

With the set-up in Fig. 5.4 a sample of the transmitted pulse is monitored on an oscilloscope. The width of the pulse at 70.7% of the peak amplitude may be taken as the pulse width. The p.r.f. trigger can be monitored on the same set-up.

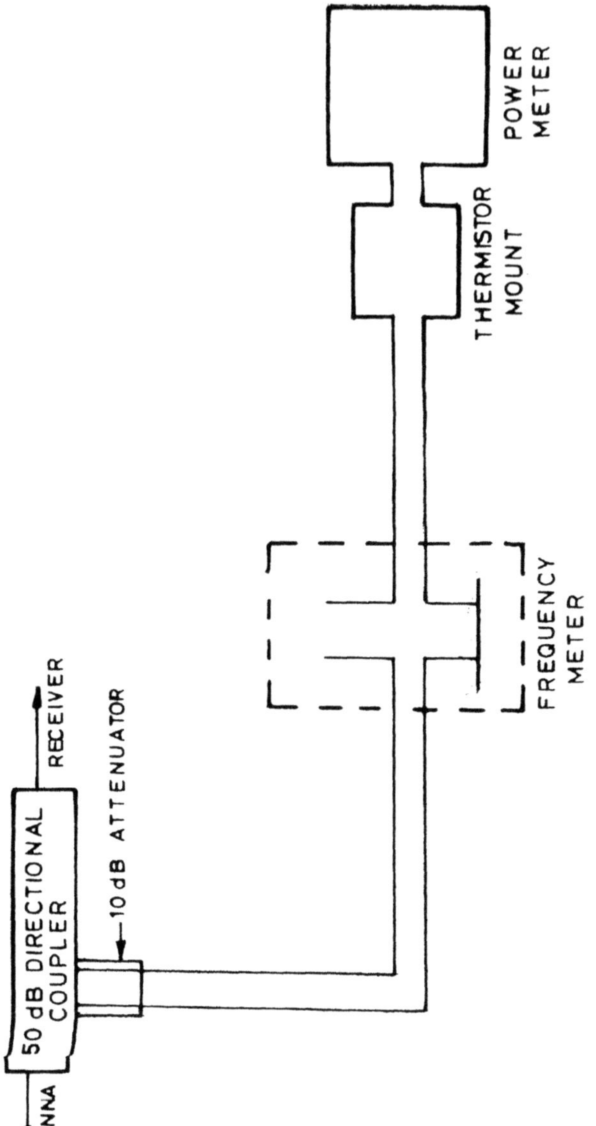

Fig. 5.3. Test equipment set-up for measurement of transmitted power.

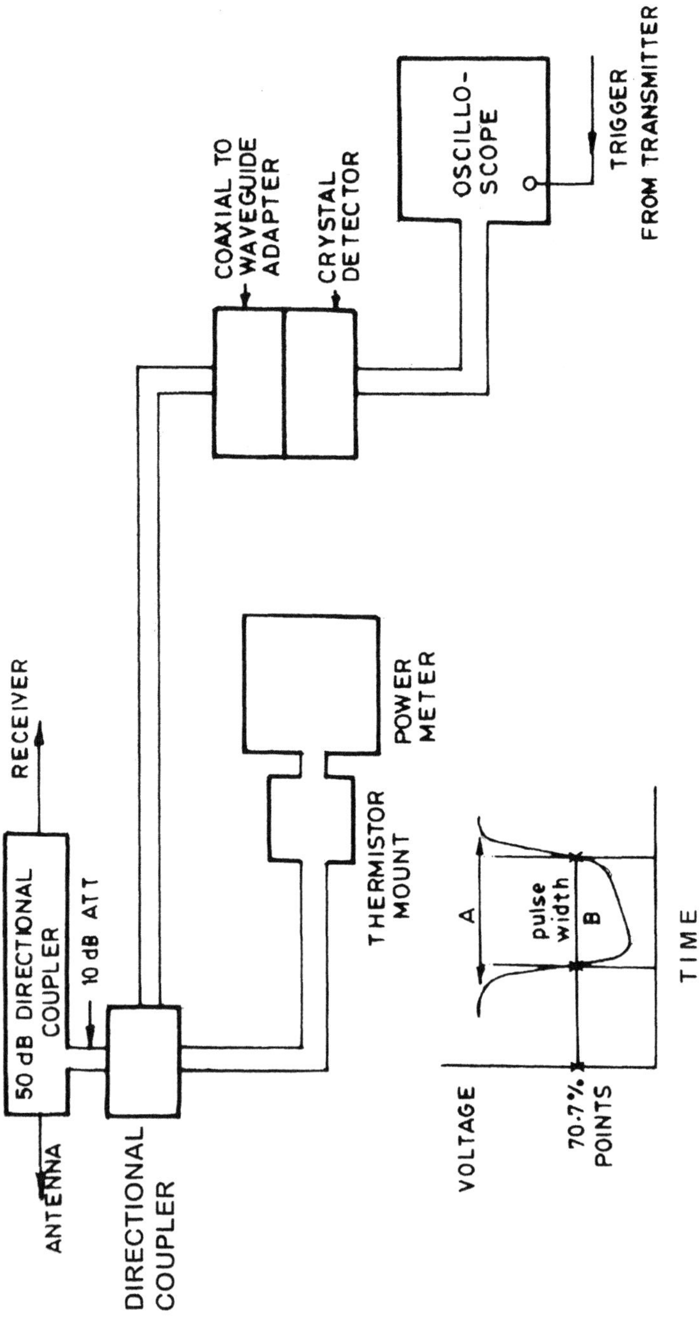

Fig. 5.4. Test equipment set-up for measurement of pulse width.

5.2.5 Receiver Parameters

Power from a signal generator is pulse modulated and fed to the receiver and the video output is monitored on an oscilloscope or printed out from the digital processor. The calibrated attenuator on the signal generator is adjusted till the signal is barely discernible amidst the noise. The attenuator reading corrected for receiver head losses and loss in connecting cables gives the MDS. The receiver noise figure, IF bandwidth, range normalisation, level thresholds and several such factors can be checked similarly with appropriate test equipment. The determination of the receiver transfer function by printing out the receiver video output for various settings of input from a signal generator (Fig. 3.10) was referred to in Chapter 3. However the signal from a signal generator is a steady one while precipitation signal is a fluctuating one from a distributed target and is averaged. The output depends on the statistical characteristics of the signal as well as the nature of the receiver response. In general, there will be a bias in the output with reference to the true average of the signal. For a square law receiver the bias is zero for a steady signal. For the logarithmic output this bias is taken as the difference between the estimates of the mean power from a logarithmic and a square law receiver. Terblanche (1996) has described a signal processing method whereby averages simulating a linear or square law receiver response can be obtained from a logarithmic receiver. Calibrating the radar with an actual weather echo in addition to the absolute calibration with a signal generator (or solar noise) has been suggested by several authors. Details may be seen in Gorgucci et al. (1984, 1989, 1991) and Schaffner et al. (1983).

Built-In Test Equipment (BITE) automatically checks the radar status almost continuously. Forward power and reflected power at the transmitter output can be monitored with a dual directional coupler and a peak power meter. The on-line noise figure of the receiver can be checked by measuring the receiver output by injecting a controlled noise trigger in the receiver chain. A simulated target at any desired range with any Doppler velocity and signal power can be created. The range is simulated by time - shifting the modulator pulse. Shifting the COHO frequency gives the desired Doppler shift. A programmable attenuator is used to set the signal strength. The output can be printed out. The system can also run diagnostic tests to identify faults up to module level. In case any measured parameters entering into the radar equation are different from the original values the new values can be used by the computer in computing the output products.

The French RONSARD radar system, which was described in Chapter 3, has a self-calibrating facility. Every 40 seconds, a known signal from a temperature-stabilised source oscillating at the transmitter frequency ±150 kHz is fed to the input of the parametric amplifiers with a fixed AGC value. The power measurement at the receiver output is sent to the computer controlling the 6-dB PIN diode RF attenuator. This feedback counterbalances the drift of various RF or IF amplifiers and keeps the overall gain constant within 0.5 dB, when external temperature varies between –20 to 50°C. (see Nutten et al. 1979, 1979b).

A remote calibration facility exists in the remote-controlled Doppler polarimetric radar of the NSSL, USA (see Chapter 3) (Zahrai and Zrnić 1993). A coherent signal source (CSS) is fed to a port in the receiver waveguide. This produces an output offset by 500 Hz. from the IF. This frequency is mixed with the RF oscillator and fed through a variable attenuator to the waveguide port. This is used to calibrate overall receiver gain and also to monitor and correct for any amplitude and phase variations in the AGC and I and Q processors. The facility can also produce and display a constant Z_{DR} value on the PPI for calibration. A phase shifter in the IF path can simulate differential phase shift and velocity.

5.2.6 Antenna settings

The manufacturer ensures that when the antenna pedestal is placed over level surface, and when the elevation angle reads zero, the radiating axis is horizontal; also that it coincides with the optical axis. At the time of installation and in periodical checking the verticality of the front plane of the antenna dish may be checked with a plumbline. It is also necessary that the antenna faces true north when the azimuth scale reads zero. The setting can be done initially in several ways.

(1) Before installation of the radar, the azimuth with respect to magnetic north, of a reference object at a distance, is determined and corrected for the current value of the magnetic declination for the particular place and the antenna optic axis aligned using a telescope. Alternatively, before the antenna is installed, a pole star observation[6] is taken and corrected for the deviation of the azimuth of the pole star from true north at the time of observation.

(2) Subsequent checks may be made by observing the plane of the antenna with a theodolite[7] placed at a distance.

(3) A more convenient procedure is to observe the echo of a distant point object of known azimuth, which gives a permanent echo. The satellite-based Global Positioning System (GPS) can give accurate position information.

---

[6] This is applicable in the northern hemisphere only.

[7] This is possible only if the antenna is not covered by a radome.

5.2.7 Solar observations

The sun (so do many other stars) emits electromagnetic radiation in a wide range of frequencies. For example, at a wavelength of 10 cm, the solar radiation flux at the earth varies from 70 to 250 flux units, depending on solar activity, one flux unit being 10^{-22} W m^{-2} Hz^{-1}. This is detectable by most weather radars. Since the angular diameter of the radio sun is about 0.5° it can be considered to be a point source for most meteorological radars, which have beam widths larger than 0.5°. The position of the sun in the sky at any time is accurately known from the Ephemeris, or it can be stored in the radar computer. If the antenna (in the non-transmitting condition) is pointed at the sun and adjusted to maximise the video output from the radar receiver, the azimuth and elevation of the antenna are accurately known and the antenna settings can be adjusted accordingly (Raghavan 1984). The procedure can be carried out any time during daylight hours and is free from errors due to ground reflections. It is preferable to take the observation at local noon as the rates of change of elevation and azimuth of the sun are minimal at this time. The method is particularly useful where dual wavelength or dual polarisation antennas have to be aligned to point exactly alike.

A more important application of the solar observation is to calibrate the radar system gain, antenna beam width, boresight error and receiver noise figure. The values of the expected solar flux are published by astronomical observatories. Since the flux fluctuates with solar activity, it is desirable to attempt the calibration in a quiet sun period. Also, the noise from the quiet sun is unpolarised; the antenna will receive only half the power as it normally receives only one polarisation. The power received will depend on the receiver bandwidth and antenna efficiency[8]. A correction to the observed beam width and gain is necessary because of the finite angular diameter of the sun. Procedures and corrections necessary are discussed in Frush (1984), Pratte and Ferraro (1989), Stutzman and Ko (1974) and Whiton et al. (1976).

5.2.8 Effective performance

The above discussion dealt only with the basic radar attributes. The actual capability of detection by radar of a given rate of precipitation or a target of given cross section depends on numerous other factors. Besides the MDS

---

[8] In the case of dual polarisation radars the polarisation switch may have different properties in transmitting and non-transmitting condition.

of the receiver the losses introduced in any data processing adjuncts e.g., averaging, range normalisation etc. will affect the effective sensitivity to weak echoes. On the other hand, increase of dwell time and averaging of signals improves sensitivity of the radar (see e.g. Smith 1993). The capability for detection also depends on the range, the propagation conditions, the height of the target, the echoing mechanism and attenuation in the path due to precipitation. Values of dBZ, which can be detected at various ranges by a typical 10 cm radar were shown in Fig. 3.27.

For comparison of clear air and precipitation echoes, an equivalent reflectivity factor Z_e was defined in equation (2.97). A comparative performance of some radars of the USA in detecting various types of echoes in terms of a "Practical Minimum Detectable Effective Reflectivity Factor (MDZ_e)" as a function of range (R km) is given in Fig. 5.5 (after Pratte and Keeler, 1986). The specifications of these radars are given in Table 5.2. The "Clear Air" radar is a hypothetical one.

5.2.9 Other Methods of Calibration

Calibration of radar with rain gauges is discussed in Chapter 7. While the space-borne TRMM radar is being validated using ground-based radars and rain gauges, the TRMM is also being used to calibrate ground-based radars (Chapter 7). Scanning radars have also been calibrated using profilers and disdrometers (Chapter 10). Atlas (2002) has reviewed calibration methods from the earliest days of radar meteorology to the present.

Table 5.2 Specifications of radars, performance of which is shown in Fig. 5.5.

Radar	WSR57M	WSR-74C	CP2	Clear Air
Antenna size m	3.7	2.4	8.5	8.5
λ cm	10.7	5.4	10.7	10.7
Beam width °	2.2	1.6	0.9	0.9
Antenna gain dB	36.5	38.0	43.0	43.0
Pulse width μs	4.0	3.2	1.0	4.0
P_t kW	410	250	1040	250
p.r.f. Hz	164	259	960	960
Scan rate $°s^{-1}$	18	18	15	3.8
Receiver noise figure dB	8.0	9.0	4.2	4.2
3 dB bandwidth MHz	0.8	0.8	1.2	0.3

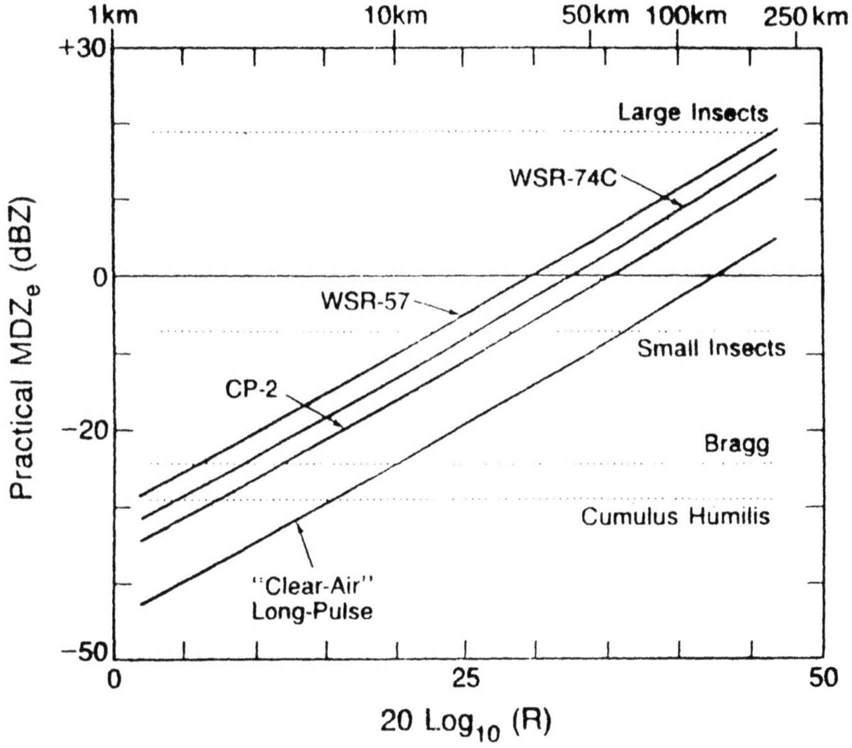

Fig. 5.5. Practical MDZ_e for different radars as function of range (from Pratte and Keeler 1986; courtesy AMS). Specifications of the radars are given in Table 5.2.

Chapter 6

RADAR OBSERVATION OF WEATHER SYSTEMS

In the first section of this chapter we discuss radar observation of thunderstorms and other mesoscale phenomena and the radar indicators and signatures of severe storms, stratiform anvil, hail, mesocyclone, tornado, gust front and downburst. This is followed in section two with use of radar data in studies of the growth of convection. Section 6.3 deals with derivation of products from radar observation directly or in conjunction with numerical models, such as cooling rates, latent heat release, turbulence and thermodynamic and microphysical retrievals. Section 6.4 deals with short-period forecasting using radar along with other data sources.

6.1 Precipitation-related Phenomena

6.1.1 Scales of Weather Systems

Atmospheric disturbances occur on various spatial scales. Classifications proposed by some authors are given in Table 6.1.

Though definitions vary, mesoscale phenomena which roughly extend from a few km to a few hundred km and last from a few minutes to several hours, are ideally suited for study by radar not only by virtue of their spatial extent and the spatial and temporal resolution achievable by radar but also because these sub-synoptic scale phenomena tend to slip through the conventional synoptic observational network. Important among the phenomena studied by radar are convective systems ranging from a tornado or its parent mesocyclone to clusters of thunderstorms or squall lines to tropical cyclones.

Table 6.1. Scales of Weather Phenomena

Author	Scale		Characteristic Dimension
Ligda (1951)	SYNOPTIC		1000 km
	MESO		Some tens of km
	MICRO		1 km or less
Orlanski (1975)	MACRO		> 2000 km
	MESO	α	200 km - 2000 km
		β	20 km - 200 km
		γ	2 km - 20 km
	MICRO	α	200 m - 2000 m
		β	20 m - 200 m
		γ	2 m - 20 m
Fujita (1981)	MASO	α	4000 km - 40,000 km
		β	400 km - 4000 km
	MESO	α	40 km - 400 km
		β	4 km - 40 km
	MISO	α	400 m - 4 km
		β	40 m - 400 m
	MOSO	α	4 m - 40 m
		β	40 cm - 4 m
	MUSO		< 40 cm

6.1.2 Convection and Thunderstorms

The first major field experiment for the study of the thunderstorm using radar and aircraft observations was undertaken in the USA (Byers and Braham 1949) although many studies and field experiments based on limited surface and upper air data were conducted much earlier. [e.g. Chatterjee and Sur (1939), IMD (1944), Sen (1931), Sohoni (1928), in India]. Byers and Braham distinguished three stages in the life of a thunderstorm cell namely the cumulus, mature and dissipating stages. By definition, a thunderstorm is one in which lightning occurs and thunder is heard. Many convective cells may not reach that stage of growth[1] but may give radar echoes. A cell may be defined as a compact

[1] Visual or aural observation of lightning and thunder may be missed. One way to identify a thunderstorm is to use a Lightning Detection System (LDS). A 41-45 dBZ or higher reflectivity factor observed by radar in a 40 km × 40 km square is taken in the USA to be a thunderstorm (Bosart and Landin 1994) irrespective of whether an observer hears thunder (which is the conventional criterion for reporting a thunderstorm). It is, however, difficult to distinguish from the radar echo alone, between a shower arising from a cumulus and a thundershower from a cumulonimbus in which electrical charge separation is involved. Some researchers have tried to do so from the height of the echo top. Tohsha and Ichimura (1961), in Japan, assumed that echo tops reaching above the −15.6°C level (between 7.5 and 8.2 km altitude in Japan) are thunderstorms and those below that level are showers. In India, Chatterjee and Prem Prakash (1990) took the altitude limit as 8 km on the ground that clouds of lower height may not have sufficient charge separation to be classified as thunder clouds.

region of strong vertical motion and is recognised on radar by the precipitation echo. In terms of the radar observables, a cell may be defined as an entity with closed isopleths in the low elevation reflectivity pattern, with a definite minimum or break in spatial continuity from other such entities. It is also characterised by significant vertical motion as seen by a Doppler radar. Though the cell is considered an individual unit, it is not isolated, as air may flow in laterally from outside the cloud (entrainment) at various levels. Thunderstorms may occur as a single cell, or multicell (i.e. consisting of a series of cells forming and dissipating) or supercell which is a giant single cell.

In the cumulus stage, the radar-detectable precipitation particles generally form near the cloud base[2] and the first radar echo may be seen. If the updraft is strong enough, hydrometeors can be supported without falling and grow to larger sizes. Ultimately they become heavy and fall; their drag as well as the cooling caused by melting and evaporation of hydrometeors initiates the downdraft in the lower portion of the cell. This marks the mature stage of the cell. At this stage the radar echo extends both below and above the initial level. It is generally considered that because of evaporation of the hydrometeors the downdraft air is cooled (see e.g. Fujita 1959). Moist adiabatic descent ensures that it is denser than the environment and it sinks rapidly. If the environmental lapse rate is nearly dry adiabatic, the acceleration of the downdraft is increased. Because of evaporation, the downdraft may usually be moister than the environmental air at the surface. However, cases of the downdraft air at the surface being warmer than the environmental air have been reported (e.g. Banerjee et al. 1969). This may be due to lack of moisture and hence dry adiabatic descent. Mull and Rao (1950) proposed a mechanism for formation of the downdraft in terms of the sudden congelation of supercooled water above the 0°C isotherm. According to them, this produces heating at that level. This air aloft will be potentially cooler than the layer below which contained the supercooled water. This results in a turbulent motion carrying ice crystals below the 0°C level. Here they melt, cooling the air and triggering the downdraft. As this process is independent of evaporative cooling, it is possible for the downdraft to be drier than the environment in some cases as already noted in Chapter 4 in connection with the occurrence of superrefraction after a thunderstorm. This reasoning of Mull and Rao in respect of the convective scale downdraft appears to be supported by the subsequent finding of Leary and Houze (1979) that cooling due to melting

---

[2] The first radar echo has in some cases been observed as high as the −30°C level (Browning and Atlas 1965) probably due to strong updrafts carrying the particles to that height in the time taken for them to grow to radar-detectable sizes.

plays an important role in the initiation and maintenance of the mesoscale downdraft.

Besides these mechanisms, a non-hydrostatic vertical pressure gradient (produced by rotation) may also drive the downdraft down. The various mechanisms may be recognisable by Doppler radar. Roberts and Wilson (1989) have listed (Table 6.2) four downdraft forcing mechanisms and the associated (Doppler) radar signatures.

The downdraft spreads at the surface giving rise to a sudden increase in horizontal wind (the squall) and in some cases to the gust front which can be seen by a sensitive radar as a refractive index discontinuity. The spread is usually asymmetric, the strongest winds being in the forward sector (Venkataraman and Bhaskara Rao 1966). De and Sen (1961) found the squall wind speed at Calcutta highly correlated with the maximum height of the radar echo and the direction of squall to be generally in the direction of movement of the radar echo. In the dissipating stage the downdraft weakens and light rain continues. The interaction of the downdraft with the environmental air often leads to development of second generation cells[3] usually but not necessarily in the direction of cell motion.

6.1.3 Severe Storms

Thunderstorms are known as severe if they cause damaging winds or hail. While very strong vertical shear may restrict the growth of thunderstorms, shear

Table 6.2. Downdraft forcing mechanisms and associated radar signatures

Mechanism	Signature
Evaporational cooling	Radial convergence below cloud base, particularly when Z_e decreases significantly towards the ground. Radial convergence above cloud base, particularly when associated with a Z_e notch and/or a very dry environmental layer.
Melting cooling	Radial convergence just below the 0°C level.
Precipitation drag	Radial convergence within a descending reflectivity core > 50 dBZ_e.
Vertical pressure gradient	Azimuthal radial velocity couplet indicative of rotation, especially when vorticity increases with decreasing height.

[3] Chatterjee and Sur (1939) were probably the first to point this out.

up to some limit helps the growth[4]. Such shear causes separation of the updraft and downdraft regions and, hence, enables the updraft to extend to greater heights and lead to a severe storm. If the updraft is stronger than the fall velocity of precipitation particles, the latter are carried upwards and radar echo is observable at higher altitudes. The updraft thereafter decelerates and, hence, the visible cloud top (which may be due to small hydrometeors not detected by the radar) may often be a few hundred metres above the radar echo top.

Single cell storms are usually not severe and they last only about 30 minutes. Multicell and supercell storms (the distinction between the two is not sharp) grow to greater heights, produce more precipitation and often have dangerous features such as hail, violent wind and turbulence.

In a multicell storm the cells form and move in an organised manner. Typically 2 to 4 cells may be there at a time and new cells may develop with updrafts of 10 to 15 m s^{-1}. Fig. 6.1 shows the vertical structure of a multicell storm according to Browning et al. (1976) inferred from radar and aircraft observations. This figure can be thought of as an instantaneous picture of a storm with cells in various stages of development or as a sequence of stages in the life of one cell. The cold outflows from the cells combine to form a gust front and new cells are triggered along the front. This regeneration enables the storm to last a long time.

Individual cells move roughly with the mean (vertically averaged vector) environmental wind in the layer occupied by the cloud (Bhaskara Rao and Dekate 1967, Byers and Braham 1949). The layer wind from 700 to 500 hPa is often taken as the wind guiding the motion and radar observations agree generally with this [e.g. Byers and Braham (1949), Chalon et al. (1976), Fankhauser (1976), IMD (1944), Rao et al. (1961), Subramanian and Sehgal[5], (1967), Verma and Awadhesh Kumar (1993)]. There is also a transfer of horizontal momentum between the lower and upper troposphere because of the vertical motion. However, the motion as seen on radar is the combined effect of translation and propagation (development of regenerated cells preferentially in certain directions). Hence, the resultant motion of the storm as a whole may be to the right or in some cases to the left of the mean wind. Right moving storms are more common in the USA, but it is not necessarily so in other areas (Houze et al. 1993).

---

[4] Bhaskara Rao and Dekate (1967, 1971) defined a Relative Top Shear (RTS) as the shear between the mean vector wind in the layers over which the growing cloud extends and the next higher layer. They found, in India, that if RTS is not more than 10 knots km^{-1}, the cloud grows further.

[5] Subramanian and Sehgal (1967) found, in Agartala (northeast India), that movement of convective cells agreed, in 30% of cases, with the wind at 10,000 feet (700 hPa) and more generally, in 60% of cases, with winds at levels from 7,000 to 15,000 feet. They however, consider that the extrapolation of the last half hour's movement of the radar echo is a more successful predictor for operational short period forecasting at the airport than the use of upper wind data.

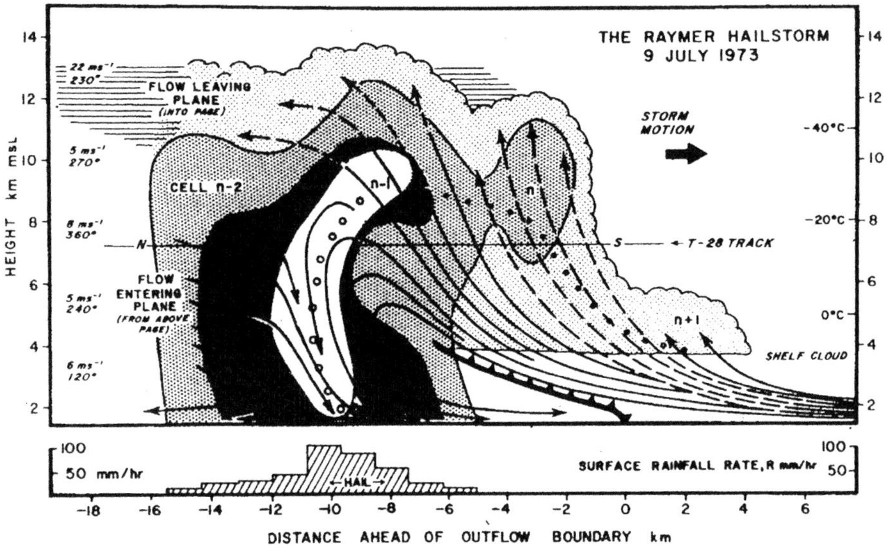

Fig. 6.1. Schematic model of a muticell hailstorm in Colorado, USA. (from Browning et al. 1976; courtesy AMS).
Top Fig: Vertical section along the storm's north to south (N-S) direction of travel, through a series of evolving cells. Solid lines are streamlines of flow relative to moving system; they are shown broken on left side of the figure to represent flow into and out of the plane and on the right to represent flow remaining within a plane a few kilometres closer to the reader. The chain of open circles represents the trajectory of a hailstone during its growth from a small particle at cloud base. Lightly stippled shading represents the extent of cloud and the darker grades of shading represent radar reflectivity factors of 35, 45 and 50 dBZ. Temperature scale on the right is that of a parcel lifted from the surface. Winds (m s^{-1}, degrees) on the left are environmental winds based on soundings behind the storm. The numbers n-2, n-1, n and n+1 indicate the different cells.
Second Fig: 2-minute average surface rainfall.

The speed of motion (usually derived from radar observation)) determines the duration of rainfall over any point. Slow moving storms may cause flooding.

6.1.4 The Supercell Storm

The supercell storm is a single giant quasi-steady thunderstorm cell. A supercell generally forms in an environment of strong low-level shear and curvature of wind shear with height. On the radar it may be seen as a solid mass of high reflectivity unlike in the multicell storm where several cells may be distinguished. The updraft insulated from dilution by environmental air, becomes so strong that the cloud particles are carried to a great height before they grow to a size detectable by radar. These precipitation particles fall elsewhere. Hence, a region of weak or no echo ($Z_e < 20$ dBZ or often < 5 dBZ) appears in the RHI or vertical section of the radar echo at low levels topped by a wider overhang of the reflectivity pattern at middle levels. This has been called a weak echo vault or Weak Echo Region (WER) (Browning and Donaldson 1963, Browning and Ludlam 1962). The vault has been explained (Browning 1965) as probably due to centrifugal evacuation of particles within a rotating column of air. If the WER is surrounded by precipitation echoes it is known as a Bounded Weak Echo Region (BWER). The presence of a WER or BWER has often been taken as an indication of a severe storm with likelihood of hail. Airflow in Weak Echo Regions of hailstorms have been studied by dropping chaff packets near the cloud base into the updraft, from aircraft (Marwitz 1973) and tracking them with radar. It was found that the inflow air has its origin near the surface. The inflow is typically negatively buoyant below cloud base and there exists a significant non-hydrostatic pressure perturbation in most storms. A vertical velocity maximum usually exists within the WER. Marwitz and Berry (1971) used a ground-based 10 cm radar and an airborne 3 cm radar while the aircraft measured the updrafts in the WER. The maximum updraft found by them was 18 m s^{-1} and the width of the WER narrowed from about 4 miles near cloud base to 2 miles at 16,000 feet. The radar echo was tilting considerably to the right of the mean environmental winds. The echo intensity above the WER reached 30 dBZ at 25,000 feet. Cases of WER observed in India were termed "dry holes" (Sharma 1965, Mathur and Kulshrestha 1966) but their association with severe weather was not clearly established.

The up - and downdrafts also develop an intertwined three-dimensional structure (Browning 1964). As a result, the updraft rotates cyclonically and a vortex known as a mesocyclone may develop and give rise to a tornado. The precipitation particles descending within the rotating downdraft may be seen on non-Doppler radar as a "hook" echo or occasionally as a spiral with an enclosed eye (Stout and Huff 1953, Fujita 1981) much like the tropical cyclone structure described in Chapter 8 but on a smaller scale. Fig. 6.2 shows a hook echo observed with an X-band radar at New Delhi. This echo was associated with a severe hailstorm which damaged a passenger aircraft. With operational non-Doppler radar the hook echo or other sharp protuberances in the radar echo known as "distinctive echoes"[6] (Forbes 1981) were looked for as the main indicators of hail- and tornado-producing storms. A Doppler radar observes a large horizontal shear or a couplet of opposite radial velocities as a vortex signature (see section 3.3 and discussion later in this chapter) at low levels and a divergence signature at the top.

---

[6] Distinctive echo means one which possesses a marked appendage on the right rear flank or has a comma shape or Line Echo Wave Pattern (LEWP, see Para 6.1.5) according to Forbes (1981).

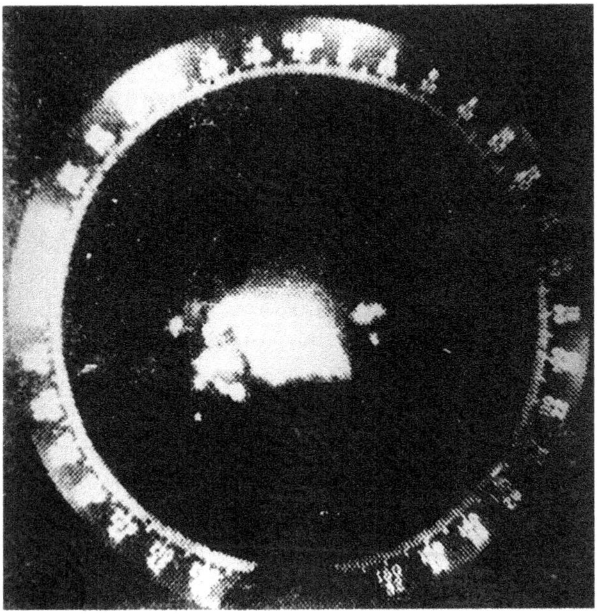

Fig. 6.2. Radar PPI picture of a hook echo from a hailstorm near Delhi (from Mull and Kulshrestha 1962; courtesy IMD).

In the USA, the following characteristics observable by radar, are considered sufficient to identify a thunderstorm as a supercell, according to Harris et al. (1994) who have proposed that some of these can be used in an algorithm for identifying severe storm structure.

(1) long life > 6 hours.
(2) quasi-steady state appearance.
(3) motion to right of mean mid-level environmental wind (this may not apply in other areas).
(4) a mesocyclone (see Para 6.1.8) observed in the velocity field.
(5) a BWER or WER at mid-levels in a region of very high reflectivity marked by strong reflectivity gradients, capped by a high reflectivity core aloft, and associated with the maximum echo top.
(6) a mid-level echo overhang extending above and beyond a low-level WER at the right-rear flank. The WER is marked by strong reflectivity gradients.
(7) very strong, precipitation-laden downdraft and resultant gust front, and
(8) frequently, a low-level hook-shaped echo.

Supercells can exist for several hours. Sometimes a supercell may split into two and one part move to the right and one to the left. Supercells can form singly or as part of a line (see below) or in a sequence, one spawning another.

6.1.5 Clusters and Squall Lines

Under favourable conditions thunderstorms occur in clusters or form along lines. Lines of thunderstorms with the width of the line small in comparison with the length are termed squall lines. A radar PPI picture of a squall line is shown in Fig. 4.1.They form along fronts or airmass boundaries or as a result of continuous generation of thunderstorms in convergence-divergence patterns produced by vertical transfer of horizontal momentum in pre-existing thunderstorms. While the line (or cluster) as a whole may move with the mean wind, the development of fresh cells preferentially at one end while others dissipate may change the curvature or orientation of the line or make it appear to move parallel to itself or at an angle to itself. The line may also elongate by addition of cells at one end. The line can also assume a wavy shape due to different acceleration in different parts of the line. This has been called a Line Echo Wave Pattern (LEWP) (Nolen 1959). According to Nolen, tornadoes often form in LEWP's. A radar study at Chennai (formerly Madras) (Raghavan 1962) showed that convective cells tend to form along lines, some of them wavy, and the lines have preferred orientations depending on the topography as well as the synoptic situation. The movement was with the mean upper winds and in most cases perpendicular to the orientation of the line. Many of the lines were not straight but concave or convex as seen by an observer towards whom the line was moving; and that concave lines yielded greater rainfall over the observing site than convex ones. A concave shape in a squall line has been linked to very strong surface winds in the USA (Hamilton 1970).

De and Sen (1961) and De (1963) studied squall lines in the pre-monsoon season over Gangetic West Bengal lasting 3 to 10 hours. These originated as isolated cells and after forming into lines of varying lengths (upto 300 km) moved at speeds upto 60 kmph, often faster than the 700 hPa wind. The direction of movement was up to 90° to the right of that wind. In the same area, Chaudhry and Rakshit (1970) found squall lines moving within 40° clockwise of the wind at 4.5 km level. Ray and De (1971) studied 60 pre-monsoon season squall lines near Guwahati. These had an average height of 9 km, duration of 3 to 4 hours, length of 45 n. miles and moved roughly with the 700 hPa wind.

Boucher and Wexler (1961) considered precipitation lines including squall lines in several geographical regions. They found that the lines are invariably oriented in a direction clockwise from the 700 hPa wind direction. The motion is better correlated with the 700 hPa wind than with winds at other levels. The lines are not necessarily associated with fronts. Based on these observations, Boucher and Wexler developed a regression line connecting the 700 hPa wind component normal to the line orientation and the average line speed in knots. They suggested a motion forecast procedure using the known 700 hPa wind and the regression line.

Pestaina-Haynes and Austin (1976) compared maritime tropical lines (GATE[7] and Barbados) with continental mid-latitude (Montreal) precipitation lines. Let us refer to these as G, B and M lines, respectively. From radar and synoptic data, they found that M and B lines have similar orientations but travel in opposite directions in accordance with the winds at their locations. The G lines which lie in the zone of convergence of the trades are oriented east-west and hardly have any motion perpendicular to their orientation. The M and G lines were longer than B lines. G and B lines have greater number of smaller cells than M lines. In B and M, orientation of the line is at a small angle counter-clockwise from the steering wind. They noted that motion is usually to the right of the steering wind because of preferential formation of new cells on the right rear flank (propagation). They found the steering levels to be respectively 1000, 850 and 700 hPa for G, B and M.

In Australia, radar echoes near Melbourne were studied by Clarke (1988) who found that the echoes are often arranged along lines. Individual echoes moved always to the right of the direction of propagation of the line. Echoes moved with speeds of 2 to 20 m s^{-1} with the mean wind over part of the cloud.

The properties of squall lines thus depend on local climatology. This is best studied by radar at each location of interest. Squall lines can exist for several hours and continue after the initial triggering mechanism ceases to exist, because of the non-hydrostatic pressure field induced by relative motions near the boundaries of a large convective system. They are important because they can produce heavy rain and severe weather for long periods and can be a serious hazard for aviation. Structure of squall lines and clusters is further discussed in section 6.2.

6.1.6 Mesoscale Convective System (MCS)

A generic term proposed by Zipser (1982) for any group of interacting convective elements is the "Mesoscale Convective System" (MCS). This includes the stratiform region which develops in association with convective cells (see section 6.2). Long-lived convective systems in mid-latitudes have been called Mesoscale Convective Complexes (MCC) (Maddox 1980).

---

[7] GATE = Global atmospheric research program (GARP) Atlantic Tropical Experiment, an international experiment carried out in the tropical north Atlantic in 1974.

6.1.7 Hailstorms

Hail grows in severe convective storms by repeated deposition of water substance on ice particles and being supported by updrafts in thunderstorms, can grow to large sizes. The typical trajectory of a hailstone in a storm is shown in Fig. 6.1. However spatial separation of hail and rain has been reported in some storms (Ulbrich and Atlas 1982).

Hail usually consists of layers of ice and liquid water. Air as well as other substances (particulate matter and even insects) may be trapped inside. Such particles with water and ice mixed have been called spongy-ice particles. Scattering by hail was discussed in Chapter 2. The operational need to identify hail is important because of the enormous damage that hail can cause. Detection of hail enables warnings for aviation and agriculture and even the possibility of suppressing hail (Chapter 11). Presence of hail can also lead to large errors in radar rainfall estimates (Chapter 7).

A qualitative indication of a severe storm is the formation of a hook echo (Fig. 6.2) or distinctive echo but it is not a necessary accompaniment of hail formation. Donaldson (1958, 1959, 1961, 1965) found a high correlation between frequency of hail occurrence (verified on the ground by observers), on the one hand, and radar echo height and observed Z_e on the other. Waldvogel et al. (1979) using 3 cm radar in Switzerland, found that if the height of the 45 dBZ contour exceeded the freezing level by at least 1.4 km, it was a sure indication of hail aloft although not all cases were followed by hail at the ground. According to Geotis (1963), Z_e exceeding 55 dBZ continuously for several minutes is a definite indication of hail in New England (USA). In South Africa, Mather et al. (1976) used echo top heights greater than 12 km and maximum Z_e greater than 45 dBZ to identify clouds to be seeded for hail suppression. In the National Hail Research Experiment (NHRE) in the USA, reflectivity factor of 45 dBZ at S-band between the −5 and −30°C levels was taken as an indicator of hail probability (Federer 1977). The height criterion is not widely accepted, as several authors found that hail occurrence was not very sensitive to echo top height (e.g. Kulshrestha and Jain 1967, Misra and Prasad 1980). The difference in Z_e at low altitude between severe and non-severe storms is slight. It is more significant at the middle and upper portions of the storm. A peak was observed by Donaldson in severe storms at about 20,000 feet. Maximum Z_e value in a hailstorm was 76 dBZ. That is, the vertical profile of Z_e (VPR) rather than the value at one level was a better indicator of hail. This is probably evidence of storage of hail near the altitude of the updraft maximum. This finding led later to the concept of high VIL[8] and rapid increase

[8] See Chapter 2 for definition of VIL.

of VIL with time as indicative of hail (Greene 1972, Greene and Clark 1972). With digital radar, rapid 3-D processing enables a VIL map to be produced in near real time (section 3.6). The formation of a WER or BWER is indicative of intense updraft and hence likelihood of hail and possible tornado formation. Donaldson (1965) gave several criteria for recognition of a severe storm and hail from the reflectivity pattern. Lemon (1978) found (with non-Doppler radar) that the following criteria predicted hail successfully in the USA.
(1) Peak Mid-level (5 to 12 km above ground level) $Z_e \geq 45$ dB.
(2) Mid-level echo overhang extending at least 6 km beyond the outer edge of (or beyond the strongest Z_e gradient of) the low level (≤ 1.5 km above ground) echo.
(3) The highest echo top located on the storm flank that possesses the overhang and lies above the low level Z_e gradient between the echo core and echo edge or lies above the overhang itself.

As the hail melts, when it comes down, a vertical profile of reflectivity (VPR) can be found. Schmid and Waldvogel (1986) using 10 cm radar in Switzerland found maximum Z_e at an altitude of $h_0 - 1$ km where h_0 is the 0° isotherm level. Z_e varied within ±1 dB between h_0 and $h_0 - 1.5$ km. Above h_0, Z_e decreased with height at a rate of −1 to −2 dB km^{-1}. Below $h_0 - 1$ km, Z_e decreased at a rate of about −3 dB km^{-1}.

Such criteria may be incorporated in algorithms for hail detection. A Severe Weather Probability (SWP)[9] algorithm has been developed in the USA, consisting of a magnitude-weighted summation of VIL values in a grid and is claimed to work well in Oklahoma (Winston and Ruthi 1986). The Waldvogel et al. (1979) criterion mentioned above, has also been used in an algorithm. A one-dimensional numerical model of a cumulus cloud was developed by Danielson et al. (1972). They claim that it generates hail and radar reflectivities in a realistic manner. They calculated reflectivity from Mie theory for water and ice spheres for comparison with radar-measured values.

The correlations of echo top height, reflectivity factor etc. with hail occurrence are generally based on observation of hail at the ground. Hence, these relationships could be vitiated by the possibility of melting of the hail after its observation by radar and before it reaches the ground. Also, as pointed out by Sulakvelidze (1968), the relationships vary with geographical conditions and from season to season at the same place.

Since the kinetic energy of falling hail is important from the damage point of view, Waldvogel et al. (1978, 1978b) derived relationships between Z_e and kinetic energy flux E (expressed in J m^{-2} s^{-1}). Their empirical relationship was of the form,

[9] Severe weather is defined in this context as presence of hail 3/4 inch diameter or larger and/or wind gusts of 50 knots or more (Greene and Clark 1972).

$$E = 5.0 \times 10^{-6} \times Z_e^{0.840} \tag{6.1}$$

They also obtained the relation between Z_e and hailfall rate H mm h^{-1} as

$$H = 6.53 \times 10^{-4} \times Z_e^{0.747} \tag{6.2}$$

They used these relations to compute E and H from 10 cm radar observations. They integrated the area having Z_e equal to or above a specified value Z_x over a period of time. This quantity was called "Area of Z Time Integrated" AZTI(Z_x) and expressed in km^2 min. According to them, if AZTI(55) exceeds 100 km^2 min, hail at the ground occurs.

With Doppler radar, the signature of upper level divergence can be detected (see Chapter 3). If the divergence pattern is axisymmetric, the magnitude of the divergence can be equated to the updraft speed and the potential for hail formation. Witt and Nelson (1991) developed a technique for estimating probable maximum hailstone size. From the single Doppler divergence signature at the storm top, the maximum Doppler velocities towards and away from the radar are noted. The algebraic difference between the two gives the magnitude of the divergent outflow. This is a measure of the maximum vertical velocity. To take account of the area of updraft also, they defined a "High Radial Shear Area" (HRSA) which could be computed. They found good correlation between these quantities and the maximum hail diameter. Based on this technique, as well as on high Z_e and high VIL, automated hail detection algorithms have been devised.

Also, as discussed in chapter 2, hail echo gives a high Z_e and a low Z_{DR} on dual polarisation radar. Figs. 3.29 and 3.30 show RHI scans of Z_H and Z_{DR} respectively in a hailstorm. At the intense portion of the storm (range 34 km) Z_H shows a maximum. The corresponding Z_{DR} values are slightly negative or near zero all the way to the ground indicating hail reaching the surface. Aydin et al. (1986, 1990) defined a hail signal (H_{DR}) using the properties of hail (isotropy and high Z_e) and rain (anisotropy). They noted that,

(1) In rain Z_{DR} is always positive (up to 4 dB generally).

(2) In hail Z_{DR} is around 0 dB in contrast to large positive values in the surrounding rain, while Z_H(or Z_V) increase or maintain a high value in the hail region. H_{DR} is defined as

$$H_{DR} = Z_H - f(Z_{DR}) \tag{6.3}$$

where the function $f(Z_{DR})$ is, according to them,

$$\begin{aligned} f(Z_{DR}) &= 27, && \text{if } Z_{DR} \leq 0 \text{ dB} \\ &= 19Z_{DR} + 27, && \text{if } 0 \leq Z_{DR} \leq 1.74 \text{ dB} \\ &= 60, && \text{if } Z_{DR} > 1.74 \text{ dB} \end{aligned} \tag{6.4}$$

$H_{DR} > 0$ indicates hail.

The larger the H_{DR}, the greater the certainty that the high Z_e is not due to raindrops. At low elevation angles below melting level, H_{DR} is a good indicator of the presence of hail. Brandes and Vivekanandan (1998) have proposed an algorithm based on H_{DR} for hail detection and estimation of hailstone size[10]. Since K_{DP} is also low for hail, the combination of Z_H and Z_{DR} or Z_H and K_{DP} can be used for distinguishing rain and hail rates (Balakrishnan and Zrnić 1990). A combination of polarimetric radar and wind profilers have been used by May et al. (2001) to deduce the presence and size of hail.

For hail of 1 to 4 cm diameter, Z_e at 10 cm is higher than at 3 cm. (Hence, a dual wavelength radar can be used to compute Z_{10}/Z_3. However Z_3 can be lower due to attenuation in heavy rain without hail being present. Alternatively Z_{10} and Z_5 can be used. Eccles and Atlas (1973)[11] suggested the use of the range derivative (d*y*/d*r*) of the dB difference *y* between Z_{10} and Z_3. According to them, a negative (d*y*/d*r*) is an indicator of the far edge of a hailshaft and a large positive (d*y*/d*r*) may be interpreted as the near edge of a hailshaft. However, this may not be true if smaller hail is present. Jameson and Srivastava (1978) found that negative d*y*/d*r* can occur within the hail region due to changes in hail diameter or the thickness of water coat of the hail. Thus, according to them, negative d*y*/d*r* indicates the presence of hail but may not be able to establish the boundaries of the hail region. Moreover, dual wavelength systems do not lend themselves to operational use because of the need for two radar systems and the requirement of accurately matching the two beams.

6.1.8 Mesocyclone and tornado

The mesocyclone is a small scale vortex which may (or may not) contain one or more tornadoes. Fig. 6.3 (from Ray 1986) shows a single Doppler mesocyclone signature with an embedded tornado. The isodop pattern of a Rankine vortex was discussed in Chapter 3. The core diameter of a mesocyclone is about 5 km. Hence, the vortex signature (couplet of opposite Doppler velocities) will be detected only if the pulse volume is smaller than this. This is one of the limitations to the range up to which a mesocyclone is detected. Early treatments of detection of the mesocyclone vortex may be seen in Burgess (1976), Donaldson (1970), Hennington and Burgess (1981) among others.

---

[10] According to Brandes and Vivekanandan, Z_H is not a good indicator of hail because Mie scattering occurs.

[11] Ulbrich and Atlas (1982) have published a "Hail Parameter Diagram" similar to the Rain Parameter Diagram discussed in Chapter 7.

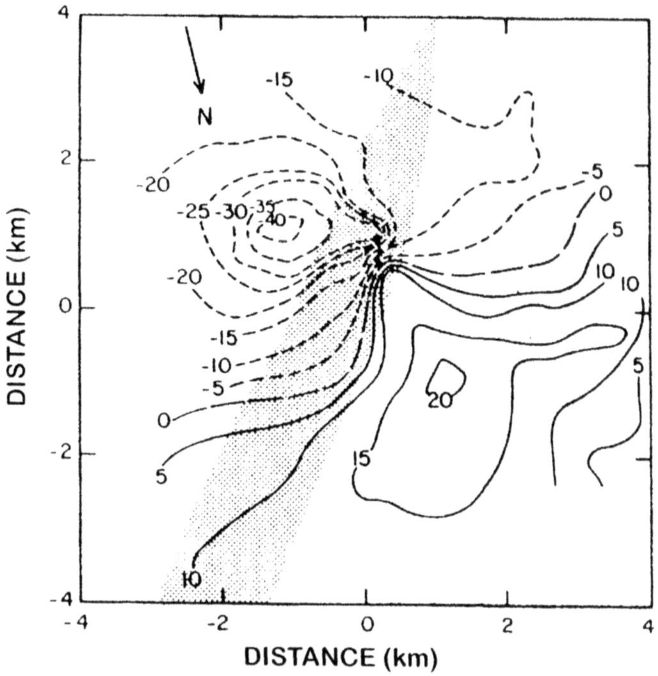

Fig. 6.3. Single-Doppler mesocyclone signature with an embedded high shear location (TVS) indicating a tornado. (from Ray et al. 1981; courtesy AMS). The mesocyclone centre is at the centre of the grid. Doppler velocities (m s^{-1}) are positive (negative) for flow away from (towards) the radar.

Criteria for mesocyclone detection were specified by these authors. The principal ones were:

(a) Significant azimuthal shear (i.e. the change in Doppler velocity at the same range with change of azimuth) (at least 0.005 s^{-1}) must exist between closed isodops of opposite sign after removing storm motion.

(b) Shear pattern should extend vertically over a height interval (say 3 km) comparable to the horizontal diameter of the vortex and to persist for a time duration greater than half the period of revolution of the feature.

It is necessary to scan at several elevations and this has to be done fast enough as the phenomenon evolves rapidly. This is best done in an automated volume scan. Hennington and Burgess (1981) proposed a pattern recognition technique which identifies a consistent increase (or decrease) of Doppler velocity with azimuth while scanning. The azimuth locations where the sign of the first space derivative of the velocity field changes (at a constant range) are the end points of a "pattern vector" representing a shear segment. A number of such vectors exist, one at each range gate over the mesocyclone (thick nearly horizontal lines[12] in Fig. 6.4). The outline of the pattern vectors across the centre line of the mesocyclone is called the pattern vector envelope. The shape shown in the figure is the signature of the mesocyclone. If the difference between the beginning and ending velocities is D_v and the distance between them is D_d, D_v/D_d (expressed in m s^{-1} km^{-1}) is the azimuthal shear and $D_v \times D_d$ is the angular momentum (m s^{-1} km). Only those pattern vectors are retained, in which these quantities exceed certain preset thresholds. A Z_e threshold is also established to reject data from very weak reflectivity regions. The beginning and ending azimuths and velocities, the range and the time corresponding to the beginning azimuth in each pattern vector are saved. An example of an azimuth-range plot (B-scan) of the Doppler velocities in a mesocyclone is shown in Fig. 6.5 (from Wieler 1986)[13]. The heavy lines in the figure are the beginning and ending points of the pattern vectors. The circled values are the extreme velocities. Scan at one elevation produces a two-dimensional feature consisting of an ensemble of pattern vectors. Scans at several elevations with checks for correlations of their locations and

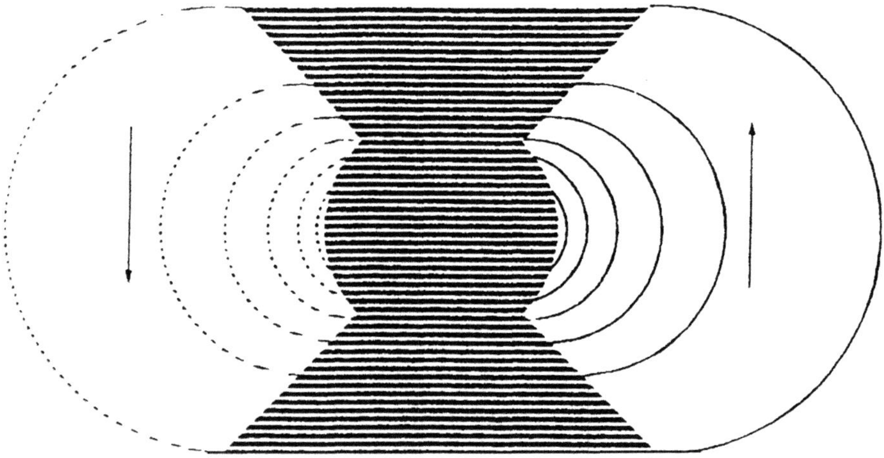

Fig. 6.4. An example of pattern vectors (thick nearly horizontal lines) in a mesocyclone and the pattern vector envelope. The thin lines are isodops. The direction of rotation is indicated by the arrows. (from Desrochers and Donaldson 1992; courtesy AMS).

[12] The radar is assumed to be south of Fig. 6.4, the range being large in comparison with the vortex diameter. That is why the pattern vectors appear as nearly straight lines.

[13] The concept of pattern vectors can be used for detection of the BWER, an indication of a supercell thunderstorm (Harris et al. 1994). The BWER appears on a PPI as a ring or ridge of high reflectivity surrounding a distinct minimum. Hence pattern vectors can be constructed from a B-scan of the reflectivity field (not the Doppler velocity field) in a manner similar to the mesocyclone case.

AZIMUTH

RANGE		325		326		327		328		329		330	
51.60	-14	-17	-22	-22	-13	-7	1	6	13	20	24	23	23
51.45	-15	-18	-22	-20	-14	-5	3	11	18	22	24	24	23
51.3	-17	-22	-26	-20	- 8	-1	4	13	17	23	25	25	23
51.15	-19	-25	-28	-19	- 6	0	5	12	18	24	26	25	23
51.00	-18	-26	-32	-21	- 6	2	8	14	20	27	28	26	25
50.85	-18	-27	-35	-26	-10	2	8	12	20	28	30	27	24
50.70	-18	-28	-32	-25	- 6	5	8	12	21	29	31	27	24
50.55	-14	-23	-31	-25	- 2	8	9	12	23	32	31	27	24
50.40	-13	-23	-29	-21	- 2	8	10	13	24	31	29	27	24
50.25	-10	-24	-24	-15	- 4	4	9	13	26	31	29	26	24
50.10	-11	-19	-16	-12	- 7	0	7	14	27	30	27	26	23
49.95	- 3	-17	-15	-10	- 6	0	8	15	27	29	27	25	23
49.80	- 2	-18	-17	-12	- 9	-2	8	18	28	28	26	24	22
49.65	- 5	-16	-17	-15	-11	-4	6	19	28	28	26	25	23
49.50	- 8	-17	-19	-17	-11	-5	9	23	28	26	24	24	23
49.35	- 5	-16	-17	-18	-14	0	14	23	28	25	24	23	21
49.20	- 5	-16	-20	-20	-14	7	15	24	28	25	22	23	21
49.05	-14	-19	-21	-19	- 7	4	12	23	27	23	21	21	20

Fig. 6.5. Azimuth-range plot of the Doppler velocities in a mesocyclone (from Wieler 1986; courtesy AMS). Range is in km. Range gates are at 150 m intervals. The heavy lines in the figure are the beginning and ending points of the pattern vectors. The circled values are extreme positive and negative velocities (m s^{-1}). The highlighted values represent one pattern vector.

dimensions to see if the features seen at different elevations represent the same mesocyclone, produce a three-dimensional feature. Algorithms based on these considerations have been developed by Wieler (1986) and Zrnić et al. (1985b) among others.

Because of the small size of the phenomenon, it may happen that random positions of the radar beam relative to the mesocyclone (or tornado) centre may cause spurious indication of changes in magnitudes and azimuthal separation of Doppler velocity peaks (Wood and Brown 1997). Also, as the mesocyclone changes range relative to the radar, the deduced core diameter and mean rotational velocity of the mesocyclone signature oscillate back and forth, even though the physical width of the radar beam changes uniformly with range (Wood and Brown 2000).

In some mesocyclones, the rotation can be anti-cyclonic. The ability to detect such meso-anticyclones should be included in the algorithms (Wieler 1986).

Donaldson et al., have, in a series of papers (Desrochers and Donaldson 1992, Desrochers et al. 1986, Donaldson and Desrochers 1990, 1991) developed the concept of Rotational Kinetic Energy (RKE) and Excess Rotational Kinetic Energy (ERKE) in mesocyclones. For a rotating cylinder of mass m, density β and radius r rotating about its central axis with a maximum rotational velocity v, the moment of inertia is $I = mr^2/2$, the mass per unit height is $m = \beta\pi r^2$ and the angular velocity is v/r. Hence RKE is,

$$\text{RKE Joules m}^{-1} = \frac{Iv^2}{2r^2} = \frac{\beta\pi r^2 v^2}{4} \tag{6.5}$$

For a mesocyclone, r and v can be obtained from the search for the couplet of velocity maxima at each elevation. The radius is half the distance between the velocity maxima and the rotational velocity at the vortex perimeter is half the algebraic difference between the velocity maxima multiplied by cos ϕ. (ϕ is the angle between a line joining the velocity maxima and a constant-range line and indicates the superposition of divergent or convergent components over pure rotation). To incorporate the condition of a minimum shear S_m (usually taken as 0.005 s^{-1} as mentioned earlier) as the threshold value to qualify as a mesocyclone, these authors introduced the concept of Excess Rotational Kinetic Energy (ERKE). Since the shear is the ratio of v to r, the rotational velocity v_m required to give a shear S_m for a given r is rS_m. If this is subtracted from the actual rotational velocity v, we get the excess $(v - rS_m)$ over that needed to maintain mesocyclonic shear. Using $(v - rS_m)$ instead of v in equation (6.5) gives the ERKE.

$$\text{ERKE} = \frac{\beta\pi r^2 (v - rS_m)^2}{4} \tag{6.6}$$

If $(v - rS_m)$ is negative, ERKE is taken as zero. They defined an average mesocyclone having an ERKE of 540 MJ m^{-1} as a Climatological Mature Mesocyclone (CMM) and expressed ERKE of other mesocyclones in units of CMM. This is called two-dimensional ERKE. Three-dimensional ERKE is obtained by integration over a vertical column (above 4 km altitude) using data from a volume scan. The 3-D ERKE was found to be a good measure of mesocyclone intensity and predictor of tornado and other severe weather probability. They developed on this basis, a mesocyclone-detection and tornado-prediction algorithm.

The tornado is a fast-rotating misocyclone in the Fujita classification. It can sometimes be anti-cyclonic. It may be visible as a funnel cloud hanging from the parent cumulonimbus but even this may be hidden in heavy rain. Hence, the detection of the tornado has long been a problem although its likelihood may be predicted from the parent thunderstorm which can be seen on radar. The presence of a hook echo has been taken as evidence of a rotating updraft and hence of the likelihood of a tornado. Fig. 3.2 (from Zrnić et al. 1985) is a Doppler spectral display relating to a tornado. The 7th range gate from the bottom exhibits the two velocity maxima and is centred on the tornado. A double-peak symmetric with respect to the mean wind spectrum is called the Tornado Spectral Signature (TSS) (Zrnić and Istok 1980). However, there is an asymmetry in this spectrum indicating that targets are being centrifuged outwards (Zrnić et al. 1977). The neighbouring range gates in Fig. 3.2 show only a widening of the spectrum. As a tornado is very small, it may often be entirely contained in the sampling volume of the radar except at close range. In such a case the vortex signature will not be seen on the isodop pattern. The couplet of opposite velocities around the vortex will be averaged and a near-zero mean Doppler velocity will be indicated. The spectral width will be widened. The radar is more likely to see (as discussed above) the circulation associated with the parent mesocyclone which is larger in extent but may indicate lower peak velocity than in the tornado. Wind in a tornado can probably go up to 142 m s^{-1} (Fujita 1981), although such high winds have not been measured because of the difficulty of operating anemometers in extreme weather. Hence, with most radars there will also be a serious problem of velocity aliasing. A concept of a Tornadic Vortex Signature (TVS) (Brown et al. 1978) was formulated with the following guidelines:

1. An azimuthal shear of at least 20 m s^{-1} over an azimuthal distance of about one beam width.
2. Signature with extreme Doppler velocity values of opposite sign, after TVS translation has been removed.
3. Anomalous shear region not more than about 1 km in range extent (otherwise it would indicate a shear line rather than a small-scale vortex).
4. Shear region at least several kilometres in vertical extent.
5. Persistent anomalous shear region at the same general heights for about 10 minutes or more.

They found that the TVS originated at storm mid-levels and took about 25 minutes to work its way to the ground. During this time the signature magnitude increased at all detectable heights until the tornado reached its maximum size. When the tornado was shrinking or dissipating, the magnitude decreased at all heights. Based on these principles, algorithms for tornado detection have been devised (see also Chapter 9). These appear to have been successful in many cases although not all tornadoes are accompanied by a TVS. A comparison between non-Doppler and Doppler radar forecasts of tornadoes in the Joint Doppler Operational Project in the USA showed (Ray 1986) a distinct improvement with Doppler, in the Probability of Detection (POD) and Critical Success Index (CSI) and to some extent in the lead time (LT) and reduction in the False Alarm Rate (FAR)[14].

Detection of the parent mesocyclone, as noted earlier, will provide a longer lead time for tornado prediction. However, not all mesocyclones produce tornadoes and not all tornadoes are produced in mesocyclones. For example, tornadoes can form along gust fronts or in downbursts. Tornadoes are most common in the USA and have been extensively studied using dual or triple Doppler radars (e.g. see Ray et al. 1981, besides the above references). A case study in Japan using Doppler radar and other means may be seen in Kobayashi et al. (1996). In India tornadoes occur in the northeastern states and rarely in other areas[15]. Radar studies [Gupta and Ghosh (1980), Mukherjee and Bhattacharya (1972), Shravan Kumar and Sen Sarma (1970)] are few and only qualitative.

6.1.9 Gust Front

A dome of cold air and therefore high pressure is formed near the base of a group or line of thunderstorms. The pressure rise is partly due to the cold air and partly due to the deceleration of the downdraft. The leading edge of the mesoscale pressure dome followed by a surge of gusty winds on or near the ground is known as the gust front. The front is formed by the spread of the cold downdraft air from the thunderstorm undercutting warmer environmental air. As the thunderstorm travels, the gust front is most prominent and moving fastest ahead of the storm. Air motion associated with the front is turbulent and is shown in Fig. 6.6 (after Wakimoto 1982). In this figure, produced using a Doppler radar, Wakimoto identified for the first time the "precipitation roll" formed by airflow deflected upward by ground friction.

---

[14] Probability of Detection (POD) = $X/(X + Z)$, False Alarm Rate (FAR) = $Y/(X + Y)$, Critical Success Index (CSI) = $X/(X + Y + Z)$, where X = Number of forecasts of severe events that occur, Y = Number of forecasts of severe events that do not occur, Z = Number of forecasts of non-severe events that occur severe. Lead Time (LT) = time between the advisory/warning and the event.

[15] Climatology of tornadoes in and around India may be seen in Saha (1966, 1971) and Peterson and Mehta (1981). Saha estimated a pressure deficit of 100 hPa corresponding to tangential wind velocity of 184 knots in an Indian tornado.

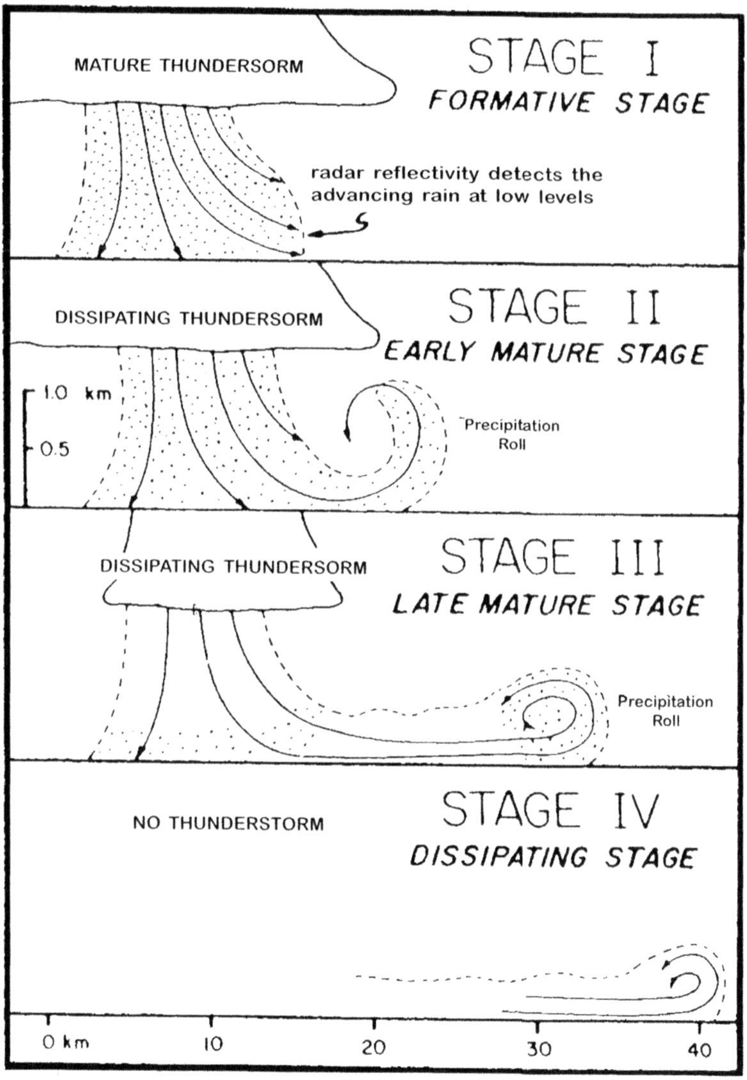

Fig. 6.6. Vertical section of a gust front in four stages (from Wakimoto 1982; courtesy AMS)

An aircraft encountering the front will experience strong horizontal shear. i.e. a sudden increase in wind speed often accompanied by a large change in direction. The front can be present even after the thunderstorm has dissipated and can travel (under the influence of the pressure difference) several km ahead of the storm and may have little or no rainfall associated with it. The front generally extends to a height of about 1 km but can sometimes go up to 4 km. Radar may detect echoes from the particles in the "precipitation roll". The radar also detects the weak echo (typically < 10 dBZ) which results from the refractive index discontinuity in the front, as a thin line, the movement of which can be tracked. An example is shown in Fig. 4.9. Doppler radar can monitor the broadened spectrum width of these velocities, which results from the turbulence associated with the gust front. Using Doppler radar and in situ observations, Fulton et al. (1986) found that the interaction of the outflow air with stable ambient air produces a series of gravity waves propagating ahead of the outflow.

A gust front associated with squall lines is likely to move faster and farther from the thunderstorm than in the case of single cells. The gust front may be as much as 30 km ahead of the storm (Joseph et al. 1980). A "nose" forms in front as the air near the ground moves more slowly (due to friction) than the air above. It is important to note that precipitation at the ground is not necessary for the gust front to be formed. Thus in andhis (see Para 6.1.12 below) which are thunderstorms without rain reaching the ground, dust and sand are carried up with the gust front.

When gust fronts from different groups of thunderstorms collide, strong updrafts are produced and fresh convective clouds develop (Wilson and Schreiber1986, Mahoney 1988). Wilson and Mueller (1993) proposed a means of nowcasting thunderstorm initiation by observation of clear air echoes from convergence lines (see section 6.4). Radar detection of the gust front is further discussed in Chapter 9 in the context of aviation.

6.1.10 The Downburst

In an investigation of tornadoes in the USA, Fujita (1981, 1985) found that trees were damaged in a "starburst" pattern as though the air had diverged in all directions radially from a central point. From this and later studies the downburst was discovered and so named by Fujita. The downburst is a strong downdraft which induces an outburst of damaging diverging winds at the surface. When the outburst wind is over 4 km in dimension it is called a macroburst. When it is less than 4 km (i.e. the misoscale) it is known as a microburst. Operationally, a microburst is defined as one having a divergent wind of at least 10 m s^{-1} (20 knots) over a distance of 4 km. The shear is 0.0025 s^{-1}. The word "micro" in the latter term should not be taken to mean that it is less intense[16]. On the contrary it can produce winds up to 75 m s^{-1} (146 knots) and the shear can be over 100 knots between the two ends of a runway; hence it is a serious aviation hazard. A simple model of a microburst is shown in Fig. 6.7. Note the asymmetry in the outflow. The winds may be either straight or curved. Twisting

[16] The main difference between a microburst and a gust front is one of scale. The former is one to two orders of magnitude smaller than the latter.

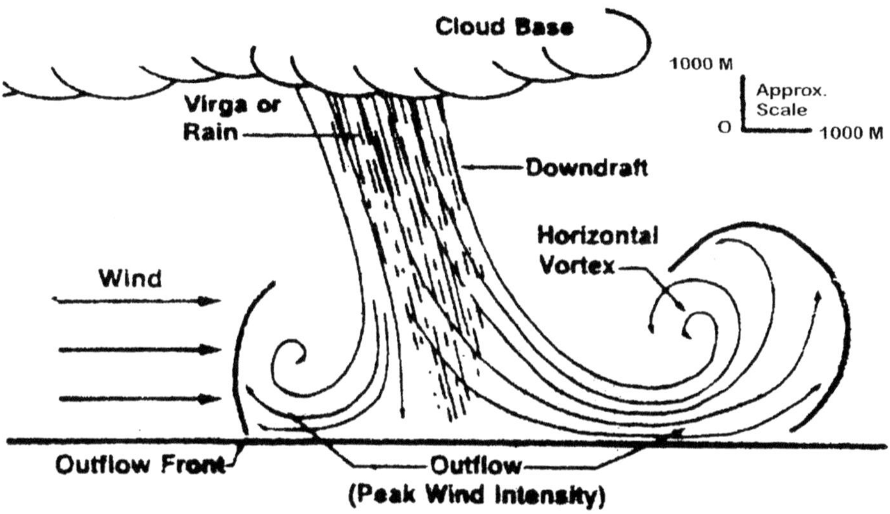

Fig. 6.7. A simple model of a microburst (courtesy Dr. J. McCarthy, Aviation Weather Associates, Inc., USA)

downbursts characterised by curved streamlines are often induced by a rotating thunderstorm. On the RHI of a non-Doppler radar a vertical shaft of descending virga and a splashing echo near the ground can be seen. On a Doppler radar, the downward velocity as well as the diverging winds near the surface can be observed (Fig. 3.13).

Using Doppler radar, Rinehart et al. (1993) measured the strengths of the microburst divergence (i.e., the volumetric rate of flow per unit depth expressed in $m^2\ s^{-1}$) in 908 microbursts in Florida and found the frequency distribution to be exponential. They also measured the rotation angle and found that a majority of microbursts rotated cyclonically but a signifcant number rotated anti-cyclonically. (The proportion of cyclonic rotation will go up at higher latitudes because of the increase in the Coriolis force). Thus, rotation is common among microbursts but the tangential velocity is small in comparison to the radial (diverging) velocity.

Roberts and Wilson (1989) identified four radar features as usable for predicting microbursts.

(1) a descending reflectivity core
(2) increasing convergence within and near cloud base
(3) a reflectivity notch
(4) rotation

Using a dual-polarisation radar, Wakimoto and Bringi (1988) found that a near-zero Z_{DR} core descended with a microburst-producing downdraft. The minimum Z_{DR} surrounded by higher values indicates the presence of ice particles which have had less time to melt than the surrounding particles. This is interpreted as a downdraft. Doppler velocities in a flare echo, if observed, also provide a measure of particle fall speeds in the downdraft (as noted in Chapter 4) and may help to identify a microburst.

Microbursts can occur even in association with showers from large cumulus; a thunderstorm is not necessary. They are termed wet or dry depending on whether the rain reaches the ground or not. In the Joint Airport Weather Studies (JAWS) Project in 1982 in the USA (McCarthy et al. 1982), 186 microbursts were observed over 86 days and of these 83% were dry (Fujita 1985). A dry microburst can come down even from innocent-looking virga. Dry microbursts can escape detection by conventional (i.e. non-Doppler) weather radar as the absence of precipitation may result in only weak echoes from the refractive index changes. Microbursts can also raise dust, reducing visibility. Microbursts have been observed visually and been photographed. (Caracena et al., 1989). Their structure has been inferred also from the pattern of the damage swaths on the ground.

Downbursts can occur in clusters or along lines. Their outflows may therefore interact and produce complicated Doppler velocity patterns. A phenomenon known as "derecho" consisting of a cluster of downbursts producing damaging winds has been identified in the USA (Johns and Hirt, 1987) as occurring within extratropical mesoscale convective complexes. A derecho often exhibits a bow echo on radar (see Paragraph 6.1.11).

The Andhis of North India are prime candidates as microburst producers. Downburst hazard to aircraft is discussed in Chapter 9.

6.1.11 The Bow echo

A type of radar echo associated with downbursts was identified by Fujita. He called it the bow echo and described it (Fujita 1981) as a bow-shaped mesoscale echo, a potential inducer of microbursts. The following description is based on his Doppler and non-Doppler analyses of a number of cases. "The initial stage of a bow echo is a tall thunderstorm with a gust front located along the leading edge of the activity line (stage A in Fig. 6.8). A downburst descends (stage B) from the tall echo, distorting the line echo into a bow-shaped bulge. The downburst intensifies at the location of the arrow attached to a bow. When the downburst reaches its mature intensity, the bow echo takes the shape of a 'spearhead', accompanied frequently by a channel (or trench) of weak echo located along the center axis of the downburst flow. Downburst-induced tornadoes are likely to form at this stage. As the downburst weakens, after

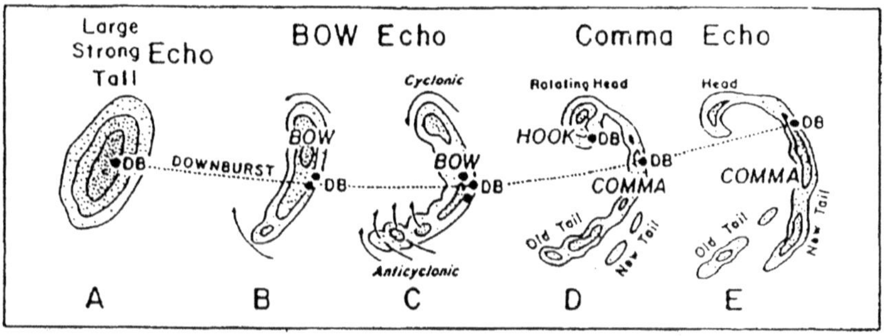

Fig. 6.8. Fujita's model of an isolated bow echo associated with a downburst. (from Johns 1993; courtesy AMS).

pushing ahead of the parent bow echo, a mesoscale circulation dominates the area of the bow echo. Then a rotating head with an appearance of a hook echo forms near the center of the mesoscale circulation. The final stage of bow echo often turns into a comma-shaped echo which gradually disappears along with the weakening mesoscale circulation". Lee et al. (1992) divide bow echoes into two categories – squall-line-type bow echo (SLBE) and single-cell-type bow echo (CBE). In the case of an SLBE, inflow from the rear (rear-inflow jet) forces the squall line to form a bulge. There is also entrainment of dry air in the rear, which results in a concavity, because of erosion of the backside of the echo by evaporation. The meteorological conditions leading to formation of a bow echo in the USA are discussed by Johns (1993). A bow echo is usually but not necessarily the precursor of a microburst. It is of operational importance because the weak echo channel is associated with extreme winds.

6.1.12 The Andhi

The andhi is a type of thunderstorm common in north India in the dry pre-monsoon months. Because of the extreme dryness of the environment the precipitation particles completely evaporate before reaching the ground. Although there is no rain at the ground, the downdraft, squall and gust front are present. Because there is a deep layer of dry adiabatic lapse rate on such an occasion, there is strong acceleration of the downdraft, which has warmed moist adiabatically for much of its descent. The wind raises dust which seriously affects visibility up to a considerable altitude. A model of the andhi[17] has been

[17] The andhi is usually also referred to as "dust storm" in India. This should be distinguished from "dust raising winds" which are due to strong pressure gradients not associated with thunderstorms. The latter are also referred to as "dust storms" in some countries.

formulated by Joseph (1982). The dust itself scatters very little radiation at the radar wavelengths used, but the precipitation echo aloft can be seen. Figures 6.9 and 6.10 respectively are PPI and RHI pictures of an andhi taken with a 3 cm radar at New Delhi. Similar dry thunderstorms have been reported from the USA e.g. by Krumm (1954).

6.1.13 Low Precipitation Severe Storms

A dryline which is a boundary between moist and dry winds is an important location for convective development (see Weston 1972, regarding the dryline in northern India). In the USA, it is a well-documented phenomenon (Schaefer 1986). It can often be seen as a thin line on radar because of the humidity gradient. The region near the dryline generates thunderstorms. In Oklahoma, USA, near the dryline, a class of thunderstorms known as low precipitation severe storms has been identified by Bluestein and Parks (1983). Because of little precipitation, they are often difficult to detect by radar. A dual-Doppler radar study of such a storm, made by Bluestein and Woodall (1990) showed an absence of a strong downdraft and the presence of long-lived rotating updrafts and of large hail which gave a radar echo.

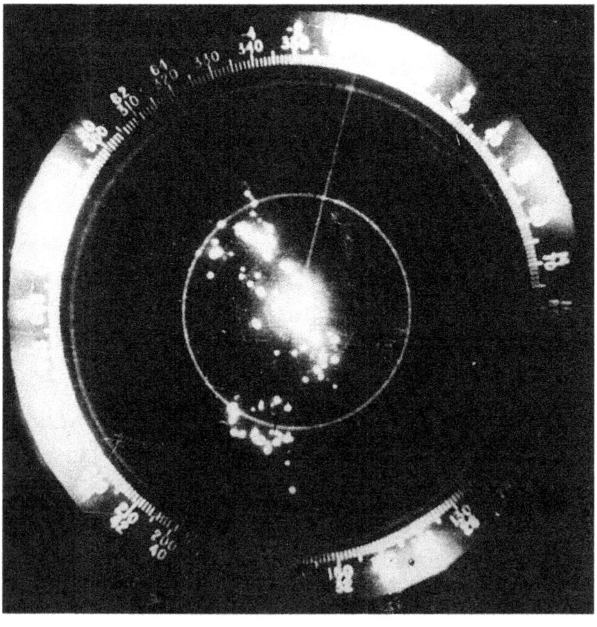

Fig. 6.9. Precipitation echo aloft from an Andhi (dust storm) on a 3 cm radar near New Delhi (PPI picture). (from Mathur and Kulshrestha 1966; courtesy IMD).

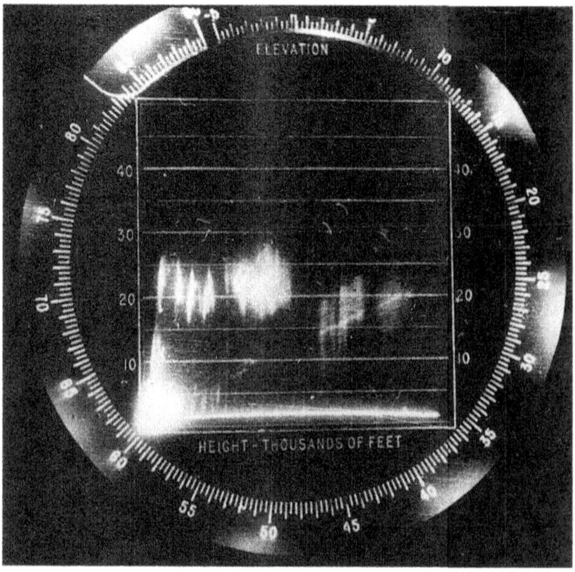

Fig. 6.10. Precipitation echo aloft from an Andhi (dust storm) on a 3 cm radar near New Delhi (RHI picture). (from Mathur and Kulshrestha 1966; courtesy IMD).

6.1.14 Nor'westers

Pre-monsoon season squall lines of northeast India and Bangladesh are called nor'westers because many of them (not all) move from northwest to southeast. They form at the interface of moist southeasterly air and drier land winds and are extremely destructive. The pressure rise associated with a nor'wester may be as much as 13 hPa (Hoddinott 1986). They often produce hail and occasionally tornadoes. These have been extensively studied with conventional data (e.g. IMD 1944). Among radar studies may be mentioned, Das et al. (1957), De (1959), Koteswaram and De (1959) and De and Sen (1961). The heights of tops in these storms have been studied with airborne radar, optical camera and ground-based radar (Cornford and Spavins 1973).

6.2 Studies of the growth of convection

6.2.1 Radar Climatologies

In the early days of radar meteorology, statistics relating to convective cells observed by radar were compiled for various areas. The statistics inclusde echo top heights, echo sizes, preferred areas of development, diurnal variation, seasonal variation and frequencies of direction and speeds of motion. This climatology was, for the first time, made possible by radar and besides being useful in aviation forecasting and short-period local forecasts, it has helped in physical understanding of the phenomena involved. For example, studies of frequencies of convective echo top heights observed with 3 cm radar at New Delhi and Calcutta (now Kolkata) (Kulshrestha 1962, De 1959, Mukherjee et al. 1972), showed echo tops exceeding 20 km on several occasions. Cornford and Spavins (1973) using airborne forward-looking 3 cm radar and optical camera determined the heights of cumulonimbus tops around Calcutta in the pre-monsoon season and compared them with data from a ground-based 3 cm radar. These were important results at the time, because (in the absence of observations) it was thought by many meteorologists at that time, that monsoon rain is mostly of warm origin and clouds rarely extend much beyond the freezing level. While the warm rain hypothesis was largely true of monsoon rain on the west coast of peninsular India, the above studies using radar and high-flying jet aircraft showed that the same thing did not apply to other parts of India[18].

The limitations and biases involved in such data should be kept in mind. Though most of these authors speak of cloud tops, these are really radar echo tops which are usually lower than the visually observed cloud tops because the top portions of convective clouds would not have sufficient reflectivity to give radar echo. No reflectivity thresholds have been specified by most of these authors. Also, the radars used were not as sensitive as more modern radars. Secondly, some of these radars had rather wide beam widths in the elevation plane and there could be appreciable error in the height determination. Thirdly, these being manually operated radars, echoes had to be selected by the operator for height finding and hence the statistics may not include all echoes occurring at a given time. Further, observational schedule was determined according to operational requirements, implying that observations may have been restricted to some hours of the day or to some seasons. Hence, the statistics of diurnal or seasonal variation would be biased towards the active periods of observation. Wherever more modern digital radars have become available, it may be worthwhile repeating the study to get an unbiased climatological record.

---

[18] Radar studies have shown that frequencies and heights of convective clouds in northern India are highest in the monsoon season (Chatterji and Prem Prakash 1990, Kulshrestha 1962, Kumar 1992). This is not true in other parts of the country. Highest thunderstorm heights around Chennai (Madras) (Jayanthi et al. 1980, Raghavan et al. 1981), Hyderabad (Thomas 1973, Thomas and Raghavendra 1977) and Calcutta (Biswas and Gupta, 1989, 1992), occur in the pre-monsoon season and there is a sharp fall in heights after the monsoon sets in.

Echo top heights of convective cells have also been sought to be related to other radar-observed parameters. A good example is the work of Gagin et al. (1985). They considered the difference (D) between the maximum height of each cell during the observation (H_{max}) and the lowest summit height (h) of the cell when it starts raining at the ground. They related D to the maximum reflectivity factor (Z_{max}), the maximum precipitation area (A_{max} km^2), the duration of precipitation (DUR in minutes), the maximum volume rate of rainfall (RVR sq km mm h^{-1}) computed using a standard Z-R relationship and the time TX for any of these parameters X to reach maximum value. In each case they arrived at a power law relationship of the form,

$$X = AD^b \tag{6.7}$$

with a correlation coefficient ranging from 0.6 to 0.8. Curiously, they found the exponent in most cases to be close to an integer. Although considerable care has been taken in this study, it is subject to the basic sources of error such as range biases created by decreasing detectability and increasing beam volume at long ranges.

A positive correlation between the height and the other parameters is to be expected, as a cloud which grows taller or covers a larger area may give rise to greater precipitation. However, this may not be true in every situation. For example, the reflectivities and rain rates in tropical cyclones are not well-correlated with echo top heights (see Chapter 8). Rosenfeld and Gagin (1989) studied the relationship of the total rain volume (RV) of convective cells and clusters of cells in Israel and South Africa (two very different climatic regimes) with the life cycle maximum echo top height above freezing level measured by radar (as well as with other parameters measured by other means). They found a very strong dependence of RV on the height in both the areas.

The rates of vertical growth of the radar echo have been found to be up to 20 m s^{-1} by Battan (1953) in the USA. De and Rakshit (1961) in Calcutta found growth rates ranging from 0.3 to 8.9 m s^{-1}. They concluded that the rain in Calcutta was mostly due to the condensation-coalescence mechanism. Cornford and Spavins (1973), using airborne radar, found a maximum growth rate of 8.6 $\pm$ 3 m s^{-1} at Calcutta. Ramana Murty and Biswas (1961) and Ramana Murty et al. (1960) divided convective echoes around Delhi into three types (warm, moderately cold and supercooled) based on the heights of their tops and found their growth rates (in m s^{-1}) to range from 1 to 2.6 for warm, about 6 for the moderately cold and 9 to 12 for the supercooled. Vertical growth rates at Mumbai (formerly Bombay) (Mukherjee et al. 1977) were found to vary from 1.5 to 16 m s^{-1} and decay rates from 0.3 to 7 m s^{-1}. More recent studies at Delhi (Chatterjee and Prem Prakash, 1990b, Chatterjee et al. 1993) give growth rates of 0.9 to 25 m s^{-1} and decay rates of 0.9 to 8.7 m s^{-1}. These are averages during the life of the echo. The growth of thunderstorms is not at a steady rate but pulsating.

The spatial distribution and diurnal changes in occurrence of precipitating clouds and their movement in various seasons within the area of coverage of the radar has been studied at various places, e.g. Frank et al. (1967) in the USA and Rao et al. (1961), Raghavan et al. (1981), in India. These studies show that there are preferred areas of formation of convective cells determined by local topography and other features. For instance, a river bed, even if it is dry, is a preferred place of initiation of convective cells compared to neighbouring land areas. The effect of the coastline i.e. the land-sea interface, on the time of occurrence of the convection is also brought out in these studies. The time of occurrence of the echoes is also dependent on the season as well as local factors such as the sea breeze.

6.2.2 Echo Properties and Growth Mechanisms

Convective clouds generally form in a random manner in an area favourable for convection, by reason of the large scale airflow and appropriate thermodynamic conditions. Assuming that an initial cloud element grows further in size while entraining environmental air in some proportion, López (1976, 1977) showed that this process would result in a log-normal distribution[19] of the sizes of the cells. In addition to the growth of an individual cumulus, clouds can grow by merger into larger elements. If the cumulative per centage frequency of sizes is plotted on logarithmic probability paper, a straight line should result. In the GATE experiment in the tropical Atlantic, the echoes observed by ship-borne radars were classified into three size scales, viz. the D scale (≤ 100 km), generally associated with pure convection and hence called convective scale, C scale (101 -1000 km) and B/C scale (> 1000km). The latter two were considered to be mesoscale. The majority of echoes were in the D scale with relatively low frequencies in the other two scales. López (1976, 1977, 1978) and Houze and Cheng (1977) found that the frequency distribution of echo sizes did fit a log-normal distribution, to a great extent. The echo top height distribution was also log-normal. A similar result was arrived at in other coastal and oceanic areas of the Western Hemisphere. However, at the large-size end there was a deviation from the log-normal. Echoes of very large sizes were fewer than the number to be expected from a log-normal distribution. This was explained by López (1977) in terms of a limit to growth by convection. A similar study was made by Raghavan et al. (1983) at Chennai, India, using radar

---

[19] For the theory of the log-normal distribution see Croxton et al. (1969) or Crow and Shimizu (1988).

data of the MONSOON-77 and MONEX-79[20] programmes. They found that while the D scale predominated, echo sizes followed a log-normal distribution only up to a size of about 200 km^2 (Fig. 6.11). Larger echoes were *more*

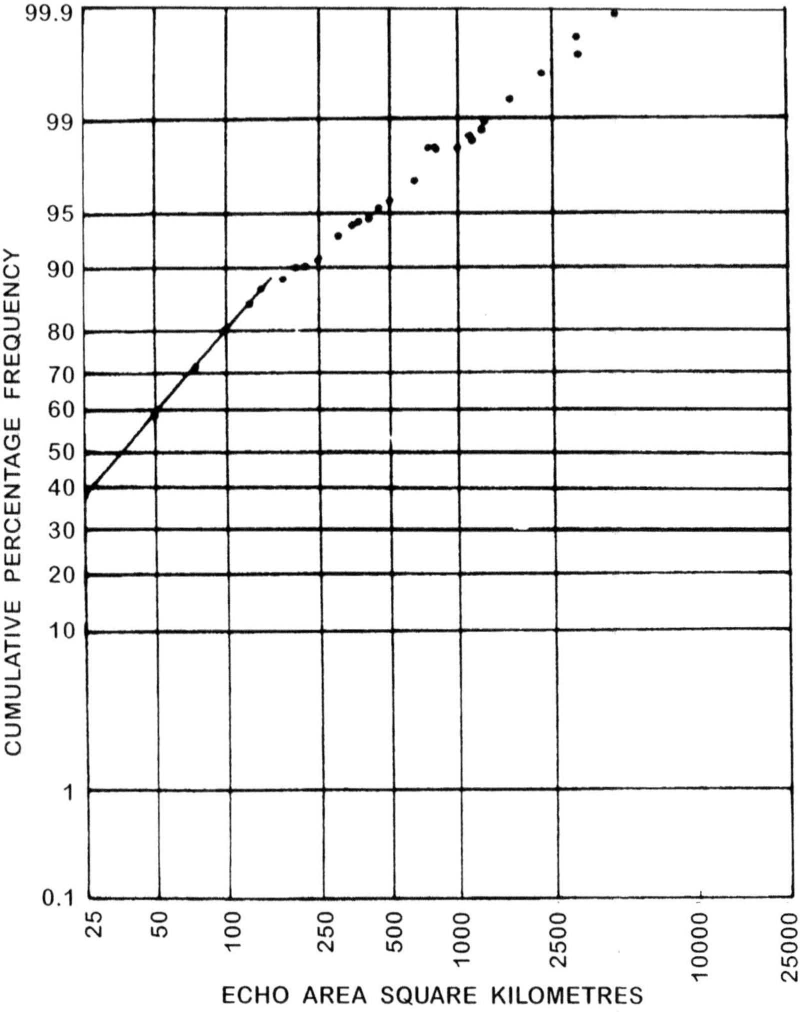

Fig. 6.11. Frequency distribution of radar echo sizes near Chennai – southwest monsoon season, 1979, sea area. (after Raghavan et al. 1983; courtesy AMS). The plot is on log probability paper. Echo area on a logarithmic scale is represented on the X-axis. Cumulative percentage frequency on a probability scale is represented on the Y-axis. Points beyond about 200 km^2 deviate *below* the straight line drawn.

[20] MONEX-79 = MONsoon EXperiment-1979 (a part of GARP). MONSOON-77: A programme of observations conducted in 1977 as a precursor to MONEX.

numerous than predicted by the log-normal distribution in the case of echoes over sea or land in the southwest monsoon season and echoes over the sea in the northeast monsoon season in the two years.. Raghavan et al., (1983) explained the more numerous large size echoes as due to the formation of a large stratiform anvil in association with the convective cluster. On the other hand, in the case of echoes over land in the northeast monsoon season the points deviated *above* the log-normal line (Fig. 6.12) i.e. the large echoes were *fewer* than predicted by the log-normal distribution. Raghavan et al. (1983) explained the

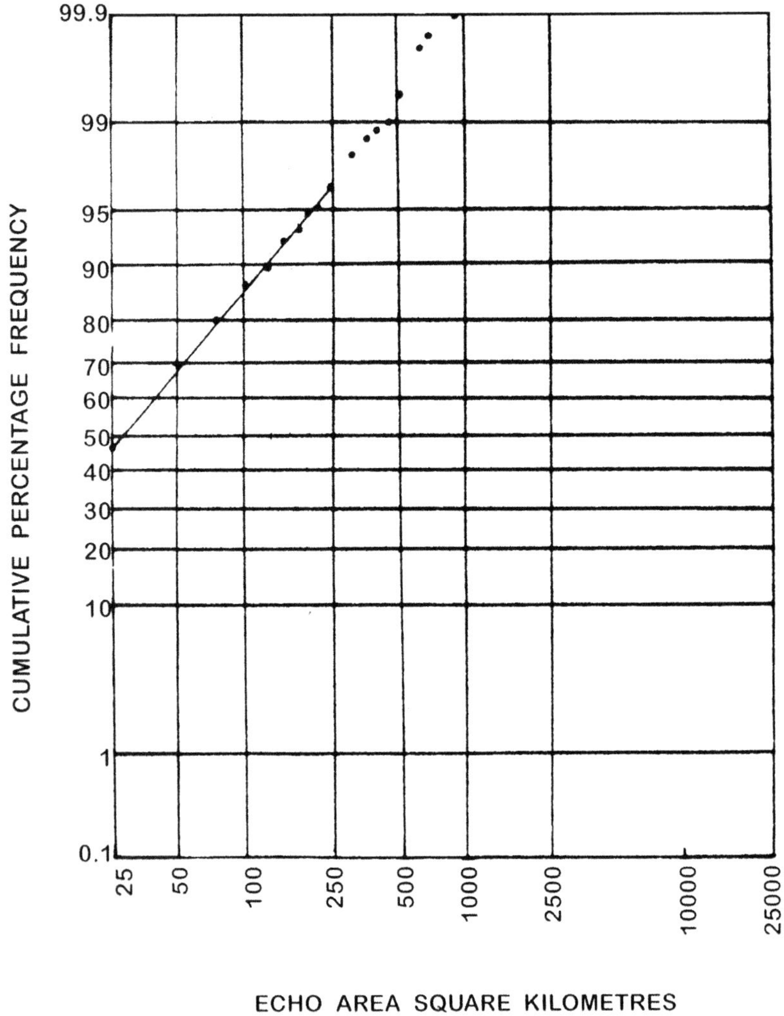

Fig. 6.12. As in Fig. 6.11 but for northeast monsoon season, 1979, land area. (after Raghavan et al. 1983; courtesy AMS). Points beyond about 200 km^2 deviate *above* the straight line drawn.

latter result as due to cessation of growth when the northeast monsoon cloud systems come over land. Chatterjee et al. (1991, 1992) also found that sizes and heights of radar echoes of convective clouds around Delhi and other continental stations in India were log-normally distributed. The higher echo tops were fewer than indicated by the log-normal distribution. This was explained by them as limitation of growth due to the tropopause. But echoes of very large area were more numerous than indicated by the log-normal distribution. This has been explained again in terms of the stratiform anvil.

6.2.3 The Stratiform Anvil

In the Atlantic studies, it was noticed that a large and thick mesoscale stratiform anvil forms around the convective towers in cloud clusters and squall lines (e.g. Houze and Betts 1981). A model of such anvil formation was given by Leary and Houze (1979) based on radar and aircraft observations in GATE (Fig. 6.13). Although further refinements based on Doppler radar observations, have been made (e.g. Smull and Houze 1987, Biggerstaff and Houze 1993), Fig. 6.13 still represents the essential features. The major features apparent from the model are,

(1) The anvil trails behind the convective echo.

(2) The convective part is characterised by high Z_e (45 dBZ or more), high rainfall rates (several tens of mm h^{-1}), vertically oriented reflectivity contours,

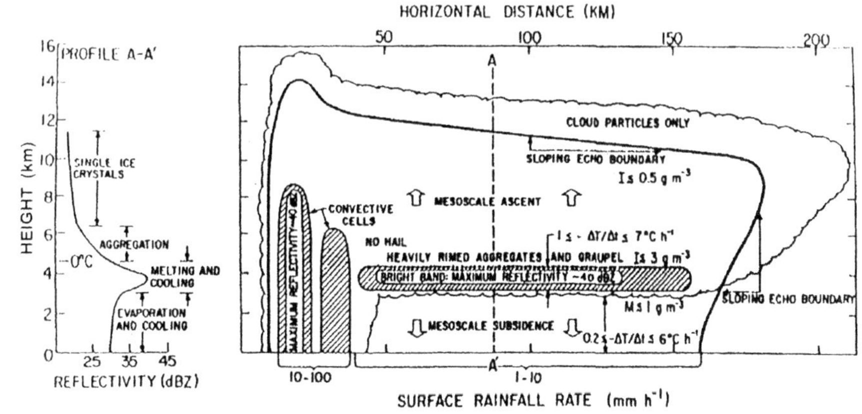

Fig. 6.13. Schematic vertical cross section and vertical profile of radar reflectivity (along dashed line AA' in the cross section) in horizontally uniform precipitation associated with an anvil cloud. (from Leary and Houze 1979; courtesy AMS). The anvil cloud occurs to the rear of intense convective cells propagating from right to left in the figure. The dark solid line is the contour of minimum detectable radar echo, lighter solid lines and shading indicate contours of higher reflectivity, and the scalloped line indicates the cloud boundary.

vertical velocities of several m s^{-1}, and small area in the horizontal plane. In some cases the mean vertical velocity is maximum in the lower troposphere; but in other cases it is a maximum in the upper troposphere.
(3) The stratiform area is very large in the horizontal plane, has a low Z_e (< 30 dBZ), low rain rates (a few mm h^{-1}), horizontally oriented reflectivity contours and a bright band and small vertical motions (several cm s^{-1}) called the mesoscale ascent and descent[21]. The vertical motions are upward in the upper troposphere and downward in the lower troposphere.,
(4) Besides the large area of precipitation in the stratiform anvil, the precipitation persists after the convective precipitation has stopped. For these reasons the precipitation volume in that part is roughly equal to that in the convective towers.

Similar stratiform anvils have been observed in many other locations e.g. in West Africa in the COPT-81 Experiment[22] (Roux and Sun 1990). Such anvils have been observed in the case of mid-latitude squall lines also. From single Doppler radar observations of such squall lines Houze et al., (1989) arrived at a conceptual model of the flow patterns at various levels. Smull and Houze (1987) made a dual-Doppler analysis of such a squall line. They noticed a narrow "transition zone" with weak reflectivity at low levels, between the convective and stratiform areas. There was a secondary maximum of reflectivity in the trailing stratiform region. Vertical profiles of reflectivity and vertical velocity obtained by them in the three zones[23] are shown in Figs. 6.14 and 6.15. Note that there is a mesoscale downdraft in the lower troposphere and an updraft above, in the transition and stratiform regions, but there is only ascending motion in the convective zone. Fig. 3.16 gives the vertical profiles of hydrometeor fall velocity, reflectivity factor, divergence of the horizontal wind and vertical air velocity in the trailing anvil of a mid-latitude squall line determined by the EVAD technique. High resolution dual-Doppler analysis (combined with other data) has made it possible to investigate the relationship between precipitation intensity and the spatial structure of the mesoscale up- and downdrafts (Biggerstaff and Houze, 1991). Thermodynamic and microphysical retrievals have also been attempted in such anvils, as discussed in section 6.3.

The case of the tropical systems is similar except that in the tropical case (1) the maximum reflectivity is lower (2) the reflectivity decreases more rapidly with height above freezing level and (3) the vertical velocity is lower

---

[21] Vertical velocities were obtained originally from aircraft observations. In subsequent studies Doppler radar observations became available [e.g. Fig. 3.16(c)]. Later, wind profilers have also been used (see Chapter 10).

[22] COPT = Convection Profonde Tropicale.

[23] Structure of the convective, transition and stratiform zones as seen by "clear air" radar is discussed in Chapter 10.

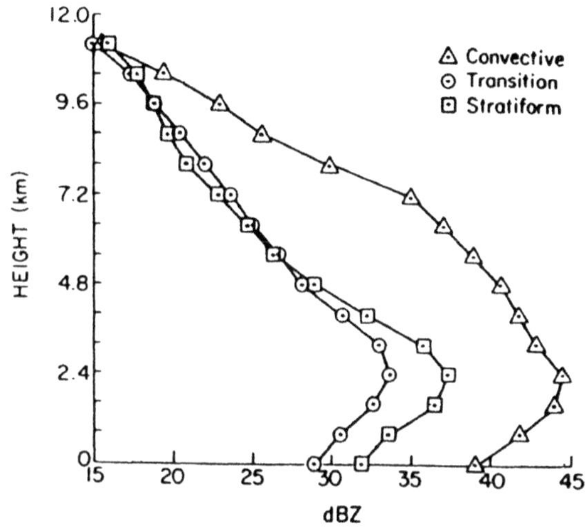

Fig. 6.14. Vertical profiles of mean radar reflectivity factor (dBZ) in the convective, transition and stratiform regions in a mid-latitude squall line with a trailing region of stratiform rain. (from Smull and Houze 1987; courtesy AMS).

(Szoke et al. 1986). In tropical mesoscale convective systems in the Darwin area of Australia, studied with microwave Doppler as well as wind profiler radar, the leading convective region followed by the transition region and trailing anvil were identified (e.g. Cifelli and Rutledge 1994). Raghavan et al. (1983)

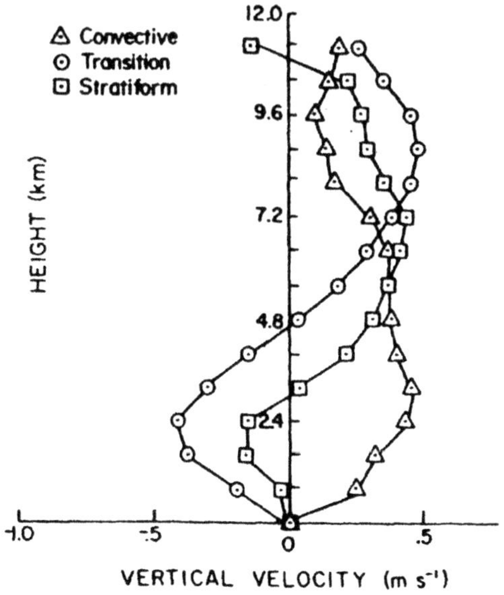

Fig. 6.15. As in Fig. 6.14 except for vertical velocity (w m s^{-1}). (from Smull and Houze 1987; courtesy AMS).

verified the formation of a stratiform anvil[24] at Chennai through the radar observation of reflectivity distribution and of melting bands. They, however, found that the stratiform anvil was not a trailing region but could be elsewhere. They attributed this deviation from the Leary-Houze model to the different wind profile in their area. Houze et al. (1981) also found an anvil in the vicinity of north Borneo and it was not a trailing cloud. Houze and Churchill (1987)[25] found in a Bay of Bengal monsoon depression that the anvil was trailing the convective line in some cases and in some others the convective cells were randomly embedded in the stratiform rain. In a GATE squall line too, Houze and Rappaport (1984) reported stratiform rain extending in front of the convective region.

The discovery of the stratiform anvil has several implications (see e.g. Houze 1997). Zipser et al. (1981) pointed out that the existence of mesoscale fields of kinematic and thermodynamic variables signifies that convective clouds go through their life cycle in an environment that is not the same as the large scale environment. The assumptions hitherto made in cumulus parameterisation that the bulk of convective transports takes place in the cumulus towers became questionable. Also, as already noted, the stratiform anvil contributes about 50% of the total volume rainfall of the weather system. Leary and Houze (1979, 1980) used the observed vertical profiles of radar reflectivity and precipitation amounts and determined the mesoscale up- and downdraft transports based on certain assumptions. They computed the heating by condensation in the mesoscale updraft and the cooling in the melting layer. Ultimately, the result of the anvil formation was found to be a net heating of the middle and upper troposphere and cooling at lower levels (Houze 1982, 1989, Johnson and Young 1983). This influences the large-scale circulation (Hartmann et al. 1984). Hence the correct identification of rainfall as stratiform or convective has assumed importance.

The question arises whether a stratiform anvil occurs in association with convective clouds in tropical cyclones. Airborne radar studies of Atlantic tropical cyclones suggest that it does (Jorgensen et al. 1985; see Chapter 8 for a further discussion).

---

[24] Raghavan et al. (1981) and Sivaramakrishnan et al. (1985) have pointed out that most of the radar echoes around Chennai are convective in origin. In most cases the stratiform echo appears to be generated from the convective cloud by the mechanism discussed above and is not a result of large-scale advection.

[25] Houze and Churchill (1987) using airborne instruments, found very little liquid water above the freezing level in the stratiform region thus verifying the radar indication that the precipitation in this region was indeed stratiform.

6.2.4 Fractal Dimensions of Convective Clouds

The study of sizes of radar echoes of convective clouds leads us to examine if these are fractals. Mandelbrot (1977) proposed that shapes occurring in nature though looking irregular and complicated, have the property of self-similarity i.e. the shapes are similar irrespective of the scale of magnitude to which they grow. He termed such shapes as fractals. The smoothness or roughness of a cloud boundary, for example, can be expressed in terms of an area-perimeter relationship. A more complicated shape involves a larger perimeter for the same area. The relationship is of the form,

$$P = SA^{(D/2)} \tag{6.8}$$

where P is the perimeter, A the area, D is a number called the fractal dimension and S is a constant of proportionality. Lovejoy (1981, 1982) plotted the areas of radar echoes as well as cloud elements seen on infrared satellite pictures against their perimeters and the best fit was obtained with $D = 1.35$ which is close to that expected for turbulent isobars in turbulence theory. He concluded that the structure remains the same irrespective of size in the size range up to 10^6 km^2. He also concluded from this that rain and cloud perimeters have no characteristic horizontal length scale between 1 and 1,000 km. Similar studies using radar echoes have been made by Rys and Waldvogel (1986) and Skoda (1987). The former found a decrease of D (smoothening of the shapes) with strong winds and the latter found variations of D according to the physical processes, including interaction with orography, involved in cloud formation. In India, Jain (1989) found the fractal dimension for radar echoes near Chennai in the northeast monsoon season to be $D = 1.30$. The largest echo sample in his case was 5350 km^2. Examining satellite pictures in the Indian region (over a much larger area) Jayanthi et al. (1990) for the same season got the same value of D. Chatterjee et al. (1994) obtained a higher value (1.42) for convective clouds around Delhi in the pre-monsoon season. No systematic study of this parameter over different seasons and regions is available.

Whether these results are consistent with the development of stratiform anvils beyond a certain stage of growth of the convective elements (which would imply a change in cloud structure) is not clearly established.

6.3 Derived Parameters relating to Weather Systems

6.3.1 Derived parameters

Besides the basic products from a Doppler weather radar which are the reflectivity factor, mean velocity and spectrum width, attempts have been made to derive several other atmospheric parameters not directly observable by radar. Some derivations usually involve assumptions which may not be realistic in all situations. Despite this limitation, radar meteorological research has yielded data of useful quality and given an insight into precipitation development processes and thus enhanced considerably the utility of the radar observation.

Some of the more obvious products derived from the reflectivity factor Z were introduced in Chapter 2. These include precipitation rate, liquid water content, and terminal velocity of precipitation particles. Measurement of precipitation will be dealt with in detail in Chapter 7. In this section examples of other parameters derived from radar observations are given.,

6.3.2 Cooling Rates

In their study of GATE convection (see section 6.2), Leary and Houze (1979) computed the cooling in the melting layer (due to melting of the falling particles) from the equations,

$$\frac{\Delta T}{\Delta t} = \frac{V_I L_f I}{C_p \rho \Delta z} \tag{6.9}$$

$$\frac{\Delta T}{\Delta t} = \frac{V_R L_f M}{C_p \rho \Delta z} \tag{6.10}$$

where T and t denote temperature and time respectively, L_f the latent heat of fusion, C_p the specific heat of air at constant pressure, ρ is the air density, Δz the depth of melting layer, I the ice water content (kg of solid precipitation particles per c.m. of dry air), M the liquid water content (kg of liquid precipitation particles per c.m. of dry air) and V_I, V_R are the terminal velocities of fall of solid and liquid particles. The equations are derived by assuming that mass flux of precipitation particles IV_I and MV_R acts as a heat source for the mass of air contained in the layer through which the melting hydrometeors pass. The cooling rates are directly proportional to the terminal velocities of the particles. Leary and Houze (1979) assumed $V_I = 1.5$ m s^{-1} for aggregate snowflakes and $V_R = 6$ m s^{-1} for rain drops. I and M were calculated from the observed reflectivity factors Z_I and Z for solid and liquid precipitation from the equations,

$$I = 8.0 \times 10^{-3} Z_I^{\,0.61} \tag{6.11}$$

$$M = 5.5 \times 10^{-4} Z^{0.80} \tag{6.12}$$

Changes in the liquid water content of precipitation in the stratiform region between 2.5 and 0.5 km levels were attributed to evaporation beneath the melting layer. This was accompanied by cooling below the melting layer in the region of horizontally uniform precipitation. The cooling rate in the 2.5 to 0.5 km layer was calculated using the equation,

$$\frac{\Delta T}{\Delta t} = \frac{V_R L_v \Delta M'}{C_p \rho \Delta z} \tag{6.13}$$

where L_v is the latent heat of vaporisation, Δz is 2 km and $\Delta M'$ the change in average precipitation liquid water content from 2.5 to 0.5 km..

The cooling rate can be several degrees per hour. Cooling due to melting thus plays an important role in the initiation and maintenance of the mesoscale downdraft (see also Paragraph 6.1.2.).

6.3.3 Latent Heat Release (LHR)

Condensation of water vapour in clouds results in the release of an enormous amount of latent heat which is utilised in driving the kinematic and dynamic processes in the precipitating system. It may often sustain the growth of larger weather systems from the initial cloud. Longley (1949) computed the Latent Heat Release[26] (LHR) in a hurricane by simply multiplying the rate of rainfall by the latent heat of vaporisation of water. Thus if 1 mm h^{-1} of rain occurs over an area of 1 square km, the Rain Volume Rate (RVR) is 1 sq km mm h^{-1} and the rate of water collection in the 1 sq. km area is 1000 tonnes per hour. Taking the latent heat to be 597 calories per gm (at a temperature of 0°C) this gives an energy release of approximately 2.5×10^{12} Joules per hour or 6.93×10^{8} W. Using this conversion, the LHR distribution in a tropical cyclone was computed from radar-estimated rainfall rates by Raghavan and Varadarajan (1981) (see Chapter 8) who went on to relate it to cyclone intensity. While this rough calculation may be adequate for order-of-magnitude computation of LHR in a tropical cyclone, a more realistic computation will have to take into account the condensation and evaporation rates in the cloud. Sikdar et al. (1974) attempted such a computation in a severe convective storm and compared it with the result of a numerical experiment. According to them, if the total radar-detected liquid water mass is M, its rate of change is given by,

[26] Latent heat is released to the atmosphere when a cloud forms. However, if the cloud dissipates without precipitating, latent heat is taken from the atmosphere again. Net gain of heat by the atmosphere occurs only when precipitation occurs. Hence we link the LHR only to precipitation and not to cloud formation.

$$\frac{dM}{dt} = G - R - E \tag{6.14}$$

where R is the rainfall rate, G the condensation rate in the updraft and E the evaporation rate in the downdraft. (dM/dt) was obtained from the measured variation with time, of the liquid water content (LWC) computed from an M-Z relationship (see chapter 2) from radar RHI profiles and R got by gauge measurements. E was taken as 2/3 of R. Then G was computed from equation (6.14). The maximum value obtained was about 8×10^6 kg s^{-1}. Multiplying by the latent heat of vaporisation, Sikdar et al., obtained a value of 4.7×10^{12} calories per second (or 1.7×10^7 MW) for the LHR for the case studied. This method assumes that all water is in liquid form. To take account of the ice water content, Chandrasekar et al. (1991) and Tong et al., (1995) have presented another method using polarimetric radar. The ice particles do not contribute much to the quantity Z_{DP} [Z_{DP} is defined by equation (2.83)]. Z_H is the sum of $Z_{H(R)}$ and $Z_{H(I)}$, the contributions from rain and ice respectively. From measurement of Z_{DP} and the relationship of Z_{DP} with Z in rainfall it is possible to get the rain portion of the reflectivity signal $Z_{H(R)}$. The fraction f of the reflectivity signal due to ice hydrometeors is then obtained as,

$$f = \frac{Z_{H(I)}}{Z_H} = 1 - 10^{-0.1(\Delta Z)} \tag{6.15}$$

where ΔZ is the horizontal deviation in dB of the data point from the "rain line" in a Z_{DP}-Z_H scattergram. Chandrasekar et al. modified equation (6.14) to include the effects of freezing and melting as follows:

$$\frac{dM}{dt} = G - R - E - F + Me \tag{6.16}$$

where F is the rate of freezing of liquid water and Me is the rate of melting of ice. They also wrote a similar equation for the rate of change of the ice water content,

$$\frac{dI}{dt} = F - Me + D - S \tag{6.17}$$

where D is the rate of deposition of ice and S its rate of sublimation. However, they neglected D and S as they were small. M was found from the M-Z relation and I by a correction procedure which amounts to adding 7.2 dB to the reflectivity from ice hydrometeors and then using a M-Z relation. Rain rate was obtained using Z and Z_{DR} (see Chapter 2, also Seliga et al. 1986). The assumption that $E = (2/3)R$ was avoided by taking $G - E$ as a single quantity and multiplying it by the latent heat of vaporisation. Similarly $F - Me$ was multiplied by the latent heat of freezing. The variation of LHR with time could also be obtained.

Tong et al. (1995) also computed LHR from the updraft field obtained by multiple Doppler analysis. According to them, the two methods agree fairly well. Since the latent heat of freezing is far smaller than that of vaporisation, the contribution of the freezing and melting to the LHR was found to be quite small. Thus the earlier methods using conventional radar were not far wrong for most purposes.

6.3.4 Turbulence – Energy Dissipation Rate

Scales of turbulent eddy motions in the atmosphere were discussed in Chapter 2. The fluctuations around the sine wave in a VAD scan (see Chapter 3) can be taken to indicate turbulent motion of hydrometeors. Lhermitte (1968) studied these fluctuations in a snowstorm to deduce turbulent air motion, neglecting the contribution of wind shear and variations in particle fall speeds. The spectral densities of turbulence can be used to find the rate of energy transfer within the inertial subrange, which is called the energy dissipation rate ε (units m^2 s^{-3}). The energy input in a convective storm is primarily from the updrafts. So the transition wavelength, L_0, also called the outer scale, between the energy input range and the inertial subrange should be less than the dimension of the up- and downdraft circulation cell i.e. within a few km. Within the inertial subrange the three-dimensional spectral energy density is (see Brewster and Zrnić 1986, Doviak and Zrnić 1993),

$$E(K) = A\varepsilon^{2/3}K^{-5/3} \tag{6.18}$$

where A is a non-dimensional constant equal to about 1.6 and K is the three-dimensional wave number magnitude. The longitudinal spectral energy density $S_l(K)$ can be expressed as a function of $E(K)$ and K. Brewster and Zrnić (1986) give the equation,

$$S_l(K) = \frac{18}{55}A\varepsilon^{2/3}K^{-5/3} \tag{6.19}$$

Provided signals are strong, the Fourier transform of the radial velocities along a radial will give a good estimate of $S_l(K)$, from which ε can be computed.

In Chapter 3 it was seen that the variance of the Doppler velocity spectrum is composed of four contributions i.e.,

$$\sigma_v^2 = \sigma_t^2 + \sigma_s^2 + \sigma_d^2 + \sigma_a^2 \tag{6.20}$$

If we neglect σ_a due to antenna movement which may be not more than 0.3 m s^{-1}, we may try to separate the others to determine the turbulence contribution and therefrom calculate ε assuming that the turbulence is homogeneous and isotropic. The following simple treatment is from Frisch and Strauch (1976). The fall speed contribution to the observed variance is,

$$\sigma_d^2 = \sigma_D^2 \sin^2\theta \qquad (6.21)$$

where σ_D^2 is the variance of the fall velocities of hydrometeors and θ is the angle of elevation at which the radar observation is made. For low elevation operation this term can be neglected. The contribution of wind shear can be divided into two parts:

(1) σ_{sr}^2 due to the shear k_r (s^{-1}) in the radial velocity along the beam i.e. the radial direction, and

(2) σ_{st}^2 due to the shear k_t (s^{-1}) in the radial velocity in the transverse direction. k_t itself can be divided into two perpendicular components (horizontal k_h and vertical k_v),

These are given by,

$$\sigma_{sr}^2 = \frac{(k_r h)^2}{12} \qquad (6.22)$$

$$\sigma_{st}^2 = (0.3 k_t r\phi)^2 \qquad (6.23)$$

$$\text{and } k_t^2 = k_h^2 + k_v^2 \qquad (6.24)$$

where r and h are the range and range resolution (metres) and ϕ the one way half-power beam width. Hence,

$$\sigma_s^2 = \left(k_h^2 + k_v^2\right)\left(0.3 r\phi\right)^2 + \frac{\left(k_r h\right)^2}{12} \qquad (6.25)$$

From Doppler velocity measurements k_r, k_h, k_v can be calculated. Except at very short ranges the range resolution is much smaller than the angular resolution. Therefore the k_r term can be neglected. The difference $\sigma_v^2 - \sigma_s^2$ gives σ_t^2 the variance due to turbulent fluctuations within the pulse volume. If the scales of motion involved are within the inertial subrange and if the k_r term is neglected, ε can be calculated from the relation (Istok and Doviak 1986),

$$\varepsilon = \frac{2.4\sigma_t^3}{r\phi A^{3/2}} \qquad (6.26)$$

In severe convective storms, spectral width can be large (up to tens of $m\ s^{-1}$) and values of ε as high as $1\ m^2\ s^{-3}$ have been obtained although more typical values are of the order of $0.05\ m^2\ s^{-3}$. The maximum values have been observed downstream of the updraft.

Meischner et al. (2001) compared ε derived from C-band Doppler radar spectrum width with in situ measurements from aircraft in thunderstorm anvils. They found values between 10^{-4} and $5 \times 10^{-2}\ m^2\ s^{-3}$ inside moderate to strong storm cells. Generally, close agreement between radar estimates and aircraft estimates were found for low shear areas.

ε is considered to be a useful quantity as it is proportional to the acceleration which an aircraft experiences due to turbulence. However, as can be seen from the above, the calculation is based on several assumptions and there are numerous sources of error especially in weak signal conditions. For a more detailed discussion including the errors involved and comparison with aircraft measurements, see Bohne (1981), Brewster and Zrnić (1986), Doviak and Zrnić (1993), Frisch and Clifford (1974), Istok (1981), Istok and Doviak (1986), Lewis (1981), Mahapatra (1999) and Srivastava and Atlas (1974).

Using conventional radar, Gorelik and Mel'nichuk (1968) noted that if a radar observes vertically upwards, the gradients of the wind velocity in the horizontal do not affect the spectrum of the radar signal. However the fall velocity differences of the hydrometeors do. Assuming that the contribution of fall velocity differences is small in comparison with turbulence effects, they computed ε from the measured mean-square width of the signal fluctuation spectrum for vertical observation of clouds without precipitation. For the precipitation case, they noted that the variance of turbulent velocities varies with the pulse volume. If two pulse widths are used, two pulse volumes with different vertical widths result. Assuming that the spectrum of fall velocities does not change in the two cases, Gorelik and Mel'nichuk could eliminate the fall velocity effect and compute ε.

Determination of ε in the clear atmosphere is dealt with in Chapter 10.

6.3.5 Thermodynamic and Microphysical Retrievals

Gal-Chen (1978) showed from continuity and momentum equations which describe the evolution of convective systems, that knowledge of the wind and its time history would suffice to determine, uniquely, the pressure and density deviations from their horizontal averages. The three components of wind velocity data are obtainable from dual- or mutiple-Doppler radar observations. Their time derivatives give the accelerations. The accelerations are caused by forces, principally pressure gradient, buoyancy and Coriolis force. Hence the equations of motion can be used to derive these forces. The method has been refined and used by several researchers with variations and case studies of squall lines including their stratiform regions, microbursts etc. have been published. A simple treatment is given below.

The radar-derived winds are substituted into the anelastic form of the equations of motion. These may be written as (Parsons and Kropfli 1990),

$$\frac{1}{c_p\theta_{ev}}\left\{\frac{du}{dt}+f_1+m_1\right\}=G=-\frac{\partial\psi}{\partial x} \quad (6.27)$$

$$\frac{1}{c_p\theta_{ev}}\left\{\frac{dv}{dt}+f_2+m_2\right\}=F=-\frac{\partial\psi}{\partial y} \quad (6.28)$$

$$\frac{1}{c_p\theta_{ev}}\left\{\frac{dw}{dt}+f_3+m_3\right\}=H=-\frac{\partial\psi}{\partial z} \quad (6.29)$$

where,

ψ = non-dimensional pressure $= (p/p_0)^{\frac{R}{c_p}}$

p = pressure,

p_0 = a reference pressure (1000 hPa)

R = gas constant

c_p = specific heat of dry air at constant pressure

$u, v, w,$ = x, y, z components of wind

t = time

f_i, i = 1,2,3, components of Coriolis force

m_i i = 1,2,3, represent sub-grid scale mixing

g = acceleration due to gravity

β represents buoyancy

θ_{ev} = environmental virtual potential temperature.

Only the variations and not the absolute values of pressure, temperature etc. are obtainable from these equations. θ_{ev} may be obtained independently from radiosonde or aircraft observation. The velocities are computed with reference to the moving storm by subtracting the storm motion. The time derivatives du/dt etc. contain the partial time derivative $\partial u/\partial t$ and the spatial gradient term e.g. $u(\partial u/\partial x)$. These can be calculated from velocity data of successive pulse volumes and successive scans[27]. The Coriolis terms can also be computed from interpolated velocities. The mixing terms can be estimated from the measured wind field using a parameterisation scheme. The pressure field in a horizontal grid can be obtained from equations (6.26) and (6.27). Using this in equation (6.28), the buoyancy field is obtained. The buoyancy term involves a knowledge of the condensate mixing ratio with contributions from precipitation and cloud particles. The precipitation part can be computed from a *Z-M* relationship [equation (2.57)]. The cloud particle contribution cannot be determined from the radar data; it has to be neglected or assumed or based on independent measurement. The temperature field is determined by solving a thermodynamic equation for the potential temperature.

Besides pressure, retrievals of temperature variations and the heat budget

[27] Error arises from the scan process taking a finite time.

also have been attempted. Brandes (1984) and Roux (1985) used a simplified thermodynamic equation, in addition, to calculate the horizontal mean values of temperature at each altitude (assuming an initial value). The horizontal mean values of pressure were also computed from the vertical projection of the equations of motion. Thus, not only the pressure and temperature perturbations but the pressure and temperature fields could be computed. Obviously, the derivations will be in error due to errors in the measured wind components. There are several assumptions and sources of error involved. Even so, the results in some of the case studies cited are claimed to compare reasonably well with independent measurements.

Single Doppler radar, which gives only one component of velocity, has been used by Bonesteele and Lin (1978) by supplementing it with data from another similar storm. Roux and Sun (1990) used single Doppler data in a West African squall line for this computation. They used the ABCD method (see Chapter 3) to combine several successive scans of a single Doppler radar as a squall line approached, to get the three dimensional wind field and thereafter carried out the thermodynamic retrieval technique. Liou et al. (1991) have proposed a method using single Doppler radar in conjunction with a numerical model. Successive values of observed Doppler radial velocities representing one component of the wind are inserted into the model and the other two components are modified by a variational method. Sun and Crook (1994), derived three-dimensional wind and thermodynamic fields with single Doppler radar and verified them with a dual-Doppler analysis.

For more detailed treatment of thermodynamic retrievals the following references may be seen: Bonesteele and Lin (1978), Brandes (1984), Chong et al. (1980), Gal-Chen (1978), Hane and Scott (1978), Hane and Ray (1985), Kessinger et al. (1988), Liou (2001), Liou et al. (1991), Parsons and Kropfli (1990), Ray (1990), Roux (1985, 1988), Roux and Sun (1990), Scialom and Lemaître (1991), Sun and Houze (1991, 1992), Sun and Crook (1994, 1996), Sun et al. (1991).

Microphysical retrieval consists of solving a time-dependent and three-dimensional set of continuity equations of heat and water substance with assimilated profiles of heat and moisture and Doppler radar-derived airflow. The technique maps a set of observational data on to a set of numerical model output data. Cloud physical processes such as coalescence or breakup of drops, rain evaporation, hail growth etc. are parameterised[28]. Suitable drop size distributions for liquid and solid hydrometeors are assumed. The method is described by

[28] Polarimetric radar products can be used for obtaining actual microphysical data. They enable better identification of particle phase but since observables which delineate particle types are not sharply defined, a fuzzy logic approach has been used by Vivekanandan et al. (1999).

Zeigler (1985, 1988). Hauser and Amayenc (1986) used also the water content and hydrometeor fall velocities derived from radar reflectivity data as inputs. The temperature field is also needed for deriving microphysical variables. This can be obtained from independent observations or from the thermodynamic retrieval technique. Hauser et al. (1988) combined the thermodynamic and microphysical retrieval techniques. Related work may be seen in Braun and Houze (1992, 1993), Marécal and Hauser (1991), Rutledge (1986) and Rutledge and Hobbs (1983, 1984).

6.4 Short Period Forecasting

6.4.1 Nowcasting

Meso-α and meso-β scale phenomena (see Table 6.1 for definitions of scales) have a small spatial extent and a life ranging from a few minutes to a few hours and are, therefore, difficult to forecast accurately by conventional synoptic methods or the larger scale numerical weather prediction models. Continuous or frequent radar observation enables extrapolation in space and time and serves as a means of forecasting for short periods. Besides radar, satellite and other observational tools may be used. Such forecasting which is based mainly on extrapolation of intensive observations (both in situ and by remote sensing) is called "nowcasting" where very short forecast periods (less than two hours) are involved. The lead time for forecast depends on the phenomenon as well as on the purpose for which the forecast is required. For instance, in the case of mesocyclones and tornadoes a lead time of only a few minutes may be achievable but this may be adequate in the case of aviation requirements. A larger and longer-lived system e.g. a tropical cyclone can be predicted with a lead time of a day or longer and such lead time is necessary for taking precautions. The prediction may be based on observation of an existing phenomenon which can be extrapolated in space and time or on the observation of some precursor phenomena.

The tracking of existing radar echo patterns over a period of time by manual or computer methods was discussed in Chapter 5. These have been used for forecasting. The SHort-term Automatic Radar Prediction (SHARP) developed in Canada (Austin and Bellon 1982) and similar correlation techniques were referred to.

Several radar-based indicators of severe weather such as hook and bow echoes, WER, BWER, hail signals, mesocyclone and tornado signatures, gust fronts and downbursts have been discussed in section 6.1. As noted in Chapter 4, the observation of 10 dBZ or stronger echo aloft precedes lightning strike by several minutes. All these indicators, if obtained in real time, help in nowcasting for periods ranging from a minute to two hours. Although these are observed after the phenomenon has started, many of them precede the occurrence of severe weather at the ground. The use of polarisation parameters may also be exploited (e.g. for nowcast of hail and microburst).

The formation of convergence lines or fronts representing boundaries of temperature, moisture or wind discontinuities, is often detectable by radar due to the refractive index change. The low level convergence can be detected by Doppler radar. As already discussed, the gust front has long been recognised as a precursor to further thunderstorm or dust storm development. Wilson and Schreiber (1986) and Mahoney (1988), among others, have shown that echoes from convergence lines are precursors to initiation of convection. Points at which two boundaries collide or intersect are particularly preferred for fresh thunderstorm initiation. A boundary can be seen as "a thin line of enhanced reflectivity and/or a line of apparent convergent flow in Doppler velocity with dimensions of 1 to 3 km wide, 10 km long and present for a minimum of 15 minutes" (Wilson and Schreiber, 1986). A thin line detection algorithm was referred to in Chapter 4. Cumulus clouds forming along the boundaries can be seen on satellite pictures. The sea breeze front tracked by radar can also indicate areas of likely convective development if supported by other data. Because of low reflectivity, it is necessary to use a very sensitive radar for detecting these phenomena.

Wilson and Mueller (1993) developed a nowcasting procedure for up to 30 minutes by making use of extrapolation of existing cells as well as use of convergence boundaries as locations for initiation of new cells. A similar technique was used in a tropical area (Darwin, Australia) by Keenan et al. (1992).

The use of squall lines as precursors in the case of tropical cyclones is discussed in Chapter 8.

6.4.2 Use of Integrated Observing Systems

Radar is not used in isolation but as part of an Integrated Observing System (IOS) with satellite, surface and upper air data because the exact location of initiation of mesoscale phenomena will depend on the temperature and moisture and wind fields. The procedures will also be dependent on local climatology and vary from region to region. From a climatological study, statistical predictors of severe storm occurrence in the subsequent hours can be developed, the inputs being current radar and other measured atmospheric parameters (from surface and radiosonde observations). The procedure of Boucher and Wexler (1961) for forecasting the motion of precipitation lines knowing the 700 hPa wind from radiosonde observations and the climatological properties of the lines was referred to in section 6.1.

Xu Yumao et al. (1988) in China, have developed regression equations for forecasting severe convective weather over a period of up to two hours. They take the values of selected parameters[29] at time t when severe weather is not occurring and compute a predictor y. According as y exceeds certain thresholds y_1 or y_2 the forecast is for a severe thunderstorm or a very severe one. Severe storm is taken as one having hail < 2 cm diameter or surface wind between 17 and 25 m s^{-1}; very severe storm is taken as one having hail > 2 cm diameter or surface wind > 25 m s^{-1}. The data of five years were used to determine the regression coefficients and the equations were applied with a high degree of success in the sixth year. Similar regressions were also developed to predict whether a severe storm already occurring at time t would continue up to $t + 2$ hours.

The FRONTIERS system in the U.K. [Browning (1979, 1986), Browning and Collier (1982)] has been mentioned in Chapter 5. This combines geostationary satellite imagery with radar data from a network of stations.. The satellite imagery is used to infer probable areas of precipitation[30] outside the area of good radar coverage (Collier et al., 1989). NIMROD, the successor to FRONTIERS produces very short range forecasts of a variety of meteorological elements including precipitation, every 15 minutes. Radar data

---

[29] The measured parameters were:
(1) maximum Z_e dBZ,
(2) maximum height (km) of echo top and environmental temperature (°C) at this level,
(3) height (km) of the echo with maximum intensity and the environmental temperature (°C) at this level
(4) maximum speed (km h^{-1}) of the echo ensemble,
(5) average speed (km h^{-1}) of the echo ensemble,
(6) the ratio of the thickness between condensation level and freezing level to the thickness between freezing level and echo top,
(7) the wind shear (s^{-1}) between lower and upper troposphere (the level representing each is chosen depending on the season),
(8) height of the 0°C level,
(9) stability criterion i.e. difference between equivalent potential temperatures at 500 and 850 hPa,
(10) departure (°C) of the maximum surface temperature from the climatological mean maximum temperature for a ten day period, and
(11) the time deviation between the time of occurrence of severe weather and the climatologically most probable time of occurrence for the place and season.

[30] The calibration of satellite data using radar estimate of precipitation is discussed in Chapter 7.

from one or a network of radars as well as geostationary satellite data are displayed together in systems such as FRONTIERS. Geostationary satellite data are available at half to one hour intervals; and they are usually received at analysis centres some 30 minutes or so after than the time of observation. Their resolution may also be inferior to that of radar though their area coverage is much larger. Hence, for very short period forecasts of small scale phenomena, radar data may be used with surface data (and wind profiler data where available) to complement them. For relatively longer periods, satellite and radiosonde data can be used in addition. The merger of the radar and satellite data and their computer processing is best done by man-computer interactive systems such as McIDAS (see Chapter 7; also see Heymsfield et al.1983). While fast processing demands the use of automated methods, human interaction is essential to reconcile any differences between various data sources and analyse the field using the forecaster's skill. Hence, the provision of a versatile workstation is as important as the observing equipment. The data may also be used with mesoscale numerical weather prediction models to introduce a forecast element instead of mere extrapolation of present weather. The problem of assimilation and interpretation of the wide variety of data available makes the human forecaster's skill and experience indispensable. Finally, fast communication of the output to the user is essential as short period forecast is obviously highly perishable.

Lightning detection equipment and the relationships between lightning and radar-observed parameters (see Chapter 4) have also been used in shot-period forecasting. Lightning detection supplements radar observation where radar network coverage is poor (Reap 1986).

6.4.3 Automated Radar Subsystem (ARSS) for Very Short Range Forecasting

Ivanov et al., (1989) have described an Automated Radar Subsystem (ARSS) for very short range forecasting developed in the former USSR. This consists of a dual wavelength radar (3 and 10 cm), a preliminary processing equipment and a minicomputer. The range is 125 or 250 km and displays have 100×100 elements with resolution 2×2 or 4×4 km. There are several modes of operation: (1) reflectivity measurement at 3 or 10 cm. (2) simultaneous operation at 3 and 10 cm. (3) a mode of obtaining information on turbulence in clouds and precipitation that allows to identify zones of higher inhomogeneity of wind field within the scanning radius of the radar. The data products can be transmitted on telephone lines to users (forecasters, aviation and agricultural interests).

Chapter 7

ESTIMATION OF PRECIPITATION

In the first section of this chapter we introduce the subject of estimation of precipitation using radar. Section 7.2 deals with the estimation of precipitation using measurements of reflectivity factor. Drop size measurements, Z-R relations and corrections needed for the effects of range, melting band and attenuation are discussed. Calibration with rain gauges is also dealt with in this section. In section 7.3, area-time integral methods are discussed. Section 7.4 deals with precipitation estimation using attenuation measurements. A treatment of multi-parameter methods follows in section 7.5. Combinations of reflectivity and attenuation, reflectivity and Doppler velocity at vertical incidence, dual wavelength measurements and polarisation parameters are considered. In section 7.6 we discuss the comparisons of radar and satellite measurements and the combined use of the two techniques. Finally, in section 7.7 we deal with the use of space-borne radar for precipitation measurement.

7.1 Measurement of Precipitation

7.1.1 Rain Gauge and Remote Sensing Measurements

Precipitation is the end product of many weather systems. A measurement of this quantity is needed not only for an understanding of atmospheric processes but also for a wide range of practical applications ranging from flood forecasting and water resource management to microwave communications. Depending on the application, it may be necessary to measure the amount of precipitation (mm), its rate (mm h^{-1}) with specified time resolution, its phase (i.e.

liquid or solid precipitation), its spatial and temporal distribution, its vertical profile, its drop size spectrum and in some cases, its chemical composition. For most purposes Areal Precipitation Estimates (APE) i.e. area-integrated values of rate or amount of precipitation are needed and for certain purposes linear integration over a path length may be needed.

The conventional instrument for rain measurement at the ground is the rain gauge which measures the volume of water falling over a known area (of the mouth of the funnel of the gauge, typically 127 to 324 cm^2). The volume divided by the area represents the height up to which the rain water would have stood if all the rain falling over an area remained there without runoff, seepage or evaporation. This height (expressed in mm) is, by definition, the rainfall. Solid precipitation is measured in terms of melted water equivalent. The collection in an ordinary rain gauge is measured manually at regular intervals. Recording gauges measure continuously and the rate of precipitation can also be obtained from them. Since a gauge measurement is a point measurement and since rainfall is not a continuously variable function in space, a measurement by a gauge is not truly representative of its surroundings. An areal average of a large number of gauges is essential for most applications. The spatial variation of rainfall depends on the characteristics of the rain (convective or stratiform) and the topography. Very large spatial gradients can occur in the case of convective precipitation and in hilly terrain. The APE obtained, therefore, depends critically on the gauge network density. The World Meteorological Organisation (WMO) has recommended (WMO 1994) certain network densities for various types of terrain for synoptic, meteorological and hydrological purposes but these may not be suitable for all applications. For discussions on reliability and errors of gauge measurements see Sevruk (1989), Huff (1970).

The operation of gauges distributed over several places and collection of data from them has several problems of logistics and communications. The majority of rain gauges in networks in most countries are ordinary gauges monitored by human observers at intervals of time and hence cannot give instantaneous or short-period rainfall rates. Where rain rates are required, fast response recording or telemetering gauges have to be employed and these are expensive. For certain applications, a vertical profile of precipitation may be needed; it cannot be obtained from a rain gauge. Finally, there is no satisfactory method of in situ measurement of precipitation over the ocean.

Use of a remote sensing method to measure or estimate precipitation would offer the equivalent of a very high density network, resolution limited only by that of the sensor. Such a high resolution device would capture isolated heavy rain rates which may be missed or smoothed out by any practical network of rain gauges. The remote sensing device can give instantaneous rates of precipitation. It will also enable real time collection of data for large areas from a few centralised locations. Data can be collected over the sea area also. Vertical profiles can be obtained. The high resolution data can be used as input to numerical models. Since relatively few sensors are needed the cost may not be higher than that of a dense rain gauge network if the telecommunication cost for the latter is taken into account.

The two common remote sensing devices used are the radar and the satellite, the latter using visible, infrared or microwave sensors or radar itself. The disadvantage of both of them is that precipitation rate or amount is not measured directly but has to be derived from certain quantities measured by them. In the case of radar, these are the radar reflectivity factor at one or more wavelengths, the attenuation or the polarisation parameters. However, in spite of its own errors, the rain gauge is the standard with which any other measurement of precipitation is to be compared. Hence we have to examine the conceptual differences between the gauge and radar methods of measurement.

Unlike a rain gauge which makes a measurement over a very small area at the ground, the radar measures Z_e (or some other parameters) integrated over the pulse volume which is a function of the range and the beam width and is usually much larger than the volume sampled by a gauge. This is, therefore, more representative of a larger area but the sampled volume is at a height above ground which depends on range. In case attenuation is measured it is integrated over a path length. The techniques of radar estimation of precipitation may be divided into

(1) single parameter methods and

(2) multi-parameter methods.

The single parameter may be Z_e (dBZ) or the specific attenuation k_R (dB km^{-1}) which can be related to the precipitation rate R (mm h^{-1}) by an equation such as (2.56) or (2.115).These methods involve gross assumptions about the Drop Size Distribution ((DSD) and ignore the effect of hydrometeor shapes. Multi-parameter methods consist of the measurement of two or more parameters (e.g. reflectivity or attenuation at two wavelengths; reflectivity in different polarisations), which will enable the DSD or the drop shapes to be taken into account. The determination of DSD is not merely a step to calculate precipitation rate, but is of interest by itself for understanding the structure and physics of precipitating systems. Measurement of Z_e alone is the method most used and is considered in detail in section 7.2. The use of other techniques will be considered in subsequent sections.

7.2 Estimation of Precipitation from Reflectivity Factor

7.2.1 Errors in measurements of Z_e

The sample monitored by the radar is an ensemble of precipitation particles of various sizes, shapes and of two phases (water and ice[1]) all of which have an effect on Z_e. (We shall use the terms Z and Z_e interchangeably below for convenience). The larger the range, the greater is the sampling volume, and the greater its height above ground. At larger ranges the volume sample is not representative of the lower portion of the precipitating cloud. Also the beam may not be filled by the particles; hence Z will be underestimated. The minimum detectable Z increases with range. This will result in error in the measured spatial distribution of precipitation. (The range normalisation corrects only for the r^2 attenuation and does not correct for these problems). In measuring Z we assume Rayleigh scattering. We can correct for attenuation in the path to the sampling volume (see Paragraph 7.2.8) and for occultation of the beam by obstacles in the path but these corrections have several uncertainties. The presence of hail or melting band in the beam will give a high value of Z, leading to overestimation of R compared to what a rain gauge measures. Assuming that the same particles fall into a rain gauge below, they might have undergone modification in size, shape and phase after detection by the radar. Besides melting, Mason and Ramanadham (1954) identified the following causes of modification of drop size distribution (DSD):

1. Coalescence within cloud,
2. Accretion with cloud droplets,
3. Coalescence between rain drops below cloud base, and
4. Evaporation below cloud base.

According to Hardy (1963) the number of small drops is depleted by all these processes while the number of large drops is increased by coalescence and accretion but decreased by evaporation. Further, the particles sampled by the radar may be blown downstream by the wind and therefore may not correspond to rain at the gauge below. For this reason an areally averaged rainfall both from rain gauge and radar will compare better than point estimates. If there is vertical motion, the actual rainfall will differ from that computed using the terminal velocity alone [equation (2.52)]. For example, if there is an updraft equal to the terminal velocity, the rainfall rate would be zero (Battan 1976).

[1] Ice can be in various forms.

To minimise errors it is necessary to choose a radar with a wavelength relatively free from attenuation due to intervening rainfall, at the same time giving a detectable power return even from weak precipitation. It is necessary to keep the beam width small and the scanning at as low an elevation as possible so that the beam height is low and also the beam does not cut the melting layer. The 10 cm (S-band) wavelength is normally chosen but in temperate countries where rainfall rates are not high, the 5 cm (C-band) offers the advantage of narrower beam width at less cost. Hence Japan and most countries in Europe have used the C-band for most of their radars while countries at lower latitudes (USA, India) have preferred the S-band. In cold climates where mainly snow is to be detected, 3 cm radar may have an advantage as attenuation in snow is low and reflectivity at 3 cm is high. (The use of still higher frequencies in space-borne radar will be discussed later).

Finally, there are radar calibration errors (see Chapter 5). Unless the radar is well calibrated there can be systematic under- or overestimation of precipitation. For example, according to Ulbrich and Lee (1999) there is systematic underestimation by several US WSR88-D radars probably due to calibration biases.

The factors introducing discrepancies between radar and rain gauge measurements are listed in Table 7.1 following Zawadzki (1984).

7.2.2 Drop Size Distribution (DSD) Measurements

If the DSD is known, Z can be computed from equation (2.55) and R from equation (2.53), assuming a drop size vs terminal velocity relationship such as that of Gunn and Kinzer (1949). A knowledge of the DSD[2] is thus important for establishing a valid Z-R relationship. Experiments have been made in several countries to measure the DSD at the ground or from aircraft at flight level or from balloons. The use of ground measurements for determining the Z-R relationship has the limitation that the DSD measured at the ground may not be the same as that at the radar pulse volume. And the sampling volumes of such measurements are small in comparison to radar sampling volumes. The measuring instruments (briefly described below) have their own errors (e.g. they may not respond well at very large or very small drop sizes) and the Z computed from them may not agree with the Z measured by radar.

Early ground measurements used a simple technique of exposing a filter

[2] For a discussion on "What is a rain drop size distribution" see Jameson and Kostinski (2001).

Table 7.1. Sources of errors in radar and gauge comparisons (after Zawadzki 1984).

Sources of random errors	Sources of systematic errors	Sources of range-dependent errors
Variability of Drop Size Distribution (DSD) at ground from storm to storm, Variability of DSD at ground within each storm, Time-space variability of precipitation under cloud base, Advection of precipitation, Variations in the radar calibration, and Residual signal fluctuations.	Change in DSD from the height of radar measurement to ground, Attenuation, Wet radome, and Time sampling by the radar.	Size of the radar sample volume coupled with the non-uniformity of the field of reflectivity, Increase in height with range of the sample volume; Minimum detectable signal, and Propagation, shadows.

paper coated with a dye to the rainfall for a measured time interval. Raindrop cameras (Jones 1959) and disdrometers (Joss and Waldvogel 1967) have subsequently been used. The latter uses a transducer in which downward displacement of a diaphragm under the influence of drop impact causes an induced voltage in a sensing coil. In the laser disdrometer (Calla et al. 1986), the rain particles are illuminated by a Helium-Neon laser and are imaged as shadow graphs into a linear photo-diode array. The particle size is determined by the number of photodiodes shadowed. Hauser et al. (1984) have described an optical spectropluviometer (used in the COPT-81 experiments) which could measure drop sizes and fall speeds. Kruger and Krajewski (2002) have described a two-dimensional video disdrometer in which two orthogonal light sheets that are projected onto line-scan cameras, form the virtual measuring area. Hydrometeors that fall through the light sheet cast a shadow that is recorded by the cameras. Detailed shape and size information for each individual hydrometeor is available. The light sheets, spaced 6 mm apart are also used to measure the vertical fall velocity. The Precipitation Occurrence Sensor System (POSS) (described in Chapter 11) consisting of a bistatic X-band CW radar also retrieves DSD from the Doppler spectrum.

Measurements have been made on board aircraft at levels equal to the radar beam height using hot wire detectors, pneumatic droplet samplers or foil samplers (e.g. Brown and Braham 1959) or particle measuring probes [e.g. Houze et al. (1979), Jorgensen and Willis (1982)]. Willis (1984) used an airborne optical spectrometer. A balloon-borne radiosonde type of instrument for measurement of drop size (and charge) has been described by Bateman et al. (1994).

There is vast literature[3] available on comparisons of disdrometer-measured

[3] References are too numerous to be listed here.

DSD's with radar measurements, the nature of precipitation being usually identified from radar [e.g. Maki et al. (2001), Schuur et al. (2001)].

7.2.3 *Z-R* Relationships

Z-R relationships derived from DSD measurements have been reported by a large number of authors. The best known is the relationship,

$$Z = 200R^{1.6} \tag{7.1}$$

(see Chapter 2), which is commonly used as a "default" relationship[4] in some countries.

An alternative method is to relate Z measured by radar with the precipitation rate R recorded at the ground by a rain gauge or a network of gauges on a large number of occasions. A *Z-R* relation can be derived by regressing observed values of Z_e or dBZ_e against those of[5] R or dBR. This method is a purely statistical one (without a physical foundation)[6] and will include the effect of errors in measurements by radar. Hence the relationship so derived may be applicable only to similar measurements by the same radar. The statistical and physical methods can also be combined.

A fairly comprehensive list of *Z-R* relationships published all over the world up to 1970 is given by Battan (1973; his Table 7.1 for rain and Table 7.3. for snow). Stout and Mueller (1968) have reviewed the *Z-R* relationships and discussed their limitations. A few selected relationships published by various authors subsequent to 1970 are listed in Table 7.2. An inspection of Battan's tables and Table 7.2 in this book will show a wide variation in the values of the coefficient A and exponent b for equation (2.56). These appear to depend on the place, season, type of precipitation (drizzle, shower, stratiform rain, snow etc.). This is to be expected because the DSD varies with type of precipitation (Mason and Andrews 1960) and may vary spatially and temporally even within a single precipitating system.

---

[4] This is often referred to as the Marshall-Palmer (M-P) relationship, although Marshall and Palmer (1948) gave the coefficient and exponent as 296 and 1.47 respectively.

[5] dBR is the value of R expressed in decibels above (or below) a threshold of 1 mm h^{-1}. Regression of Z_e against R and regression of dBZ_e against dBR may give different *Z-R* relationships (Campos and Zawadzki 1999). The latter gives weightage to the lower values. In hydrometeorology the higher values of R are more important. Hence the *Z-R* regression is preferable.

[6] Jameson and Kostinski (2001b) have argued that since *Z-R* relationships are usually based on statistically inhomogeneous data they are simply statistical correlation fits and have no physical justification.

Table 7.2. Z-R relationships $Z = AR^b$ obtained by various methods (after 1970).

A	b	Area or project	Precipitation type	References and Remarks
24.7	2.37	South Dakota (USA)	Convective storms	Miller (1972). Climatological Z-R relation (see equation 7.2).
300	1.4	Florida Area Cumulus Experiment (FACE) (USA)	Convective	Woodley et al. (1975), Gagin et. al. (1985). Also used as default value for WSR-88D network.
230	1.25	GATE		Hudlow and Arkell (1978)
185	1.22	GATE		Atlas et al. (1990b). Climatological Z-R relation. (Atlas et al. give different sets of constants for different ranges and for different radars in GATE. Other constants have been given for GATE by other authors. They are based on drop size measurements on board ships or aircraft).
55	1.6	USA (National Weather Service)	Convective (flash flood situations)	Jorgensen and Willis (1982)
300	1.35	USA (National Weather Service)	Hurricane	Jorgensen and Willis (1982)
300	1.4	-do-	Default values for US WSR-88D	Fulton et al. (1998)
250	1.2	-do-	-do-	-do- For tropical environments
340	1.53	COPT-81 (West Africa)		Hauser et al. (1988); from surface measurements of precipitation from optical spectropluviometer.
300	1.5	Switzerland	All rain	Joss and Waldvogel (1970), Schiesser et al. (1995)
316	1.5	Swiss Meteorological Institute	for rain	Joss and Lee (1995)
100	1.2	Chennai (India)	Seasonal mean for northeast monsoon	Raghavan and Sivaramakrishnan (1982). By comparison of radar- and rain gauge-derived areal averages of precipitation.

Table 7.2 continued.

A	*b*	Area or project	Precipitation type	References and Remarks
100	1.3	Chennai (India)	Seasonal mean for southwest monsoon	Raghavan et al. (1987). By comparison of radar- and rain gauge-derived areal averages of precipitation.
230	1.4	Massachusetts (USA)	Average - ordinary rain	Austin (1987)
400	1.3	- do -	Intense convective cells	- do -
100	1.4	- do -	Non-cellular rain	- do -
650	1.36	Bauru (Brazil)	all rain for near range	Calheiros and Zawadzki (1987). From comparison of probabilities of Z_e and R in non-simultaneous measurements.
50	1.5	Darwin (Australia)	convective	Steiner and Houze (1993). Convective and non-convective parts were separated according to reflectivity criteria and then compared with rain gauges
150	1.5	-do-	non-convective	
82	1.47	-do-	convective	Steiner et al. (1995); after adjustment with rain gauges
143	1.5		stratiform	
334	1.19	Kapingamarangi	convective	Atlas et al. (1999).
147	1.34	Atoll	transition	Based on disdrometer and 915 MHz profiler
278	1.44	TOGA	stratiform	measurements.
865	1.08	COARE[7]	stratiform	Classification based on variation of median volume diameter and rain rate. Values shown here are for one occasion. Atlas et al. obtain widely different values for other occasions.
178	1.51	National MST	convective	Narayana Rao et al. (2001).
162	1.44	Radar Facility,	transition	Z and R from disdrometer data.
251	1.48	Gadanki (India)	stratiform	Classification from MST radar and Lower
155	1.39		northeast monsoon	Atmospheric Wind Profiler. Seasonal values are irrespective of
407	1.32		southwest monsoon	classification.

[7] TOGA COARE = Tropical Oceans-Global Atmosphere Coupled Ocean Atmosphere Response Experiment

The *Z-R* relationship will also depend on altitude as the terminal velocity of the hydrometeors varies with altitude.

Fujiwara (1965) classified data taken from the USA, Japan and Germany into six types of precipitation echoes and plotted domains of *A* and *b* (Fig. 7.1) for each type. For example, in this figure, growing convective cells tend to have a low value of *b* while stratiform echoes or those from decaying convective cells have higher *b* values. Stout and Mueller (1968) attempted to classify the *Z-R* relations by rain type or according to synoptic situations. They assumed a constant *b* of 1.5 and varied only *A*. Joss and Waldvogel (1970b) proposed that from the vertical profile of *Z*, the type of precipitation may be recognised by an automated program and the values for the appropriate type may be used. Steiner and Houze (1993) and Steiner et al. (1995) in Darwin, Australia, separated the convective and non-convective portions of a radar echo by reflectivity criteria and determined separate *Z-R* relationships for each by comparison with rain gauges. Several authors have attempted classification of precipitation based on the drop diameters, rain rates, convective, transition or stratiform characteristics, horizontal gradient of reflectivity, vertical profile of reflectivity, altitude of freezing level, cloud depth and the bright band in the sampling volume, and

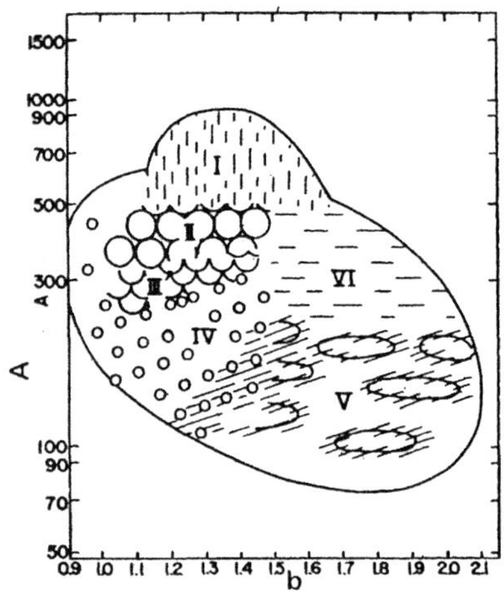

Fig. 7.1. Variation of *A* and *b* with type of precipitation echo (after Fujiwara 1965; courtesy AMS). I. diffuse and heavy part of thunderstorm echo, or tall echo evaporating. II. core of thunderstorm or intense solid echo (which is sometimes diffuse). III. generating or developing stage of intense cells. IV. small solid "airmass" echoes either scattered or lined up with rainband. V. stratified plane echo or weak diffuse echo. VI. completely diffused echo from a thunderstorm. (Between IV and V there is an intermediate region of stratified echoes with convective cells among them).

tried to work out *Z-R* equations for each case [e.g. Atlas et al. (1999), Narayana Rao et al. (2001), Rosenfeld et al. (1995, 1995b), Steiner and Houze (1993), Steiner et al. (1995), Tokay and Short (1996)]. The classification has been made from radar or profiler observations and drop sizes at the ground were measured by disdrometers. Atlas et al. (1999) consider that *Z-R* relationships must be physically based; they base the relationship on the D_0 (median volume diameter), Z and R variations using measurements by disdrometer and UHF profiler. Their values for one occasion are shown in Table 7.2 and they have given widely different values for other occasions. Chandrasekar and Bringi (1987) made simulation studies using various values of N_0, D_0 and m of a gamma distribution and also varied the sampling volume and showed that discrepancies between radar-measured reflectivity factor and that computed from surface measurements are not due to physical causes alone.

Miller (1972) proposed that "climatological *Z-R* relationships" for particular areas can be derived. He prepared Cumulative Frequency Distribution (CFD) curves of Z and of rainfall rates for a particular area in the USA. If, for example, in the CFD curve for Z, 80% of observed values of Z fall below 50 dBZ, and the CFD curve of rainfall rate shows that 80% CF corresponds to $R = 41.4$ mm hr^{-1}, he took 50 dBZ as equivalent to 41.4 mm hr^{-1}. Plotting a number of such *Z-R* pairs he derived a *Z-R* equation,

$$Z_e = 24.7R^{2.37} \tag{7.2}$$

This concept was further developed by Calheiros and Zawadzki (1987) and Atlas et al. (1989, 1990b). Pairs of values R_i and Z_i may be taken such that the Cumulative Distribution Functions (CDF) match i.e.

$$\int_{R_i}^{\infty} \mathrm{P}(R)dR = \int_{Z_i}^{\infty} \mathrm{P}(Z)dZ \tag{7.3}$$

Then the pairs R_i and Z_i determine the climatological *Z-R* relation provided that the probability distribution functions of R and Z are representative of the season and region. Since Z measured at long range is likely to be an underestimate for reasons discussed earlier, different *Z-R* relationships for different ranges were formulated by Calheiros and Zawadzki (1987). Atlas et al. (1990b) also stratified the *Z-R* relations by range. Their coefficient (A) decreases with increasing range while the exponent (b) does not change appreciably.

For forms of precipitation other than rain, *Z-R* relationships are even more uncertain.

7.2.4 Observational Procedures

The basic observational procedure is to make a scan at a low elevation and record the signal over each range-azimuth bin, while averaging samples as discussed in Chapters 3 and 5. Low elevation is important to make the radar sample representative of the low level precipitation and to avoid cutting the melting band. Zero elevation is usually unsuitable as there is likely to be too much ground reflection, ground clutter or occultation by intervening obstacles. It is difficult to ensure that the beam remains below the melting layer, at higher latitudes where the melting layer altitude is low. However, in view of possible occultation in some azimuths, scanning at higher elevation is also necessary. If a volume scan is being taken, the first three or four scans at intervals of one beam width can be used for the rain estimation. Higher elevation data are inserted in those bins which are affected by clutter or occultation at the lowest elevation and a composite reflectivity distribution obtained. The use of a low antenna scan speed (not more than 3 rpm; preferably less) is required for maximising the number of samples averaged. However, this speed in conjunction with the need for several elevations increases the total acquisition time for one set of data and results in assuming that rainfall rate is constant for that period. Corrections as discussed in Chapter 3 are applied[8]. Correction may also be made for partial occultation of the beam in some bins. Bins contaminated by clutter or anomalous propagation echoes may be assigned values interpolated from neighbouring azimuths[9]. Manual intervention may be necessary to identify clutter and anomalous propagation as automated systems may not be entirely effective. The signals in each bin are then converted into Z_e values, which are further converted into R, assuming a Z-R relationship, before integration over Cartesian squares[10]. The Z-R relationship is usually a mean for the season or for a type of rainfall obtained by prior comparison with rain gauges; or different relationships for different ranges may be used. Default values of A and b are usually provided but the operator can choose other values. If separate Z-R relationships are to be applied for convective and stratiform parts of the precipitation, criteria for automatic recognition using values of reflectivity or reflectivity gradient should be used.

If contours of various dBZ levels are available (as in the case of manually operated radars), the area between two successive contours can be multiplied

[8] Correction for attenuation by intervening rainfall may be necessary in the case of radars operating at C-band or shorter wavelengths (see Para 7.2.8).

[9] This will be valid only if spatial gradients of precipitation rate are small.

[10] It may not be meaningful to integrate Z_e over a large area. Hence R is integrated.

by the mean of the rainfall rate values of the two contours to yield volume rain rate in km^2 mm h^{-1}. If the acquisition of successive data sets are centred at times T_1, T_2 etc., at regular time intervals t, the rain rate at time T_n is assumed to be constant over the period $T_n - (t/2)$ to $T_n + (t/2)$. With this assumption the rain volume (km^2 mm) over an extended period of time may be obtained by adding the successive volume rain rates and multiplying by t. The rain volume calculated by summation over a long period is the cumulative rainfall. With modern digital radars a volume scan can be completed in about 5 to 10 minutes and a map of rain volume obtained every 5 or 10 minutes.

The procedures used in the US WSR-88D radar may be seen in Fulton et al. (1998) and those in Europe in Meischner et al. (1997).

7.2.5 Use of Vertical Profiles of Reflectivity (VPR) - Correction for Range and Melting Band

The simple procedure given above is inadequate in many circumstances. As the range increases the radar beam sees the higher part of the cloud and therefore the observed reflectivity is in many cases less than what is expected at the surface. In stratiform precipitation the reflectivity is nearly constant as height increases in the rain portion, becomes a maximum in the bright band and rapidly decreases at higher levels. Hence the radar may show an increase of reflectivity with range up to the point when the beam intersects the bright band and a decrease in reflectivity at farther ranges. In convective precipitation the reflectivity increases with height up to some altitude and thereafter decreases but not as rapidly as in the stratiform case. Fig. 7.2 shows vertical profiles of reflectivity (VPR) in convective and stratiform precipitation. Errors caused by using the reflectivity aloft to represent precipitation at the ground are often far larger than errors due to inappropriate Z-R relationships. These problems are aggravated in hilly terrain[11]. If we can determine the VPR it should be possible to derive from the observed reflectivity aloft, the reflectivity near the ground which is the quantity to be compared with the surface rain gauge measurement (Joss and Waldvogel 1990).

The reflectivity $Z(x, h)$ at a geographic location x, at an altitude h can be related to the reflectivity $Z(x, 0)$ by the relation,

$$Z(x,h) = Z(x,0)z_D(h) \tag{7.4}$$

where the factor $z_D(h)$ has been called the vertical profile of reflectivity (VPR)

[11] A study of the problems in radar estimation of precipitation in mountainous terrain may be seen in Young et al. (1999).

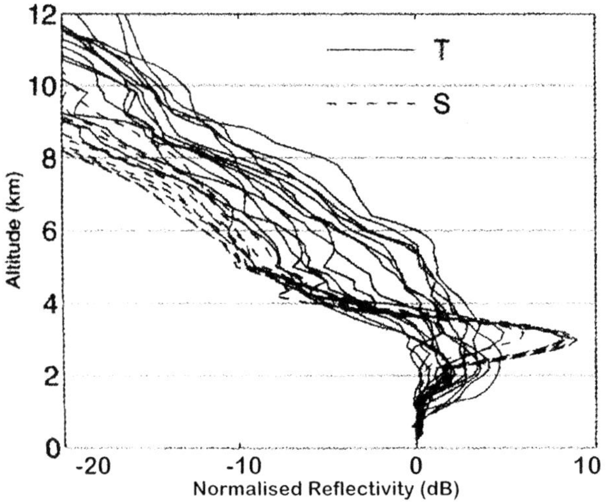

Fig. 7.2. Vertical profiles of reflectivity in convective (T) and stratiform (S) precipitation (from Vignal and Krajewski 2001; courtesy Dr. Krajewski and AMS).

(Andrieu and Creutin 1995). The radar observing the region near x yields a reflectivity averaged over the sample volume (Vignal and Krajewski 2001),

$$\bar{Z}(x,\theta) = \int_b^t f^2(\phi,h) Z(x,h)dh \tag{7.5}$$

where θ is the angle of elevation, ϕ the beam width, b and t are the altitudes of the base and top of the beam at location x and f represents the one-dimensional power distribution of the beam at altitude h. The radar value is thus a function of the real VPR and the radar parameters. If we can assume the VPR to be homogeneous in the region considered[12], we can put,

$$Z(x,h) = Z(x,0) z_{Da}(x,\theta) \tag{7.6}$$

where,

$$z_{Da}(x,\theta) = \int_b^t f^2(\phi,h) z_D(h)dh \tag{7.7}$$

If the VPR is known and assumed to be constant over the whole area of measurement, a correction factor can be applied to each radar measurement to obtain $Z(x, 0)$ and using a Z-R relationship, the corresponding surface rain rate can be obtained. The VPR can be determined from a volume scan of the radar, using nearby echoes where the bright band beam height and beam spread are not important and the same VPR can be assumed to apply over the whole domain of interest. Koistinen (1991) developed daily mean vertical profiles from nearby

[12] Note that we are only making an assumption that the rate of change of reflectivity with height is constant and no assumption is made about variation of reflectivity in the horizontal.

echoes and applied them to ranges up to 150 km. However, besides the assumption of homogeneity of VPR, this procedure, which is operationally used in Switzerland (Joss and Lee 1995), depends on the availability of nearby echoes. Alternatively, a climatological profile based on past data and averaged over time and space can be used. Climatological data may not be available from the same radar for areas, which are occulted at low elevation; these may have to be determined from a radar at another site which can see these areas. Andrieu and Creutin (1995) proposed a method involving determination of the ratio of reflectivities at two (or more) elevations at various distances. They call this the "ratio evolution" and developed an "inverse" method to identify the VPR corresponding to the observed ratio. Vignal et al. (1999, 2000) and Vignal and Krajewski (2001) extended this to determination of "local" VPR's for areas of about 20 km × 20 km (or sectors of 15° width in azimuth and variable range intervals) using a volume scan. Fig. 7.2 shows such local VPR's on a single occasion. Germann and Joss (1999) determined VPR's, integrating reflectivity over 600 km^2 × 60 min for making profile corrections in shielded regions in the Alps. Seo et al. (2000) proposed a procedure for real-time adjustment of range-dependent biases in radar rainfall estimates due to non-uniform vertical profile of reflectivity. Using volume-scan measurements, their procedure estimates conditional mean and variance profiles of point reflectivity in the vertical and calculates, as a function of elevation angle and slant range, a multiplicative adjustment factor to be applied to the apparent radar rain rate. It must be mentioned that all these techniques require large computer power.

Harrold et al.(1974) showed that errors due to the bright band could be reduced significantly by applying a statistically derived correction factor which was a function of the height of the melting layer and the range of the area of interest. This correction factor may apply only to the area over which it is derived. To overcome this problem, Harrold and Kitchingman (1975) and later Smith (1986) developed a method, which can apply a case-by-case correction in real time. The algorithm detects the bright band (on non-Doppler radar). It also derives values for the characteristic parameters of height and intensity of the bright band which are used to calculate a correction. The bright band causes a maximum in the profile of apparent rainfall rate against range. Maxima due to the bright band are separated from those due to other features by comparing equivalent heights of maxima obtained at two different elevations. Peaks due to local precipitation maxima should be coincident in range and not in height while the reverse is true for the bright band peaks. This has been introduced operationally in the UK. Hardaker et al. (1991, 1993, 1995) have developed a microphysical model of the melting layer from which they derive a VPR, which is used to generate correction factors for the bright band in terms of surface rainfall rate. With surface-based radars in the tropics, it is often possible to avoid cutting the bright band by restricting the range and the beam elevation.

Kitchen et al. (1994) devised a method of correction of radar data for the effects of bright band, range and orographic growth by diagnosing at each pixel, the variable parameters in an idealised reflectivity profile in real time using other meteorological observations besides radar. Thus surface observations and assumed lapse rates are used to calculate the wet-bulb freezing level. Precipitation top heights are determined from geostationary satellite infrared observations combined with radiosonde data. Smyth and Illingworth (1998) distinguished between melting snow in the bright band layer and graupel in areas of vigorous convection, using LDR measurements[13]. From this they derived separate VPR's for convective and stratiform precipitation.

7.2.6 Calibration with Rain Gauges

The map so obtained may give a realistic representation of the spatial distribution of precipitation subject to range limitations, but the magnitude of the rainfall is likely to be in error because of the uncertainty of the *Z-R* relationship. One possible method of overcoming this problem is to calibrate or "adjust" the radar values with rain gauges (Hitschfeld and Bordan 1954). The objective is to combine the point measurement accuracy of the gauges with the spatial resolution and coverage of the radar. It should be clear that this calibration with rain gauges is only to allow for uncertainty of the *Z-R* relationship and is not a substitute for the calibration of the radar parameters.

If at a point P the rain rate measured by the gauge and radar respectively are G and R, the radar estimate is made to agree with the rain gauge value[14]

---

[13] This is based on the finding of Frost et al. (1991) that LDR in melting snow in the bright band was − 16 dB while in graupel it was − 25 dB. Frost et al. verified these results from aircraft observations.

[14] The radar value represents the instantaneous rain rate or an average of successive scans if a volume processing at several elevations is combined. The rain gauge value may be typically a one minute average if a fast response gauge is used. But often, in conventional recording gauges it may be possible only to evaluate rain rate over a period of 15 minutes or more. The T minute average rain rate R_T (mm h^{-1}) can be converted if necessary into a shorter period t minute (say one minute) rain rate R_t by a power law (e.g. Ajayi and Ofoche 1984),

$$R_t = a.R_T^b$$

where a and b are constants. In India, Calla et al. (1987) derived similar power law relationships between available 60 minute rain rates and t minute (t = 1,5,10,15 or 30) rain rates. They have listed the values of the coefficients and exponents for various Indian cities.

by multiplication by the ratio $F = G/R$. The same ratio is used to adjust the magnitude of the radar-estimated rain rate at all other points. This assumes that the rain gauge value is the correct one. The *Z-R* relationship used for the purpose may be a mean value obtained from previous experiments and is not critically important.

Wilson (1970) found that for a 1000 square mile area the r.m.s. error of the radar estimate of the areal total is reduced by 39% by calibration with one gauge and that the radar calibrated with one gauge is equivalent to 4 gauges placed uniformly over the area.

The data of a rain gauge at a point may not correspond to the radar-measured value for the various reasons stated earlier. A better result can be obtained by using data from a number (*N*) of rain gauges over a finite area (Brandes 1975, Wilson and Brandes 1979). The establishment of a number of gauges and data collection from them is expensive and may seem to defeat the very purpose of using the radar as a centralised rain measuring device. Hence, only a few gauges may be used in what may be called a target area to calibrate the radar over its entire area of coverage including sea areas where there are no gauges.

The gauge to radar ratio *F* can be computed for each gauge site *i* and the average ratio used for adjustment i.e.,

$$F = \frac{1}{N}\sum_{i=1}^{i=N}\frac{G_i}{R_i} \tag{7.8}$$

An alternative is to average the *G* and *R* values spatially over the whole of the calibrating area. Then,

$$F = \frac{\sum G_i}{\sum R_i} \tag{7.9}$$

In this case the values receive a weight proportional to the rainfall at each site.

Averaging of precipitation over a specified area covered by a number of gauges can be done by the following methods (see Abbi 1972, WMO 1994).

1. Taking an arithmetic mean of the rainfall collected in all the gauges over the area. This method is suitable where there is a uniformly spaced network and spatial variation of precipitation is not appreciable. If there are *N* stations and P_r is the precipitation measured by the *r*th rain gauge, the areal average P_{av} is given by,

$$P_{av} = \frac{1}{N}\sum_{r=1}^{r=N} P_r \tag{7.10}$$

2. Drawing isohyets (isopleths of equal rainfall) and taking an area-weighted mean. If the areas measured between successive isohyets are A_r ($r = 1$ to N)

and P_r (r = 1 to N) are mean values of precipitation between two successive isohyets, the average P_{av} is given by,

$$P_{av} = \frac{\sum_{r=1}^{r=N} P_r A_r}{\sum_{r=1}^{r=N} A_r} \tag{7.11}$$

3. Drawing "Thiessen polygons" which are formed by the perpendicular bisectors of the lines joining adjacent gauges. Each rain gauge value is taken to represent the polygon around it and is weighted by the area of the polygon. If the area of the polygon around the rth gauge is A_r, the average P_{av} is again given by equation (7.11).

More refined methods have also been devised to overcome shortcomings of the above simple methods. For comparison with radar estimates a particularly suitable method is that of English (1973). His model is aimed at a computer-based computation of rain amounts at points of a regular grid from a scattered distribution of rain gauge values over areas such as river sub-catchments. The model is a combination of inverse distance-squared weighting (i.e. each gauge is assigned a weight in inverse proportion to its distance from the grid point for which the value is computed) and the fitting of a linear surface. This has the advantage of dominance of the observations close to the interpolation point and a general dependence upon the distribution of a number of observations. If one or more gauges do not have valid data the rest can be used without changing the weighting factors.

If the number of gauges used is considered inadequate to define the areal rainfall field accurately, the radar-derived spatial distribution itself may be used to interpolate between the gauges (Harrold et al. 1974) and derive an optimum rain gauge estimate G to compare with the radar estimate. This procedure assumes that radar measures the spatial distribution correctly and requires correction only of the magnitude.

More complex schemes have been devised. For example, a gauge may be assigned a weight varying inversely with distance of the gauge from the grid point at which it is compared with the radar. Woodley et al. (1975) who deployed 220 gauges in a 510 km^2 area in Florida, USA, used the daily G/R ratio for adjustment and reported a 15% improvement in accuracy. The time integration improves the agreement.

Instead of using a constant multiplying factor F, Creutin et al. (1988) have proposed a multivariate objective analysis technique to combine radar and gauge data and obtain a more accurate estimate of true ground rainfall. Anagnostou and Krajewski (1998) have proposed a technique of minimisation of the r.m.s differences between hourly rain gauge network and radar accumulations.

The first major quantitative precipitation measurements with a network of radars was the one organised in the GATE experiment in 1974 (Hudlow et al.1979, Patterson et al. 1979). Ships with C-band radars, rain gauges and drop size measurements made on board were used to derive *Z-R* relationships (Table 7.2).

Hildebrand et al. (1979) analysed the data of a 5 cm radar in Montana, USA, with 109 gauges in a 1500 km network centred at about 60 km from the radar, and also data of a 320-gauge network in a 105 km area with two 10 cm radars at the edge of the gauge network. They found that the combinations of radar and gauge data are no more accurate than gauge-only data for gauge densities greater than 1 gauge per 100 km. This emphasises the point made earlier that it is appropriate to use a small network of gauges to calibrate the radar estimates.

Other studies and analyses of this type may be seen in Barnston and Thomas (1983), Barnston et.al. (1983), Collier et al. (1975, 1983), Harrold et al. (1974), Raghavan and Sivaramakrishnan (1982), Raghavan et al. (1987), Woodley et al. (1975, 1982, 1983) and Zawadzki et al. (1986).

If the gauges telemeter their data to the radar site, a real time correction of the radar precipitation map is possible. This is being done in several countries. In the USA, hourly precipitation data from automated rain gauges are combined with radar data (Fulton et al. 1998). The receipt of the gauge data without undue delay at the radar site or analysis centre is crucial.

Anagnostou and Krajewski (1999) have developed an algorithm for real time use, which combines radar and gauge data. It carries out beam height-effect correction, vertical integration, convective-stratiform classification, conversion from radar observables to rainfall rate, range-effect correction, and transformation of the estimated rainfall rates from polar coordinates to a Cartesian grid. Additionally, the algorithm applies advection correction to the gridded rainfall rates to minimise the temporal sampling effect and subsequently aggregates the corrected rainfall rates to 1-hourly, 3-hourly, and storm-total accumulations. The system applies different parameter values for convective and stratiform regimes.

7.2.7 Neural Network Method

Xiao and Chandrasekar (1997) have introduced an entirely new approach viz. a neural network (NN)[15] based technique. Without assuming any *Z-R*

[15] The neural network concept was discussed in Chapter 3 in the context of identification of AP echoes.

relation, this technique maps the three-dimensional radar measurements (such as Z or Z_{DR}; see section 7.5 regarding Z_{DR}) to the ground rain gauge measurements directly. Using gauge measurements as the target set of observations on the ground, the neural network approximates the relation between the radar measurements and ground rainfall observations based on a training data set. When the neural network is trained appropriately, it can be applied to new data sets. They expect this technique to generate more accurate rainfall estimates than the use of Z-R or Z_{DR}-R relationships. Also according to them the neural network estimates using both Z and Z_{DR} are better than the neural network estimates using Z alone. A neural network can be "tuned" to a specific kind of storm or for several storms. But, to generalise it to other storms, it is necessary to collect new data and retrain the neural network. Liu et al, (2001) have developed an "adaptive" neural network that is easy to train and can continuously update the structure by incorporating the latest information into an existing neural network without having to retrain from the beginning. A further discussion of estimation of rainfall from neural network may be seen in Bringi and Chandrasekar (2001).

7.2.8 Correction of Rain Estimates at Attenuating Wavelengths

Hitschfeld and Bordan (1954) examined the question of a correction for attenuation due to intervening rainfall, of measurements at an attenuating wavelength. They proposed a correction scheme but came to the conclusion that such corrections are highly sensitive to small errors in radar calibration. They recommended that an attenuation correction should not be attempted unless the radar calibration is highly accurate. For this reason they preferred radar calibration with rain gauges as discussed in Para 7.2.6. Schemes for correction of 5 cm radar observations (this being a widely used wavelength) have been tried on the basis of Z-R and Z-k_R relationships (see e.g. Hildebrand 1978, Patterson et al.1979). If Z measured (i.e. attenuated value) at range r is $Z_a(r)$, it can be converted into corresponding $k_{Ra}(r)$. Then the corrected value $Z'(r)$ is given by,

$$\log Z'(r) = \log Z_a(r) + 2\sum_{x=1}^{x=r-1} k_{Ra}(x) \tag{7.12}$$

Using the corrected value $Z'(r)$, a new value $k'_{Ra}(r)$ is found from the Z-k_R relation. Using this we get a second step value $Z''(r)$,

$$\log Z''(r) = \log Z_a(r) + 2\sum_{x=1}^{x=r-1} k'_{Ra}(x) \tag{7.13}$$

These iterative steps are continued until the last Z value shows a difference of less than a dB with the previous one. R is then computed from Z.

The use of attenuation itself as a measure of precipitation rate is discussed in section 7.4.

7.3 Area-Time Integral Method of Precipitation Estimation

7.3.1 Cumulative Precipitation over Large Areas and Long Periods

Comparisons with gauge networks show that error in the radar estimate tends to decrease, if integrated over a larger area (say, a few hundred km^2) and over a longer time (say, a few hours). Collier (1989) showed that the error of a radar calibrated with one rain gauge is about 30% for integration over less than 100 km^2 /2 hrs, while it is less than 10% for 400 km^2 /4 hrs. There are hydrological applications (e.g. total water availability over a river catchment) for which a large area-long time estimate of rainfall is useful. Because of the averaging in space and time the Z-R relationship assumed is not critical. An early method which was tried was to photograph the radar scope (at constant settings) on the same film by multiple exposure at frequent intervals. Densitometric measurement on the film was used to evaluate the cumulative rainfall over the area. Alternatively, successive MDR plots (see Chapter 5) could be used to integrate rain rate over a period of time. [See Kulshrestha (1966) for a review of some early methods].

7.3.2 Area-Time Integrals

Several authors have found (e.g. López et al. 1983, 1989) that the total rain volume integrated over several hours over a weather system is well related to the area of the system. In Florida, USA, over an experimental area of 1.3×10^4 km^2, López et al. (1983) obtained a correlation coefficient of 0.92 between rain volume and area for times of the order of 1 hour. This suggests that by integrating the radar echo area (above a set threshold value of reflectivity) over a period of time we can get an Area-Time Integral and multiplying this by an average rainfall rate we can get the rain volume. The advantage of the method, if it works, would be its simplicity and avoidance of the use of Z-R relationships every time[16] and the radar calibration will not be critically important.

[16] This is akin to methods of rain estimation from satellite visible and infrared imagery over large areas. See section 7.6.

Following Doneaud et al. (1984) we may put the rain volume V(km^2 mm) over area A during time T as,

$$V = \int_T \int_A R da dt \tag{7.14}$$

where R is the precipitation rate and a and t are the variables corresponding to A and T. If a constant rain rate R_c can be assumed

$$V = R_c \int_T \int_A da dt \tag{7.15}$$

The double integral in equation (7.15) is called Area-Time Integral (ATI) and is expressed in km^2 h. It can be expressed as the sum of areas A_i over which rain above a specified Z_e (or R) threshold was detected in the ith observing period with a time interval of δt_i between observations i.e.,

$$\text{ATI} = \int_T \int_A da dt = \sum A_i \delta t_i \tag{7.16}$$

The rain volume (RV) (km^2 mm) over the area, derived from a gauge network or estimated by radar using some Z-R relationship can be divided by the ATI to get the average rainfall rate. This can be used with subsequent measurements of ATI to obtain a rain volume estimate.

Doneaud et al. obtained a power law relationship between RV and ATI of the form,

$$\text{RV} = K(\text{ATI})^b \tag{7.17}$$

Since R_c = (RV)/(ATI) we have,

$$R_c = K(\text{ATI})^{b-1} \tag{7.18}$$

The measured area will evidently depend on the Z_e threshold assumed. For various thresholds, Doneaud et al. obtained different values of K and b. If $b < 1$, R would decrease with increasing ATI, which is contrary to experience. Hence, they chose a threshold which would make b very close to 1, making the RV-ATI relationship a linear one. This threshold was 25 dBZ for convective rainfall[17] in North Dakota, USA.

In the GATE ship radar data, Marks and Hudlow (1976) found an almost linear relationship between the volumetric water flux from echo base (m^3 /5 minutes) and the echo area (km^2) and there was little variation with type of echo.

Geotis (1991), in a study in Darwin, Australia, with C-band radar, obtained power law relationships of the form $F = aA^b$, where F is the rain flux in kg s^{-1} and A is the area in km^2 of echo exceeding 15 dBZ. F was calculated from a Z-R relationship. The coefficient and exponent varied widely and were generally different for monsoon and non-monsoon rains. The monsoon

[17] 25 dBZ corresponds to a rain rate threshold of 1.33 mm h^{-1} assuming $Z = 155R^{1.88}$

rains, which were probably largely stratiform, had low values of F but with large areas, so exhibited lower values of the coefficients. However, he found that increasing the threshold value to 25 dBZ does not change the exponent significantly, while the coefficient increases as is expected. Atlas et al. (1990) and Rosenfeld et al. (1990) considered a general case where there is a total rain area A_0, while the area with precipitation rate R above a threshold value p, is $A(p)$. The volume rain rate $V(p)$ (considering only rain above the threshold) is,

$$V(p) = A(p)R_c(p) \tag{7.19}$$

$R_c(p)$ is the mean rain rate above the threshold p and is given by,

$$R_c(p) = \frac{\int_p^\infty RP(R)dR}{\int_p^\infty P(R)dR} \tag{7.20}$$

where $P(R)$ is the percentage of the area where the rain rate is between R and $R + dR$. The volume rain rate measured in equation (7.19) is a fraction ϕ of the total volume rain rate V_t.

$$\phi = \frac{\int_p^\infty RP(R)dR}{\int_0^\infty RP(R)dR} \tag{7.21}$$

$$V_t = \frac{A(p)R_c(p)}{\phi} \tag{7.22}$$

Dividing by A_0, we get the average area wide rain rate $<R>$,

$$<R> = \frac{F(p)R_c(p)}{\phi} = S(p)F(p)$$

$$\text{where } F(p) = \frac{A(p)}{A_0} \text{ and } S(p) = \frac{R_c(p)}{\phi} \tag{7.23}$$

$S(p)$ may be defined in terms of the probability distribution function (PDF) of rain rates by dividing equation (7.20) by (7.21),

$$S(p) = \frac{\int_0^\infty RP(R)dR}{\int_p^\infty P(R)dR} \tag{7.24}$$

and can be taken, according to Atlas et al. (1990), as determined entirely by the climatological PDF for the regime (season/region) in question. We can expect a PDF without large variations to be applicable, if the rainfall is integrated over a long time or large area. The area should be large enough to include storm cells at various stages of their life cycle so that a single snapshot of the fractional area coverage serves as a good indicator of the average rain rate.

Thus <*R*> can be inferred by measuring the fractional area covered by rates above the threshold, provided *S*(*p*) is known. If we average *A*(*p*) in equation (7.22) over time *T*, the total rain volume *V* over time *T* is,

$$V = \frac{\left[A(p)\right]_{\text{av}} .TR_c(p)}{\phi} = \left[\text{ATI}\right]S_p \tag{7.25}$$

i.e. if *S*(*p*) is known, *V* can be determined from the ATI. *S*(*p*) for any region can be obtained as the slope of the regression between *V* and ATI for the life of one storm or as the slope of the regression of <*R*> and *F*(*p*) for many storms at one instant. Rosenfeld et al., (1990) also used the echo top height information from radar to determine the "effective efficiency" E_e of a storm, which is the fraction of water vapour carried from surface through the cloud base to the top. With increasing E_e the PDF is shifted towards higher rainfall rates. They call this technique the Height-Area Rainfall Threshold (HART) method.

7.4 Precipitation Estimation from Attenuation

7.4.1 Relationship between Attenuation and Rain Rate

As seen in Chapter 2, the relationship between rainfall rate (R mm h^{-1}) and attenuation due to rainfall (k_R dB km^{-1}) can be expressed as a power law,

$$k_R = A''R^{b''} \tag{7.26}$$

The exponent is close to unity at some wavelengths (e.g. 0.9 cm; see Table 2.4). The average rain rate over a path from radar (range zero) to a range *L* is

$$R_{\text{av}} = \frac{1}{L}\int_0^L R(r)dr \tag{7.27}$$

where *R*(*r*) is the precipitation rate at range *r*. The total attenuation (dB) along the path will be,

$$A = A''\int_0^L \left[R(r)\right]^{b''} dr \tag{7.28}$$

If *A* is deemed to be caused by an average precipitation rate R_x,

$$A = LA''R_x'' \text{ or } R_x = \left(\frac{A}{LA''}\right)^{(1/b'')} \tag{7.29}$$

Substituting for *A* we get,

$$R_x = \left\{ \frac{1}{L} \int_0^L \left[R(r)^{b''} \right] dr \right\}^{(1/b'')} \tag{7.30}$$

Therefore if $b'' = 1$, $R_x = R_{av}$; i.e. the average rain rate derived from the attenuation measurement will be equal to the true average rain rate in the path irrespective of the actual distribution of $R(r)$ along the path.

7.4.2 Measurement of Total Attenuation

If the attenuation over a transmission path is measured at a wavelength at which b'' is nearly unity, the average rain rate over the path can be computed. This has been done by many communication engineers using communication links. It can also be done using a radar and passive reflectors installed a few kilometres away. The received echo powers from two reflectors are compared in dry conditions and also while there is precipitation between them. The total attenuation in the path between the reflectors is thus obtained and this is independent of the radar calibration. Collis (1964) used a 8.7 mm wavelength radar and related the attenuation to the rainfall measured at a few gauges along the path. An advantage of this wavelength is the appreciable attenuation over a short path which makes it measurable. However he found that the relationship was not consistent from storm to storm. He attributed this to variability of the drop size distribution. But he concluded that the measurement would be useful in detecting heavy rainfall in river catchments. Harrold (1967) in a similar experiment, used an 8.6 mm radar and tetrahedral reflectors placed in the same line of sight. The nearer reflector was 26.5 km from the radar and the spacing between the reflectors was 6.9 km. The path average rain rate was obtained from 8 gauges between the reflectors. Assuming $b'' = 1$, he obtained A'' ranging from 0.20 to 0.39 (mean value 0.26 or 0.27) which agreed fairly well with theory. But he found that with heavy rain the signal went below the MDS.

Atlas and Ulbrich (1977) found that while b'' increases from 1.04 to 1.16 from 0.86 cm to 3.2. cm wavelength, the estimated error in derived R at the two wavelengths were 9 and 21% respectively. Even at 3.2 cm therefore the errors were less than those incurred from use of Z-R relations. However, thcy also found that in heavy rainfall, measurement was not possible because of excessive attenuation.

Besides variation in drop size distribution the effect of vertical motion will also vitiate the comparison with rain gauge catch as in the case of Z-R measurements. Moreover, the technique gives only path averages along the radial where the reflectors are situated. This method is thus suitable for communication engineers who are interested in attenuation along specific paths. Areal rainfall measurements will need reflectors all around the radar. The method is also not suitable for measurements at long ranges.

7.4.3 Total Attenuation from Air- or Space-borne Radar

In the case of a radar looking down or at an angle at the ground or sea surface from an aircraft, a strong echo from the surface will be received in addition to the precipitation return. Fig. 7.3 from Meneghini et al. (1986) shows the range normalised signal power (dBm m^2) received at two wavelengths (X-band and Ka-band) by a nadir-looking airborne radar as a function of range. The aircraft was flying at an altitude of 4.5 km. Fig. 7.3a shows the signal obtained during precipitation and Fig. 7.3b when there was no precipitation (a few minutes later). Considering only the Ka-band data (dashed curves in the figure)[18], the ratio of the sea surface signals (strong spikes at 4.5 km range) without and with rain should give the total attenuation. That is, instead of using passive reflectors we use the surface signal as reference for determining the attenuation. This is called Surface Reference Technique (SRT). The reference signal can be obtained from the same spot when precipitation is not occurring or in a neighbouring area[19]. In either case, it is assumed that the back-scattering cross section per unit area (σ) of the surface is unchanged despite the spatial or time difference. But the signal power from the surface alone in the presence of rain P'_{Gn} will be lower than the return P_{Gn} in the absence of rain. The suffix n denotes the nth range bin. It was shown by Meneghini et al. (1983) that,

$$P_{Gn} = \frac{C_G \sigma_1}{r_n^3} \text{ and } P'_{Gn} = \frac{C_G \sigma_2 A_n}{r_n^3} \tag{7.31}$$

where C_G is a constant and A_n is the total rain attenuation given by,

$$A_n = \exp\left(-0.46 \int_{r_0}^{r_n} k_R \, ds\right) \tag{7.32}$$

where s represents path length, k_R is the attenuation coefficient in dB km^{-1}, r_0 is the range to the bin at which the precipitation starts and r_n the range

[18] The dual wavelength aspect of this figure will be discussed in section 7.5.

[19] The surface reference technique has also been used with ground-based radar. From echoes from a mountain with and without rain in the path, the path-integrated attenuation between the radar and the mountain is obtained and therefrom the rain estimate is made (Delrieu et al. 1997, 1999). Testud et al. (2000) developed an algorithm called ZPHI to derive rain profile along a radar beam using polarimetric parameters.

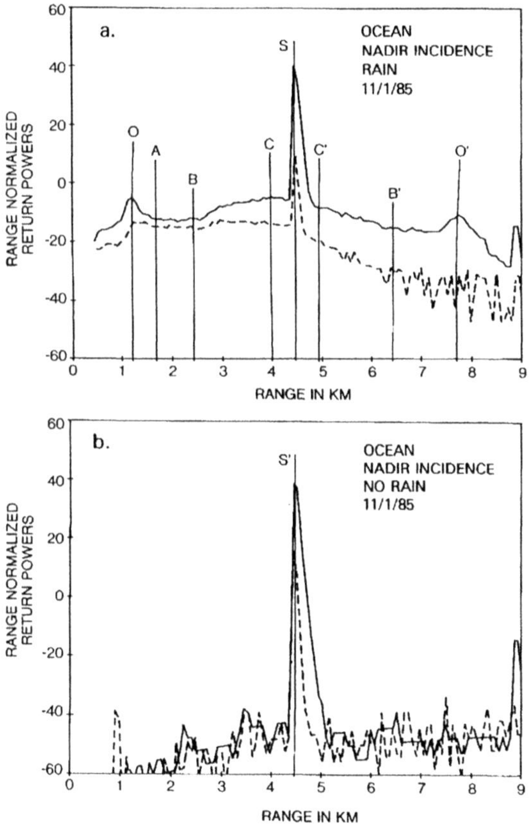

Fig. 7.3. Range normalised signal powers received by an airborne nadir-looking dual-wavelength radar plotted against range. (from Meneghini et al. 1986; courtesy Dr. R. Meneghini of NASA, USA and AMS). a. Signal in presence of precipitation. b. Signal in absence of precipitation. Units on Y-axis dBm m². Continuous line is for X-band and dashed line for Ka-band. Points S, S' refer to sea surface 4.5 km below aircraft. Point O is bright band. The echoes beyond 4.5 km are images (See text for discussion).

to the centre of the nth bin. The σ values for the two cases (σ_1 and σ_2) are assumed unequal. If the received power from rain at the nth range bin is P_{Rn} the total received power from the nth bin while there is precipitation will be not P'_{Gn} but ,

$$P_{1n} = P'_{Gn} + P_{Rn} + P_{N1} \tag{7.33}$$

and in the absence of precipitation,

$$P_{2n} = P_{Gn} + P_{N2} \tag{7.34}$$

P_{N1} and P_{N2} represent noise power at the receiver in the two cases. The estimate A_{nc} of A_n is then,

$$A_{ne} = \frac{P_{1n}}{P_{2n}} \tag{7.35}$$

From equations (2.115), (7.32) and (7.35) we have,

$$\int_{r_0}^{r_n} R^{b''} ds = -\frac{1}{0.46A''} \ln \frac{P_{1n}}{P_{2n}} \tag{7.36}$$

If we define the average rain rate in the path, R_{av} as,

$$R_{\text{av}} = \left[\frac{1}{r_e} \int_{r_0}^{r_n} R^{b''} ds \right]^{(1/b'')} \quad \text{where } r_e = r_n - r_0 \tag{7.37}$$

$$R_{\text{av}} = \left[-\frac{1}{0.46A''} \cdot \frac{1}{r_e} \ln \frac{P_{1n}}{P_{2n}} \right]^{(1/b'')} \tag{7.38}$$

For the Ka-band $b'' = 1$ and equation (7.38) is simplified.

7.5 Multi-parameter Methods of Precipitation Estimation

7.5.1 Parameters of the Drop Size Distribution (DSD)

The power law relationships between Z and R and k_{R} and R are uncertain largely because of variation in the DSD. If the DSD is known, R, Z and k_{R} can be uniquely determined without depending on empirical relationships. Without knowing the DSD the efforts to relate Z and k_{R} to R are only statistical exercises lacking in physical elegance and in accuracy as well.

We can look at the problem in another way, following the treatment of Jameson (1991). Suppose we wish to relate R to any one radar-measured quantity X. R and X can be expressed as ,

$$R = C_R N_T \left(D^{3+p}\right)_{\text{av}} \text{ and } X = C_X N_T \left(D^q\right)_{\text{av}} \tag{7.39}$$

where C_R and C_X are constants, N_T is the total concentration of drops, D is drop diameter, p is the exponent of the relation between terminal velocity and diameter (see Chapter 2) and the symbol $(\,)_{\text{av}}$ represents averaging over the DSD. Empirical power law relations are of the form

$$R = AX^b \tag{7.40}$$

From these equations we get estimated R denoted by $E(R)$ as,

$$E(R) = C_R C_X^{-b} N_T^{1-b} X^b \text{ where } b = \frac{\ln(D^{3+p})_{\text{av}}}{\ln(D^q)_{\text{av}}} \tag{7.41}$$

In practice, empirical relations like equation (7.40) are derived after sampling over many different DSD's and b represents an average value.

If $3 + p = q$, then regardless of the DSD, $b = 1$ and $E(R)$ is proportional[20] to X. In general the more $3 + p$ departs from q the greater is the uncertainty in $E(R)$ due to variation in DSD. To establish the DSD we need to measure more than one parameter. If we assume that the DSD is exponential, we have from Chapter 2,

$$N_D = N_0 \exp(-\Lambda D) \text{ or } N_D = N_0 \exp\left(\frac{-3.67}{D_0} D\right) \tag{7.42}$$

There are two variables here (N_0 and Λ or N_0 and D_0). Hence it is necessary to measure two independent parameters to define the DSD. An example is the determination of N_0 and D_0 from measurement of Z_H and Z_{DR} discussed in Chapter 2.

Ulbrich and Atlas (1978) expressed Z, R, k_R, Σ (optical extinction) and M (liquid water content) as functions of N_0 and D_0. These expressions are reproduced in Table 7.3. They developed a "Rain Parameter Diagram"[21] (Fig. 7.4) which has R and Z along its axes and carries isopleths of N_0, D_0, and M. In the version of the diagram given in Fig. 7.4, the symbol W is used for liquid water content (broken lines); isopleths of $\log_{10} N_0$ are drawn as lines with short and long dashes and isopleths of D_0 are solid lines. The symbol μ is used for the exponent m of the gamma distribution [equation (7.61)]. $\mu = 0$, written at the top left signifies that an exponential distribution is assumed.

The diagram shows that if any two of the above-mentioned quantities are measured, all the others can be uniquely determined subject to the assumption that DSD is exponential. Ulbrich and Atlas (1978) also show that Fig.7.4 is applicable to any DSD (not only exponential) for which two of the moments are known. Any of the Z-R relations listed in section 7.2 can be plotted as straight lines on the rain parameter diagram and will not be parallel to the N_0 isopleths as they should be if N_0 is a constant. Ulbrich and Atlas (1978) have also presented additional overlays which can be laid on the base diagram of Fig. 7.4 to evaluate k_R, and Σ at different wavelengths and different temperatures. Z_{DR} which is a function of only D_0 (Chapter 2) for an exponential distribution can also be included.

7.5.2 Multi-parameter Methods Available

We shall now consider the combinations of parameters which can be

[20] An example is the case of attenuation at a wavelength of 0.86 cm.

[21] The diagram is an extension of an earlier one by Atlas and Chmela (1957).

Table 7.3. Expressions for Z, R, M, k_R and Σ (from Ulbrich and Atlas 1978).

$$Z = 10^6 N_0 \int_0^\infty D^6 e^{-\Lambda D} dD = 8.03 \times 10^4 N_0 D_0^7 \tag{7.43}$$

$$R = 0.6\pi N_0 \int_0^\infty v(D) D^3 e^{-\Lambda D} dD = 0.6\pi N_0 F_n(D_0) \tag{7.44}$$

where $v(D)$ is the fall speed m s^{-1} of a drop of diameter D. The integral $F_n(D_0)$ is evaluated numerically using rain drop fall speeds of Gunn and Kinzer (1949).

$$M = \frac{\pi}{6} N_0 \int_0^\infty D^3 e^{-\Lambda D} dD = 1.73 \times 10^{-2} N_0 D_0^4 \tag{7.45}$$

$$k_R = 0.4343 N_0 \int_0^\infty Q_D e^{-\Lambda D} dD = 0.4343 N_0 F_A(D_0) \tag{7.46}$$

where Q_D (cm^2) is the total attenuation cross section (see Chapter 2) and $F_A(D_0)$ is an integral to be evaluated using known attenuation cross sections.

$$\Sigma = \frac{\pi}{20} N_0 \int_0^\infty D^2 e^{-\Lambda D} dD = 6.35 \times 10^{-3} N_0 D_0^3 \tag{7.47}$$

Eliminating N_0 between equations (7.43) and (7.44) we get an expression for R, in terms of D_0 and Z.

$$R = (2.35 \times 10^{-5}) D_0^{-7} F_n(D_0) Z \tag{7.48}$$

Using equation (7.48), isopleths of D_0 are plotted on the diagram. Using other combinations of these equations, the other isopleths are also plotted.

measured by radar to estimate R. To the parameters mentioned in the rain parameter diagram we may add measurements at more than one wavelength. Besides Z_{DR} we can also add the other polarisation parameters discussed[22] in Chapter 2. Good reviews of multi-parameter methods may be seen in Atlas et al.(1984), Hall (1984) and Rogers (1984).

7.5.3 Reflectivity Factor and Attenuation

Rain estimation by (1) measuring Z at an unattenuating wavelength (say, 10 cm) and k_R at an attenuating wavelength (say, 3 or 1 cm) and (2) measuring k_R at two attenuating wavelengths (say 1 and 3 cm) has been studied by several authors e.g. Sulakvelidze and Dadali (1968), Eccles and Mueller (1971), Goldhirsh and Katz (1974) and Goldhirsh (1975).

Eccles and Mueller (1971) developed a method of determining the specific attenuation over a short range interval, excluding the attenuation in the intervening path, using the ratio of returned powers at two wavelengths. They

[22] We are not considering non-radar methods such as passive radiometry or measurement of optical extinction.

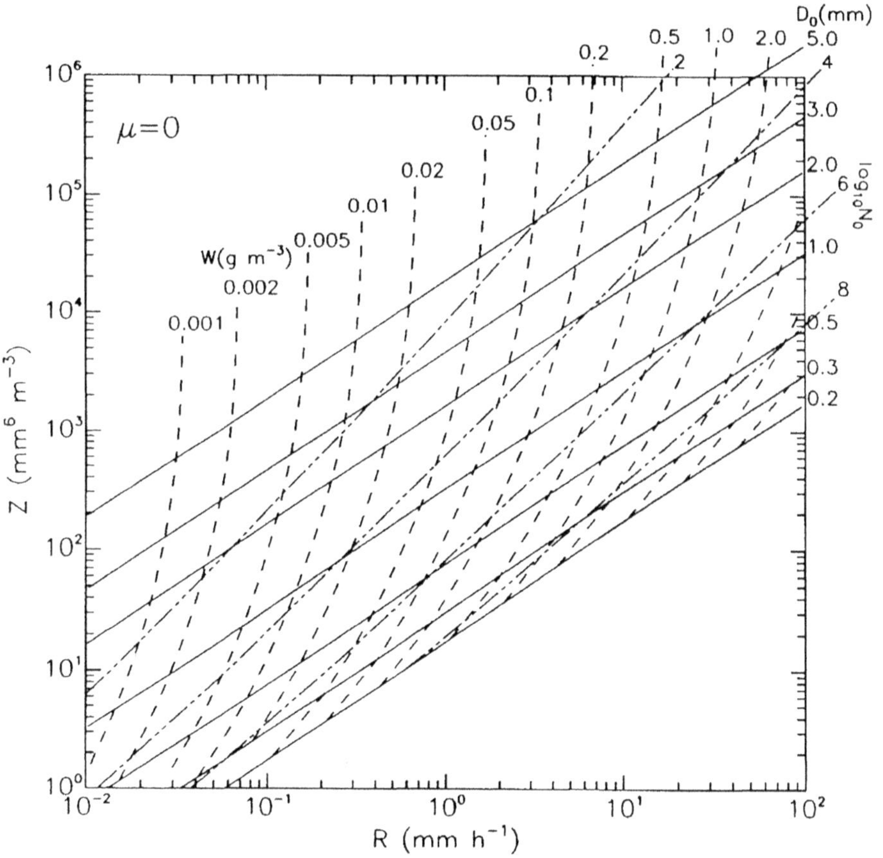

Fig. 7.4. Rain Parameter diagram of Ulbrich and Atlas. [from Ulbrich and Atlas 1978. Copyright by the American Geophysical Union (AGU). Reproduced courtesy Dr. Ulbrich and AGU]. Base diagram: Radar reflectivity factor Z mm^6 m^{-3} versus rainfall rate R mm h^{-1}. Isopleths are:
1. median volume diameter D_0 mm (solid lines),
2. $\log_{10}N_0$ (lines with short and long dashes), N_0 being the distribution parameter m^{-3} cm^{-1}
3. liquid water content W g m^{-3} (broken lines)

considered the average echo powers P_{10} and P_3 at 10 cm and 3 cm wavelengths from matched beams of a dual wavelength radar, at range $r = r_0 + s$ in an area of precipitation starting from range r_0.

$$P_{10} = \frac{C_{10} Z_{10}}{r^2} \tag{7.49}$$

$$P_3 = \frac{C_3 Z_3}{r^2} 10^a \text{ where } a = -0.2 \int_{r_0}^{r_0+s} k_{R3} dr \tag{7.50}$$

k_{R3} being the attenuation rate (dB km^{-1}) at the 3 cm wavelength and C_{10} and C_3

are radar constants. If hail is absent, Z_{10} can be equated to Z_3. Dividing one equation by the other and taking logarithms we get,

$$10\log\frac{P_{10}/C_{10}}{P_3/C_3} = 2\int_{r_0}^{r_0+s} k_{R3}dr \tag{7.51}$$

If, at a farther range $r_0 + s + s'$ within the rainfall area, the received powers at the two wavelengths are P'_{10} and P'_3 and the incremental attenuation rate between s and s' is k'_{R3} we have,

$$2\int_{r_0}^{r_0+s} k_{R3}dr + 2\int_{r_0+s}^{r_0+s+s'} k'_{R3}dr = 10\log\frac{P'_{10}/C_{10}}{P'_3/C_3} \tag{7.52}$$

From the last two equations we get,

$$2\int_{r_0+s}^{r_0+s+s'} k'_{R3}dr = 10\log\frac{P'_{10}P_3}{P'_3P_{10}} \tag{7.53}$$

Here, the effect of attenuation up to the range $r_0 + s$ has vanished. If the average attenuation rate over the incremental distance s' is k_{R3av}

$$2k_{R3av}s' = 2\int_{r_0+s}^{r_0+s+s'} k'_{R3}dr \tag{7.54}$$

From the last two equations,

$$\begin{aligned} k_{R3av}s' &= (0.5)10\log\frac{P'_{10}P_3}{P'_3\,P_{10}} \\ &= (0.5)(10\log_{10}e)\ln\frac{P'_{10}P_3}{P'_3\,P_{10}} \end{aligned} \tag{7.55}$$

which can be approximated as,

$$k_{R3av}s' = 2.1715\left[\frac{P'_{10}P_3}{P'_3\,P_{10}} - 1\right] \tag{7.56}$$

The rain rate R (or M) over the range interval can then be determined from a power law relationship. The advantage of the method is that the radar calibration cancels out and the method is less sensitive to DSD variations than the standard *Z-R* method.

Alternatively, assuming the exponential DSD, the averaged Z_{10} at 10 cm wavelength over a path of uniform rain from range r_0 to $r_0 + s$ can be written, using equation (2.42), as,

$$Z_{av} = N_0 6!\Lambda^{-7} = N_0 6!\left(\frac{D_0}{3.67}\right)^7 \tag{7.57}$$

The average k_{R3} will be, from Equations (2.113) and (2.114)[23],

[23] C and n are defined by equation (2.112).

$$k_{\text{R3av}} = 0.4343\Gamma(n+1)CN_0\Lambda^{-(n+1)}$$
$$= 0.4343\Gamma(n+1)CN_0\left(\frac{D_0}{3.67}\right)^{n+1} \tag{7.58}$$

Equations (7.57) and (7.58) can be solved for N_0 and Λ (or N_0 and D_0) using appropriate values for C and n.

If two attenuating wavelengths (say, 1 and 3 cm) are used, two equations similar to (7.58) for k_{R1av} and k_{R3av} (the suffixes 1and 3 indicating the wavelengths) are available for evaluating N_0 and Λ or D_0. Also by analogy with equation (7.53) we can get,

$$2\int_{r_0+s}^{r_0+s+s'} (k'_{\text{R1}} - k'_{\text{R3}})dr = 10\log\frac{P'_3P_1}{P'_1P_3} \tag{7.59}$$

where the symbols and suffixes have meanings similar to those used earlier. From this, Nakamura et al. (1990) expressed the difference D_{Att} between the specific attenuations at the two wavelengths[24] as,

$$D_{\text{Att}} = \frac{1}{2s'}10\log\frac{P'_3P_1}{P'_1P_3} \tag{7.60}$$

From D_{Att} the rain rate can be derived using an appropriate DSD.

Berjulev and Kostarev (1974) used equation (7.53) to determine attenuation over a path length at 8.6 mm wavelength. They used a 3 cm radar for the first wavelength and neglected the attenuation at this wavelength. They deployed rain gauges along the path. They found that attenuation in rain is a rapidly varying parameter.

Goldhirsh (1975) considered the errors in these techniques. Assuming $\pm$ 2 dB uncertainty in Z and $\pm$ 25% in k_{R}, he concluded that the combination of Z_{10} and k_{R1} produced acceptable uncertainties in estimation of R and M. The combination of k_{R1} and k_{R3} involved larger errors; and the combination k_{R3} and Z was impractical.

For a 10 cm - 3 cm combination of wavelengths, the dual wavelength method has to be applied over a long path interval. In the 10 cm - 1 cm combination saturation will occur at 1 cm wavelength if a long path or a high rain rate is involved. A large number of independent samples is also needed to obtain acceptable variance in the P_r measurements. The assumption of equal Z at the two wavelengths may not be valid in the case of 1 cm, as Mie scattering can occur at this wavelength even from rain drops (Nakamura and Inomata 1991). The method assumes uniform rain over the path which may be the case only in light

---

[24] They used Ka (0.87 cm) and X (3 cm) band radars; hence, here, the suffixes 1 and 3 refer respectively to these two wavelengths.

stratiform rain. The assumption of absence of hail is also a limitation. The problem of matching of beams at the two wavelengths is also quite a serious one. The techniques therefore do not appear promising for operational use with ground-based radars. However, Nakamura et al. (1990) point out that the method is useful to get a vertical profile of rain rate which is useful to communication engineers, besides giving the structure of precipitation. They have also suggested that D_{Att} can be used to calibrate Z measurement by the radar instead of comparing with rain gauge values.

We have seen in Chapter 2 that N_D can be expressed as a 3-parameter distribution.

$$N_D = N_0 D^m \exp\left(\frac{-3.67 + m}{D_0} \cdot D\right) \tag{7.61}$$

This equation has 3 parameters N_0, D_0 and the exponent m which can be solved for, if three independent measurements are made. Sulakvelidze and Dadali (1968) proposed that measurements be taken of reflectivities at two wavelengths and attenuation at a third wavelength. The wavelengths should be so chosen that the scattering and attenuation properties are distinctly different at the three wavelengths. They proposed the wavelengths of 7 and 0.9 cm for reflectivity and 2 cm for attenuation measurement. But considering the difficulties encountered with even two wavelengths, the practical viability of a 3-wavelength measurement is doubtful. Ulbrich (1983) has proposed a method of determining the parameters of the gamma distribution with dual measurements only. By working back from the numerous empirical Z-R relationships found all over the world (see Battan 1973 and Table 7.2 of this book) he found that N_0 and m are not independent but related by an equation,

$$N_0 = C_N \exp(3.2m) \tag{7.62}$$

where C_N is a constant. Hence measured Z and k_R can be expressed as functions of m and D_0 only; in principle m and D_0 can be evaluated but the practical limitations mentioned above still apply.

Meneghini et al. (1986) have obtained interesting results with an airborne dual wavelength radar. They used X- and Ka-band (34.45 GHz) radars looking down over the ocean surface on an aircraft flying at an altitude of about 4.5 km. In Fig.7.3a, which was discussed in section 7.4, the signal peaks at the bright band O (the peak is smaller at Ka-band). Thereafter the Ka-band signal decreases compared to the X-band because of the attenuation. The measured difference in dBZ values at the two wavelengths ($\mathrm{dBZ_X} - \mathrm{dBZ_{Ka}}$) called the measured Dual Frequency Ratio (DFR), is partly due to Mie scattering and partly to differential attenuation. Then there is a sharp spike (S) at both wavelengths at 4.5 km range. This is the return from the sea surface. The spike at the Ka-band is attenuated more by the rain. Beyond 4.5 km range "image" signals were obtained by a process of scattering of the incident radiation from the surface to the rain where a portion of the energy is radiated back to the surface where it is scattered towards the antenna. These signals decreased with range due to attenuation except for a small peak at the image of the bright band. In Fig. 7.3b, for observations taken over the same spot a few minutes later when there was no rain, the peaks due to the sea surface unattenuated by rain were alone observed.

Meneghini et al. suggest that rain rate over the path can be estimated in several ways.

(1) The total attenuation SRT method discussed in section 7.4.

(2) Z at each range gate from point A to point C (Fig. 7.3a) is computed from the returned power and converted to R using a Z-R relationship. The region AC is chosen to avoid the bright band. The R values are averaged over this region; attenuation is neglected.,

(3) Eccles-Mueller method. Subtract the difference between the ordinates at point A from the corresponding difference at point C. This is proportional to the differential attenuation between A and C. Again the melting region is avoided.

(4) The differences between the surface returns at S at the two wavelengths in rain conditions is taken. From this is subtracted the corresponding quantity (at S') in no rain conditions to get a result proportional to the differential attenuation. These results are of interest in relation to satellite-borne radar (see section 7.7).

The airborne radar mentioned above was subsequently modified to be able to receive H and V polarisations. Using it, Kumagai et al. (1993) and Meneghini and Kumagai (1994) made LDR measurements in the X-band besides the DFR in stratiform and convective rain in the western Pacific. In the former, measurable LDR values (maximum -12 dB) were confined to the melting layer. The maximum DFR occurred at heights above the LDR peak. The dBZ and measured DFR are relatively insensitive to particle shape but are determined by the phase state of the hydrometeors and their size distribution. On the other hand, LDR is relatively insensitive to particle size and depends on shape. Noting this they have modelled the melting layer. This is of interest again for satellite-borne radar measurements of rainfall, as the melting layer cannot be avoided in such measurements. In convective rain the maximum LDR was much smaller but extended both above and below the freezing level. This was interpreted as due to an extended region containing liquid, frozen and partially melted hydrometeors.

7.5.4 Reflectivity and Doppler Velocity at Vertical Incidence

In Chapter 3 we mentioned the possibility of obtaining particle fall speed, vertical air motion and DSD from Vertical Incidence Doppler (VID) measurements. If DSD can be determined, obviously, rainfall rate can be derived. Considerable work has been carried out subsequent to the early methods mentioned in Chapter 3, but these techniques are mainly directed to determination of DSD and are of research interest. The measurement, being restricted to the vertical direction over the radar site, is not suitable for operational measurements of rain rates over an area.

It was seen in Chapter 3 that assuming the Marshal-Palmer DSD, Rogers (1964) obtained V_t, the mean Doppler fall speed measured by a vertically pointing Doppler radar.

$$V_t = A''Z^{b''} = 3.8Z^{1/14} \tag{7.63}$$

A correction for air density at higher altitudes can be made as discussed in Chapter 2. Subtracting V_t from the observed mean Doppler velocity, the vertical air velocity w can be obtained. Alternatively, instead of using equation (7.63), an empirical V_t-Z relationship can be obtained using DSD measurements at the ground or from aircraft, and w computed as above. If the vertical air velocity can be assumed to be zero, the fall velocity can be equated to the observed Doppler velocity. Then the Doppler velocity spectrum represents the spectrum of the fall velocities. The reflectivity factor contributed by drops having diameter D to $D + dD$ and corresponding terminal velocities v to $v + dv$ can be expressed as $N_D D^6 dD$ and also as $Z(v)dv$. Hence,

$$N_D = \frac{Z(v)}{D^6} \cdot \frac{dv}{dD} \tag{7.64}$$

Taking dv/dD from the data of Gunn and Kinzer (1949) (see Chapter 2), N_D for any D can be obtained from the Doppler spectra, i.e. the DSD can be determined. If there is a known vertical air velocity w and if it is uniform over the sampled volume, the observed Doppler spectrum can be displaced along the velocity axis by an amount w, to yield the spectrum of terminal fall velocities (Rogers 1967). The drop sizes can be inferred as before. This is usually called Rogers' method. While the vertical velocity w can be determined from a knowledge of Z and mean Doppler velocity, a knowledge of the complete Doppler spectrum is needed for determining DSD. The DSD obtained is very sensitive to the value of w used. This means that the DSD estimate could have large error even for small errors in w. On the other hand, a rough knowledge of the DSD, such as can be obtained if there is an independent rain rate or LWC measurement, is sufficient to derive a fairly accurate value of w.

Sekhon and Srivastava (1971) used Rogers' method to derive DSD in a thunderstorm. They estimated the errors by working out the DSD using the vertical velocity w obtained by the Rogers' method and by using the values $w + 1$

and $w - 1$ m s^{-1}. From this, they concluded that an error in w causes a change of slope of the DSD but not its form i.e. it continues to be exponential. Srivastava (1990)[25] suggested the use of the equation (2.46) for the fall velocity, reproduced here as equation (7.65).

$$V_D = a_1 - a_2 \exp(-600D) \text{ m s}^{-1} \quad (7.65)$$

At vertical incidence and with no vertical air motion this equation can be integrated assuming an exponential DSD to yield the mean Doppler velocity V_0.

$$V_0 = a_1 - a_2 \left(\frac{\Lambda}{\Lambda + 600} \right)^7 \quad (7.66)$$

From this equation Λ can be determined for stratiform rain. Knowing Λ, N_0 can be calculated from equation (2.42) using measured Z. Thereafter M and R can be computed. The method does not require the full spectrum to be seen. If vertical air velocity is to be allowed for, it can be determined by the EVAD technique (Srivastava et al. 1986) discussed in Chapter 3.

Richter (1993) has proposed that the vertical air motions in the same range gates as the particle motion can be determined independently by a Radio Acoustic Sounding System (RASS) (see Chapter 10). Hauser and Amayenc (1981, 1983, 1984), in an attempt to avoid assuming the M-P relationship, introduced a "3P method". They assumed an exponential DSD but used a least square fit of a theoretical spectrum to each experimentally observed Doppler spectrum using the French RONSARD C-band radar. From this, they could determine simultaneously N_0, Λ and w. Thus the DSD was determined and therefrom M and R could be obtained. They concluded that the technique would apply to stratiform rain but not convective precipitation or snow. The method requires a knowledge of the complete Doppler spectrum. The 3P method avoids the large error in DSD which results from errors in w in the Rogers' method.,

Klaassen (1989) used some characteristic parameters of the Doppler velocity spectrum instead of looking at the entire spectrum. These were:

(1) Z,

(2) mean Doppler velocity,

(3) maximum Doppler velocity which is taken as the velocity at the upper part of the spectrum where Z falls 10 dB below its maximum value and,

---

[25] Ice particles of the same mass can have different fall speeds because of variations in shape and bulk density. This is an additional source of variance in VID spectra. Passarelli and Srivastava (1980) considered the spectral variance to be made of two components – that due to mean fall speed-mass relationship and that due to fluctuations about the mean. These two can be of the same order.

(4) median Doppler velocity which is the velocity dividing the spectrum into two equal reflectivity parts.

A gamma distribution [equation (7.61)] is postulated. Since four observed parameters are available, w, N_0, D_0 and m can be calculated. This method is called the 3V (three velocities) method. Klaassen concluded that the method is suitable where large drop sizes are involved. Thomson and List (1996) in Canada, used an X-band Doppler radar in a vertically pointing mode in conjunction with a disdrometer at the ground. The radar-measured Z was calibrated using values of Z calculated from disdrometer data. They have described a technique of deriving both the drop size spectra and the updraft velocities from these measurements.

Walker and Ray (1974) proposed a technique using dual wavelength vertical incidence Doppler radar to estimate vertical velocity, DSD and turbulence. The average power from particles with velocities in the range v and $v + dv$ is,

$$S(v)dv = C\sigma(v)N_v(v)dv \tag{7.67}$$

where C is the radar constant, $\sigma(v)$ the backscattering cross section and $N_v(v)$ the number density. The ratio of $S_1(v)$ and $S_2(v)$ at two wavelengths, when normalised by the ratio of the radar constants is,

$$\frac{S_1(v)}{S_2(v)} = \frac{\sigma_1(v)}{\sigma_2(v)} \tag{7.68}$$

The vertical wind is found by fitting the measured ratio with that expected from still air, a determination independent of DSD and radar calibration. DSD is obtained from,

$$N_D(D) = N_v(v)\left(\frac{dD}{dv}\right)^{-1} \tag{7.69}$$

where $N_v(v)$ is obtained from equation (7.67) after the spectra have been corrected for the effects of vertical winds.

For more details on this topic the reader is referred to Atlas et al. (1973), Hauser and Amayenc (1981, 1983, 1984), Joss and Waldfogel (1970), Klaassen (1983, 1989), Pasqualucci (1984), Sekhon and Srivastava (1971), Srivastava (1990), Srivastava et al. (1986), Ulbrich (1983, 1992), Walker and Ray (1974) and Wiley et.al. (1970).

7.5.5 Use of Polarisation Parameters

Determination of precipitation rate from measurement of polarisation parameters was discussed in Chapter 2. Seliga et al. (1981) compared the rain

rate R_{ZDR} obtained from measurement of Z_H and Z_{DR} with estimates of R from three other methods viz.,
(1) Area-averaged rainfall rate R_0 from a dense network (20 km^2 per gauge) of rain gauges,
(2) Rain rate R_{ZR} from a conventional Z-R relationship,
(3) Rain rate R_{CZR} from a Z-R relationship calibrated with rain gauges over a sub-area.
They found that R_{ZDR} gave rise to lower average differences (22%) with reference to R_0 than the other two estimates. Bringi et al. (1982) compared R_{ZDR} estimates with rainfall-range profile derived from a disdrometer at the radar site and found excellent agreement. Goddard et al. (1982) compared measured Z_H and Z_{DR} with values derived from disdrometer measurements. They found an average difference of 1.6 dB in the case of Z_H and 0.3 dB for Z_{DR}. Seliga et al. (1986) compared observed Z_H (mm^6 m^{-3}) and Z_{DR} (dB) with values calculated from measurements of R (mm h^{-1}), M (g m^{-3}) and D_0 (mm) by a disdrometer and proposed (empirical) power law relationships between these quantities. Zhang et al. (2001) used a video disdrometer to find a relation between m and Λ in the gamma distribution and used this along with radar-measured Z_H and Z_{DR} to determine the DSD and therefrom the rain rate.

Other authors [e.g. Chandrasekar et al.(1990), Gorgucci et al. (1994), Sachidananda and Zrnić (1987)] have also found power law relationships connecting these quantities either by using actual radar data or by postulating a gamma distribution and calculating from the expressions for Z_H, Z_{DR}, and R. These relationships are of the forms [with Z_H usually expressed in mm^6 m^{-3} and Z_{DR} in dB; Z_H is in dB in equation (7.72)],

$$R = CZ_H^a Z_{DR}^b \tag{7.70}$$

$$R = CZ_H^a 10^{bZ_{DR}} \tag{7.71}$$

$$R = C \cdot 10^{(aZ_H - bZ_{DR})} \tag{7.72}$$

The relationships have been derived with various assumptions regarding axial ratios, canting angle, exponential or gamma DSD and maximum drop diameter. A standard error not exceeding 0.1 dB for Z_{DR} and 1 dB for Z_H measurement is needed. This can be obtained by averaging 60 to 70 contiguous sample pairs at S-band (Sachidananda and Zrnić 1986).

For calibration of Z_{DR} measurements, the radar antenna is pointed vertically. In this case measured Z_H and Z_V should be identical. Long term averages of any differences can be used as a correction for Z_V. Even so, at low values of Z_{DR}, the error in measurement may give rise to zero or negative values of Z_{DR}. In such a case, equation (7.70) breaks down but equation (7.71) or (7.72) can be used. For zero Z_{DR}, these equations reduce to Z-R relationships.

Errors in dual-polarisation measurements can arise also from non-matching of sidelobe patterns in the two polarisations. Detailed discussions on errors may be seen in Chandrasekar et al. (1990), Chandrasekar and Bringi (1988), Gorgucci et al.(1994) and Sachidananda and Zrnić (1987).

Also, as seen in Chapter 2, R can be derived from K_{DP} (degrees per km) from power law relationships of the form of equation (2.88). Seliga and Bringi (1978) drew a curve relating K_{DP}/N_0 and D_0 on Fig 2.6, for an exponential distribution. The figure could then be used to derive D_0 from measured Z_{DR}; then K_{DP}/N_0 is read off and knowing measured K_{DP}, N_0 is determined. Thus from measurement of Z_{DR} and K_{DP}, the DSD and rain rate can be derived. Using Z_{DR} and K_{DP}, relationships of the form,

$$R = CK_{DP}^{a} Z_{DR}^{b} \text{ or } R = CK_{DP}^{a} 10^{bZ_{DR}} \qquad (7.73)$$

have been derived (e.g. Ryzhkov and Zrnić 1995). These usually assume a drop shape vs. size relationship given by equation (2.64) reproduced here as equation (7.74).

$$\frac{b}{a} = 1.03 - 0.62 D_{eq} \qquad (7.74)$$

Gorgucci et al. (2000) pointed out that the value 0.62 of the coefficient (designated as β) of D_{eq} is not always valid and they formulated an algorithm for deriving β from measurements of Z_H, Z_{DR} and K_{DP}. Using this, Gorgucci et al. (2001) formulated a rainfall algorithm making the constants in equation (7.73), functions of β.

Jameson (1994) has suggested a method of deriving R from a measurement of specific differential attenuation k_{Dif} (see Chapter 2) and K_{DP} and taking their ratio. Balakrishnan et al. (1989) simulated rain rates from a disdrometer-measured DSD using various polarimetric algorithms using Z_H only, Z_H and Z_{DR}, K_{DP} only, and Z_H, Z_V and K_{DP} together. The r.m.s. errors of these algorithms were in decreasing order. The first of these was expected to give optimum results at $R <$ 20 mm h^{-1}, the second between 20 and 40 mm h^{-1} and the third above 40 mm h^{-1}.

K_{DP} has to be determined over a finite range interval over which the rain rate is assumed to be uniform. This assumption will be valid only if the interval is very small. However, in a small range interval, the estimate of K_{DP} will be more affected by the Doppler spread of the signal. Also, since K_{DP} is very small it is useful only for measurements of rain rates greater than about 50 mm h^{-1}. Sidelobe contamination will also affect the phase measurements (Sachidananda and Zrnić 1987). The differential propagation phase shift can be extracted for circular polarisation radar also and used to estimate rainfall rate (e.g. Tan et al. 1991).

Ruf et al. (1996) have described a small, low power, high p.r.f., Ka-band dual-polarisation radar which they have used with a corner reflector at a range of about 1 km. They determined k_{Dif} and used a linear relation[26] between k_{Dif} (dB km^{-1}) and R (mm h^{-1}),

$$k_{\text{Dif}}\ \text{dB km}^{-1} = 0.035R\ (\text{mm h}^{-1}) \tag{7.75}$$

to derive R. The measurements were verified with a tipping bucket rain gauge. It should be possible to instal a ring of reflectors around the radar and get measurements in all directions.

Sauvageot (1996) has proposed the estimation of rain rate from measurements of differential reflectivity and differential attenuation (k_{Dif} integrated over a path length) at an attenuating wavelength. According to him, this being a differential measurement is independent of radar calibration and from attenuation by undetected clouds. The method can also be used as a detector of hail and anomalous propagation.

Ryzhkov and Zrnić (1995) compared the dual-polarisation estimators of rain and found that the one using K_{DP} and Z_{DR} yielded a standard deviation 2 to 3 times less than one using Z_H and Z_V, in a squall line in the USA[27]. According to Zrnić and Ryzhkov (1996) [see also Vivekanandan et al. (1999), Brandes et al. (2001)], the advantages of using K_{DP} are:

(1) immunity to beam blockage,

(2) immunity to ground clutter cancelling and

(3) ease of isolating effects of anomalous propagation.

Straka and Zrnić (1993) recommend a simple Z-R relation when $R < 20$ mm h^{-1}, a combination of Z_H and Z_{DR} when $20 < R < 70$ mm h^{-1} and an R-K_{DP} relation for $R > 70$ mm h^{-1}. While this may be suitable for research experiments this does not seem simple enough, at present, for operational radars giving real time outputs.

The major merit of dual polarisation measurements is the ability to distinguish the precipitation phase i.e. distinguish between rain, hail and snow and thereby contribute to more accurate estimates of rainfall[28] [Balakrishnan and Zrnić (1990), Bringi and Chandrasekar (2001), Ryzhkov and Zrnić (1998)]. Thus, Direskeneli et al. (1986) used the hail signal H_{DR} (defined in Chapter 6) to identify regions of hail and omit them from their rainfall calculations using a relation similar to equation (7.70). Since the polarimetric radar signatures for different hydrometeors is not mutually exclusive and unique, use of neural network and fuzzy logic

---

[26] The average of the H and V specific attenuations can also be linearly related to the rainfall rate. But the differential attenuation k_{Dif} offers the advantage of reducing the dependence of estimated R on changes in drop shape.

[27] Comparisons were made with a rain gauge network.

[28] Application of such identification is not confined to rainfall estimation but contributes to study of precipitation processes and is useful as input to numerical weather prediction models.

methods have been proposed for automatic identification [Liu and Chandrasekar (2000), Vivekanandan et al. (1999b), Zrnić et al. (2001)]. Another merit is in the estimation of vertical profiles of instantaneous precipitation rate which are particularly useful in communication applications. Hence, while dual-polarisation techniques have produced exciting results for researchers they are also much more promising for operational use than other multi-parameter methods and their induction into operational radar networks is being seriously considered [Doviak et al. (2000), Illingworth and Zrnić (1995), Zrnić and Ryzhkov (1999)].

7.6 Comparison of Satellite and Radar Measurements of Precipitation

7.6.1 Satellite Estimation of Precipitation

Satellite remote sensing is attractive for precipitation measurement as large areas including the oceans can be covered. There are techniques based on sensing by satellite-borne radiometers, of the solar radiation reflected in the visible region or the emitted infrared (IR) or microwave radiation. The methods used with visible or IR wavelengths relate the area of cloud above a threshold brightness (visible imagery) or brightness temperature (infrared imagery) and rate of variation of this area to the rain volume [e.g. Arkin and Meisner (1987), Bhandari et al. (1987), Doneaud et al. (1987), Griffith et al. (1978), Martin and Howland (1986), Rao et al. (1989) and Stout et al. (1979)]. An early treatment of the whole subject of satellite rainfall estimation may be seen in Barrett and Martin (1981). Since these methods do not measure precipitation directly, the estimate has to be calibrated with reference to surface measurements i.e. "ground truth"[29]. Rain gauge networks and radars have been used for the purpose. The radar is equivalent to a gauge network of high density and an additional advantage is that a coastal radar has coverage over a part of the ocean. Some areas such as the UK and Japan have dense radar networks which can be used to validate satellite observations. The satellite and radar are complementary. While using the radar data to validate satellite observations, the satellite data

[29] Satellite IR radiometers see high clouds which may not be supported by precipitation below. Therefore the agreement with radar may not be good on a point to point basis. Hasler and Morris (1986) who studied hurricanes using stereoscopy from two geostationary satellites and compared with radar observations concluded that at least in tropical cyclones, rainfall estimates from satellite IR data may be accurate only for large spatial and time scales.

can be used to correct possible errors (e.g. due to anomalous propagation or due to shielding by hills) in the radar field (e.g. Fiore et al., 1986).Thus an estimate obtained by these devices over a limited area can be used to extend precipitation measurements to much larger areas including the oceans. Good matching of the radar and satellite imagery in space and time is needed.

7.6.2 Calibration using radar

In the Griffith et al. (1978) technique, geostationary satellite imagery (visible and IR) and (10 cm) radar PPI products were superposed on the McIDAS[30] system to determine echo area A_e exceeding a threshold rain rate of 1 mm h^{-1} and cloud area A_c above a preset threshold. A_e was related to the volumetric rain rate by a suitable coefficient. If the cloud area is A_c and the maximum cloud area attained by the cloud system is A_m, curves or tables relating (A_e / A_m) to (A_c / A_m) were prepared. Then for each satellite picture the (A_c / A_m) was read and the corresponding (A_e / A_m) determined from the curves or tables. The volumetric rain rates were verified with radar and gauge data. The errors in the method were found to decrease with increasing period of estimation. Fig. 7.5 gives a comparison of daily isohyets from satellite and radar in GATE (see section 7.2). The radar data were obtained from the GATE ships. The method is considered suitable for real time estimation of rainfall.

In the method of Arkin and Meisner (1987), measurements are made of the monthly mean fraction of 2.5° latitude-longitude squares covered by clouds whose equivalent blackbody temperature is below 235°K as seen in geostationary satellite imagery. This fraction is multiplied by a constant K expressed in mm per day, to obtain a Quantitative Precipitation Estimate (QPE). The constant was determined as 71.2 mm per day by Arkin (1979) using the rain gauge and radar data of GATE. The method is useful for estimates of large scale convective rainfall over the oceans and has been adopted for the Global Precipitation Climatology Project (GPCP) of the WMO. It has been used for mapping Indian Ocean rainfall by Rao et al. (1989)[31] using the Indian National geostationary Satellite (INSAT). But the method is not suitable for short-period or small-scale estimation of precipitation.

Martin and Howland (1986) formulated a "grid-history technique" using

---

[30] "McIDAS" stands for "Man computer Interactive Data Access System" at the University of Wisconsin, USA. See Suomi et al. (1983) for a description.

[31] Rao et al. (1989) have used the GATE value of the constant K. It is desirable to make comparative measurements in various areas using gauges and radar as was planned for the GPCP.

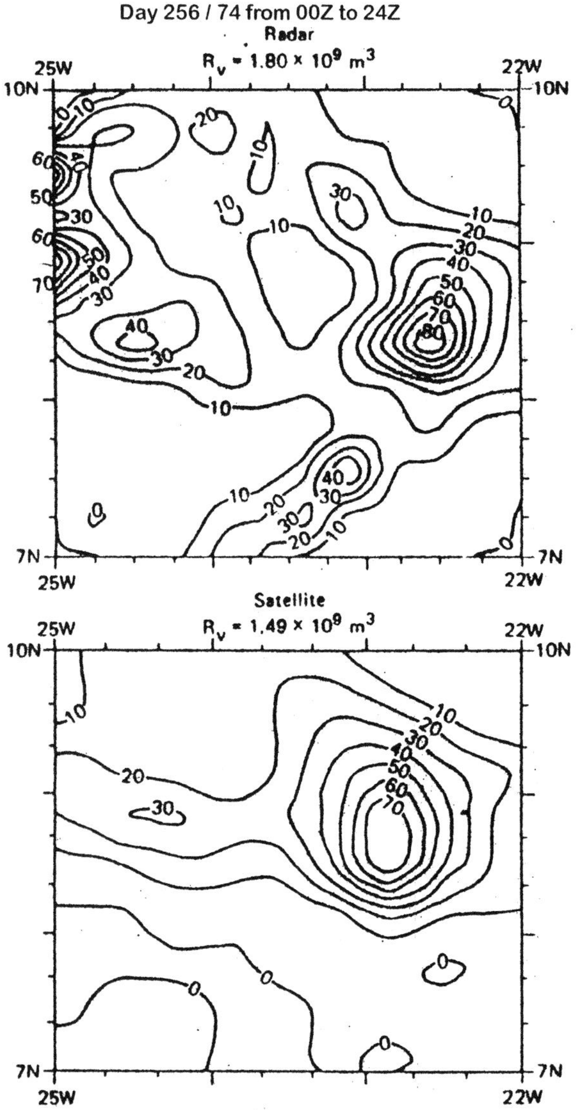

Fig. 7.5. Radar- and satellite-inferred daily rain estimates (mm/day) from GATE (from Griffith et al. 1978; courtesy AMS). Top figure: Radar isohyets from four GATE ships; data composited over (1/4) degree latitude-longitude squares. Bottom: Satellite isohyets from digital IR data; (1/3) degree squares. R_v is time-integrated rain volume m^3 for 24 hours.

visible and IR geostationary imagery. Using the McIDAS, a sequence of images, say, at hourly or half-hourly intervals is viewed. Comparing the image at each point of a grid [the grid spacing was 0.2° or 0.1° (latitude/longitude)] with certain threshold values of brightness or brightness temperature, rain rate values of zero, light, moderate or heavy with weights 0, a_1, a_2, a_3, are assigned to each grid point. The rain rates corresponding to the three designations are derived by calibration with rain gauge data. The daily rainfall R_d at each grid point is

$$R_d = f_1 a_1 + f_2 a_2 + f_3 a_3 \tag{7.76}$$

where the f's are frequencies of occurrence of the a's during a 24 hour period. Bhandari et al. (1987) applied the Martin-Howland technique to the data of the American geostationary satellite (GOES)[32] stationed over the Indian Ocean in 1979, in conjunction with data of a 10 cm radar at Chennai (Madras) and a network of rain gauges yielding 24-hour rainfall amounts. They calculated the fractional area covered by various rainfall rates as determined by radar and from this, assigned the weights for the light, moderate and heavy categories in the satellite brightness temperatures. Using these weights they computed the satellite estimate of daily rainfall in an area adjacent to Chennai. Fig. 7.6(a) shows an example of the isohyet pattern from the satellite. Figs. 7.6(b) and 7.6(c) respectively give the gauge- and radar-estimated rainfall distribution for comparison. The radar estimate was based on the seasonal *Z-R* relationship of Raghavan and Sivaramakrishnan (1982) given in Table 7.2. There is a broad agreement in the spatial distribution of daily rainfall but there are mismatches in sub-areas. The method was found suitable for convective precipitation over small areas but rather laborious as it involved partly subjective assignment of the rain category.

Vicente et al. (1998) have developed an automated rainfall estimation technique which computes in real time, convective rainfall rates based on a nonlinear, power law regression relationship between cloud top temperatures from the GOES satellite and radar-derived rainfall estimates. Gourley et al. (2002) developed a "multisensor" technique which combines radar data in Cartesian coordinates with GOES data through GIS software. In this case also, the GOES rainfall is based on a regression of cloud top temperatures with radar-derived rainfall. They identify the melting layer from the radar volume scans and reject data above the melting layer. They consider this method to be suitable for mountainous regions where the disadvantages of the radar are made up by the satellite data.

7.6.3 Estimates from Satellite Microwave Radiometers

Multichannel microwave radiometers on board polar orbiting satellites have been used for determination of precipitation as well as liquid water

---

[32] GOES = Geostationary Operational Environmental Satellite.

content and water vapour distribution (see e.g. Wilheit 1986; Wilheit et al., 1977). These use a wide range of frequencies (6 to 185 GHz) to monitor the upwelling radiation from the ocean surface. The attenuation of this radiation by the hydrometeors in the path to the satellite changes the brightness temperature sensed by the radiometer. At low rain rates the brightness temperature is proportional to rain rate. At frequencies up to about 35 GHz the attenuation is mainly due to absorption. In this case, the ice portion of the cloud is neglected. At higher frequencies scattering is more important and gives rise to lower brightness temperatures than would otherwise be expected. Here, the ice portion also contributes significantly. The microwave radiation (unlike the visible and IR) is directly related to the hydrometeors. It can penetrate the clouds above but suffers from poor spatial resolution (i.e. a footprint 25 to 140 km on a side and area of 600 to 6000 km^2 at the lower frequencies, but better at higher frequencies) due to the relatively long wavelength, compared to the visible and IR methods. Hence, the footprint may be made up of rain and no rain areas and the maximum rain rates smoothened out. Regression equations connecting the brightness temperatures at two or more wavelengths with the rain rate have been formulated. The radiance from the precipitation shows good

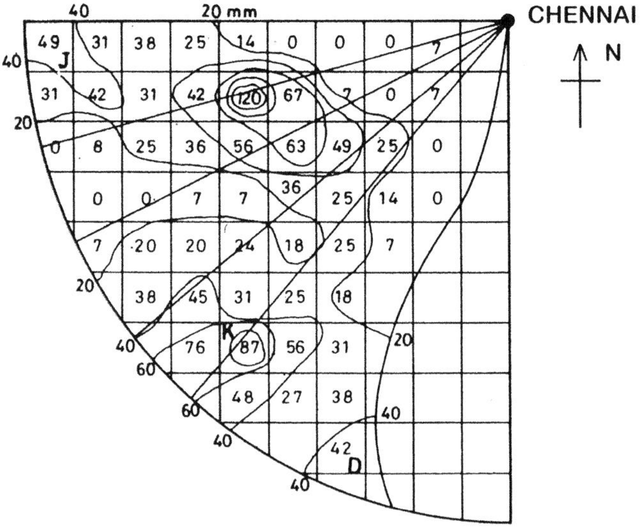

Fig. 7.6. Cumulative rainfall (mm) November 13-14, 1979 in a quadrant of radius 200 km over land near Chennai, India (from Bhandari et al. 1987; courtesy ISRO)
Isohyets at 20 mm intervals. The four radials to the southwest are limits of radar shadows hatched in Fig. 7.6c.
(a) Rainfall distribution from GOES satellite, 0.2 degree Grid.

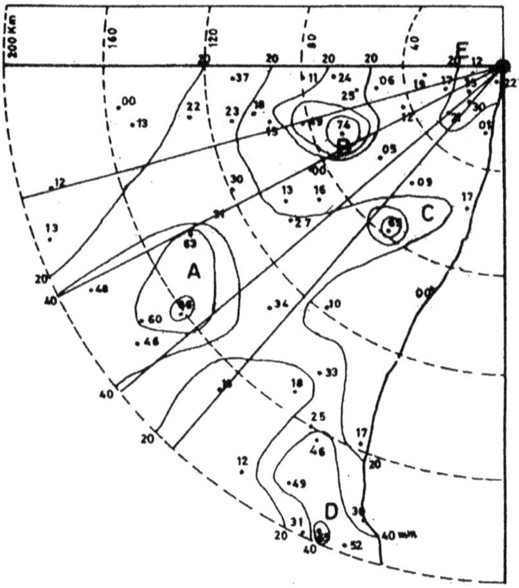

(b) Rainfall distribution from rain gauges

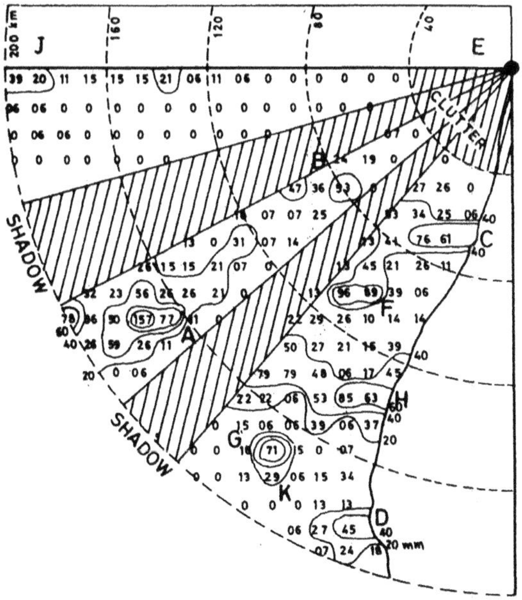

(c) Rainfall distribution from radar located at Chennai.

contrast with the sea surface which has low emissivity and therefore the method is suitable for oceanic rainfall if the footprint is not contaminated by land. Over land the precipitation return is masked by variations in the land surface emissivity which is quite high.

Validation of the technique is by radar. The US Defense Meteorological Satellite Program (DMSP) satellites carry a Special Sensor Microwave Imager (SSM/I) which is a multi-channel microwave radiometer. Rainfall algorithms developed for these have been calibrated using radar network measurements in the USA, UK and Japan to develop instantaneous rain retrieval algorithms over land and ocean (as distinct from the time-integrated rainfall product). Details may be seen in Ferraro and Marks (1995).

7.6.4 Area-Time Integral Methods

Doneaud et al. (1987) applied the ATI method (discussed in section 7.3 for the case of radar) to satellite (visible and IR) data. The thresholds for satellite data were chosen to match the satellite-derived ATI to the radar-derived ATI. Thus, the radar acted as a reference, using which, satellite estimates could be derived for larger areas. Atlas and Bell (1992)[33] have examined the relation between the (visible and IR) satellite-based and radar-measured ATI and pointed to the usefulness of such a relationship in combining satellite-borne radar observations with geostationary satellite visible and IR observation for precipitation estimates.

7.7 Space-borne Weather Radar

7.7.1 Requirements of Space-borne Weather Radar

We have seen in the last section that passive remote sensing from satellites by visible or IR radiometers is only an indirect method of precipitation estimation and that microwave sensing has several limitations particularly, poor resolution and lack of contrast with surface emissivity over land. Will active

[33] Visible and IR satellite methods are indeed better at estimating rain areas than in estimating rain amounts. In an earlier study in Canada, Lovejoy and Austin (1979) estimated the contribution to the r.m.s error in satellite measurement of rainfall to be about 5% for gauge to radar adjustments, 22% for radar area to satellite area and 44% for satellite area to satellite rain amount, yielding a total error of 49% in the whole process.

remote sensing by a satellite-borne weather radar be a more satisfactory method? The experience with airborne radar suggested an affirmative answer to this question. Space-borne radar has been studied for the past two decades and became a reality in 1997 in the form of the Tropical Rainfall Measuring Mission (TRMM) satellite.

The minimum altitude of an orbiting satellite will be a few hundred kilometres. The use of weather radar on a geostationary satellite does not seem feasible in the near future because of the enormous range involved. A polar orbiting satellite has a large speed of motion with reference to the earth resulting in severely reducing dwell time. A space-borne radar needs a very narrow beam to ensure filling by precipitation, of the instantaneous field of view. A wide swath of observation is necessary to avoid gaps in coverage and this, in turn, implies fast scanning reducing the number of independent samples over a given volume. It has to detect precipitation at a long range but extending over a relatively short range interval close to the earth's surface[34]. While, as in the case of airborne radar, it can yield a vertical profile of precipitation, it has to discriminate between the precipitation echo and the stronger return from the earth's surface (see section 7.4). Since the weight and power consumption are severe constraints in a satellite payload, it is desirable to use wavelengths shorter than the conventional 10 to 3 cm range of ground-based radars. At shorter wavelengths, the signal return for a given transmitted power is higher. Good spatial resolution at long range is also achievable with short wavelengths. However, short wavelengths are subject to high attenuation due to precipitation (atmospheric attenuation is also higher at some wavelengths). Wavelengths less than 1 cm may suffer excessive attenuation. A method of precipitation estimation using attenuation measurement may therefore be suitable at these wavelengths. Besides rainfall estimate, the radar can hopefully yield information on precipitation and melting layer structure and drop size distribution. A complete treatment of space-borne weather radar may be seen in Meneghini and Kozu (1990).

7.7.2 TRMM Satellite

The Tropical Rainfall Measuring Mission is a joint project of the USA and Japan with participation by some other countries. TRMM launched in November 1997, is an experimental satellite to evaluate a multi-sensor

---

[34] This, incidentally, means that the receiver dynamic range need not be as large as in ground-based radars.

(including the first weather radar in space) system of rainfall measurement in the tropics. The satellite at an altitude[35] of 350 km has an orbit taking it between 37°N and 37°S. It has a number of radiometers in the visible, IR and microwave wavelengths besides the radar. Other mission objectives are to advance the understanding of the global energy and water cycle from a knowledge of the distribution of tropical rainfall and to understand the mechanism through which tropical rainfall affects the global circulation. This experiment will hopefully pave the way for an operational satellite-borne radar in the future. An introduction to the TRMM mission may be seen in Simpson et al. (1988). For a description of the sensors including the precipitation radar (PR) on board, see Kummerow et al. (1998).

7.7.3 The TRMM Precipitation Radar (PR)

The radar operating at a frequency of 13.8 GHz has a 2 m × 2 m phased-array antenna facing nadir and scanning ± 17° in the cross track direction. The plane of polarisation is horizontal. The peak transmitted power is about 0.6 kW, the beam width 0.7° and sidelobe level −30 dB to suppress surface clutter. The spatial resolution at the earth's surface is about 4 km at nadir. The pulse width is 1.67 μs corresponding to a range resolution of 250 m. A logarithmic receiver with dynamic range better than 70 dB is used. Simulation with airborne radar has been done extensively to decide the specifications of the TRMM radar. For example, Durden et al. (1994) have described an Airborne Rain-MApping Radar (ARMAR) operating at 13.8 GHz and simulating the scan geometry of the TRMM radar.

7.7.4 Techniques for Derivation of Rainfall

A number of studies (including simulations with airborne radars) have been made to determine the possible techniques and develop algorithms [e.g. Amayenc et al. (1996), Haddad and Im (1993), Iguchi and Meneghini (1994), Im and Haddad 1993, Kozu and Nakamura (1991), Kozu et al. (1991, 1991b), Kumagai et al. (1993b), Kummerow et al. (1989), Marzoug and Amayenc (1991, 1991b, 1993, 1994), Meneghini (1993), Meneghini and Kozu (1991), Meneghini et al. (1983, 1986, 1992), Testud et al. 1996), Weinman et al. (1990)]. The methods available are:

---

[35] The altitude has been raised subsequently.

1. The conventional *Z-R* method.
2. The Surface Reference Technique (SRT) (see section 7.4.)[36]
3. Dual wavelength method (see section 7.5.)[37]
4. Area Time Integral Method (see section 7.3 and Meneghini and Jones 1993).
5. Combined use of dual or single wavelength radar and passive radiometers. [Kozu and Nakamura (1991), Kumagai et al. (1993b), Meneghini (1993), Meneghini et al. (1997)]. This method uses the radar data to construct a set of range-profiled drop size distributions. Among these, that distribution is chosen, which gives values closest to the measured brightness temperatures from the radiometers. This method is essentially applicable to the ocean but perhaps it can be extended to land with the use of multi-frequency radiometers.

All these methods (except the dual wavelength) are being explored in the TRMM (which does not have dual wavelength radar). Since the bright band cannot be avoided, a model of the melting layer can be used to improve radar rainfall estimates (Thurai et al. 2001). The TRMM can also use non-radar methods (i.e. based on the visible, IR and microwave radiometers).

The TRMM gives out several products. In the "2A25" product, which is the main output from the radar, the algorithm selects range gates between the rain top and the lowest height above the surface that is free from surface clutter. The bright band height and climatological surface atmospheric temperature are used to define the regions where the hydrometeors are primarily liquid, solid and mixed phase[38]. The initial values of the coefficients in the k_R - Z_e and Z_e -R relationships at different altitudes are defined accordingly. The true effective reflectivity factor Z_e at 13.8 GHz at each radar resolution cell is estimated from the measured vertical profiles of reflectivity factor Z_m, correcting for path attenuation using the surface reference technique. The rainfall rate R is calculated from the estimated Z_e. Details may be seen in Iguchi et al. (2000) and Meneghini et al. (2000).

Comparison with ground truth in the form of radar and rain gauge observations is being carried out in several countries [e.g. Steiner and Houze (1993), Bolen and Chandrasekar (2000)][39]. An account of the role of radars is given by Thiele (1991). A review of the status of the TRMM two years after launch is given by Kummerow et al. (2000).

---

[36] It has been suggested (Kostinski et al. 1993) that a suitable polarisation can be chosen to eliminate the surface echo and detect the hydrometeor echo.

[37] A variation is the Dual wavelength Surface Reference Technique (DSRT) (Meneghini et al. 1987). In this method the difference of the differences between backscattered powers at two wavelengths is used to determine the differential rain attenuation.

[38] The Steiner et al. (1995) technique (see Paragraph 7.2.3) is used to separate convective and stratiform areas of precipitation. See also Biggerstaff and Listemaa (2000).

[39] TRMM PR observations have also been used to determine calibration biases of ground-based radars (e.g. Anagnostou et al. 2001).

7.7.5 Other Configurations

A number of other configurations of space-borne weather radar have been proposed but none has been realised yet. Meneghini and Kozu (1990) have discussed the proposed methods. A few are briefly discussed below.

Adaptive pointing: Atlas (1982) proposed that instead of frittering radar energy over a vast area only a small part of which may be covered by precipitation, an "adaptive pointing" technique may be used. This will consist of a forward-looking microwave passive imager which would identify areas of probable precipitation ahead of the satellite track. This information will be used to control the radar antenna so that it samples the identified clouds when the satellite comes over them. An alternative is for the radar itself to do the preliminary function of identification.

Synthetic Aperture Radar (SAR): Atlas and Moore (1987) proposed the use of satellite-borne Synthetic Aperture Radar (SAR) for precipitation measurement. SAR consists of a side-looking antenna carried on an aircraft or satellite. As the platform moves, a number of returns observed by the antenna in its successive positions are stored and coherently integrated. The successive positions of the antenna form a linear array. The integrated signal is equivalent to that which would have been received if all the elements of the array had together transmitted a pulse. The radar is meant to study ocean surface parameters and was first used in the SEASAT satellite of the USA in 1978. For precipitation measurement, beam filling would be a problem for which Atlas and Moore have suggested the use of a microwave radiometer to identify the precipitation dimension as in the previous paragraph.

Stereoradar: Testud and Amayenc (1989) and Testud et al. (1991, 1992) have proposed what they call a "stereoradar" for airborne and space-borne applications. This would consist of two radar beams at an attenuating frequency (they propose 24 MHz) directed at equal angles (about 20°) fore and aft of the aircraft or spacecraft. They propose a method of using the observed apparent reflectivities from the two beams to determine the attenuation field and the true reflectivity. If a Doppler capability is added it should be possible to retrieve two components of air velocity viz. the along-track component of horizontal velocity and a component directed at about 20° to the vertical. The problems and prospects are discussed by Amayenc et al. (1993). A dual-beam X-band antenna was installed in the ELDORA/ ASTRAIA airborne Doppler radar (see Chapter 3) and the technique tried out and validated using ground-based radar as ground truth. On the basis of this, a new formulation of the technique has been proposed by Kabêche and Testud (1995).

Chapter 8

RADAR OBSERVATION OF TROPICAL CYCLONES

Radar has contributed significantly to the operational monitoring as well as the study and understanding of tropical cyclones (TC's), which are one of the most important meteorological phenomena. We describe the various features observable by radar in tropical cyclones and the methods of determining the centre of a TC from radar reflectivity and Doppler velocity patterns, in the first section of this chapter. Section 8.2 deals with the determination and possible prediction of the track of a TC. This is followed by section 8.3 dealing with determination of winds in TC's. Section 8.4 discusses in detail the derivation of various characteristics of a TC from airborne and ground-based radars. These include the wind field, the eye structure and asymmetries observed in the TC. The relationship of eye characteristics to the intensity of the TC is also discussed. Section 8.5 deals with the distribution of rainfall. In section 8.6 we consider the inputs which radar can give for storm surge prediction.

8.1 Radar-observed Features in the Tropical Cyclone

8.1.1 Tropical Cyclone as Observed by Radar

The tropical cyclone[1] (referred to in this chapter as cyclone or as TC for

[1] While "tropical cyclone"is a generic term, the phenomenon is called "cyclonic storm" or "cyclone" in India, "cyclone" in south Indian ocean and south Pacific, "typhoon" in northwest Pacific and "hurricane" (when maximum sustained surface wind speed is 63 knots or more; otherwise " tropical storm") in the western hemisphere.

short) is a meteorological phenomenon particularly suited for observation and study by microwave radar, because of the spatial and temporal scale of the TC as well as the organised pattern of precipitation in it, which the radar can detect. The first published radar observation of a TC was made from aircraft by Maynard (1945) of the US Navy. Though Henry Piddington (1848) in India, realised as early as the mid-nineteenth century that it was coiled up like a serpent and christened the phenomenon as a "cyclone", detailed observations over the ocean became possible only with the advent of aircraft reconnaissance. It was only from the radar that the spiral structure of the rain bands could be observed for the first time (Simpson 1956). The spatial distribution of precipitation was observed in several TCs in the Atlantic and the Pacific from airborne and ground radar and a structural model was given by Rockney (1956) and later by Kodaira (1964). Fig. 8.1 gives the typical horizontal structure of a well-developed TC obtained by Raghavan (1985) by compositing radar echoes from a number of cyclones in the Bay of Bengal. These authors classified the radar echoes from a TC as below:

(1) Pre-cyclone squall lines,

(2) Outer convective activity,

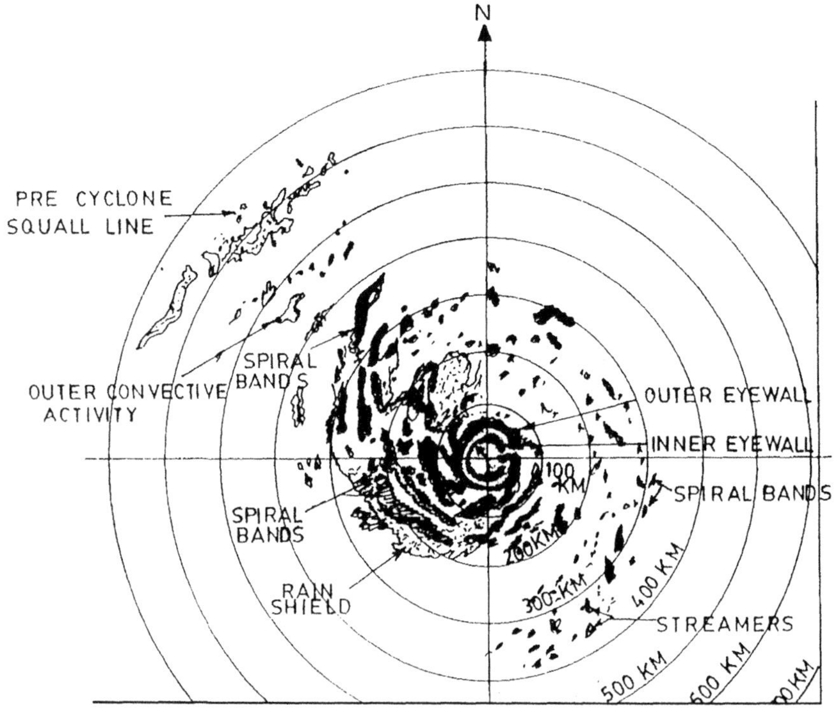

Fig. 8.1. Horizontal structure of a tropical cyclone as seen by radar (after Raghavan 1985; courtesy WMO).

(3) Spiral rain bands and rain shield,
(4) Eyewall (sometimes a double wall), and
(5) Streamers.

Willoughby (1988) who used observations with airborne radars including Doppler, has given the typical reflectivity (hence precipitation) distribution. It is possible to derive therefrom the air motions even in the absence of direct Doppler measurements. Willoughby (1988) has also illustrated the "secondary" (i.e. radius-height plane) circulation. From the observed reflectivity pattern the position of the centre of the cyclone can be determined, and by frequent or continuous observation, the motion of the TC can be tracked. Attempts have been made with varying degrees of success to predict further motion and to estimate the intensity of the TC. It is also possible to assess the spatial distribution of rainfall and Latent Heat Release (LHR). The detailed observation of the structure of the TC can also be used to infer the Radius of Maximum Winds (RMW) and the areas of maximum wind damage in the core of the TC. Doppler radar enables determination of the wind field and the maximum wind which is, by definition, the intensity of the cyclone. Thus radar is one of the most important operational aids for cyclone observation and forecasting.

8.1.2 Radars for TC Observation

The features listed, refer to a typical well-developed TC and may not be present in all TCs. As the TC develops from an initial disturbance - usually a diffuse low pressure area in the ocean - some of these features may be seen to develop gradually and undergo changes from time to time. Even so, since the TC is several hundred km across, a single stationary radar is unlikely to see all the features at one time. A composite of pictures from several radars with overlapping coverage or a time composite of observations with the same radar will give a more complete description of the entire TC. Land-based radars will still not be able to see parts of the TC which are far away from them because of range, height and signal strength limitations. Intervening heavy precipitation may mask the features beyond. Radio-wave propagation conditions may be favourable when the TC is far away and weather nearer the radar is clear. When the TC comes closer the propagation may become subnormal because of the precipitation and the rise in humidity at all levels. Thus it may sometimes happen that the TC is seen better when it is farther away than when it is near (see Paragraph 8.1.8 for an example). Radars may therefore sometimes present a biased view of the TC. In view of heavy precipitation extending over long distances and the need for quantitative measurements not vitiated by attenuation,

10 cm radars are considered ideal for use in detecting TCs from land sites although some countries use 5 cm radars because of the advantages mentioned in earlier chapters. In India, a network of 10 cm radars designated[2] as "Cyclone Detection Radars" is established along the coast. In the USA, as noted in Chapter 3, the WSR-88D, a 10 cm Doppler system is used. This wavelength has the additional advantage of a higher maximum unambiguous velocity which is important in TC observation. In Japan, the long range Mt.Fuji radar operates at 10 cm while the rest of the network is made up of 5 cm radars. On board cyclone reconnaissance aircraft, 5 or 3 cm radars are used. Most of the recent knowledge of TC core structure has come from airborne radar observations.

8.1.3 Pre-cyclone Squall lines

Pre-cyclone squall lines (Fig. 8.1) are lines of convective cells appearing about 300 to 500 km ahead of the TC centre and usually well separated from the spiral rain bands of the TC. They may be straight lines or in zigzag shape but do not follow the spiral pattern of the rain bands. An example is shown in Fig. (4.1). Each line may be up to a few hundred km in length and there may be several lines roughly parallel to each other. The set of squall lines may undergo changes due to decay and regeneration of component cells as in other mesoscale systems (see chapter 6) but do not appear to form large mesoscale anvils. Often they move with the rest of the TC at the same speed. Hence, usually, they arrive over land before the core of the TC does and cause heavy convective precipitation over narrow swaths. Senn et. al. (1965) found that in Atlantic and Caribbean hurricanes the squall lines are often, but not always, oriented perpendicular to the direction of motion of the TC. In the Bay of Bengal, Raghavan et. al. (1980) found that in all cases in which they were observed, their orientations were roughly perpendicular to the direction of TC motion at that time. The orientation is, therefore, a possible predictor of TC motion. When the TC track changes, the squall line orientation also may change correspondingly. No conclusion as to whether one is the cause of the other has been offered. Nor has any relationship been established between the characteristics of the squall line and the intensity, strength or size of the TC.

8.1.4 Outer Convective Activity

Outer convective activity (Fig. 8.1) or outer bands are names assigned by

---

[2] These radars can, of course, be used for other applications as well.

some authors to echoes, largely convective but poorly organised, which are observed in some TC's behind the pre-cyclone squall lines but ahead of the organised spiral bands.

8.1.5 Spiral Bands

The spiral bands are precipitation elements arranged along spiral arcs which appear to converge to the centre of the TC. These occur in the region between the outer bands mentioned above and the centre of the TC. These arcs correspond approximately to logarithmic spirals [Senn and Hiser (1959), Raghavan (1985)] which may be defined by the equation,

$$r = Ae^{\theta \tan \alpha} \tag{8.1}$$

where (r, θ) are polar co-ordinates of any point on the spiral with reference to the centre of the TC as origin, α is the angle of incurvature of the spiral at the point (r, θ) i.e. the angle between the tangent to the spiral at this point and the tangent at the same point to a circle centred at the origin and of radius r, A is a constant and e is the exponential base. An example with $\alpha = 15°$ is shown in Fig. 8.2. According to Senn and Hiser (1959), spiral bands propagate radially outwards from near the centre of the TC. The band thus maintains a quasi-conservative position relative to the quadrants in which it is formed and does not rotate around the TC centre.

Early ideas on cyclone spiral bands characterised them as cyclonically advected inward-propagating internal gravity waves (e.g. Abdullah 1966). However, subsequently, observations[3] from airborne non-Doppler and Doppler radars in several TC's have shown that in many storms there is a group of prominent spiral rain bands that maintain a fixed position relative to the vortex (Willoughby 1988). This feature which is not an internal gravity wave has been called (Willoughby et al. 1984) the Stationary Band Complex (SBC). It appears to act as a partial barrier to the inflow of air to the vortex. The SBC's occur near the centre in weak cyclones and near the periphery in strong ones; and in intense hurricanes the multiple convective rings (eyewalls; see Paragraph 8.1.10) probably evolve from the SBC's. According to Willoughby et al. (1984) the SBC's consist of three components: (1) the principal band, (2) connecting band and (3) secondary bands. The principal band is a prominent spiral band extending outward on one side of the vortex. It is separated from the eyewall in intense systems (see Paragraph 8.1.7). The connecting band joins the

---

[3] In these observations the bands were identified by setting threshold reflectivity factors of 25 to 35 dBZ.

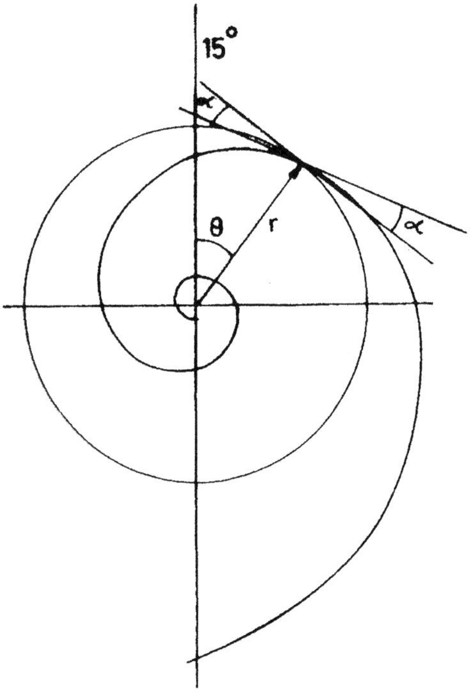

Fig. 8.2. Logarithmic spiral with crossing angle of 15°.

principal band to the eyewall (except in intense systems) and is mostly stratiform in character. The secondary bands are usually less prominent ones located on the concave side of the principal band. They mostly exhibit weak convection but in some cases may contain an outer wind maximum and eventually contract and become a convective ring.

In a Japanese typhoon, Shimazu (1997) studied rain bands using two Doppler radars and the Mt. Fuji radar. He classified the bands as (1) slow-moving wide ones and (2) fast-moving narrow ones. The former had a width of 50 to 150 km and an angular velocity of 10° per hour or less and lasted six to ten hours. The latter had a width of 25 to 50 km, angular velocity of 30° per hour and lasted one to four hours.

The crossing angles correspond to the angle of inflow at various points. The crossing angle generally increases from near zero at the radius of maximum winds (RMW) of the TC to a maximum of about 30 degrees at three times the RMW (Schwerdt et. al. 1979). The crossing angles tend to be small in intense cyclones and relatively large in weaker ones. The bands may often change their organisation and tend to break up or distort in shape when they come close to land. Each band is typically 5 to 50 km wide and may be up to several hundred km in length. Some TC's have a large number of bands packed densely; in others, there are few bands with large spaces (of very little precipitation) in between. The spiral shape of the bands enables them to be extrapolated to find their centre of convergence and thus the centre of the TC approximately when the eye of the TC is not seen by radar. A simple way to do this is to place on the PPI (reflectivity) picture, transparent overlays with the spiral of a particular crossing angle printed on each one. By adjusting the overlays the approximate point of convergence is found. Starting with a small crossing angle near the TC centre the successive bands are fitted with overlays of larger angles. It may happen that all the bands may not converge to the same point. If the points of convergence are not widely scattered the mean of the points may be taken as the centre. In weak TC's the bands are not well organised and may yield widely different centres. The choice of crossing angles for fitting is a subjective choice of the analyst. Fig.8.3 shows an example of fitting spirals to the reflectivity pattern to infer the centre. The orientations of the bands may not be clear because weaker echoes may mask them. By setting a higher threshold of reflectivity or by using the colour code of the digital picture, they can be identified.

Spiral band orientations may change with time depending on the growth or decay of individual elements. To avoid mistaking any curved line as a spiral band, he following subjective criteria are useful.

(1) An arc of say 90 degrees or more should be visible and be stable over a period of time. A long enough band can sometimes be obtained by combining PPI's from two or more radars in the network.

(2) The existence of cyclonic circulation can be verified by observing movement of individual echoes roughly along the band[4].

(3) The TC centre position arrived at by extrapolating the bands should agree reasonably well with estimates from satellite pictures or synoptic evidence.

Fig. 8.4 shows several curved arcs associated with a weak TC. Each converges to a different centre there being apparently several rival vortices. This may be the reason for the TC not intensifying. This shows that radar cannot track weak TC's or depressions although it can give other useful information such as rainfall distribution. Radar may sometimes show banding associated with a vortex looking very much like a TC and even exhibiting an eyelike feature. An example (discussed by Gupta et al. 1990) which was not a TC according to synoptic analysis is shown in Fig. 8.5.

Gall et al. (1998) have observed, in some hurricanes, spiral bands of a new kind. These are about 10 km across and occur within 100 km of the TC centre.

[4] Centre finding by the extrapolation of spiral bands is subject to large error and is resorted to only when the TC is far away from the radar and the eyewall cannot be seen. In such cases it is unlikely that reliable Doppler velocity information will be available.

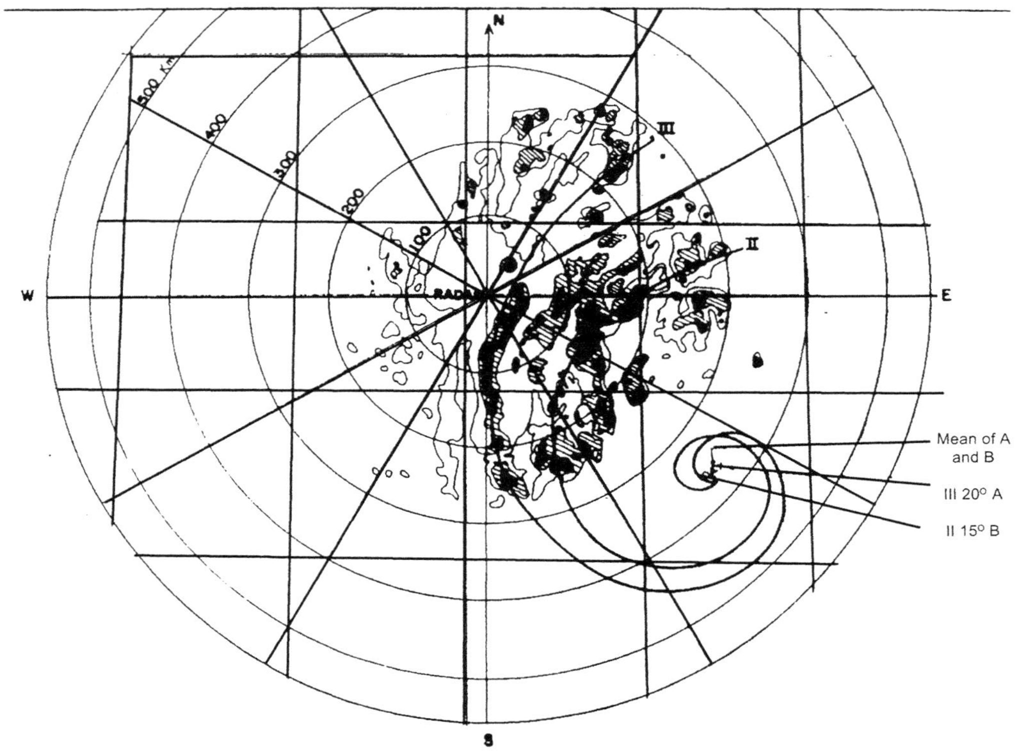

Fig. 8.3. Estimation of TC centre by spiral overlays (after Raghavan 1985, courtesy WMO). Echoes shaded black are above 37 dBZ, hatched area above 32 dBZ and outlined echo is below 32 dBZ. Use of a 15° crossing angle spiral for the band marked II and a 20° spiral for the outer band III, gives two centres of convergence A and B which are close to each other. Mean of the two is taken as the centre of the TC.

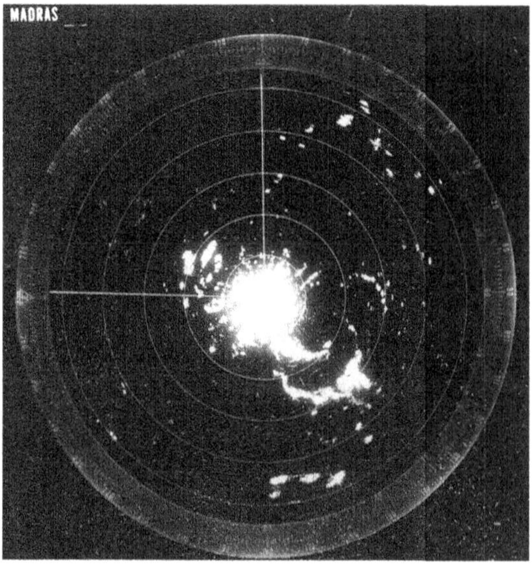

Fig. 8.4. Spiral echo patterns from a weak TC converging to different centres of circulation. (from Raghavan et al. 1980; courtesy IMD). Range rings at intervals of 40 km.

They appear to move around the hurricane with speeds close to the tangential wind at the level of the band. They induce wind speed variations of at least 8 m s^{-1} across the bands. Their role in hurricane dynamics is not known.

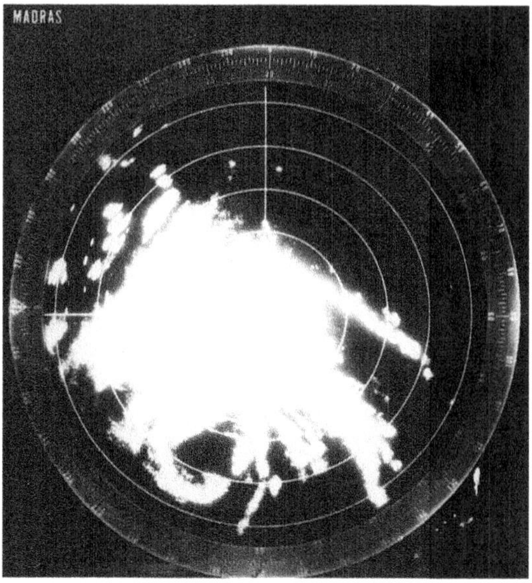

Fig. 8.5. PPI showing a vortex (courtesy IMD). Range rings at intervals of 20 km. See text for details.

8.1.6 Rain Shield

Interspersed with the spiral bands are relatively weak echoes without specific organisation which have been termed the rain shield by some authors. The precipitation is stratiform, while that in the spiral bands may be convective or stratiform. In Fig. 4.3. the echo on the left is from the stratiform rain shield and exhibits horizontally oriented reflectivity contours while the echo on the right which is from a spiral band is convective and has vertically oriented contours.

8.1.7 The Eye

When the TC intensifies, the innermost spiral band tends to reduce its crossing angle and become a circle. At the end of the band an eyewall forms approximately in the form of an arc of a circle. Gradually this may separate from the band into a distinct ring or part of a ring with an echofree area inside known as the eye which is an area of descent (Raghavan 1990). Figs. 8.6 and 8.7 with an interval of about five hours between them show this process. This normally occurs at about the time when a TC becomes "severe"[5]. Note the extremely small eye diameter in Fig. 8.7.

[5] The term "severe" here refers to the intensity of the TC, defined as the minimum sea level atmospheric pressure (MSLP) or the maximum sustained surface wind (MSSW). Terminology varies with ocean basin; the current classification in the North Indian Ocean (NIO) basin is given in Table 8.1. Other important attributes of a cyclone are:
(1) its *size* defined as the linear dimension of the zone covered by gale force (17 m s^{-1} or more) winds,
(2) its *strength* defined as the average tangential wind in the gale force area,
(3) *diameter of the eye*,
(4) *radius of maximum winds* (RMW) which is the average distance from the centre to the point in the eyewall where the wind is strongest.
The first three may not necessarily be correlated with each other. A TC may be intense but its wind speed may drop rapidly with distance outside the eyewall; hence its size and strength may be small. Conversely the maximum wind at the eyewall may not be very strong but gale force winds may extend over a large area; i.e. intensity is low but strength and size are large.

Table 8.1. Classification of cyclonic disturbances

Weather System	Maximum Sustained Surface Wind speed (MSSW)
1. Low (L)	< 17 kt (< 9 m s^{-1})
2. Depression (D)	17 to 27 knots (9 to 14 m s^{-1})
3. Deep Depression (DD)	28 to 33 knots (14 to 17 m s^{-1})
4. Cyclonic Storm (CS)	34 to 47 knots (17 to 24 m s^{-1})
5. Severe Cyclonic Storm (SCS)	48 to 63 knots (25 to 32 m s^{-1})
6. Very Severe Cyclonic Storm (VSCS)	64 to 119 knots (33 to 61 m s^{-1})
7. Super Cyclone (SuCS)	120 knots (62 m s^{-1}) and above

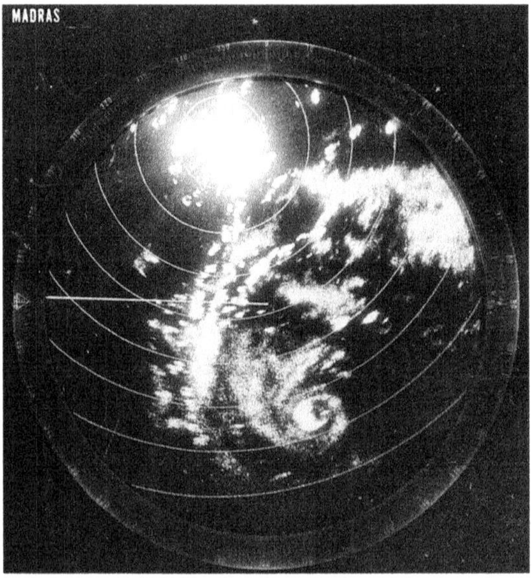

Fig. 8.6. Eye forming as an extension of the innermost spiral band. 2040 hrs. IST, 11 November 1977 (from Raghavan and Veeraraghavan 1979; courtesy IMD). Range rings at intervals of 40 km.

The process of eye formation is not fully understood. Holland (1987) has discussed some possible mechanisms. According to the radar studies of Senn

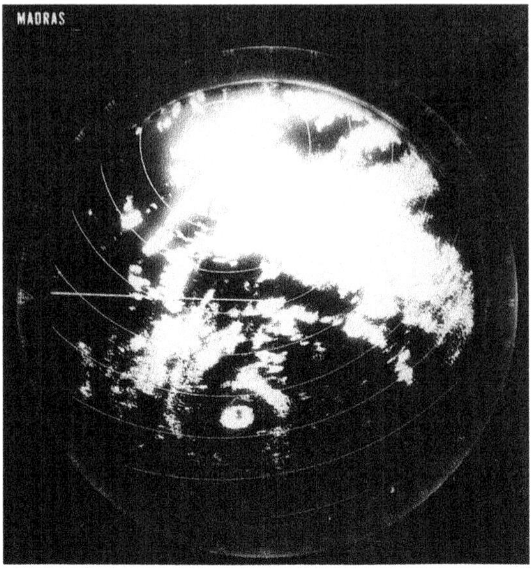

Fig. 8.7. Same TC as in Fig. 8.7 but at 0133 IST, 12 November. Eye has separated into a ring (from Raghavan and Veeraraghavan 1979; courtesy IMD. See text paragraph 8.1.9. for explanation of elliptical eyewall and its rotation. Range rings at intervals of 40 km.

and Hiser (1959) the eyewall is made up of one or more spiral band fragments in the process of generation. A distinction can be made between convective bands that spiral around the centre, but do not encircle it and convective rings that do. Since individual TC's differ widely in structure, there can be cases of intense TC's without forming an eye (Weatherford and Gray 1988). Fig. 8.8 shows a TC with several spiral bands but no eye. On the other hand a "false eye "can develop at the end of a spiral band in a weak TC without being at the centre of the system and may contain a separate pressure minimum on the high pressure side of the spiral band.(Simpson 1956, Barclay 1972, Raghavan 1977). This also represents a rival vortex which may inhibit intensification of the TC. Fig. 8.9 shows an example. In a dual-Doppler radar study Ishihara et al. (1986) observed a circulation centre in a rain band 300 km away from the typhoon centre. Multiple vortices in weak cyclones have been documented from satellite imagery too (Holland 1993, Kalsi 2002). Mukherjee et al. (1977) have described the formation of secondary vortices in an Arabian sea Cyclone. According to them, these may exhibit high winds, marked shift in wind direction, pressure fall and copious rainfall.

8.1.8 The Eyewall

The eye consists of a small, wholly or nearly echo-free area surrounded

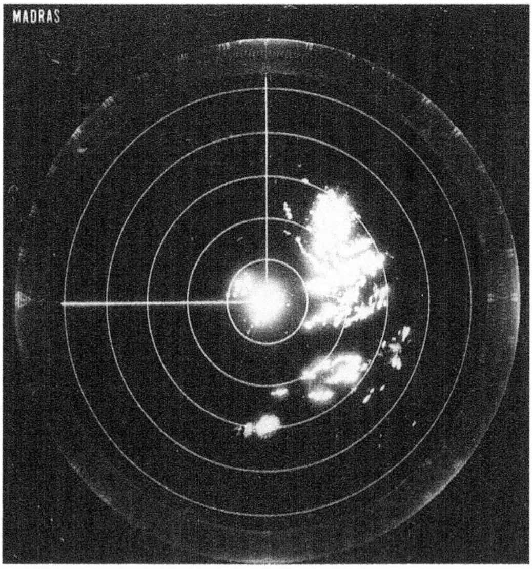

Fig. 8.8. A well-developed cyclone showing only spiral bands but no eye. (courtesy IMD). Range rings at intervals of 100 km.

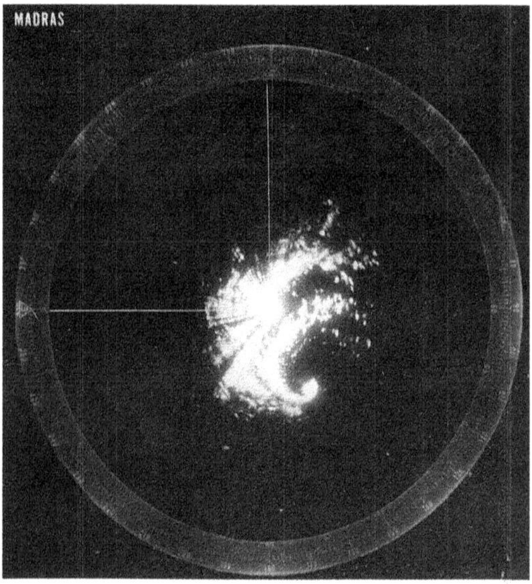

Fig. 8.9. False eye in a weak TC (after Raghavan 1977; courtesy IMD). Range exhibited is up to 500 km.

wholly or partly by a ring of convective precipitation known as the eyewall. Being the most spectacular and significant feature of the TC it has been studied very extensively by radar, satellite and other means. Land-based and airborne Doppler radar observations show the eyewall to be a tall convective cloud associated with the most intense precipitation in the TC, the strongest updrafts and tangential winds. The eyewall often slopes outwards with height i.e. the eye diameter usually increases with height [Willoughby (1988), Raghavan (1993), Tabata et al. (1993)]. This can be clearly seen in Figs. 8.10 and 8.11[6]. Precipitation particles fall out of the sloping eyewall into a precipitation shaft and form a region of high radar reflectivity. Precipitation-driven downdrafts are also present in this region. Eyewalls of some TC's are shown in Fig. 8.6 onwards. In some of these cases only a partial eyewall is seen. If, only the nearer side is seen it is possible that the farther side is not seen merely because its reflectivity or height is too low. It is however also possible that the wall is only partly formed. This is definitely the case in Fig.8.12, where the farther side of the eyewall only is seen. Three hours later (Fig. 8.13) the partial eyewall has rotated in an anti-clockwise direction and the rotation is accompanied by an anticlockwise rotation of the direction of movement of the TC. In Fig. 8.14 the eyewall of a TC is visible as a comma at a range of

[6] See also Fig. 8.40.

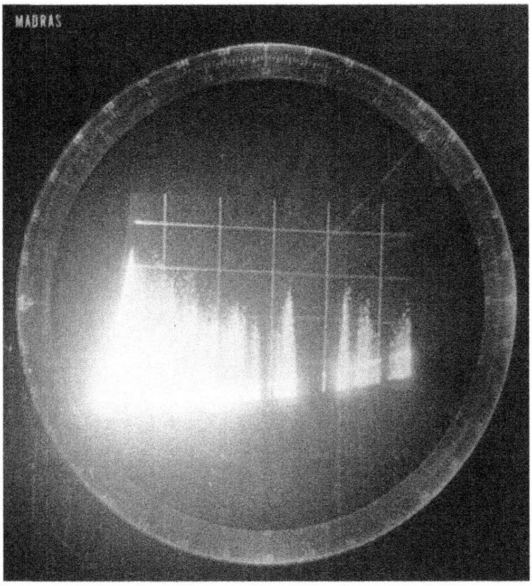

Fig. 8.10. RHI picture showing double eyewall of a TC (courtesy IMD). Range/height marker intervals are 40 km/5 km. Note that the eyewall tapers outwards with height. See corresponding PPI in Fig. 8.23.

425 km[7]. This was possible because of anomalous propagation but it is not possible to say whether the eyewall is completely formed. On the next day when the TC had come closer, it was possible to see the complete eyewall (Fig. 8.15); but a day later when the TC was still closer, only a partial eyewall could be seen (Fig. 8.16), because of subnormal propagation conditions setting in (Raghavan and Rajagopalan, 1980). When the TC comes over land, the eyewall tends to break up. Fig.8.17 shows an exceptional case in which a well-formed eye persisted several hours after landfall (Bhattacharjee and De 1976).

8.1.9 Geometry of the eye

The eyewall is often an unsymmetric structure around the eye. Even when it appears as a perfect ring (Fig. 8.18) the reflectivity distribution is likely to be highly unsymmetrical as can be seen from the pictures of the same TC in Figs. 3.6 to 3.8. While the eye of an intense TC tends to be a circle, ellipses, polygons or irregular shapes are quite common (Lewis and Hawkins 1982, Muramatsu 1986b). Fig. 8.19 and 8.20 are examples. These shapes may vary with time

[7] The eyewall because of its height and high reflectivity may be detectable at such long range but other echoes may not be.

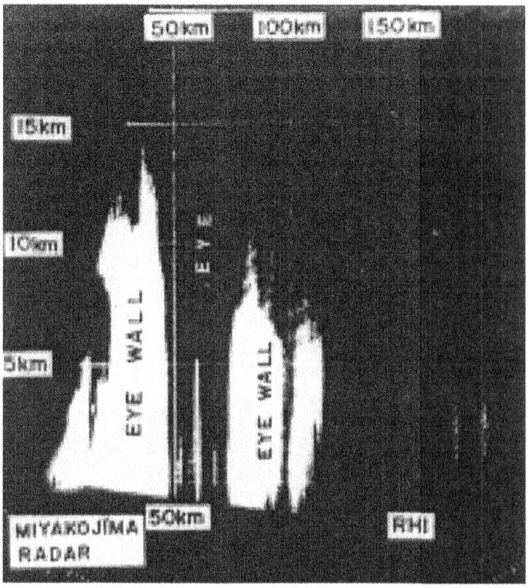

Fig. 8.11. RHI across eye of a typhoon (from Muramatsu 1986b; courtesy Meteorological Society of Japan). Note the increase, with height, of the eye diameter. See also corresponding PPI in Fig. 8.20.

and also rotate around the centre. The shape may also appear distorted due to radar observational limitations. For instance, in Fig. 8.21, ABCD and

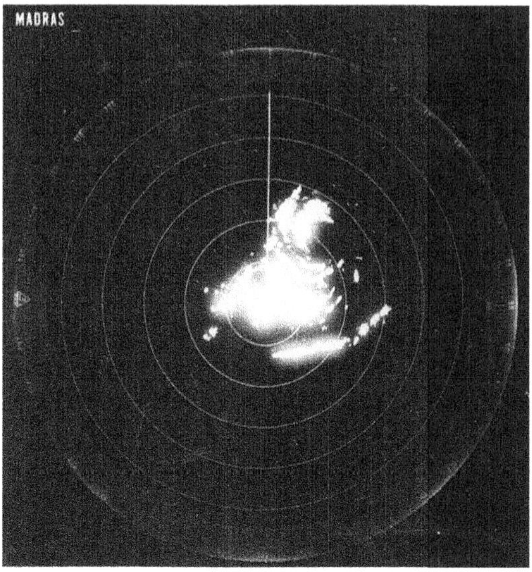

Fig. 8.12. Partial eyewall of TC, PPI picture, range displayed 500 km, time 0536 IST (from Raghavan et.al. 1980; courtesy IMD). Only the northwest sector of the eyewall is developed. The TC was moving northwest.

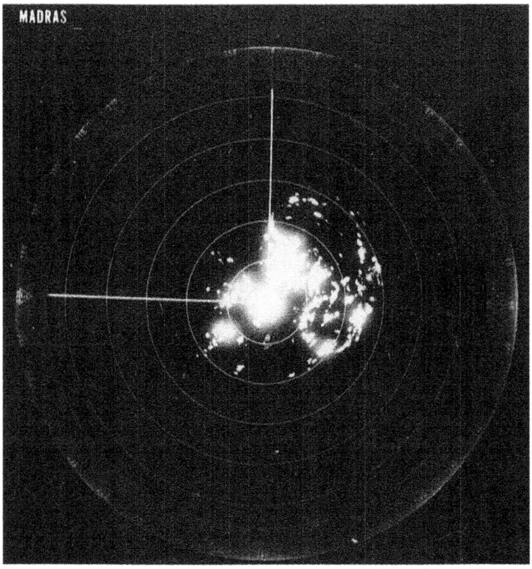

Fig. 8.13. Same as Fig. 8.12 but at 0843 IST (from Raghavan et.al. 1980; courtesy IMD). Note cyclonic rotation of eyewall. This was accompanied by a cyclonic change in direction of movement.

EFGH are the inner and outer edges of a small circular eyewall at a distance of 300 km from the radar. Owing to pulse width and finite beamwidth effects

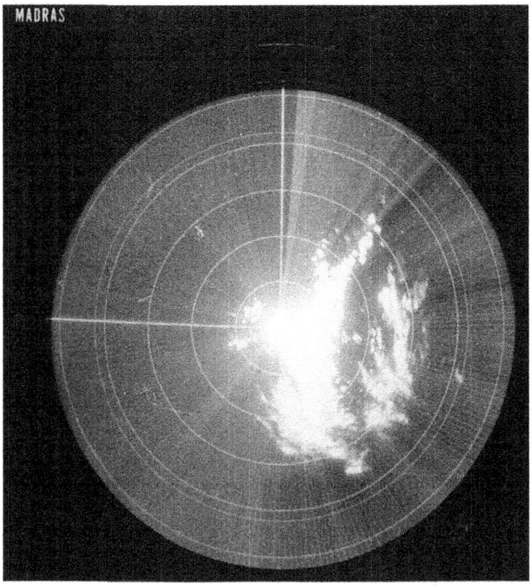

Fig. 8.14. Eyewall of a TC seen as a comma at 425 km range. PPI picture; range marker interval 100 km. 1728 IST 10 May 1979. (from Raghavan and Rajagopalan 1980; courtesy IMD). Long range detection was possible because of anomalous propagation common in that season.

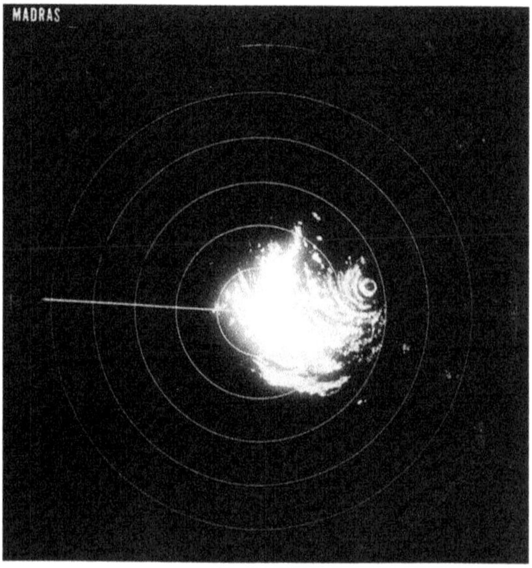

Fig. 8.15. Same as Fig. 8.14 but at 1134 IST, 11 May (courtesy IMD). Complete eyewall is seen.

ABCD will appear as an ellipse A'B'CD' with major axis along the line of sight. The outer edge EFGH will appear as an ellipse EF'G'H' with major axis normal to the line of sight. If now, the eye moves from Azimuth 150 degrees

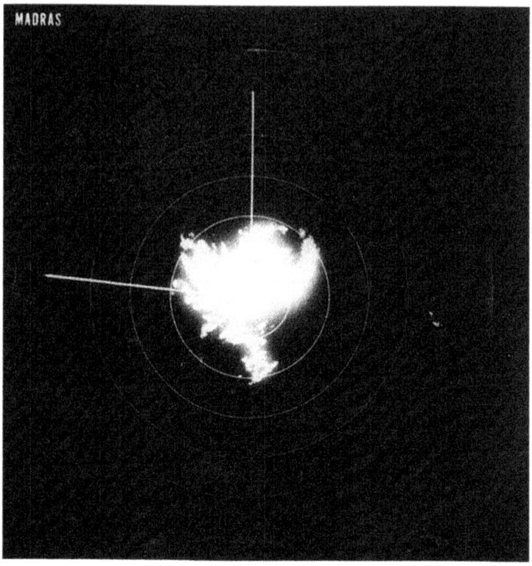

Fig. 8.16. Same as Figs. 8.14 and 8.15, but at 0719 IST, 12 May (from Raghavan and Rajagopalan 1980; courtesy IMD). Note partial eyewall at 165 km range. The deterioration in detectability as the cyclone came nearer is due to setting in of subnormal propagation associated with precipitation caused by the cyclone.

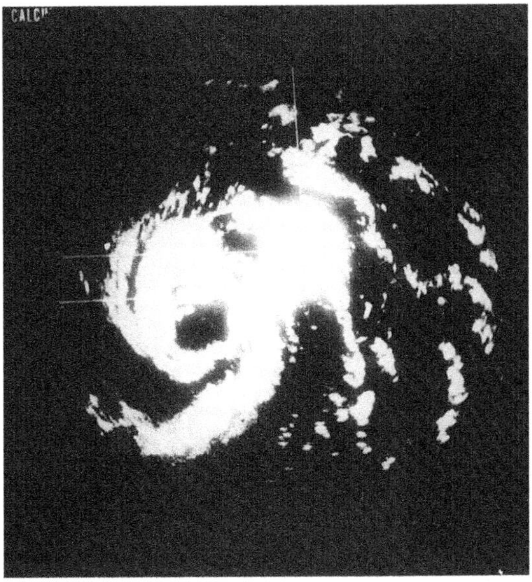

Fig. 8.17. Cyclone eye over land in West Bengal, India (courtesy IMD). Range displayed is 200 km.

to 180 degrees, the elliptical eye seems to rotate as shown in the figure. The actual appearance of the eyewall in the two positions is shown in Figs. 8.6 and 8.7. Both the shape and rotation are spurious in this case.

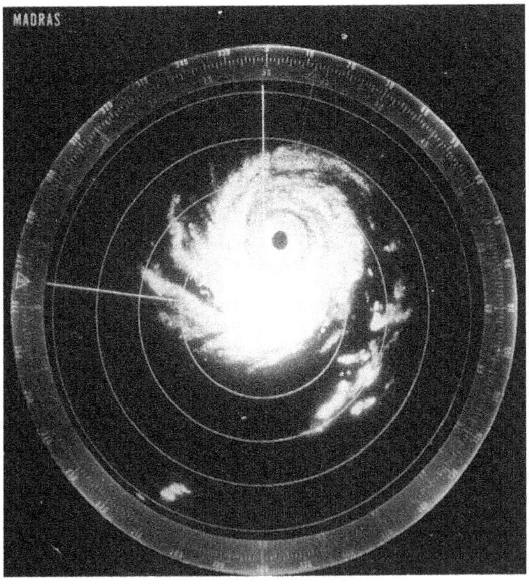

Fig. 8.18. TC eye looking perfectly symmetrical (from Raghavan et.al. 1989; courtesy IMD). Range rings at intervals of 40 km. Pictures of the same TC in Figs. 3.6 to 3.8 at different dBZ thresholds show the asymmetry in reflectivity distribution in the eyewall.

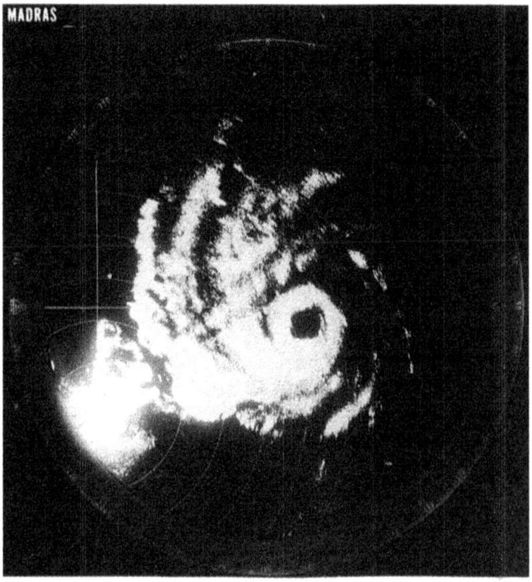

Fig. 8.19. TC with a polygonal eye. (from Raghavan et.al. 1980; courtesy IMD).

8.1.10 Double Walled Eye

In intense cyclones a second eyewall sometimes forms around the existing one. The outer wall may be merely a spiral band wrapping itself around the inner eyewall. Examples of this may be seen at Figs. 8.22 and 8.23. Fig. 8.10 an RHI, corresponds to Fig. 8.23. and shows the two walls in vertical section. In such cases, the two eyewalls need not be concentric. This phenomenon is known as an asymmetric double eye[8]. There are other cases, however, where two concentric rings known as symmetric double eye appear. Continuous air-borne radar observation has shown that a cyclical formation and dissipation of the symmetric double eye structure occurs in intense TC's of the Atlantic and Pacific (Willoughby et. al. 1982, Willoughby 1990). The asymmetric double eye does not undergo such cycles; the reason for this difference is not understood (Houze and Hobbs 1982). A double wall of either type occurs only in deep TC's but not in all of them. In the case of the symmetric double eye, wind maxima at both eyewalls corresponding to the two radii of maximum radar reflectivity have been observed by aircraft [Marks (1985), Willoughby and Black (1996)]. Raghavan et al. (1989) have documented a double wall over land.

Willoughby et al. (1984b) simulated the hurricane vortex numerically,

[8] The term "double eye" to denote one eyewall inside the other may confuse a non-specialist to think that there are two eyes separated from each other. The use of the term "double-walled eye" may remove this ambiguity.

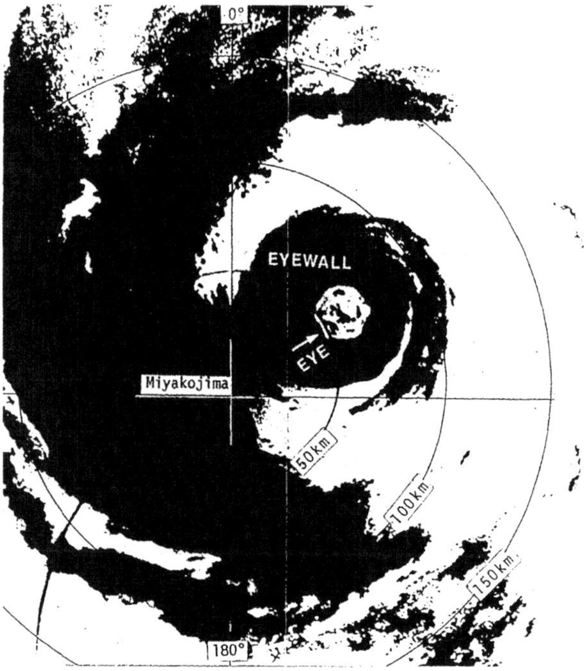

Fig. 8.20. Polygonal (hexagonal) eye of a typhoon (from Muramatsu 1986b; courtesy Meteorological Society of Japan). See corresponding RHI at Fig. 8.11.

by a nonhydrostatic model with microphysical parameterisation of moist convection. Their model was able to reproduce the concentric rings of convection with local maxima of the tangential wind in each. It could reproduce also the tapering eye wall. Several other parameters simulated were consistent with the airborne radar observations of Jorgensen (1984, 1984b) (see section 8.4).

8.1.11 TC Centre Determination from Reflectivity Observation of Eye

The observation of the reflectivity pattern of the eye by radar enables determination of the centre of the TC with considerable accuracy[9]. Where a complete circular eyewall is observed, the geometric centre of the ring is easily identified and is usually taken as the centre of the TC although the point of MSLP

[9] Fixing the centre of a cyclone within a few km is usually adequate for most operational purposes. For studies of the structure of the cyclone and for determining the wind field from Doppler radar, higher accuracy is needed.

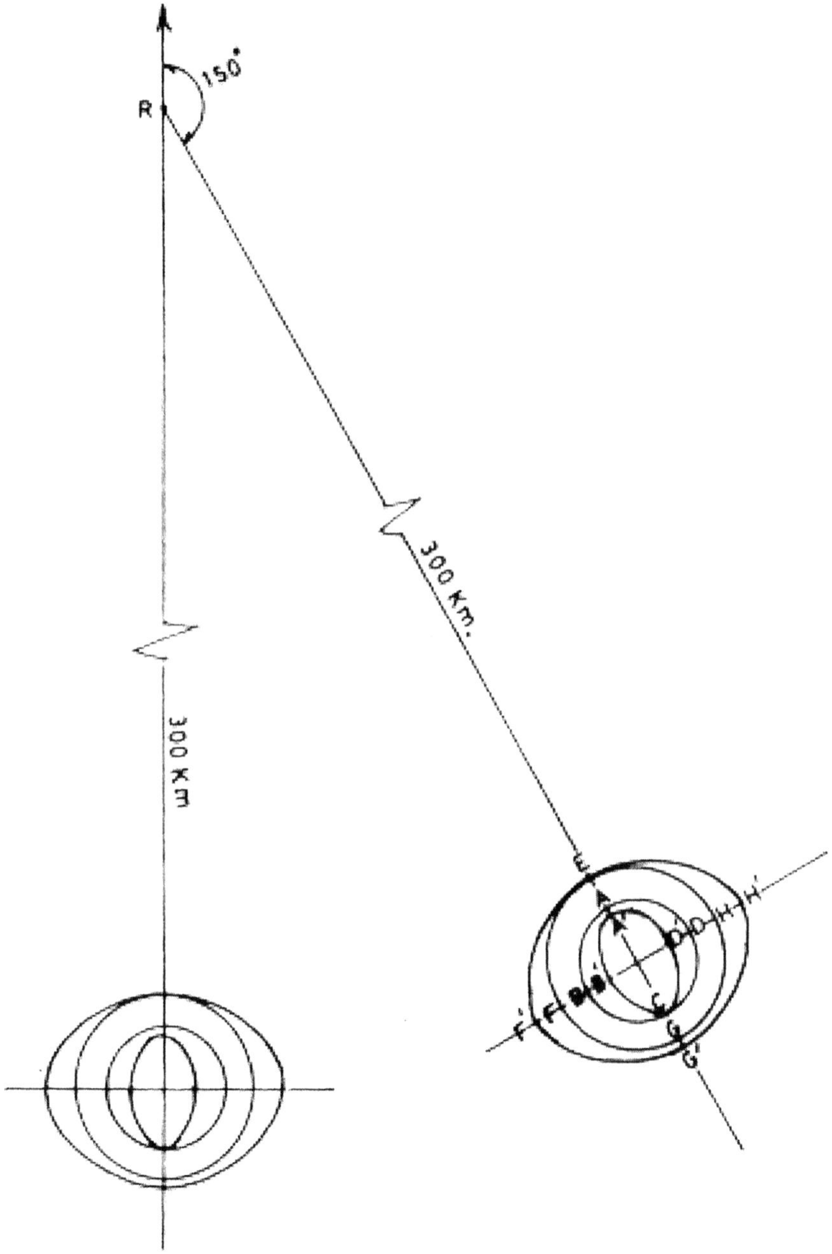

Fig. 8.21. Distortion of TC eye shape and apparent rotation of eye due to beam width (from Raghavan and Veeraraghavan 1979; courtesy IMD). R is the location of the radar.

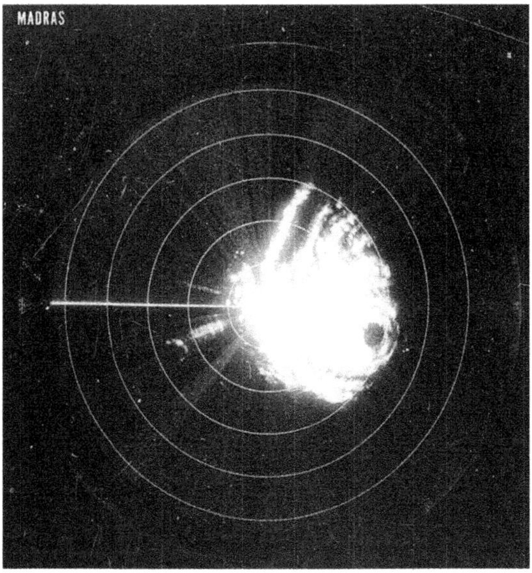

Fig. 8.22. Double-walled eye of an intense TC (after Raghavan et.al. 1980b; courtesy IMD). Range rings are at 100 km intervals.

or the "wind centre" may not necessarily coincide with it. When the eyewall is partial or of irregular shape the determination becomes a little ambiguous. The partial eyewall may be completed subjectively to obtain a circle or ellipse and the

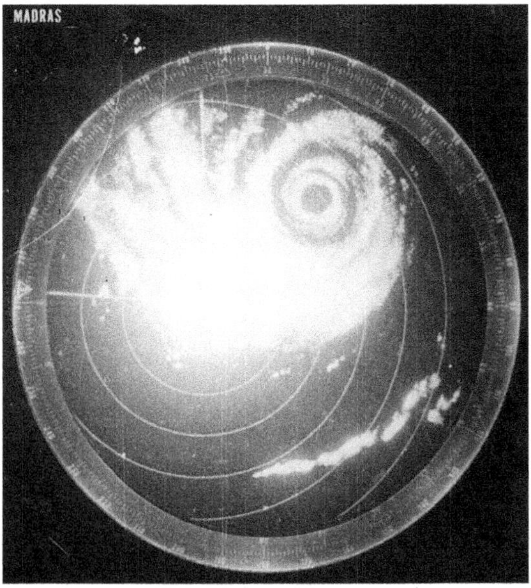

Fig. 8.23. Double-walled eye of an intense TC (after Raghavan 1997; courtesy IMD). Range rings are at 40 km intervals. Corresponding RHI showing the double wall is at Fig. 8.10.

centre of that figure taken as the centre of the TC. To reduce subjectivity in this process Meighen (1976, 1985) recommended placing of a small circle as overlay with as large an arc of the circle as possible superposed on the prominent portion of the eyewall. In Fig. 8.24 the "operational fix"[10] is the estimated geometric centre of the irregular echo-free area. The "post-analysis fix" is obtained by Meighen by fitting the circle to a part of the visible eyewall. To choose this part, the location of the arc of maximum reflectivity can be used as a guide. Even so, the method may appear subjective as the diameter of the circle to be fitted is left to the judgement of the analyst. However, continuous observation of the evolution of the echoes gives clues to the choice of the circle. In Fig. 8.25 the echo configuration at the later time (0531 UTC) decides the choice of the circle although this is not apparent from the earlier configuration at 0518 UTC. Meighen's method obviously yields not only a different centre position but a different eye size too from the conventional method. Meighen showed that fixes obtained by his method give a smoother track (Fig. 8.26). The physical basis for this may be that the wind centre tends to be located closer to the arc of maximum reflectivity in the eyewall (Marks 1990). Thus Meighen's fix is only the reflectivity-weighted centroid of the eyewall. Hence the track formed by this

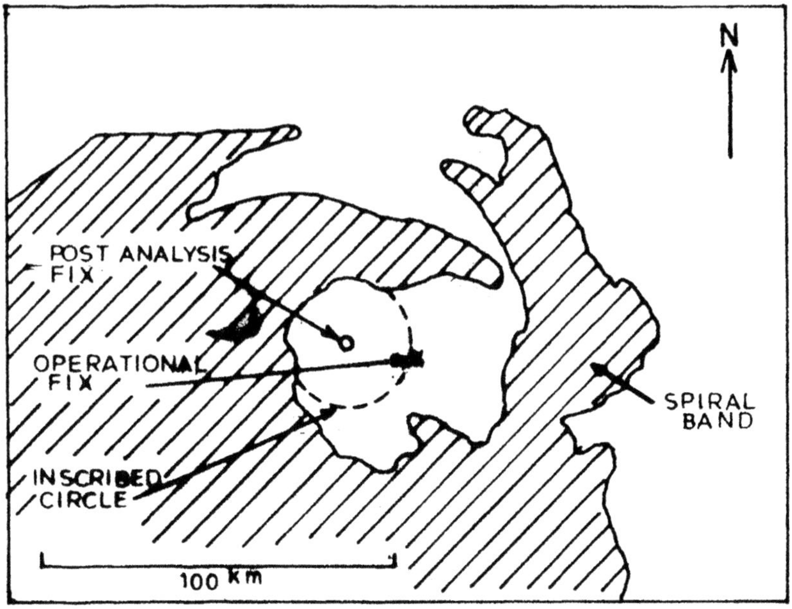

Fig. 8.24. Radar echo configuration of the eye of cyclone Trixie showing operational and post analysis centre fixes (after Meighen 1985, reproduced courtesy Dr. Meighen and WMO).

[10] The centre position estimated from the radarscope is called a "fix".

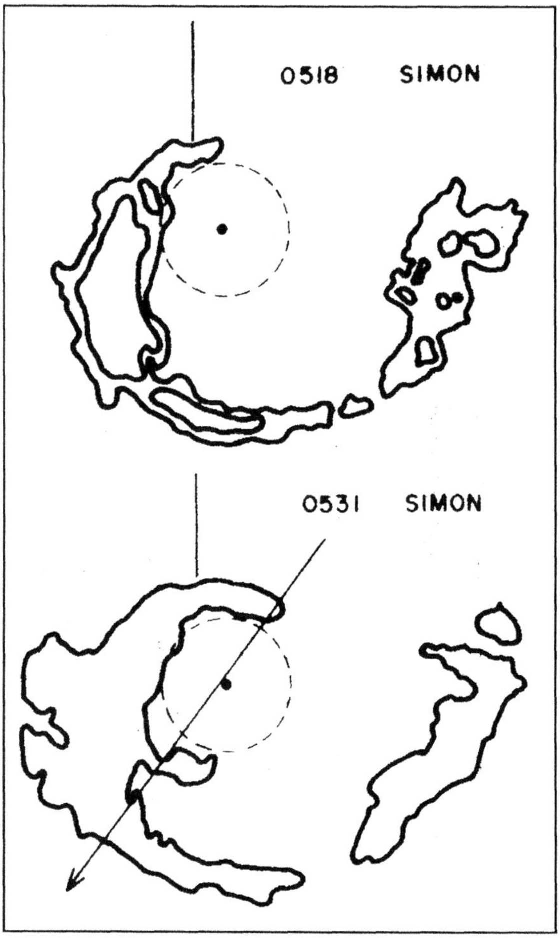

Fig. 8.25. Eye centre location TC Simon (after Meighen 1985, reproduced courtesy Dr. Meighen and WMO).

method should be smoother than the track from the geometric centres (see also section 8.2). The radar fix obtained should always be checked with satellite, aircraft and synoptic observations whenever possible.

In a semi-automated system used in the USA with airborne radar (Griffin et al. 1992) the operator subjectively determines a centre point. From this point the software estimates the location of the inner edge of the eyewall along radials at different azimuths, the inner edge being defined as the closest point at which the reflectivity is 4 dBZ higher than at the centre. The centre of the circle that fits the locations of the inner edge points is taken as the objective estimate of the centre of the eye.

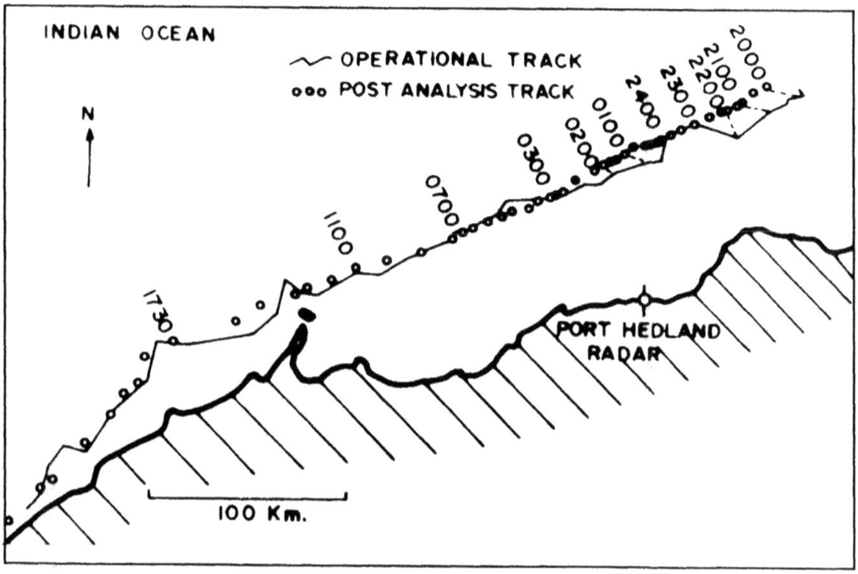

Fig. 8.26. Operational and post analysis radar tracks for cyclone Trixie (from Port Hedland radar, Australia) on 17 and 18 Feb. 1975, times in UTC (after Meighen 1985, reproduced courtesy Dr. Meighen and WMO).

The eye configuration may be different at different heights. Since the eyewall usually tapers outward with height, the eye seen at a longer range may appear larger. Error can therefore arise in the determination of its successive centre positions while the eye moves with respect to the radar. Differences between radar centre and surface synoptic positions will usually arise because the eye centre may not coincide with the pressure centre. For this reason, meteorologists usually take a TC as having made landfall only after confirmation from surface observations even if radar shows a landfall earlier.

8.1.12 Streamers

As we move beyond the eyewall the pattern of spiral bands and rain shield repeats although not symmetrically. Behind the central mass of echo there are usually more echo-free areas but further away, the rear ends of the spiral bands often extend as nearly straight convective lines associated with heavy precipitation. Rockney (1956) called them streamers. These move with the TC and are often responsible for narrow strips of heavy rainfall in the rear sector of the TC well removed from the main belt of rainfall (Raghavan and Ramakrishnan 1981). Examples of streamers may be seen at Figs. 8.18 and 8.23.

8.1.13 Centre Fix from Doppler Radar Observation

The methods of determination of TC centre discussed above are all based on the reflectivity display. It is also possible to estimate the centre from the Doppler velocity display as discussed in Chapter 3 for an axisymmetric circulation[11]. This has been discussed specifically for the tropical cyclone by Wood (1994) who has used the pattern recognition technique developed for mesocyclones (see Chapter 6) for identifying the centre. A simulation of the Doppler velocity pattern of the Bay of Bengal cyclone of November 1970 on a hypothetical 10 cm Doppler radar located at Khepupara in Bangladesh, made by Baynton (1978) is reproduced in Fig. 8.27. The two Doppler velocity maxima on either side of the eye give the maximum tangential velocities around the eye. The two maxima may differ not only because of asymmetries in the eyewall but because a component of the translational velocity of the TC gets added to or subtracted from the rotational velocity. They may also not be at the same range.

In practice there are some limitations:

(1) The maximum unambiguous Doppler range is usually much less than the maximum range of the radar for reflectivity observation. Hence it is possible to determine TC centre from the Doppler pattern only when the TC is close to the radar. It is likely that the reflectivity pattern enabling centre determination will be available long before the TC comes within Doppler range.

(2). Even within the Doppler range, range-aliased echoes may confuse the centre-finding algorithm.

(3). The simulation is that of an idealised vortex. The TC vortex may be asymmetrical and eyewall may be incomplete. In such cases it may not be possible to locate the two points N and P (in Fig. 3.20)[12]. Even if both are available. the centre of an symmetrical circulation may not be at the point C_A.

(4). The algorithm for centre detection may detect the centre of a false eye or a localised shear zone instead of the TC centre, while looking for the couplet of opposite Doppler velocities.

(5). As explained in Chapter 3, the apparent centre will be nearer the radar than the true centre when the diameter of the maximum winds becomes comparable to the range.

(6). When both the reflectivity- and velocity-based centres are available there may be a discrepancy between the two[13]. As already noted, the Doppler centre

---

[11] In the case of tropical cyclones, the value of the exponent p in equation (3.61) when r is $> r_c$ is found to be –0.4 to –0.6 instead of –1 (Riehl, 1954).

[12] However, see Paragraph 8.4.4.

[13] In a Pacific typhoon with a small eye, Stewart and Lyons (1996) found differences of 5 to 7 km between the reflectivity- and velocity-based centres.

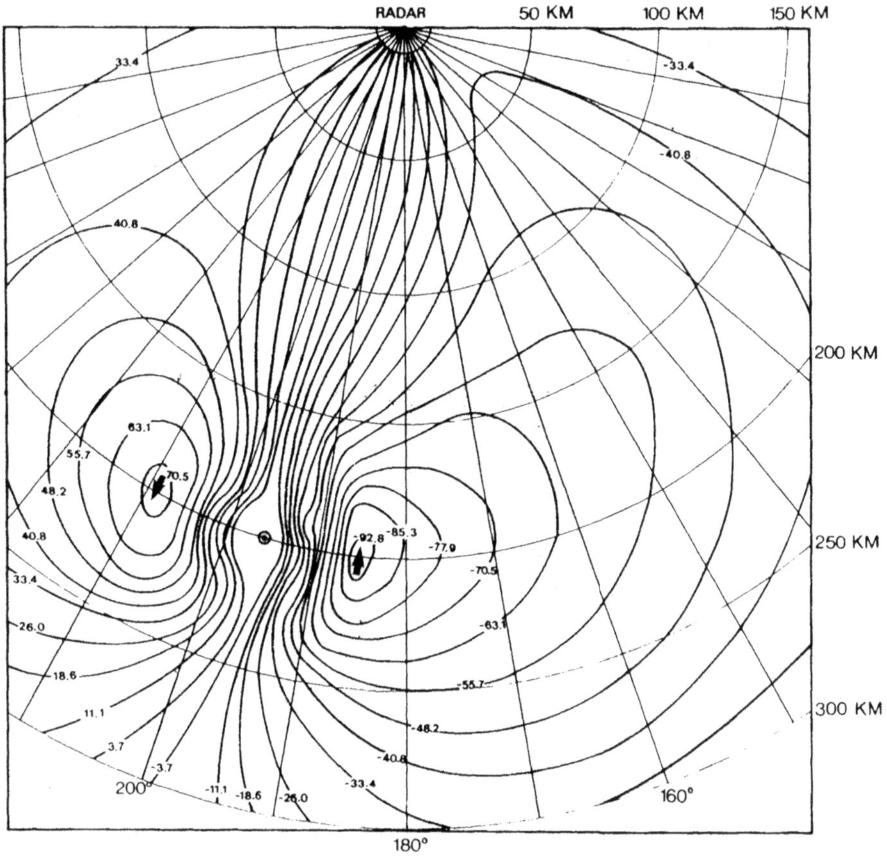

Fig. 8.27. Simulated radial (with respect to radar) velocity distribution in a Bay of Bengal cyclone located at range 200 km and azimuth 196 degrees from a Doppler radar. Velocities in knots., positive away from the radar (from Baynton 1978; courtesy AMS).

may not be the true wind centre in an asymmetrical vortex and even if it is, it may not coincide with the centroid of the reflectivity pattern. Also, the wind centre probably rotates and produces a trochoidal oscillation of the track (see Paragraph 8.2.2).

In the case of airborne radar, independent methods of deriving the centre from aircraft navigational data are likely to be more accurate than the determination from the Doppler velocity pattern. The relative positions of Doppler-based and reflectivity-based centres have not been studied statistically. It appears likely that the reflectivity-based techniques for centre determination will continue to be important for operational use.

8.2 Cyclone Motion

8.2.1 Plotting of a Radar Track

Radar has been widely used to track the motion of TC's and with varying degrees of success to predict the future motion over short periods. The centre positions determined by radar either by observing the eye reflectivity or Doppler pattern or by extrapolating the spiral bands can be plotted at regular intervals of time and the track of the TC determined. Examples of such plots can be seen at Figs. 8.26, 8.28, 8.29 and 8.30. The accuracy of such tracks will depend on the accuracy of each fix.

If a perfect circular eye is seen on radar the geometrical centre can be determined with only a small error with an automated radar. The error will mainly arise from the beam width and height considerations mentioned in section 8.1. With

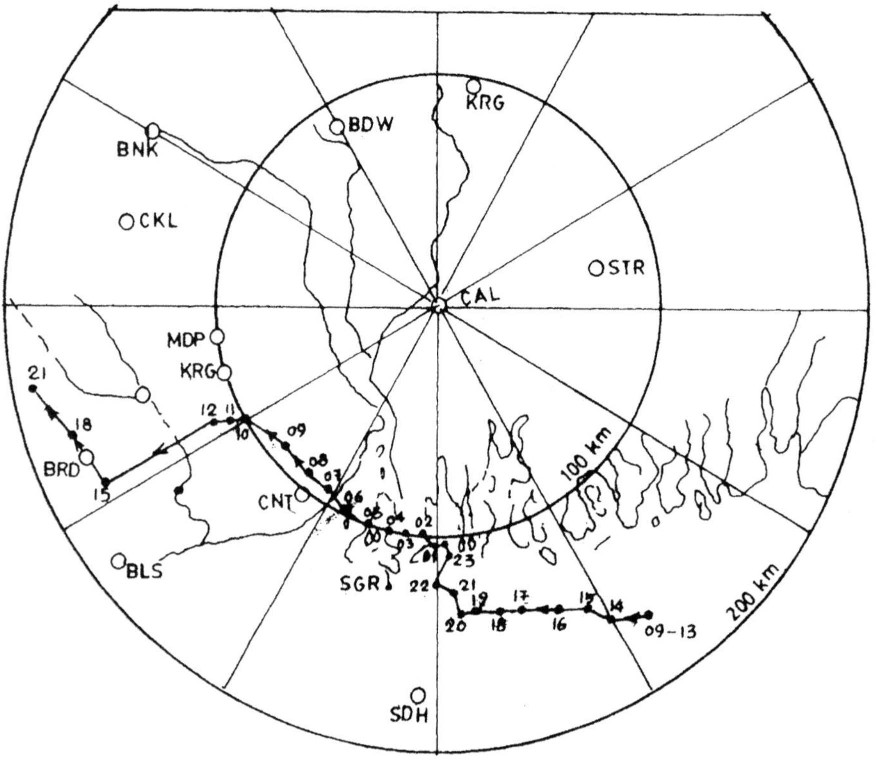

Fig. 8.28. Track of TC 14 to 15 August 1974 determined by a radar at Calcutta (Kolkata), India. (from Bhattacharjee and De 1976; courtesy IMD). From 07 UTC of 15th, the eye was over land; see Fig. 8.17.

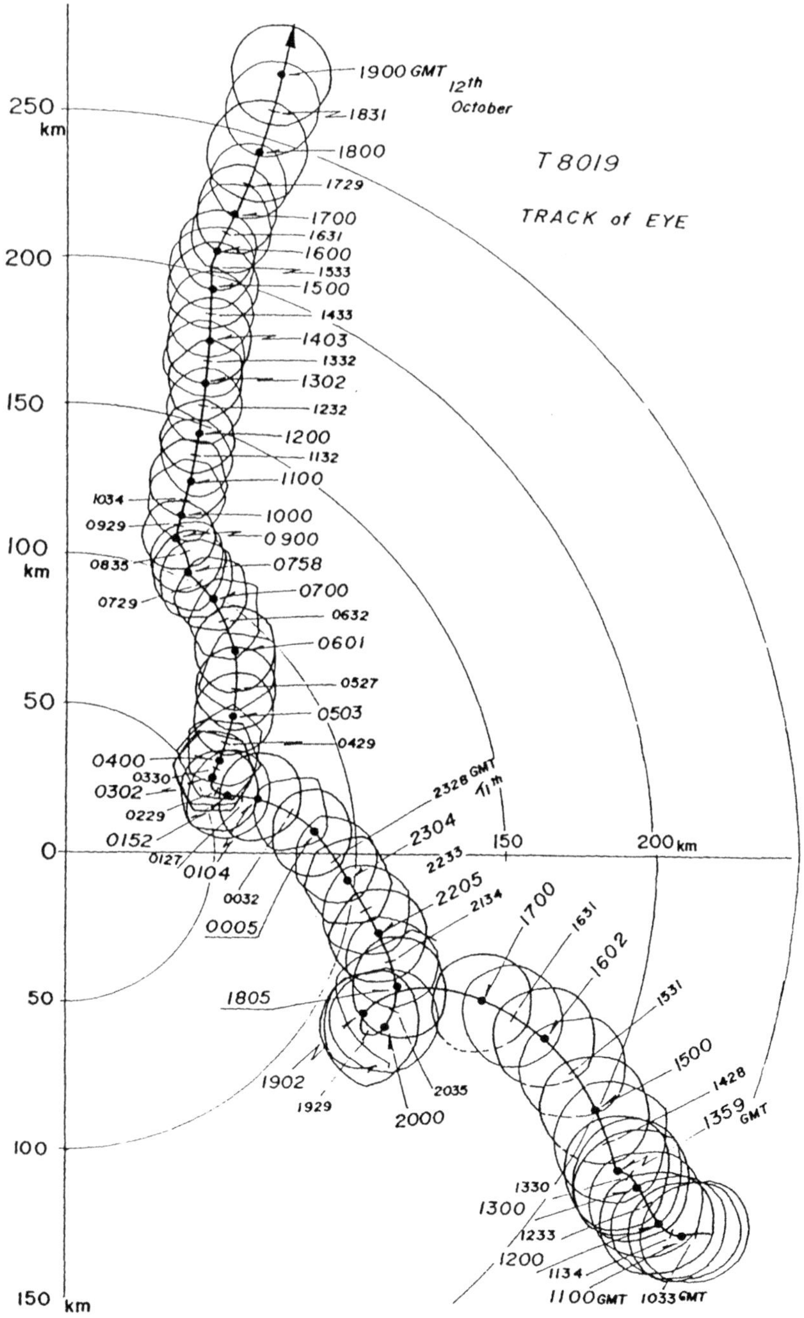

Fig. 8.29. Trochoidal track of Pacific typhoon WYNNE, 8019. Plots are at 30 min intervals from 1033 UTC of 11 October 1980 to 1900 UTC of 12th. (after Muramatsu 1986; courtesy Meteorological Society of Japan).

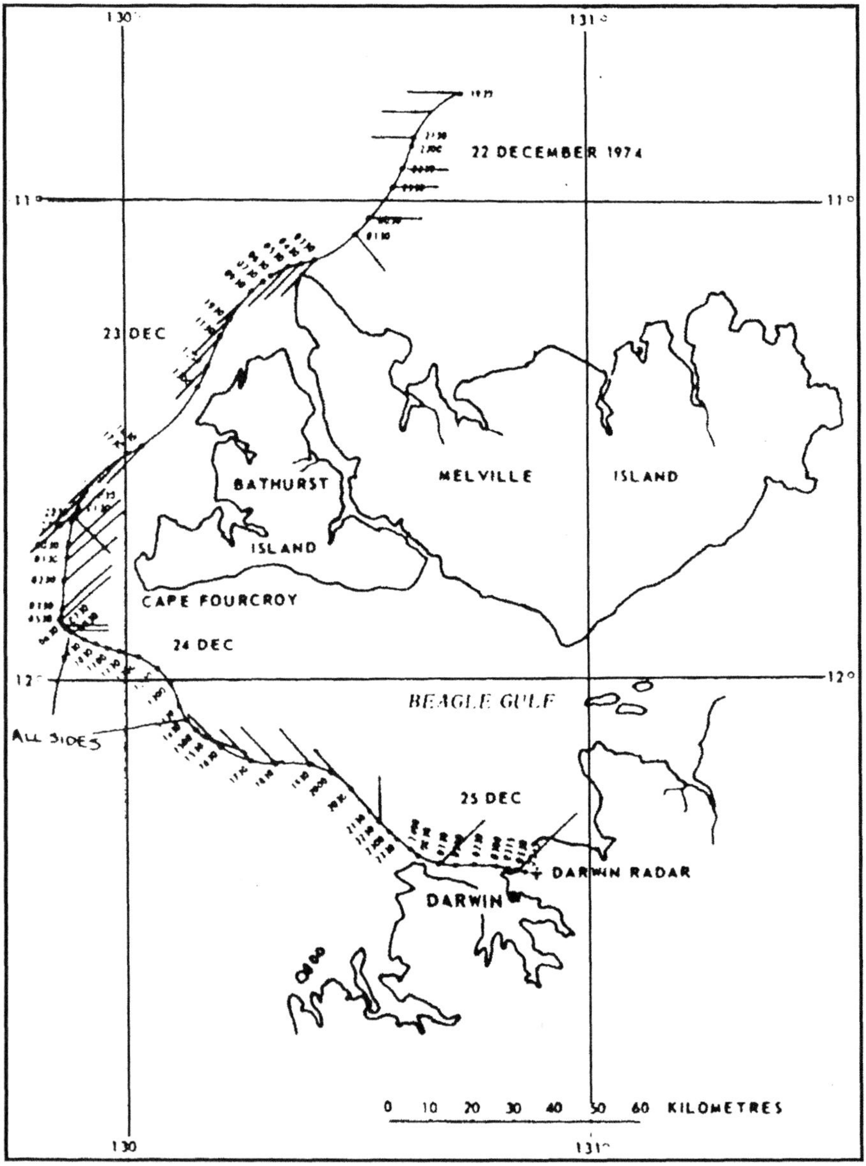

Fig. 8.30. Radar track of TC TRACY, 1974 (after Meighen 1985, reproduced courtesy Dr. Meighen and WMO). Note the oscillations. The arrows at each fix indicate the direction of the prominent convective region in the eyewall.

a manually operated radar, additional personal errors in viewing or tracing the scope or superposing overlays are inevitable. If a partial or deformed eye is seen, the determination of its centre is subjective and an error of several km is possible.

The error in cases when a major segment of the eyewall is clearly seen may be within 10 km. It is therefore of crucial importance to see as much of the eyewall as possible. At long ranges, the MDS of the radar is critical. Wherever attenuating equipment is used in the receiver for signal processing it may be worthwhile to disconnect them while trying to detect an eye at long range (Raghavan et al. 1980). If more than one radar detects the eye, all radars may not see the same pattern owing to differences in range, path attenuation, and height limitations. The one nearer to the TC centre will have a lower detection threshold and therefore likely to see more of the eye and be capable of greater accuracy. Several studies have been made by comparing fixes of the same TC from different land-based (manual) radars [e.g. Raghavan et al. (1985), Naidu et al. (1992), Rajesh Rao et al. (1992)] and differences of the order of 20 km were found (in the cases when a well-formed eye is observed) between radars and between radar and geostationary satellite. The differences are much larger in the case of spiral band fixes. All the radars may not see the same spiral band configuration. Reporting of centre positions in operational messages is in degrees and decimals of latitude or longitude i.e. rounded off to about 10 km. The code form (Chapter 5) usually provides for reporting the expected accuracy in various types of fixes.

While plotting successive positions to obtain a track it may be useful to draw circles around each point showing the probable error in the fix. If an envelope is drawn tangentially to the circles a realistic trend of the motion is obtained. It is to be expected that these circles of confusion will be larger in the case of weak TC's where the eye may not be well-formed or stable or the positions may have been obtained from spiral band extrapolation.

8.2.2 Trochoidal Motion

While some of the apparent oscillations in the track can be attributed to radar observational errors there is often a real oscillation or meandering of the TC centre about a mean path. Figs. 8.28, 8.29 and 8.30 are examples. According to some scientists the centre of the eye rotates cyclonically around the centre of the TC system resulting in a trochoidal[14] motion of the eye centre [Barclay (1972), Muramatsu (1986), Rao (1967), Subramanian (1981)]. Theoretical considerations leading to the expectation of trochoidal motion may be seen in Yeh (1950) and Kuo (1969). Horn (1951) found in the Pacific (from aircraft fixes) that the oscillations agree in amplitude with Yeh's (1950) results but the period of oscillation is shorter. Some authors have also linked the trochoidal motion with

[14] A trochoid is the locus of a point at a specified distance from the centre of a circle which rolls without slipping, along a straight line.

the changes in the shape of the eye. Muramatsu (1986) shows that the system centre follows a smoother track than the trochoidal one from radar eye centres. A study by Holland and Lander (1993) attributes track oscillations to interactions with mesoscale systems within the cyclone's circulation. Oscillations in the track can also arise if the TC motion is stalled or it makes a loop especially when changes in upper wind regime cause recurvature.

8.2.3 Smoothening and Extrapolation of Track

The fine structure of motion detected by the radar helps to pinpoint the exact path over the coastal terrain. In Fig. 8.28 for instance, short period forecasts of where exactly the eyewall will pass and create the maximum damage would have been greatly facilitated. However, a meteorologist interested in evaluating the overall motion of the system and forecast the motion over a longer period, may prefer to smoothen out some of the short period oscillations. This is particularly important when TC positions are included in warning messages sent to the public. Apparently erratic motion may lead recipients to draw wrong conclusions and undermine the credibility of the forecasts. Hence it is normal practice to smoothen out short period changes in track in bulletins to the public.

When a smoothened track or an envelope of the track is available for several hours, the average speed of motion can be determined. The track can then be extrapolated to predict motion for several more hours and even predict the point and time of landfall [(Raghavan et al. (1980), Meighen (1985)]. In Fig. 8.30 the reader can try linear extrapolation of any chosen 3 hour,6 hour or 9 hour segment of track. For example if the track from 0830 to 1130 of 24 December is extrapolated it indicates a landfall at 1730 hrs of 25 December well to the north of Darwin radar. Note that the position error of landfall is small in this case too, but the forecast time of landfall is considerably in error. Accuracy may be improved by taking a longer duration segment, but because of the trochoidal track the choice of the segment will influence the error. The speed of motion may also change considerably with time. Considering that errors in 24 hour track prediction from computer models using climatology and persistence considerations can be over one degree (in latitude/longitude), the linear extrapolation is often remarkably successful. But evidently it can be applied only if there are no meteorological indications of radical changes in direction or speed of motion of the TC. Instead of linear extrapolation, fitting of a polynomial to available radar fixes has been tried (e.g. Jayanthi and Asokan 1994). Better results can be obtained if the radar observations are combined with other data. Sheets (1985) has tracked the "mass field envelope" as measured from aircraft and used it with a radar track. Interactive displays can be used to combine imagery of a network of radars and of a satellite to obtain the best past track and extrapolation can be attempted from such a track.

8.2.4 Predictors of TC motion

Some radar-observed features have been tried as short period predictors of TC motion. The orientation of pre-cyclone squall lines has already been mentioned in Paragraph 8.1.3. Its use is limited as the squall lines usually dissipate well before landfall of the TC. Senn (1966) found an echo concentration mainly in the forward right quadrant and therefore considered it a good indicator of TC motion. But others do not find it a reliable indicator [Meighen (1985), Raghavan et al. (1980)]. In the case of unsymmetrical eyewalls, the direction (with respect to the TC centre) in which the eyewall reflectivity is highest was seen in some cases to be the preferred direction of motion (Senn 1966). In Fig. 8.12 the eyewall is seen only in the northwest sector of the TC core.. The TC was moving northwestwards. Three hours later (Fig. 8.13) the eyewall rotated to the southwest sector. The direction of motion also rotated anti-clockwise correspondingly (Raghavan et al. 1980). But this again does not happen in all TC's and hence is not a reliable predictor. For instance in Fig. 8.30 the direction of the prominent convective region in the eyewall at various times is indicated. Evidently this has little relation to the track.

8.3 Winds in Tropical Cyclones

8.3.1 SPAwinds in Cyclones

The derivation of wind vectors (SPAwinds) from the observed movement of radar echoes from precipitation is discussed in Chapter 5. With early radars it was necessary to plot the positions of individual echoes manually at frequent intervals and determine their motion. Later, this could be done by preparing movie loops (Muramatsu 1986) or by digital techniques. The SPAwind technique pioneered by Ligda (1953) attracted particular interest in the case of TC's because of the traditional paucity of wind data in TC's. Ligda (1955) found that SPAwinds in TC's correlated well with the 700 hPa level wind data.

According to Watanabe (1963) the echo movement in typhoons agrees well with the mean of the winds at cloud top and cloud base. The wind which the echo motion represents will evidently depend on the echo height. Since the radar sees echoes at different ranges at different heights the range will also be a factor. However, in TC's the wind (outside the core region) increases with height in the first 1 km and thereafter there is not much vertical variation (Meighen 1985). Hence, the height assignment is not critical. Since the wind field is generally asymmetrical in the horizontal, the wind obtained from radar in one sector (a land-based radar cannot usually observe all parts of the TC) may not be representative of other sectors.

Interest in the SPAwinds technique has diminished because of the availability of satellite-derived cloud motion vector winds and the advent of Doppler radar observations yielding the wind field.

8.3.2 Doppler Winds

Within the Doppler range, a velocity field is available subject to the problems of range and velocity aliasing. Examples of such fields from airborne Doppler radar observations are given in section 8.4. Airborne Doppler radar observation made it possible for the first time to get winds in tropical cyclones at other than flight level. The most important velocity observation is the couplet of maximum winds at the eyewall. If this can be observed it gives a direct indication of the intensity of the TC which is operationally very important. The limitations mentioned in Paragraph 8.1.13. are relevant. In the case of an incomplete eyewall it is possible that the maximum wind is underestimated.

Wind profilers and MST radars (see Chapter 10) have been used to profile the winds in parts of tropical cyclones [(e.g. Sato et al. (1991), May et al. (1994)]. The profiler suffers the disadvantage that it can see only overhead and, therefore, the limited part of the TC which comes over the radar station. But in this area the vertical motion and turbulence can be studied.

8.3.3 Derivation of Surface Wind over Land

The wind measured or estimated by radar is not the surface wind but the wind at an altitude depending on range. The wind occurring over the sea may also change at landfall (Powell et al. 1991; see also section 8.4. for the effects of landfall). Attempts have been made to relate lower tropospheric winds measured by radiosonde or wind profilers to the surface wind over land in tropical cyclones (see e.g. Dobos et al. 1995) by using suitable models. Using such relationships, it may be possible to derive from the radar measurements, the surface wind at landfall.

8.4 Cyclone Characteristics from Radar

8.4.1 Types of TC Attributes Studied

In section 8.1. the features of a cyclone seen by radar were listed. In this section we shall consider the features observed by radar in further detail and try to relate them to the structure, as well as intensity, size and strength of the TC. One approach is to measure the properties (e.g. heights, horizontal sizes, reflectivities, vertical motions) of echoes of individual precipitation elements and see how they compare with those in non-TC precipitation systems. Another is to measure group properties e.g. number and horizontal width of spiral bands, dimensions of the eyewall, total size of the echo mass, sectors of concentration of echoes, distribution of precipitation, latent heat release and mass transport, partition of convective and stratiform precipitation and asymmetries in the various distributions. Some of these measurements are possible with conventional land-based radars; others have been achieved with air-borne (non-Doppler and Doppler) radar and ground-based Doppler radar.

8.4.2 Airborne Radar Observations

C-130 aircraft and WP-3D research aircraft of the U.S.A. equipped with radars and other instrumentation have been flown through several hurricanes and typhoons. Characteristics of radars in WP-3D aircraft are given in Chapter 1 (Tables 1.5 and 1.6) and airborne radars were discussed in Chapter 3. The nose and lower fuselage C-band radars scanned horizontally and the tail X-band radar scanned vertically. The C-130 had a single horizontally scanning nose radar similar to that of the WP-3D. The flights consisted of several radial traverses oriented north-south or east-west on approximately the 700 hPa (about 3 km altitude) pressure surface [Jorgensen (1984, 1984b), Marks (1985)]. The received power data were digitised and tape recorded in 256 gates. By integration, the variance of the received power was kept below 1 dB. The antennas were scan-stabilised against small variations in pitch and roll to keep the beam at the desired tilt angle. The operator could minimise sea clutter by adjusting the tilt. The radars were calibrated several times during the hurricane season. The main limitations of the air-borne radar observations were inadequate beam filling and attenuation by intervening precipitation and atmospheric gases. The former source of error could alone cause a loss of up to 30 dB at long ranges. The loss due to intervening precipitation was about half of this. Use of time composites minimised the errors. The horizontal structure of precipitation was composited from the various radars. A typical vertical cross section of reflectivity in the core region is shown in Fig. 8.31 along with wind and other profiles from other instruments.

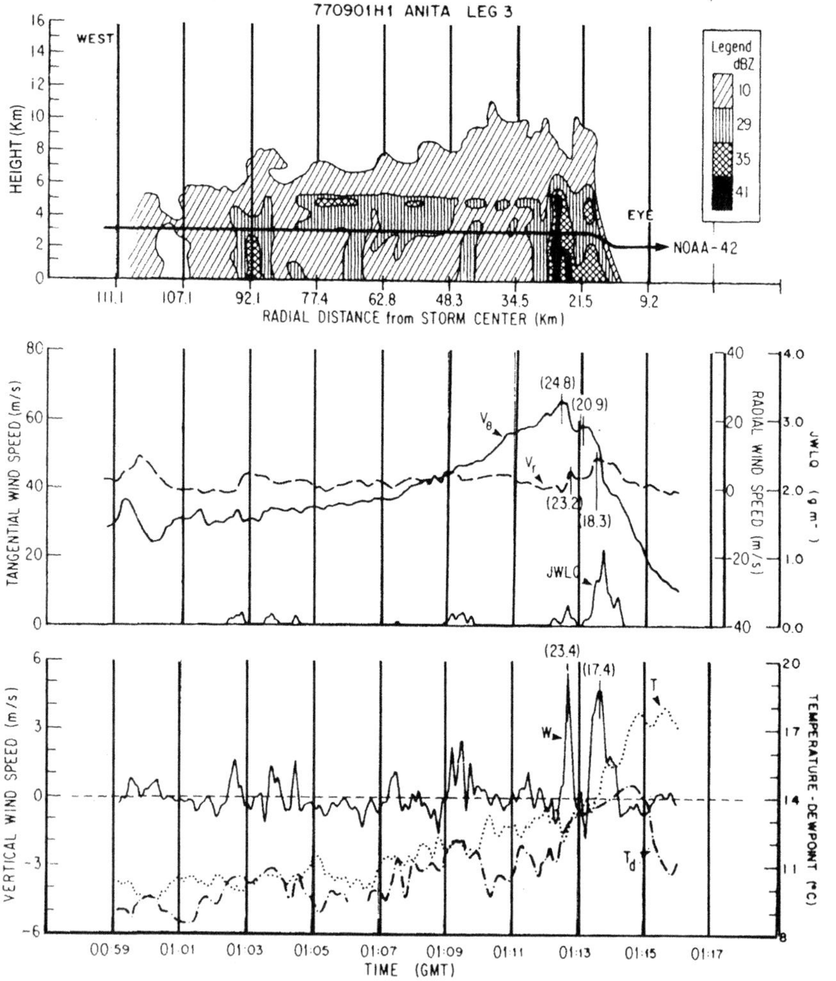

Fig. 8.31. Profiles of radar reflectivity, tangential wind V_t, radial wind V_r, cloud water content JWLQ, vertical wind W, temperature T and dew point T_d in a hurricane (from Jorgensen 1984; courtesy AMS). Wind data are relative to the moving storm. The radar reflectivity cross section was generated by compositing vertical rays from the tail radar at four samples per minute. Locations of the peaks of the horizontal and vertical wind are indicated in parentheses. The aircraft flew from west of the centre inward to the eye.

8.4.3 Airborne Doppler Radar

The tail radar of the WP-3D was converted into a Doppler system (characteristics in Table 1.6) and the first comprehensive Doppler observations were made in 1982 (Marks and Houze, 1984) in tropical storm Debby and later in other hurricanes [e.g. Marks and Houze (1987), Marks et al. (1992)]. The methods of observation and data processing are dealt with in chapter 3 [see also Jorgensen et al. (1983), Hildebrand et al. (1983]. Briefly, the same echo volume is sampled in two orthogonal legs of the flight resulting in pseudo-dual-Doppler data. The vertical incidence data provide the vertical Doppler velocity. Vertical air velocity is obtained by subtracting the estimated precipitation particle fall speed. The latter is taken as a function of the measured reflectivity (see chapter 2). The flight track and "boxes" in which data were analysed in Debby are shown in Fig. 8.32. A plot of horizontal velocities is shown in Fig. 8.33.

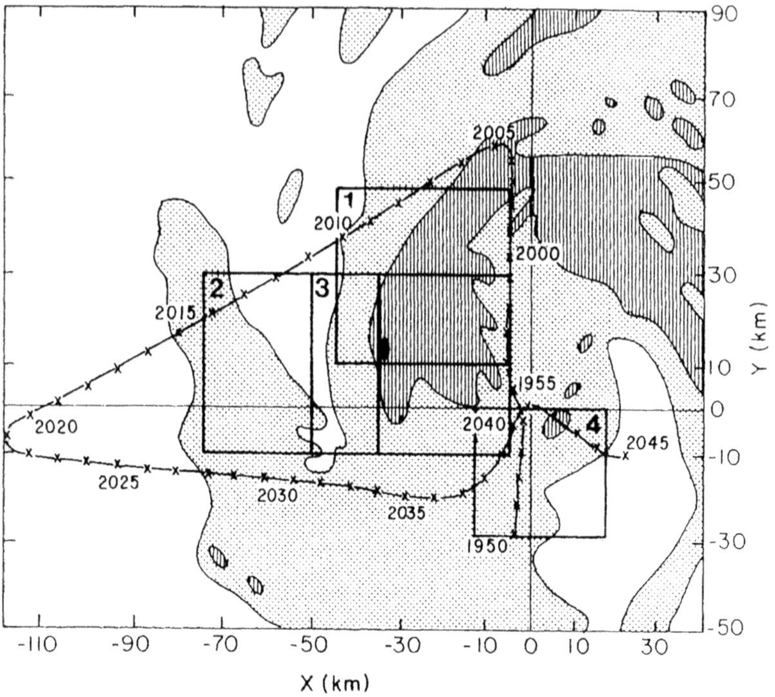

Fig. 8.32. Time composite of horizontal distribution of reflectivity in Tropical Storm Debby (1982) from air-borne Doppler radar (from Marks and Houze 1984; courtesy AMS). Reflectivity contours are for 20,30 and 40 dBZ. Aircraft flight track is indicated by the thin solid line with time indicated. The analysis boxes 1, 2, 3 and 4 are shown by thick solid lines. The origin of the co-ordinate system is at the storm centre, and the aircraft positions are plotted relative to storm centre. X and Y represent east-west and north-south axes.

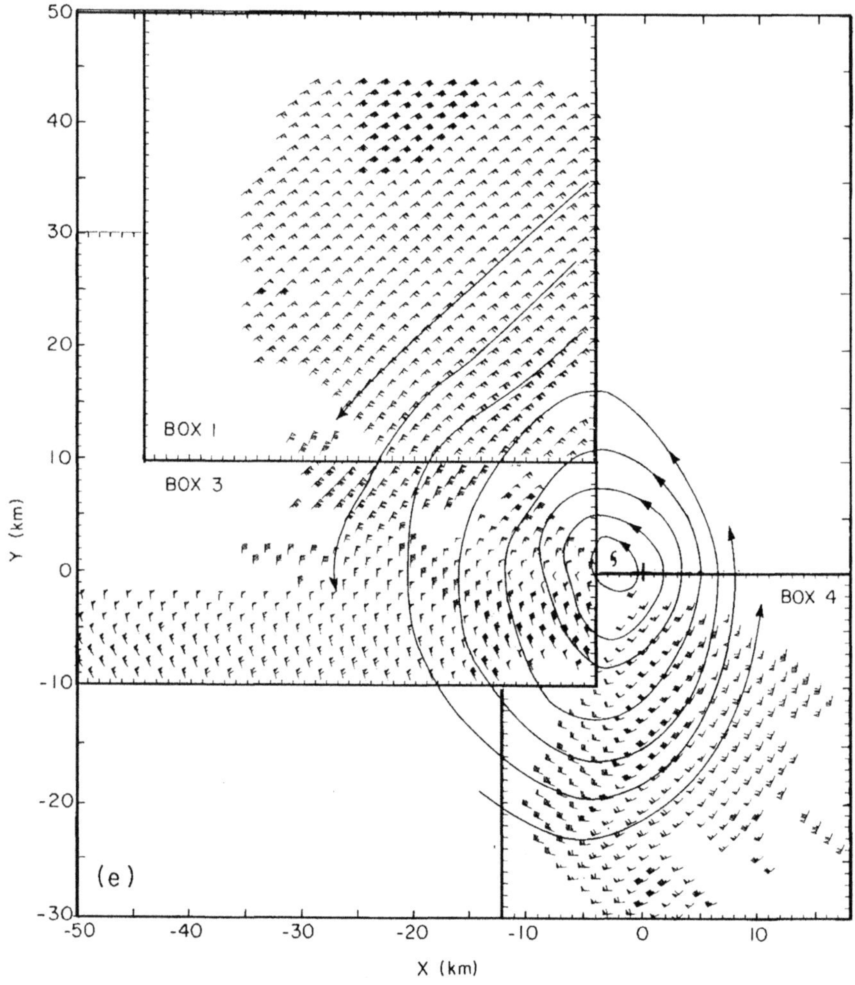

Fig.8.33. Mosaic of Doppler-derived winds in Boxes 1, 3 and 4 of Fig. 8.32 (after Marks and Houze 1984; courtesy AMS). Streamlines and storm-centre location implied by the streamlines of the Doppler winds also indicated. Note that this centre does not coincide with the origin. Wind plots use standard convention i.e.flag 25 m s^{-1}, barb 5 m s^{-1}, half-barb 2.5 m s^{-1}.

This technique assumes the wind field to be stationary for over an hour. It is also computation-intensive. It is difficult to process the mass of data on board or to transmit in real time to the analysis centre. The Velocity-Track Display (VTD; Carbone and Marks 1989, Lee et al. 1994b) technique seeks to overcome these problems. With a radar scanning in a vertical plane orthogonal to direction of flight, a radial penetration of the TC enables the tangential wind velocity V_T at any radius R (from TC centre) and height z to be computed from the observed Doppler velocity V_r at radar range r and zenith angle θ.

$$V_T(R,z) = \frac{V_r(r,\theta)}{\sin\theta\cos\phi} \tag{8.2}$$

where θ is measured with respect to the zenith as zero and

$$\phi = \sin^{-1}\left(\frac{r\cos\theta}{R}\right) \tag{8.3}$$

The VTD algorithm decomposes, by harmonic analysis, Doppler velocities on cylindrical rings into tangential, radial and mean cross-track components of the wind velocity, using data from a single penetration by one aircraft. Hence, it assumes wind field to be stationary only for about 15 to 20 minutes. The Fourier coefficients can be transmitted in real time over a satellite link. However, asymmetric component of velocity and storm motion cannot be resolved without a second leg. Roux and Marks (1996) developed an Extended VTD (EVTD) algorithm which can give additional information by combination of data from successive legs.

The use of air-borne Doppler radar enabled, for the first time, routine acquisition of winds at other than flight levels. This is particularly important for acquiring near-surface data close to land where it is not possible for the aircraft to fly low.

The National Aeronautics and Space Administration (NASA) of the USA has a high-altitude (20 km) aircraft ER-2 on which Doppler radar and in situ instrumentation have been installed. The radar known as EDOP (ER-2 Doppler radar) (Heymsfield et al. 1996) installed in the nose of the aircraft is an X-band radar with two non-scanning beams[15] one looking at the nadir and one looking about 33 degrees forward. Vertically polarised radiation is transmitted. The nadir beam measures Z, v and σ while the forward beam yields LDR in addition to these. The two beams are used together to obtain time-height sections vertical and along-track[16] air motions. The aircraft motion (independently obtained) and the hydrometeor fall velocity (derived from reflectivity) are subtracted to obtain vertical air velocity. The reflectivities are corrected for attenuation using the surface reference technique (see Chapter 7). The ER2 flights were used by Heymsfield et al. (2001) to make measurements of the vertical precipitation and thermodynamic structure in the eyewall region of a hurricane. They superposed satellite IR images on low level radar reflectivity patterns and observed convective bursts in the eyewall and a broad current of subsiding air with vertical velocity of several m s^{-1} within the eye as well as in one of the "hot-towers" in the eyewall.

---

[15] The use of two beams will also enable implementation of the stereoradar technique.

[16] Cross-track velocity cannot be obtained as there is no scanning.

8.4.4 Ground-based Doppler Radar

Use of ground-based Doppler systems has been confined in earlier years mostly to phenomena of smaller scales than TC's, mainly because of range and maximum velocity limitations. Baynton (1978) suggested a radar design for observing a TC, his main objective being to get the wind field and thereby distinguish a "killer storm" from a more ordinary one. His simulation of the Bangladesh cyclone of November, 1970 was shown in Fig. 8.27. Theoretically, Doppler radar should be able to obtain the intensity of the TC by observation of the couplet of oppositely directed winds in the Doppler signature of the eyewall. But the limitations in this have already been mentioned in section 8.1. More recent simulations may be seen in Wood (1991, 1994), Wood and Brown (1987, 1989, 1992) and in Wood and Marks (1989). Despite the limitations, the circulation in the core and outside can be mapped. A technique similar to the VTD called the Ground-Based Velocity Track Display (GBVTD) has been developed for use with ground-based Doppler radar [Lee et al. (1993, 1999, 2000), Lee and Marks (2000)] to map the winds around circles centred on the TC centre with a radar located away from the circle. By harmonic analysis, Fourier coefficients are determined for vortex circles for each radius and altitude using CAPPI data. Thus, the three-dimensional structure of the TC can be determined. The Doppler velocity pattern for an axisymmetric Rankine vortex was discussed in Chapter 3. If a mean wind flow is added and if there is asymmetry, the isodop pattern gets complicated. Since the tangential and radial velocities are usually asymmetric, the Fourier components from the harmonic analysis contain information on the higher wavenumbers of the tangential and radial winds at each radius. Fig. 8.34(a) (from Lee et al. 1999) shows the tangential velocities (V_T) of an assumed axisymmetric Rankine vortex with an asymmetry in one direction superposed; the corresponding isodop pattern is shown in Fig. 8.34(b). Fig. 8.35 shows the GBVTD-derived V_T which compares reasonably well with the analytic wind pattern in Fig. 8.34 (a). Lee et al. (2000) have tested this on a real TC. An important advantage of the technique is that even when the radar does not directly see the wind maxima (points N and P in figure 3.20) it still retrieves the wind maximum from the tendency of the Doppler velocity pattern. Lee and Marks (2000) have also formulated a TC centre-finding algorithm which estimates the vorticity centre by maximising GBVTD-retrieved mean tangential wind. They claim an accuracy of better than 2 km for this algorithm.

The inner core structure and vertical velocities in typhoons have been studied with a single Doppler radar [e.g. Wang and Tseng (1989), Ishihara et al. (1991), Lee and Tsai (1991), Tabata et al., 1993)] while in Japan, dual Doppler

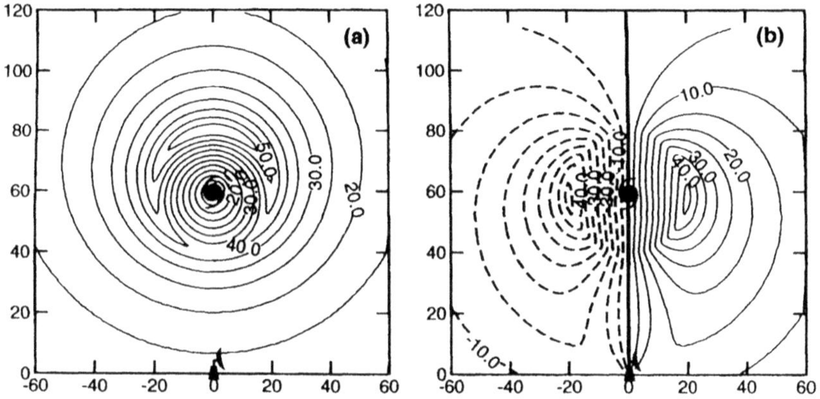

Fig. 8.34. (a) Tangential velocity (V_T) pattern of an axisymmetric Rankine vortex with asymmetry added in one direction and (b) the corresponding Doppler velocity pattern (from Lee et al. 1999; courtesy AMS). Positive (negative) Doppler velocities are contoured in solid (dashed) lines. The thick solid line represents the zero Doppler velocity contour. The hurricane and radar symbols indicate the TC centre and the radar location, respectively.

studies of typhoon rain bands and eye wall were made. [Ishihara et al. (1986), Tabata et al. (1992), Shimazu (1997)]. Vertical wind profiles by the VAD technique were obtained in a weakening Atlantic TC by Glover and Forsyth (1986) and in the eyewall region of a typhoon by Tabata et al. (1993). Though experience of use of ground-based Doppler radars operationally for TC observation and tracking is limited, a notable operational radar is the CKS airport radar in Taiwan [Wang and Tseng (1989), Ray et al. (1991)].

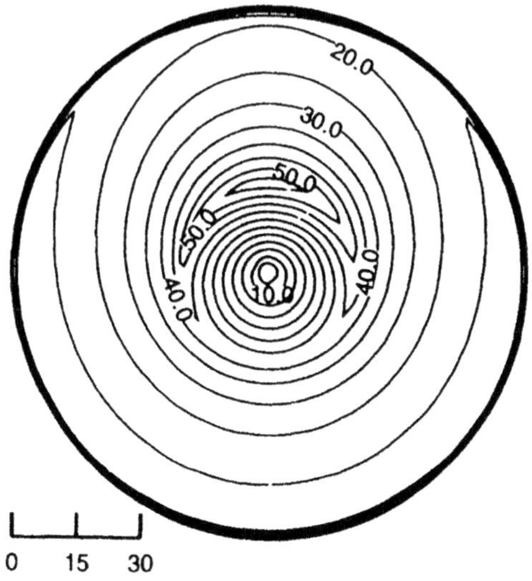

Fig. 8.35 The GBVTD-retrieved tangential winds for the vortex in Fig. 8.34. (from Lee et al. 1999; courtesy AMS).

8.4.5 Wind Field Derivatives

It was seen in Chapter 3, that in a VAD scan the horizontal wind can be expressed as the sum of Fourier terms [equation (3.51) reproduced here as equation (8.4)],

$$V_h = a_0 + a_1 \cos\phi + b_1 \sin\phi + a_2 \cos 2\phi + b_2 \sin 2\phi \tag{8.4}$$

and that in terms of the Fourier coefficients, some kinematic properties of the wind field (horizontal divergence, stretching deformation and shearing deformation) can be derived. In a series of papers, Donaldson and his colleagues [Donaldson (1989, 1991), Donaldson and Bals (1993), Donaldson and Harris (1984,1989), Donaldson and Ruggiero (1986), Harris et al. (1989,1993), Ruggiero and Donaldson (1987), Smalley and Harris (1993)] in the USA have developed this further for the case of a tropical cyclone. They identified four independent first order derivatives of the wind field in a VAD scanning circle taken outside the core of the TC. These were, downwind shear, crosswind shear, curvature and diffluence.

The kinematic properties, divergence, stretching deformation, shearing deformation and vorticity are given by the sums and differences of these first order derivatives.

$$\text{Divergence} = \text{downwind shear} + \text{diffluence} \tag{8.5}$$

$$\text{Streching deformation} = \text{downwind shear} - \text{diffluence} \tag{8.6}$$

$$\text{Shearing deformation} = \text{curvature} + \text{crosswind shear} \tag{8.7}$$

$$\text{Vorticity} = \text{curvature} - \text{crosswind shear} \tag{8.8}$$

The first three of these can be expressed in terms of the coefficients in equation (8.4). Hence downwind shear and diffluence can be computed. Vorticity cannot be determined and curvature and crosswind shear cannot be separately estimated.

These authors also defined a quantity Cyclone Intensity Indicator (CII) which is a function of (1) the wind speed measured in the VAD scan circle, (2) the mean deviation of the maximum and minimum Doppler velocities from diametric opposition and (3) the radius of the scan circle. Although this is called intensity indicator it cannot represent the true intensity because the index depends on the relative location of the radar and the TC. To overcome this problem they defined a new index, Storm Strength Indicator (SSI) by subtracting the ratio of the radius of the scan circle to the distance of the TC centre from the radar. The latter is known from conventional radar observation or other means (e.g. satellite observation). They also introduced a ratio known as Potential Vortex Fit (PVF). It is the quotient of the calculated shearing deformation divided by twice the estimated curvature term. According to them, a value of PVF = 1 indicates that both curvature and cross-wind gradient of the tangential component vary inversely with distance from TC centre. PVF of mature TC's may average 0.7 or 0.8. If PVF < 1 and is increasing, the TC intensity should be increasing. If PVF is decreasing the TC intensity should be decreasing. Persistent values well below unity may indicate a weakening cyclonic circulation. These parameters were evaluated in two hurricanes. The successful evaluation of these parameters depends on the availability of adequate echo coverage in the VAD circle without large gaps. In the absence of data from more TC's it is not possible to say at present whether these parameters evaluated in one small VAD circle away from the core would correctly represent trends in the TC as a whole and can be used by the operational meteorologist as reliable indicators of intensity and intensity change in the TC. If they can be so used, they will be valuable in cases where the eyewall itself cannot be observed by radar.

8.4.6 Variation of Eye Configuration and Size

The variation of the eye configuration and size is of great interest. The inner edge of the eyewall may have a variety of shapes but in most cases it can be treated as an ellipse with semi-major and semi-minor axes a and b which can be measured on the radar. The area of the eye can be computed as πab. The axial dimensions may range from a few km to over 50 km. Fig. 8.36 shows the variation of the eye dimensions of a TC with time[17]. Another possible measure of the size of the eye is the radial distance from TC centre to the point of maximum reflectivity in the eyewall. Raghavan et al. (1989) defined this as the Radius of Maximum Reflectivity (RMR) or R_m (Fig.8.37). If the eyewall is unsymmetrical a line can be drawn through points of maximum reflectivity and an average value of R_m obtained. To determine R_m it is not necessary to know the actual value of reflectivity; so it can be done even by a qualitative radar observation. It was seen from Fig. 8.31 that the maximum reflectivity coincides roughly with the Radius of Maximum Winds (RMW) of the TC determined independently of the radar. Jorgensen (1984) found in several hurricanes that the zone of maximum eyewall radar reflectivity was several km radially outward from the RMW at low levels while at higher levels the locations of both coincided. Marks and Houze (1987) who had the benefit of winds from Doppler radar found in hurricane Norbert (1986) that the tangential wind maximum was 4 km radially outward from the reflectivity maximum and both

[17] The eye radius was only about 2 or 3 km just before landfall. According to Raghavan (1999) this is probably the smallest eye of a TC ever documented, the previous smallest being 6 km in cyclone TRACY in Australia (Holland 1993b).

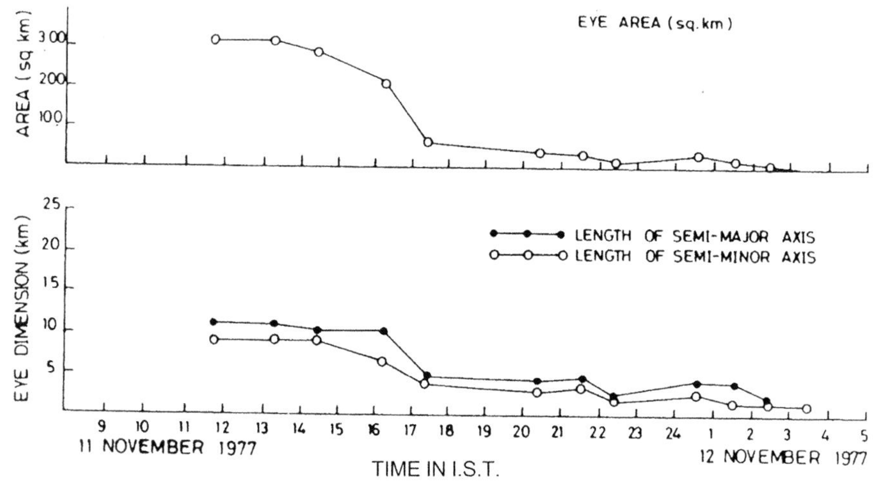

Fig. 8.36. Time variation of eye dimensions of a Bay of Bengal cyclone (after Raghavan and Veeraraghavan 1979; courtesy IMD).

sloped radially outward with height[18]. By comparing radar observations with post-cyclone wind damage surveys, Raghavan and Veeraraghavan (1979) and Raghavan et al. (1984) found that the tracks of the wind damage and eyewall reflectivity maximum were approximately co-located. Hence they concluded that RMW can be taken as approximately equal to the RMR[19]. However, since the height at which radar measures the RMR varies with range it is difficult to establish whether, in general, the RMR is outside the RMW or vice versa.

8.4.7 Relation between Eye Size and TC Central Pressure

Bell (1975) found statistically that smaller eyes were associated with deeper typhoons in the Pacific. Shea and Gray (1973) found similar inverse relationship between RMW and TC intensity (represented by MSLP) but there was a wide scatter of points. Correlation coefficients presented by Schwerdt et al. (1979) between RMW and central pressure never exceed 0.40. Hence in the absence of measurements of pressure or wind in the eyewall it is not possible to infer the intensity of a TC from the known eye diameter or RMR. However, in a given TC, changes with time, of eye size, RMR or RMW should, from

[18] In a Japanese typhoon Tabata et al. (1992) found that the RMW and maximum reflectivity coincided in the developing stage, but in the decaying stage the RMW was completely outside the eyewall.

[19] RMR is being used operationally in India as a measure of RMW for input to storm surge computation.

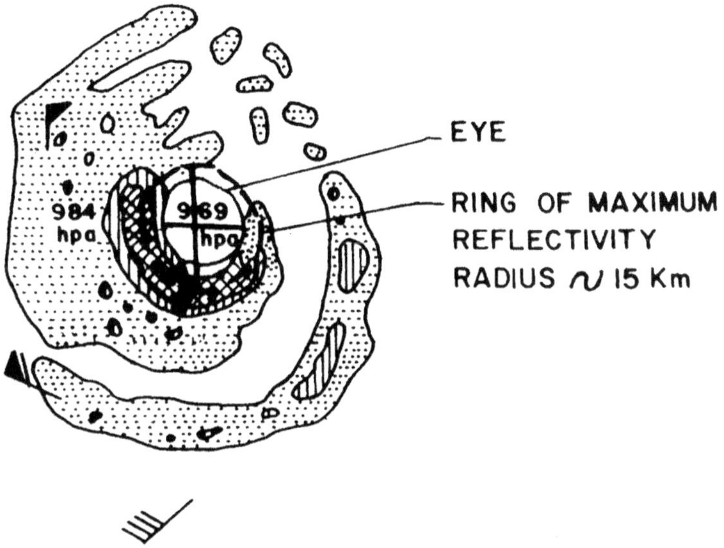

8.37. Surface structure of core area of a cyclone (after Raghavan et al. 1989; courtesy IMD). The dashed circle of radius 15 km is the ring of maximum reflectivity in the eyewall. Cross-hatched area reflectivity is 42 dBZ or more, vertical hatching 37 dBZ and stippled area 32 dBZ. Estimated central pressure, pressure reported by a ship outside the eyewall and winds from other ships are plotted.

theoretical considerations, correspond to changes in TC intensity. This has been found to be so in many cases. A very good correlation between eye radius and central pressure in an Atlantic hurricane can be seen in Fig. 8.38. The rapid decrease of eye size in the TC of Fig. 8.36 was accompanied by intensification of the TC. According to Raghavan et al. (1989), who documented variation of RMR with time, RMR changes are better than eye size changes as indicator of intensity changes. Thus, decreases in eye size or RMR with time is taken to imply intensification; conversely if the radar indicates an increase in these parameters it should signify weakening of the TC. While this may be true, there are two pitfalls in such an inference. There can be a lag of several hours between TC structure changes and wind changes (Dvorak 1984). Secondly, deterioration in seeing conditions of the radar due to range attenuation or sub-normal propagation can lead to an apparent widening of the eye. There can also be diurnal variation in the echo configuration (Mukherjee et al. 1977), which should not be interpreted as change in cyclone intensity.

8.4.8 Stormfury Hypothesis

The linkage between eye radius, RMW and TC intensity was one of the

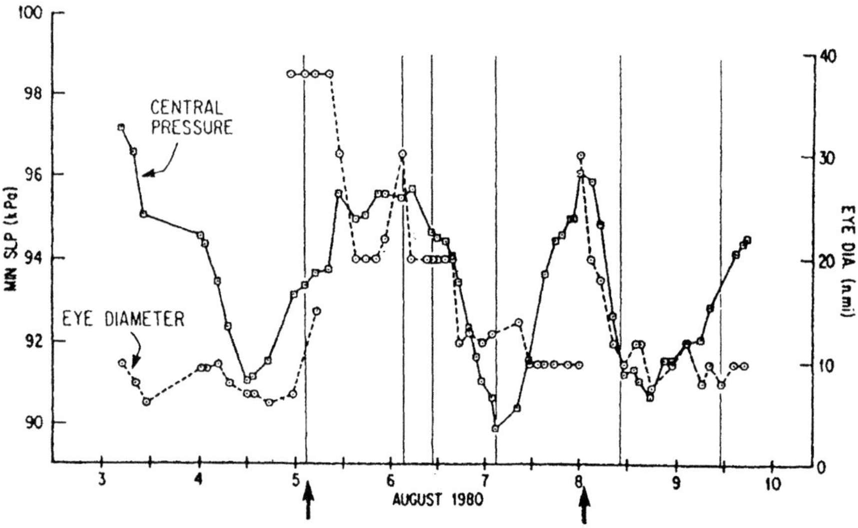

Fig.8.38. Time variation of eye radius (dashed lines) and central pressure (solid lines) for Hurricane Allen (1980) (after Marks 1985; courtesy AMS).
The date is plotted at 1200 UTC on each day. The vertical lines indicate times of aircraft research flights. The bold arrows indicate periods when the storm had two concentric eyewalls.

arguments favouring the "Stormfury" hypothesis of the 1960's (Gentry 1969). It was thought that if the clouds just outside the eyewall are seeded they would grow at the expense of the eyewall cloud leading to an effective increase of eye radius and RMW. This in turn should result in decrease of maximum wind. It was actually observed in a seeding experiment (Black et al. 1972) that an increase of eye radius and decrease of winds occurred after each seeding but a reversion to the previous state occurred later. But Willoughby et al. (1985) have argued that this could have resulted from the natural process mentioned in Paragraph 8.1.10.

8.4.9 Echo Heights and Reflectivities

Fig. 8.39 shows the radar-determined echo top heights in the core and periphery of a Bay of Bengal TC[20] over the sea. Although some errors due to the observations being at different radar ranges may be involved, it is generally seen that the heights in the core or eyewall region are lower than in peripheral areas. This is common in many TC's although the eyewall is predominantly convective and very tall convective clouds do occur in the eyewalls of some TC's. As noted in Chapter 6, tops of severe thunderstorms in Bengal often

[20] The same TC with an eye over land is at Fig. 8.17; the track is shown in Fig. 8.28.

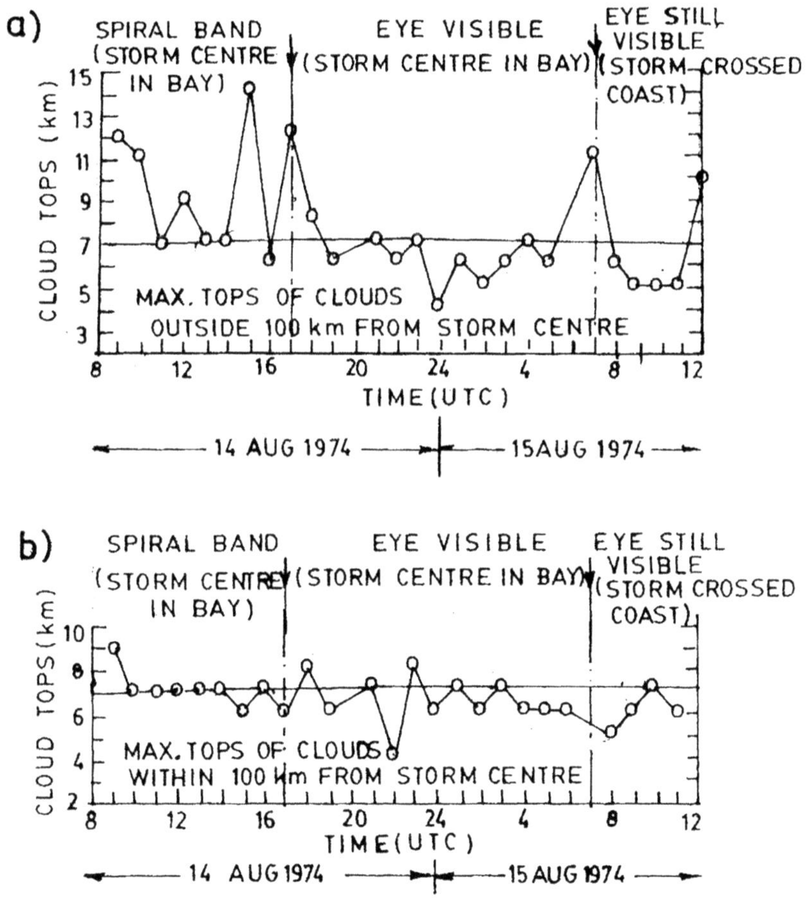

Fig. 8.39. Maximum echo tops in Bay of Bengal TC of 14 to 15 August 1974 (from Bhattacharjee and De 1976; courtesy IMD.) a) beyond 100 km from TC centre b) within 100 km from TC centre.

extend beyond 20 km. Thus the echo top heights in TC eyewalls are often lower than in local severe storms. Heights obtained by stereoscopy or from IR imagery from satellites are relatively higher (Fig. 8.40 and Hasler and Morris 1986) perhaps because of a blanket of cirrus which shows up on the satellite; but at the same time the convection does not penetrate the tropopause.

The reflectivity in the eyewall region has typically a maximum value of 40-45 dBZ (see Fig. 8.37) at low levels (850-700 hPa) and decreases rapidly beyond the freezing level (Fig. 8.40) (Hasler and Morris 1986, Jorgensen et al. 1985) whereas much higher dBZ values are found in mid-latitude thunderstorms. Thus, the TC reflectivity profiles are similar to those found in the GATE convective systems (Szoke et al. 1986). The eyewall is dominated

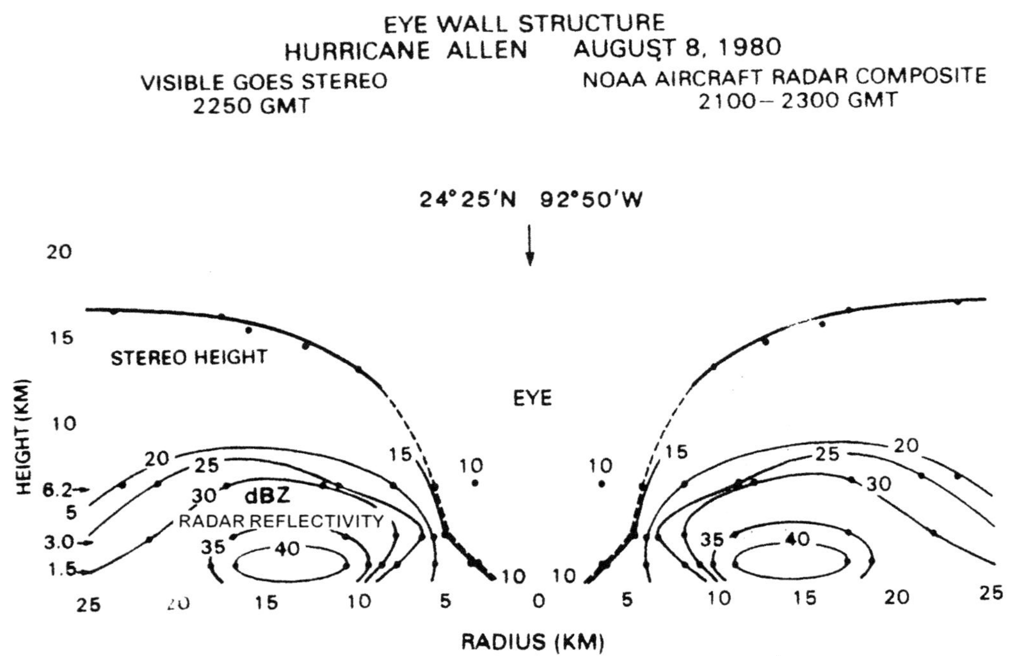

Fig. 8.40. Cross section of Hurricane Allen (1980) in the Atlantic basin (from Hasler and Morris 1986; courtesy AMS). Cloud top heights above 12 km were determined from stereoscopy of images from two geostationary satellites. The lower portion was defined by composited aircraft radar reflectivity at three flight levels (1.5, 3.0 and 6.2 km). The cross section is the average of east-west and north-south cross sections.

The outward slope of the eyewall (see Paragraph 8.1.8) with height is clearly seen. The radar reflectivity has a maximum of 40 dBZ but decreases rapidly with height. The height determined by satellite was appreciably higher than what the radar indicated.

by updrafts but their velocities are generally weaker than in local severe storms. Judging from the heights, reflectivites and vertical velocities, convection appears to be generally weaker in TC's than in local severe storms.

8.4.10 Asymmetry in Reflectivity, Velocity and Rainfall Distribution

The reflectivity distribution in the eyewall is generally asymmetric. At latitudes higher than about 20 degrees north, the eyewall echo concentration and reflectivity are highest in the right and front, but several cases have been documented in the Atlantic with the highest reflectivity to the left of the track [e.g. hurricane Norbert (1986) at 22.5 deg.N Lat. (Marks and Houze 1987), hurricane Hugo (1989) at 15 deg.N Lat. (Golden 1990)]. The reflectivity maximum coincides, as already noted, with maximum wind and rain rate. The concentration of these quantities in the left sector is considered unusual by some scientists but in most southern Bay of Bengal cyclones (south of about 20° N) maxima of these quantities were found (see e.g. Fig. 3.6) in the left sector [Biswas et al. (1988), Raghavan (1990, 1997), Raghavan and Veeraraghavan (1979), Sivaramakrishnan and Sridharan (1989)]. This has been verified by post-cyclone damage survey and rainfall analysis[21] which indicated maximum wind damage and highest intensities of rainfall in the swath traversed by the left sector of the eyewall, in these cases [Raghavan and Veeraraghavan (1979), Raghavan et al. (1984)]. Simpson and Pelissier (1971) documented the case of Hurricane Celia (1970) which exhibited the isotach maximum in the left sector of the eyewall. They attributed it to a cyclonic rotation around the vortex centre which occurs in a non-steady state cyclone. The sector of maximum wind and reflectivity appears to be generally to the left in slow moving TC's and to the right in fast moving ones as predicted by the model of Shapiro (1983). For example Black and Burpee (1987), who studied four Atlantic hurricanes, found that one which moved at 3 m s^{-1} had a core region rain maximum in the left front sector and the three which moved at 6 m s^{-1} had the maximum in the right front sector. Since the speed of motion of the TC usually increases with latitude, the apparent dependence of the position of the reflectivity and isotach maxima on latitude is explained. According to Meighen (1985) over 80% of

[21] Early analyses (Ramakrishnan 1937, Koteswaram and Gaspar 1956) of rainfall in TC's of the Bay of Bengal before radar and satellite were available have shown that (1) in the cyclone as a whole, the precipitation, on land, extends much farther to the right of the track than to the left (see also Table 8.3) and (2) in the core region, there is an area of intense precipitation immediately to the left of the track. Similar findings have been in made in more recent rainfall analyses [IMD (1970), Mukherjee et al. (1981), Mooley and Mohile 1986).

Australian cyclones have a better developed eyewall to the right of the track (right in the southern hemisphere corresponds to left in the northern). The sector of asymmetry as determined by radar is of practical importance in forecasts of wind and rainfall in the core of the TC.

Colón et al. (1970) found from aircraft reconnaissance of an Arabian sea TC (at a latitude of about 15° North), that the precipitation bands seen on airborne radar were concentrated in the 'left and front quadrants "but the isotach field showed a more typical distribution with a band of strong winds bordering the eye around the right semi-circle". They observed an eye of diameter 12 n. miles with an eyewall width of about 8 n. miles. The central pressure was 947 hPa. The maximum wind at 900 hPa was 96 knots at 12 n. miles radius. The wind decreased to 65 knots at 18 miles radius but peaked again to 95 knots at 32 miles radius. This may be a case of a double walled eye. Their RMW coincides approximately with the middle of the radar-observed eyewall. The process of intensification of the TC was characterised by a reduction in RMW.

Direct observation by Doppler radar has also shown asymmetries in the Doppler velocity in the eyewall region (Tabata et al. 1993). Black et al. (1995) using vertically pointing mode in airborne radar found asymmetry in the vertical velocities too. The strongest upward motion was found upwind of the eyewall reflectivity maximum and the strongest downward motion downwind of the maximum.

8.4.11 Mesoscale Structure

Airborne radar observations (Jorgensen 1984, 1984b) also showed that the rain bands do not exhibit well defined wind maxima or highly organised vertical motion. The strongest updraft cores (up to 16 m s^{-1}) and most (over 60%) of the vertical mass transport was in the eyewall. The most significant finding was that outside the eyewall, stratiform precipitation predominated with imbedded convective cores, the structure being similar to the mesoscale structure found in tropical cloud clusters and squall lines (Chapter 6). Rain bands in some TC's consisted of both stratiform and convective rainfall and persisted for several hours in a constant storm-relative position, while in other TC's they rapidly developed into secondary eyewalls which were primarily convective. Frequency distribution of average and maximum vertical velocities, draft dimensions, and vertical mass transport were log-normal. Even in the convective cores, updrafts were relatively weak and cloud liquid water content (measured by other instruments) was low compared to mid-latitude convection. Ice particles mostly prevailed above freezing level making it unlikely that seeding of the Stormfury type just outside the eyewall would succeed (see Paragraph 8.4.8.). Stratiform rain was found to contribute over 60% of inner core rainfall. The implications of this type of mesoscale structure are discussed in Chapter 6.

However, this mesoscale structure may not be typical of all TC's in all basins. Several other authors have observed strong convective features moving down the spiral bands. In the principal band, Willoughby et al. (1984) found cellular convection along the band axis and stratiform precipitation on its outer side. Powell (1990) also found intense convective cells on the inner side of the band. Barnes et al. (1983) found, in one hurricane, cellular convection far from the storm centre and stratiform precipitation nearer the centre. Ryan et al. (1992), who studied a developing cyclone in the southern hemisphere, found an unusually wide rain band with predominantly convective cells intermingled with stratiform precipitation, the proportion of the latter increasing down the band (see also Barnes et al. 1991). Many of these authors also noticed an outward tilt of the convective clouds with height.

Tabata et al. (1992) made a ground-based dual-Doppler radar (5 and 3 cm radars of the Meteorological Research Institute, Tsukuba) study of a typhoon. They located an outer band about 150 km from the typhoon centre, which was mainly stratiform. It exhibited a bright band with wind perturbations (see section 4.1). There was an inner band which was convective. In another Japanese typhoon, Shimazu (1997) found that both slow-moving wide bands and fast-moving narrow ones were stratiform with a few imbedded convective cells. Raghavan (1990) points out that in Bay of Bengal TC's the eyewall and spiral bands are essentially convective and no large meso-scale anvils have been observed. Thus there seems to be considerable variation in the proportion of convective and stratiform precipitation from cyclone to cyclone. These perhaps depend on the intensity of the cyclone and whether it is in a developing, mature or decaying stage.

8.4.12 Effect of landfall

The structure of a TC undergoes modifications on landfall. These are due to increased surface friction over land and the reduced water surface available for evaporation of moisture. The added surface friction reduces the wind speed (Powell et al. 1991) and increases the cross-isobaric flow towards the low pressure centre. The eyewall usually collapses[22], the banded structure

---

[22] An exception was noted in Fig. 8.17.

is distorted. The rainfall and wind distribution also should be expected to change (Parrish et al. 1982). The increased inflow may however cause increase in mass convergence and upward motion. As a result, it is possible that in some cases stronger rotational winds can develop (Wakimoto and Black 1994). The eyewall over land in Fig. 8.17 and the observation of a double wall over land mentioned in Paragraph 8.1.10. may perhaps be explained in this manner. In the TC depicted in Fig. 8.17, there were higher echo tops in the core region than outside it after landfall while the reverse was the case over sea.(Fig. 8.39).

In the USA, Bluestein and Hazen (1989) made a synthetic dual-Doppler (SDD; see Chapter 3) study of a hurricane over land. This technique makes use of a single Doppler radar (a 10 cm radar in this case) observing the moving precipitation system at two points in its trajectory. This hurricane had travelled over land for 36 hours after landfall, yet retained a circular cloud mass with well-defined spiral bands (as seen by satellite) though not the eye. The SDD technique assumes that except for motion, the system has not changed on the mesoscale between the two successive observations separated by a large azimuth angle with reference to the radar. The Doppler winds agreed well with surface and upper air observations. They found three types of mesoscale areas of precipitation viz., the central area, principal band and outer band. The last one was convective and the other two largely stratiform.

Mesocyclones and tornadoes may develop either in the eyewall region or the rain bands on landfall of the cyclone, showing localised high wind maxima on radar. The cause is probably frictional drag leading to strong vertical shear (Gentry 1983). In the USA, the preferred sector of tornado formation is the right front sector (Novlan and Gray 1974). A well-documented case of a tornado in a hurricane may be seen in Wakimoto and Black (1994). Willoughby and Black (1996) reported intense echoes orbiting around the centre on opposite sides of the eyewall. As the storm approached, cyclonic meso-vortices appeared in the eyewall adjacent to the reflectivity maxima. When the eyewall reached the coast, the vortex pair stopped rotating and intense convection erupted in the onshore winds north of the centre. Snell and McCaul (1993) observed a number of mesocyclones, many of them containing tornadoes, along a shear line in the trailing rain band of a hurricane, using Doppler radar[23]. They noted that these mesocyclones were smaller in size than those in mid-latitude severe local storms. Suzuki et al. (2000), observed with Doppler radar, several "mini supercells" some of them spawning tornadoes, in a landfalling typhoon.

The modification to the cyclone structure on landfall is particularly

---

[23] In India, no cases of tornadoes in tropical cyclones have been documented although damage patterns suggest that they were perhaps present.

important if the coastal area is mountainous (Uyeda et al. 1991). A notable example is that of Taiwan, a large mountainous island situated off the Chinese coast in the path of several typhoons. As the typhoon passes over the island it weakens due to surface friction; the eye may shrink or disintegrate, new asymmetries develop and the precipitation pattern may change considerably [Chang (1982), Gao Ding Xi (1985), Tang (1985), Wang and Tseng (1989), Liu et al. 1991)]. After the TC emerges into the sea again, it may reorganise and intensify.

8.4.13 Extra-tropical Characteristics

At higher latitudes a TC may move into an extra-tropical zone and interact with extra-tropical airmasses and fronts. This leads to change in structure as well as generation of heavy rainfall at the frontal boundary (see e.g. Sakakibara et al. 1985). There is usually no eye observed in such cases but spiral bands are observed. A case of a TC going over Canada in this fashion has been documented by Joe et al. (1991).

8.4.14 Monsoon Depressions

According to Sarker and Chowdhury (1988) monsoon depressions originating in the Bay of Bengal belong to the class of tropical systems comparable to hurricanes. This should lead us to expect that the radar echo pattern would be similar to that of tropical cyclones. Kulshrestha and Gupta (1964) and Kulshrestha (1970, 1988) have indeed noticed distinct spiral bands and in some cases, a clear vortex in radar and satellite imagery in monsoon depressions over land in northern India. These depressions have however a cold core structure in the lower troposphere; very few of them intensify into tropical cyclones. The rainfall and the lower tropospheric winds tend to be a maximum in the southwest quadrant of the depression. The radar echo pattern is highly asymmetrical and the precipitation is concentrated in some areas. The echo patterns in depressions are not well organised as in tropical cyclones and therefore are not of much help in locating the depression centre.

Johnson and Houze (1987) and Houze and Churchill (1987) studied a Bay of Bengal depression observed with airborne radar (besides various other observational tools) during the MONEX experiment. A WP-3D aircraft[24] was

[24] Specifications of the radars on board are given in Table 1.5.

used. The radar echoes from the lower fuselage radar (Fig. 8.41) show three main features (marked A, B, and C in the figure) mainly to the south and west of the depression centre. This asymmetry agrees with that noted above over land. Feature A is a narrow arc-shaped line about 500 km long with maximum reflectivity over 40 dBZ. It has a trailing stratiform region. Features B and C show stratiform precipitation with convective cells imbedded. It is not clear whether the formation of the stratiform rain is by large scale advection or by the same mechanism as in the case of squall lines and clusters described in Chapter 6.

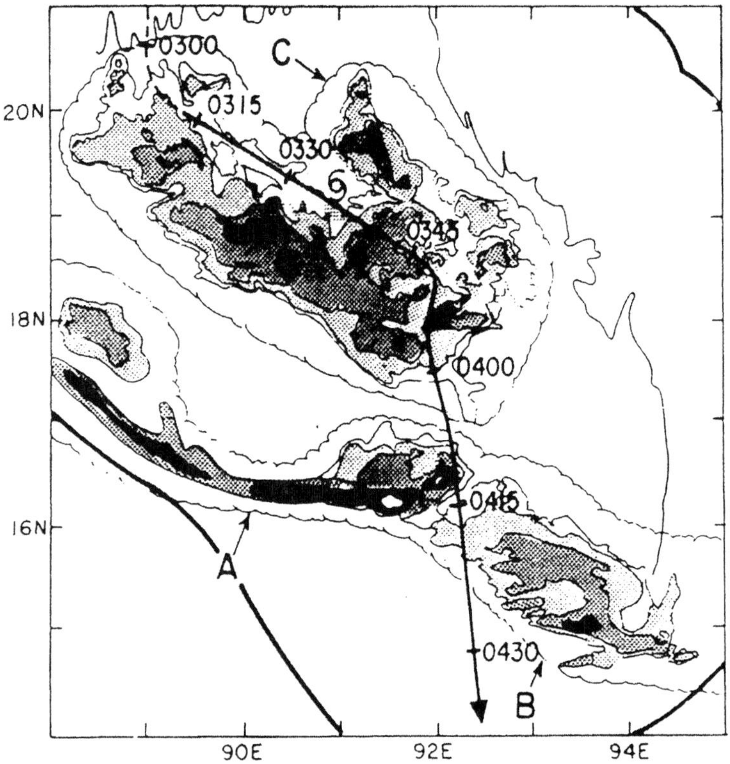

Fig. 8.41. Airborne radar observations in a monsoon depression in northeast Bay of Bengal on 5 July 1979 (from Houze and Churchill 1987; courtesy AMS). Depression centre is shown by the storm sign. Radar echoes are shown by shading thresholds of 1, 20, 30 and 35 dBZ. There are three features marked A, B and C, which are discussed in the text. The flight track is the thick line in the middle of the picture. The solid and dashed portions of the line show, respectively, the portions of the flight where radar data were and were not collected. Numbers indicate times in UTC corresponding to the hash marks on the track. The heavy bold curve outlines the area over which radar data were obtained. The scalloped line indicates approximate cloud boundaries based on satellite imagery and microphysical observations. The thin solid line is the coastline.

8.4.15 Radar-observed Features and Intensity of TC

With a Doppler radar, ideally, the intensity (i.e. the maximum wind in the eyewall) is directly determined. Several attempts have been made to correlate radar reflectivity features observed by non-Doppler radars with cyclone intensity and intensity change. While a similar correlation in the case of satellite imagery (Dvorak 1984) has been successful, the results in the case of radar have been ambiguous. The features considered are eye diameter, radius of maximum reflectivity, eye shape and configuration, double eye, thickness (i.e. horizontal width) of eyewall, heights of echo tops in the eyewall, number, thickness, length and crossing angle of spiral bands, echo area and reflectivity, and latent heat release (LHR). The eye diameter and RMR have been discussed and they were seen to be well correlated with intensity in individual cyclones but not useful while comparing one TC with another. Relationships were established by Meighen (1985) in Australian TC's between TC intensity and number of spiral bands (positive correlation), crossing angles (negative), thickness of spiral bands (negative), eyewall thickness (poor negative correlation) and eye diameter (negative, individual TC's only). According to Raghavan (1985), besides eye diameter, the shape of the eye (approximating to a complete circle), formation of double eyewall, small crossing angles of spiral bands are all good qualitative indicators of intensity. A negative relationship in the case of eyewall thickness was found in one TC by Raghavan et al. (1989)[25]. Sharma (1999) found a good negative correlation between eye diameter and eye wall width in a Bay of Bengal cyclone but neither of these parameters appeared to be well-correlated with intensity as inferred from satellite imagery.

Echo top heights in the eyewall may often be lower than those in severe thunderstorms and therefore high correlation of heights with TC intensity is not to be expected. Raghavan and Varadarajan (1981) computed, from radar data, the spatial distribution of LHR (in GW km^{-2}) and its variation with time, in a TC and found that the LHR within the core radius (taken as twice the eye radius) was well-correlated with intensity but the rainfall and LHR in the whole TC had no such relationship with the intensity. A similar conclusion was drawn by Griffith et al. (1978) from satellite imagery. Also, the relative contribution of higher rain rates to the total storm rainfall increases as the TC intensifies. There is probably also a time lag between maximum LHR and maximum intensity (Adler 1981).

A quantitative correlation of some of the quantities mentioned above with typhoon intensity near the Chinese coast was attempted by Zhou Ducheng

[25] Schwerdt et al. (1979) determined correlation coefficients between various pairs of parameters in Atlantic and West North Pacific cyclones. The correlation coefficient between RMW and central pressure ranges from 0.20 to 0.39; the correlation in the case of other parameters is much poorer.

(1985). The factors considered by him are shown in Table 8.2. Each parameter is assigned a contributing weight in m s^{-1} according to the value of the parameter as shown in the table. Typhoon intensity, expressed as maximum surface wind in m s^{-1} is the sum of the eight contributing weights. Ducheng claims that this table can be used when the typhoon is within 400 km of the coastal radar station. The mean absolute error in intensity was 4 m s^{-1}. In another method developed by the same author, two multivariate regression equations are developed between typhoon intensity y_t in m s^{-1} and the radar features. The first one applies when the eye is seen by radar.

$$y_t = 31.6613 - 0.1501X_{1t} + 1.4710X_{2t} + 0.1033X_{3t} - 0.3375X_{4t} \quad (8.9)$$

where X_{1t} = diameter of eye in km, X_{2t} = height of eyewall in km, X_{3t} = width of eyewall, and X_{4t} = minimum crossing angle of rain band in degrees.

If the eye is not seen, the second equation is used

$$y_t = 37.3686 - 0.3904X_{1t} + 0.0630X_{2t} \quad (8.10)$$

where, X_{1t} = crossing angle of spiral rain band in degrees, and X_{2t} = width of spiral rain band in km. These equations were used when the typhoon was within 350 km from the coast. The mean absolute error was 3 m s^{-1}. The correlations of some of the parameters with intensity is of opposite sign to those found by the authors mentioned earlier.

Table 8.2. Calculating table for cyclone intensity (after Zhou Ducheng 1985; courtesy WMO).

Weights	Height of eyewall km	Width of eyewall km	Diameter of eye region km	Eye Pattern	Distance typhoon centre to radar km	Feature of spiral band	Crossing angle of spiral band degrees	Compactness of typhoon echoes
1		no eye or filled				sporadical		
2	no eye or filled		no eye or filled	no eye or filled	≤ 100	cell composition	> 25	loose
3	1 to 4	≤ 10	≥ 56	broken		lumpy striped	21 to 25	
4	5 to 6	11 to 20	46 to 55	polygon	101 to 200			medium
5	7 to 8		36 to 45				16 to 20	
6	9 to 10	21 to 30		half ring	201 to 300	striped		
7			26 to 35					
8	11 to 12	> 30		ellipse close		striped cilia	11 to 15	compact
9			16 to 25		301 to 400			
10	13 to 14			round close				
11			≤ 15			striped annex cilia		
12	15 to 16			tautological eye close			≤ 10	
14	> 16							

Some of the terms used are not clear perhaps due to difficulty in translation.

8.5 Rainfall Distribution in Cyclones

8.5.1 Pattern of Rainfall in a TC

Precipitation in a TC extends over a very large area, typically 500 km in diameter. The total precipitation in this area may be 1 to 2 Gigatonnes per hour (Watanabe 1963) i.e. an average rate of 1.3 mm h^{-1}. The precipitation rate is, however, not distributed evenly. The rain rate in the eyewall and some other convective cores may be over 100 mm h^{-1}, while in the stratiform precipitation areas it will be very low. As already noted, the amount of precipitation in the stratiform region can be about 60% of the total because of the large area of the stratiform precipitation. Not all of the precipitation can be seen even qualitatively by any one radar. The quantitative range being even smaller, actual measurements will be confined to parts of the cyclone. Compositing of data from a network of radars will enable reasonably good mapping of precipitation close to land areas. Since there are identifiable features such as the precyclone squall lines, eyewall and streamers where high rates of precipitation are concentrated, a distinct spatial pattern of precipitation exists and usually persists for several hours. Since the track of the TC is predicted, the entire spatial pattern can be extrapolated over several hours, and the future position of each feature can also be predicted, the main source of error being in the track forecast. Extrapolation of the oceanic precipitation pattern is convenient as the radar sees better over the sea than over land. This forms a basis for short period forecasting of rainfall (and floods) likely to occur over land areas. Depending on the particular feature which is expected to come over a specific coastal area, localised warnings of heavy rain can be given for specific times.

8.5.2 Estimation of Precipitation

The methods of estimation of precipitation and the extrapolation of precipitation areas for short-period forecasting are discussed in Chapters 6 and 7. These apply to TC's except that the *Z-R* relationships used for other situations may not necessarily apply (see Table 7.2). In the absence of quantitative estimates, the identifiable features referred to above can still be extrapolated in space and time and useful qualitative forecasts can be produced. An interesting example is that in the case of a TC moving from the south and hitting the Saurashtra coast in western India, the streamers, which happen to be in the right rear sector, usually come over Mumbai (Bombay) city; this results in heavy rainfall over Mumbai with relatively little rainfall in areas between the storm centre and Mumbai [Mukherjee and Padmanabham (1980), Raghavan and Ramakrishnan (1981)]. This has been used for issuing heavy rainfall warnings for that city. Because of heavy precipitation in the TC extending over long distances the possibility of the radar underestimating the rate or extent of the precipitation has to be kept in view.

The precipitation represents release of latent heat as discussed in Chapter 6. The latent heat provides the energy for the maintenance of the TC according to theories of interaction between cumulus scale convection and the large scale synoptic motion field [Charney and Eliassen (1964), Yamasaki 1988). The rainfall rate can be converted into energy release (Longley 1949). One Gigatonne per hour of rainfall is approximately equivalent to 2.5×10^{18} J h^{-1}, or 7×10^{14} W. As noted already, the LHR computed in the core area appears to be related to the intensity.

8.5.3 Echo Size

The mass of precipitation echo seen by radar can be considered to be a measure of the size[26] of the system. Several early studies noted (see e.g. Kulshrestha and Gupta 1964) that the dimension of the radar echo was about one tenth of the satellite cloud imagery seen[27] in association with the TC. The ratio will evidently depend on the brightness temperature and reflectivity factor thresholds set for the satellite imagery and the radar echo respectively. The ratio of the estimated size (D1) to the echo dimension (D2)[28] for a few mature TC's in the Bay of Bengal striking the same part of the Indian coast, is shown in Table 8.3.The average ratio is 1.5. Table 8.2 also shows the extent (distance from centre) of rainfall of 2 cm per day or more in the coastal strip on either side of the track (rain gauge estimates). The tracks in these cases were east to west or southeast to northwest. As noted earlier, rainfall extends farther to the right of the track than to the left. The ratio of the total extent of rainfall (X1) to the echo dimension should be unity if the radar sees all the precipitation. The average ratio in the table is 1.2.

---

[26] See Paragraph 8.1.8 for definition of size.

[27] The satellite may see high cloud over a large area with no corresponding precipitation seen by the radar.

[28] The threshold for the echo in these cases was the minimum detectable signal of the radar.

Table 8.3. Size, rainfall extent, and radar echo dimension of some cyclones (after Raghavan 1991).

S.No.	Date of Landfall	Size D1 km	Echo Dimension D2 Km	Extent (km) of rainfall 2 cm or more per day			Ratio X1/D2	Ratio D1/D2
				Total X1	Left of track X2	Right of track X3		
1	19 Nov. 1977	800	500	470	120	350	0.9	1.6
2	12 May 1979	700	400	510	210	300	1.3	1.7
3	14 Nov. 1984	250	200	240	100	140	1.2	1.3
4	8 Nov. 1989	200	150	150	50	100	1.0	1.3
5	9 May 1990	600	400	630	150	480	1.5	1.5
	Mean Ratio						1.2	1.5

8.6 Storm Surge Prediction

8.6.1 Storm Surge

The storm surge is an abnormal rise of sea level near the coast caused by a TC. Storm surge inundation in coastal areas may go up to a height of several metres in some cases and is the most destructive feature associated with a TC. Prediction of the storm surge involves forecasting the peak height, its position and time of occurrence. Prediction is done using numerical models. The inputs needed for the model are, the intensity of the cyclone (i.e. the pressure deficit), its radius of maximum winds (RMW), the topography of the seabed near the point where the cyclone crosses the coast and the angle of the storm track with reference to the coast. The peak storm surge occurs to the right (in the northern hemisphere) of the landfall point at a distance equal to the RMW, because at this point the wind and the translatory movement of the TC add up. The probable maximum surge heights for various points on the coast are derived from a model taking account of coastal bathymetry (e.g. Ghosh 1977).

8.6.2 Radar-derived Parameters

Radar can contribute the following inputs for storm surge prediction. (1) Maximum wind in the eyewall (which can be converted by suitable relationships to a figure of pressure deficit in the TC) from Doppler radar[29].

[29] Otherwise the intensity estimate is obtained from satellite imagery.

(2) Radius of Maximum Winds (RMW) derived from the Doppler pattern and/or Radius of Maximum Reflectivity (RMR) from the reflectivity pattern. The latter is more often likely to be available and can be used in place of RMW in cases where direct measurement, by aircraft reconnaissance or Doppler radar is not possible (Raghavan 1987). The RMR is only approximately equal to the RMW. But this will not lead to much error in storm surge prediction because the dependence of surge height on RMW is not critical.

(3) Expected track and therefrom the landfall point and the angle of track with reference to the coast.

(4) the position of peak surge one RMW away from the predicted landfall point.

(5) the time of occurrence of peak surge, if the time of landfall is predicted from the radar-observed track.

Chapter 9

USE OF WEATHER RADAR IN AVIATION

In this Chapter we discuss the applications of weather radar to aviation. The types of radar used and the detection of severe weather phenomena such as thunderstorm-related hail, gust fronts, microbursts, tornadoes as well as the incidence of turbulence, icing and wind shear are dealt with.

9.1 Aviation Meteorology

9.1.1 Meteorology in Aviation.

One of the most important applications of meteorology in the last century has been aviation and indeed much of the rapid development of meteorological services has been due to aviation requirements. Monitoring and prediction of a number of meteorological parameters in all stages of a flight as well as climatological knowledge are essential for aviation safety as well as economy. Meteorological requirements for aviation are unique in the sense that a three-dimensional coverage of the atmosphere is necessary and all phenomena are to be monitored, interpreted and predicted over very short periods and information is to be communicated fast to the fast-moving aircraft. The information is also to be processed and displayed in a manner which can be quickly assimilated by pilots and ground control personnel who work under tremendous pressure. The high benefit-to-cost ratio of aviation forecasts (in terms of safety and economy) has led to intensive research on the subject.

The requirements may be broadly divided into two categories (1) for terminal operations i.e. the take-off and landing phases as well as ground operations (taxiing and parking) and (2) in-flight conditions. The first of these is the most demanding.

The meteorological parameters of interest are

(1) wind at the surface as well as higher levels. The upper level wind is important for choice of cruising level for fuel economy while the wind near the surface is important for safety in landing and take-off.

(2) temperatures which determine efficiency and therefore economy.

(3) visibility lowered by precipitation, fog, dust, lowering of cloud base or other surface phenomena may be critical for take-off and landing.

(4) wind shear and turbulence. Horizontal and vertical wind shear at low levels called Low Level Wind shear (LLWS) is a serious hazard for take-off and landing. Shear and turbulence at all levels can seriously affect safety in all phases of flight. Turbulence may occur in clear air as well as in storms.

(5) icing (i.e. deposition of ice on the aircraft) which can affect the aerodynamic performance.

(6) rain which can cause ingestion of water into the engine, affect aerodynamic performance and lower visibility.

(7) snowstorms which can cause poor visibility and snow deposition on aircraft and runways.

(8) hail which can cause damage to aircraft.

(9) lightning which may cause malfunctions of aircraft electronics and sometimes explosion of fuel. This is more serious in modern aircraft which use non-metallic structural parts losing the Faraday-cage effect.

(10) convective storms and associated phenomena such as downbursts, gust fronts, mesocyclones and tornadoes.

(11) fog which disrupts flights by loss of visibility.

A comprehensive discussion of sensors, including radar, to detect weather phenomena of interest to aviation, may be seen in Mahapatra (1999).

9.1.2 Role of Weather Radar

Radar, complemented by other observing equipment, is an ideal observational tool for continuous monitoring and short period forecasting of most of the parameters listed above. While the pilot usually has a radar on board for seeing the weather immediately ahead, he is provided, both before and during the flight, information from ground-based equipment. Reviews of the role of weather radar in aviation meteorology may be seen in Fujita and McCarthy (1990), Mahapatra and Zrnić (1991) and Mahapatra (1999).

In the early phase of radar meteorology, conventional 3 cm wavelength radars were located at airports to monitor weather in the airport area and the approach region. They were used qualitatively to issue airfield warnings for parked and moored aircraft and to provide information relating to landing and take-off. The wavelength was chosen because of easy availability of equipment and also to maximise the returned power. By tracking precipitation echoes, short period forecasts and warnings in respect of thunderstorm-related phenomena could be given. The development of Doppler radar, of algorithms to compute various derived parameters in real time and of improvements in communication techniques has now enabled thinking in terms of radar as an integral part of the "Aviation Weather System" consisting of a number of weather sensors (McCarthy and Serafin 1990) and at the same time having considerable flexibility in siting and use of the radar. Weather radar networks designed for various other operational applications can also be used for aviation. For example, the US WSR-88D radars (described in Chapter 3) operating at 10 cm wavelength have the advantages of operation free from rainfall attenuation and higher unambiguous Doppler velocity. Such a radar can usually not be located at an airport because of frequency and siting clearance problems but this need not be a disadvantage as the radar data can be transmitted in real time and be used by the aviation community. The siting of the radar away from the airport may indeed be an advantage in terms of detection of local phenomena free from clutter.

9.1.3 Terminal Doppler Weather Radar

In the USA an additional radar network known as the TDWR (Terminal Doppler Weather Radar) has been developed specifically for aviation use. This radar is similar in specifications to the WSR-88D except for operation at 5 cm and for shorter range (Michelson et al. 1990). The radar is highly automated and has algorithms for deriving various parameters in real time. The data are displayed in a "Geographical Situation Display (GSD)", (Turnbull et al. 1989, Sand and Biter 1989, Moore et al. 1991), (Fig. 9.1), at the Terminal Radar Control (TRACON) along with data from other instruments e.g. the LLWAS (see Paragraph 9.1.7). A list of TDWR products is given in Table 9.1.

These radars are sited about 5 to 15 km from the airport, as far as possible in line with the principal runway. The scanning strategy is usually to make sector volume scans of about 60° width in azimuth in the direction of the

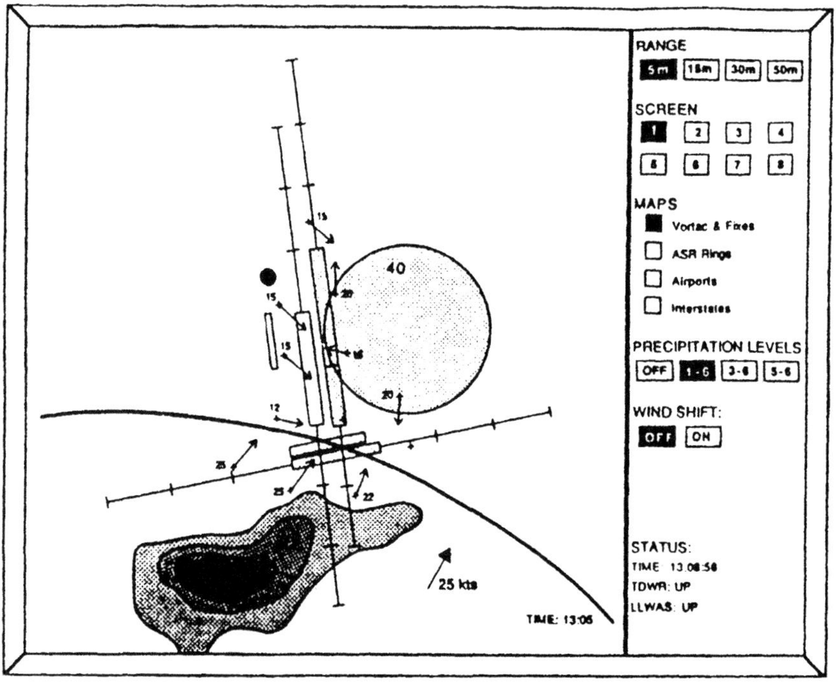

Fig. 9.1. Geographical Situation Display (GSD) used in the control tower and TRACON (from Turnbull et al. 1989; courtesy AMS). The runways are shown. The circle represents a microburst affecting runway 35-17. The number inside the circle is the strength of the microburst. The curved solid line depicts a gust front wind shift. Levels of intensity in a precipitation cell are shown by various shades. Wind vectors shown are from the LLWAS sensors at 12 sites.

airport[1]. The data set is updated every 2.5 minutes. In addition, two low altitude 360° scans updated every 5 minutes are taken for gust front detection. The algorithms cover only a limited range near the airport. The algorithm outputs go directly to the GSD without human intervention. Hence their performance has to be very reliable.

9.1.4 The Air Traffic Control (ATC) Radar

The Air Traffic Control (ATC) radar which is primarily intended for tracking aircraft, can also be used with some limitations, as a weather radar. The limitations arise because the optimum specifications for ATC and weather

[1] The radar is being upgraded to facilitate a 360° azimuth scan while maintaining an update rate sufficient to support its LLWS detection mission (Weber 2000).

Table 9.1. TDWR Products (from Turnbull et al. 1989; courtesy AMS)

S. No.	Product	Function
1	Microburst Detection	Identifies surface outflow divergence pattern that indicates microburst has reached the ground and provides microburst location with respect to runways and runway component wind-speed loss.
2	Gust Front Detection	Identifies strong horizontal convergence zone, indicating boundary of large-scale thunderstorm outflow, wind change across front, and location with respect to runways and runway component wind-speed gain.
3	Wind Shift Prediction	Predicts when a gust-frontal wind shift will affect the airport terminal and gives wind change after gust-frontal passage and predicted time of change.
4	Precipitation	Provides graphic display of precipitation for tower and TRACON supervisory positions in terms of six levels of reflectivity
5	Storm movement prediction	Predicts positions of significant storm cells using a cell-tracking algorithm.
6	Turbulence	Estimates turbulence in precipitation areas, using the second moment of the Doppler spectrum and other storm features.
7	Tornado detection	Provides location of tornadoes, using the tornado vortex signature.
8	Microburst prediction	Gives warning of microburst impact to the surface, using techniques that look at features well above the surface.
9	Convection initiation	Predicts where and when a thunderstorm will form with respect to the airport location.

radars are widely different. The ATC radar has a fan beam (i.e. wide in the vertical and narrow in the horizontal plane) with fast scanning[2] at a constant elevation angle. A weather radar needs a narrow beam width in both the planes and relatively slow scans at several elevations. Further an ATC radar is intended to reject weather echo. The operational availability of the radar for the two purposes may also clash. The ASR-9 (Airport Surveillance Radar) of the USA has a weather channel (Weber, 1989). It attempts to overcome some of these problems by having linear as well as circular polarisation. The former is used in clear weather for both the ATC and weather channels. During precipitation, polarised transmission is used. The ATC function uses the co-polar signal which rejects most of the weather echo. The weather function uses the orthogonal signal which contains the weather echo. See Mahapatra (1996), Mahapatra and Zrnić (1991) and Weber (1989) for details. The ASR-9 is being upgraded to include a Doppler weather capability (Weber 2000). Another relatively low power surveillance radar, the ASR-11 used at several US airports, also has a weather channel. It has been proposed that it could supplement WSR-88D network coverage (Bieringer et al. 2001).

[2] Anderson (1987) has developed a signal processing technique to deal with the fast scan.

9.1.5 Thunderstorms

Thunderstorms are the single largest source of severe weather affecting aviation and cause of many of the other severe weather phenomena listed in Paragraph 9.1.1. The pilot is always advised to avoid flying through a thunderstorm. This is relatively easy to do in the case of isolated thunderstorms but obviously more difficult when a line or cluster of thunderstorm cells are present. Methods of short period prediction of thunderstorms and the radar detection of conditions favourable for thunderstorm initiation have been discussed in Chapter 6. Although specific phenomena associated with the thunderstorm such as gust front, microburst, turbulence, hail and icing may be more difficult to predict, the very prediction of a thunderstorm should make the pilot anticipate the other phenomena and try to avoid entering them. Avoidance of the thunderstorm theoretically implies that all the thunderstorm-related hazards are avoided. In practice however, due to crowded airspace, turbulence in the vicinity of the thunderstorm and other considerations total avoidance is not possible. It may be necessary to look for clearances between active thunderstorms and fly though them.

9.1.6 Wind and Wind Shear

Winds in the upper air are supplied 2 to 4 times a day by tracking radiosonde balloons but increasingly, wind profilers (see Chapter 10) are becoming available for near-continuous monitoring of upper winds. A vertical time section from a single profiler or better a cross section from a network of profilers can detect the jet stream wind maximum. The vertical velocity disturbances and the turbulence associated with the jet stream can also be detected.

Wind shear at flight level can be detected from radiosonde and profiler observations. The latter provides high temporal resolution. In thunderstorms, vertical motions are recognisable by Doppler radar. Low Level Wind Shear is discussed in the next paragraph.

Low level jet streams can cause large vertical shear of the horizontal wind. This can be detected by a VAD scan (see Fig. 3.15 for an example of vertical profile of wind from a VAD scan with a microwave radar). This output can be obtained from a wind profiler also except for the lowest levels near the ground.

9.1.7 Low Level Wind Shear (LLWS)

Occurrence of wind shear at a low level, called Low Level Wind Shear[3] (LLWS), in the vicinity of an airport, can be a serious hazard to aircraft landing or taking off. If a point A and another B vertically below A have different horizontal wind velocities, an aircraft which is landing will encounter the wind at A first and then B. So it will experience a vertical shear of the horizontal wind also simply called vertical wind shear. A plane taking off will encounter these winds in the reverse order. If the points A and B are at the same height but separated by a horizontal distance, the shear between them is horizontal shear of the horizontal wind simply called horizontal shear. Wind can also have an appreciable vertical component in thunderstorms where there are updrafts and downdrafts. Change in the vertical wind is termed shear of vertical wind and should be distinguished from vertical wind shear mentioned above. Magnitude of wind shear is expressed as the vector difference between the velocities at two points A and B divided by the distance between them. The unit of wind shear is therefore s^{-1} or knots per 100 feet.

Wind shear can occur over several scales: (1) microscale (less than 100 metres), (2) mesoscale (several kilometres), (3) synoptic scale (hundreds of kilometres). Microscale shear is of the same order of dimension as an aircraft and will be felt by the craft as turbulence. Synoptic scale shear does not affect aircraft. Hence only the mesoscale shear which is a hazard for aircraft landing or taking-off, is important in the present context.

From the aviation point of view we can divide LLWS into two categories: (1) non-transitory and (2) transitory shear depending on its cause. Non-transitory shear is from phenomena[4] which may occur at specific places repeatedly and last for long periods.Transitory shear is from gust fronts, downbursts/microbursts and tornadoes, all short duration phenomena associated with convective clouds or thunderstorms. It is more intense and difficult to forecast with precision. We shall be mostly concerned here with the latter category.

LLWS causes the flight path of an aircraft which is landing or taking off, to be below (loss of performance) or above (performance gain) the intended one. Increase of tail wind or decrease in head wind will cause a transient loss

---

[3] Low level here refers to low altitude. To avoid it being misunderstood as shear of low magnitude it has been suggested that the term 'Low Altitude Wind Shear (LAWS)' should be used instead of 'Low Level Wind Shear'.

[4] Airmass frontal surfaces, seabreeze frontal surfaces, mountain waves, obstacles to prevailing wind flow e.g. due to buildings, low level jet streams.

of air speed and loss of lift and will affect the equilibrium of the aircraft. One of the consequences is overshooting or undershooting of the runway. If the pitch angle is reduced in order to gain airspeed there will be further reduction of lift causing a dangerous situation. A downdraft reduces the angle of attack which in turn reduces the lift and causes rapid sinking of the aircraft. In the case of crosswind shear, the flight path swerves to the right or left of the runway.

In the case of a gust front the aircraft will experience an increase in head wind or a decrease in tail wind i.e. a gain in performance. In the case of a microburst where air is coming down and diverging (see Fig. 3.13) both horizontal and vertical shear are involved. The aircraft landing or taking off will initially experience an increase in head wind followed by a downdraft and finally an increase in tail wind (Fig. 9.2), the latter two being a sure recipe for a crash. [For a detailed treatment of this subject see Frost et al. (1983), Fujita (1985), ICAO (1987)].

Observation of wind shear at airports has been attempted by several means but the most common is a special anemometer array called the Low Level Wind Shear Alert System (LLWAS). It consists of a set of wind instruments at and around the airport; the shear between them is computed and displayed in real time at the ATC and/or communicated to the pilot (ICAO 1987). It has been widely used in several countries with appropriate algorithms and has been largely successful with all forms of wind shear except microbursts. But it can only monitor the wind shear and not predict it. The Doppler weather radar

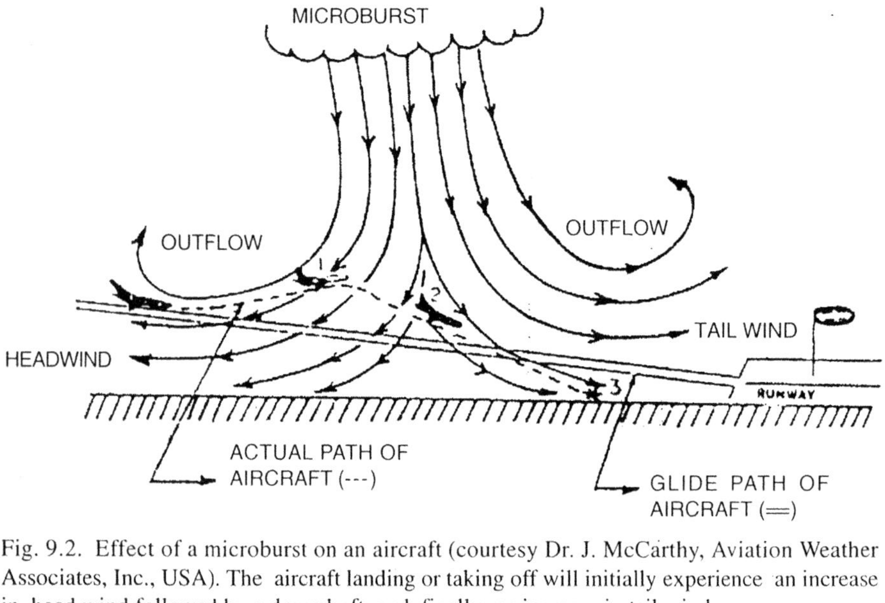

Fig. 9.2. Effect of a microburst on an aircraft (courtesy Dr. J. McCarthy, Aviation Weather Associates, Inc., USA). The aircraft landing or taking off will initially experience an increase in head wind followed by a downdraft and finally an increase in tail wind.

detects gust fronts and microbursts better and has also a predictive value in that it can detect the phenomenon well before it reaches the ground. The operational requirement is to get a prediction lead time of at least one minute. And of course, the radar detects the weather system which causes the microburst as well as other aviation hazards.

A description of a microburst detected with LLWAS and radar at Hong Kong may be seen in Lam (1996). Integration of the LLWAS and Doppler radar into a common system with appropriate software combines the merits of both and has been recommended as the best solution. The USA has implemented such a combination at the new Hong Kong airport.

Gust fronts and downbursts and the radar features used to identify them have been discussed in Chapter 6. The TDWR has algorithms for (1) Microburst detection (2) Gust front detection and (3) Wind shift prediction.

(1) The microburst detection algorithm examines radials of velocity data to locate runs of increasing radial velocities to identify small-scale low altitude divergence signatures. (This is similar to the pattern vector analysis discussed with reference to mesocyclones in Chapter 6). It also uses microburst precursor signatures aloft, e.g. convergence, rotation or descending core aloft.

(2) The gust front can be observed on the Doppler radar, (Klingle et al. 1987), as a thin line in the reflectivity field or a linear pattern of radial convergence in the radial velocities or as variation of wind velocity with azimuth or as lines of enhanced spectral width or all of these. Each feature has its limitations and therefore they are complementary. The thin line may not be present in every gust front or being a weak echo, may be difficult to identify amidst clutter. Radial convergence of the radial velocity arises because the wind behind the front is roughly perpendicular to the front while the environmental wind in advance of the front is in some other direction. It is readily recognised if the front is oriented perpendicular to the line of sight. Portions of the front which are oriented along a radial can be recognised from azimuthal shear i.e. variation of the radial velocity from azimuth to azimuth. The broadening of spectral width can be due to causes other than the gust front.

The Gust Front Detection Algorithm (GFDA) identifies runs of decreasing radial velocity which indicate radial convergence. To get the azimuthal shear, pseudo-radials of velocity data are built up at an angle from the actual radial. The thin line is detected by another algorithm. The algorithms locate features using data at two elevations (0.5° and 1.0°) and then vertically associate them to ensure that the detection is valid. Details and performance assessment may be seen in Uyeda and Zrnić (1986); Klingle-Wilson et al. (1989), Eilts et al. (1991) and Hermes et al. (1993).

(3) The Wind Shift Algorithm (WSA) (Hermes et al. 1993) estimates the wind behind the gust front from the radial velocities.

Detection of microbursts by an automated system has been implemented in Australia (Dance and Potts 2002). Agent networks (i.e., networks of cooperating asynchronous software entities) detect reflectivity cores and high divergent shear zones. Their output is integrated by higher-level agents to detect microbursts and to track microbursts through time.

There is a large number of studies of application of radar in gust front and microburst detection and prediction in the aviation context. Some selected references other than those cited already in this chapter are: [Campbell (1991), Campbell et al. (1989), Cornman and Mahoney (1991), Eilts (1992), Evans (1991), Hjelmfelt (1988), Mahoney (1988), Mahoney et al. (1989), Potts (1989), Roberts and Wilson (1989), Wakimoto (1982), Weber and Noyes (1988), Weber et al. (1991) and Wilson et al. (1988).

9.1.8 Turbulence

Turbulence in relation to aviation is defined as light, moderate or severe in terms of the acceleration that the aircraft is subjected to, as follows:

Light turbulence $0.15g \leq g' \leq 0.5g$

Moderate turbulence $0.5g \leq g' \leq 1.0g$

Severe turbulence $g' > 1.0g$

g' being the deviation from normal acceleration due to gravity g taken as 1.0.

Turbulence can produce oscillatory motion of various frequencies which can cause discomfort, or in extreme cases, structural damage. Turbulence occurs in association with thunderstorms and also in clear air e.g. at the periphery of jet streams. Detection of turbulence by Doppler weather radar is discussed in Chapter 6 and by wind profilers in Chapter 10.

Spectral width consists essentially of contributions from shear and turbulence[5]. Since both are adverse to aviation, it may not be necessary to estimate the contributions separately for aviation operational purposes. A spectrum width threshold of 4 m s^{-1} is taken as indicative of discomfort and 6 m s^{-1} as a potential hazard (Mahapatra and Zrnić 1991).

Using vertical incidence Doppler radar, Donaldson and Wexler (1969) measured the variance of the vertical velocity in a thunderstorm in which no hail fell on the ground. At least 5% of the variance measurements were too large to be accounted for by the spread of precipitation fall speeds and by vertical gradients of vertical velocity. On this basis, turbulence and/or shear was inferred in regions of high velocity variance. They concluded that vertical velocity

[5] Differences in fall velocities of hydrometeors also contribute; see Chapter 3.

variance of 2 $m^2 s^{-2}$ in a storm suggests small hail or or light to moderate turbulence. Greater than 4 $m^2 s^{-2}$ indicates damaging hail or severe turbulence or both.

With conventional radar, a reflectivity of 40 dBZ in a storm is taken as an indication of severe turbulence (Lewis 1970,1978). Turbulence occurs also in the vicinity of and above the top of the thunderstorm (Burnham 1970, Pantley and Lester 1990). Since visible cloud is larger than radar echo, some clearance around the storm is necessary. A distance of 5 n. miles beyond the 10 dBZ contour may be taken as the limit for safe flight according to Lewis (1970, 1978). Similarly a clearance above the echo is advised. In the case of squall lines, there could be severe turbulence in echo-free regions between cells.

The rules for safe flight are, according to Pantley and Lester (1990):

1. Do not fly within the anvil of a thunderstorm.
2. Do not fly within 20 n. miles of a thunderstorm.
3. Do not fly in thunderstorm tops.
4. For a wind velocity of 100 knots, an aircraft should be at least 3 km above the thunderstorm. Add (or subtract) 300 m for every 10 knot positive (or negative) departure from 100 knots.

Since the movement of the storm can also be evaluated, it is always desirable to divert behind the storm rather than in front of it.

In the NEXRAD radar (see Chapter 3) there is a Combined Moment (CM) product (Anderson and Green 1991, Klazura and Imy 1993) which displays the three base products (reflectiviy, mean radial velocity and spectrum width) in a single B-scan format display (azimuth on the X-axis and range on the Y-axis) simultaneously over a selected small area. The velocity at each grid point (0.5 km range × 1° azimuth grid) is represented by a vector whose angular orientation indicates the magnitude and sign of the radial velocity and the size of the arrowhead the spectum width. The reflectivity is portrayed in eight colours. The likelihood of turbulence can be deduced from looking at the three products.

Airborne weather radar in passenger flights can apply these criteria. But since airborne radars generally use a 3 cm radar, rainfall attenuation may make the echo appear weaker than it is, smaller than it is and its centroid nearer than it is (Metcalf 1980). Hence, airborne radar is generally used only for qualitative indication. Clark (1980) has proposed a method of correcting the observed reflectivity contours. This works by increasing the receiver sensitivity progressively with range to compensate for the two way attenuation due to rainfall; the attenuation is computed from the observed reflectivity using a k_R-Z relationship (see Chapter 2).

Meischner et al. (2001) who compared eddy dissipation rates derived from

C-band Doppler radar with in situ measurements from aircraft found good agreement between the two. According to them, although radar senses turbulence at scales smaller than the radar resolution volume, it can be related to turbulence over larger scales influencing aircraft flight performance.

9.1.9 Icing

Icing normally occurs in flights through convective precipitation containing supercooled water above the freezing level. The water freezes on contact with the aircraft. Radar which gives the height of the convective cloud and the vertical motions should help in predicting icing when combined with radiosonde and other data. Icing can also occur if a stratiform cloud is lifted by orography, leading to presence of supercooled water above the freezing level. Meischner et al. (2001) point out that small scale turbulence may force cloud droplet growth and increase the probability of formation of large supercooled drops. Hence the detection of turbulence may help in the issue of icing warnings. Ellis et al. (2000) and Vivekanandan et al. (2001) have proposed using polarimetric or dual-wavelength radar for icing detection.

9.1.10 Lightning

Relationship between lightning and radar echoes from hydrometeors was discussed in Chapter 4. By combining data of radar echo centroids and cloud-to-ground lightning strokes detected by lightning detectors within 10 n. miles of the centroid, Stern et al. (1994) developed an algorithm for positive identification of occurrence of thunderstorms[6]. This information can be used in surface weather reports for aviation in place of the thunderstorm reports based traditionally on thunder heard by a human observer.

9.1.11 Hail, Precipitation, Mesocyclones and Tornadoes

The detection and prediction of these phenomena depend on algorithms linking radar-observed parameters or signatures with these phenomena. These have been discussed in Chapter 6. The particular need of aviation is rapid computation and display of information in a form which is readily

---

[6] Lightning can also be triggered by the aircraft itself, if sufficient potential gradient is established.

assimilated.The maximum velocity difference in the region of divergence at the storm top can be obtained from the CM product mentioned above, for use in the Witt and Nelson hail algorithm (Chapter 6). Vasiloff (1991) has developed a TVS (Tornadic Vortex Signature) algorithm specifically for the TDWR. This is similar to the mesocyclone algorithm discussed in Chapter 6 except that the search for velocity differences is limited to adjacent radials instead of extending over several azimuths. This gives higher resolution without which wind maxima would be smoothed out and underestimated.

9.1.12 Choosing and Siting of the Radar

There are special problems in choosing and siting radars for detection of LLWS-inducing phenomena at airports [see Wilson et al. (1984) for a detailed discussion]. The radar must have high sensitivity to detect clear air echoes at low levels. Because of the very short lead time involved, fast data acquisition and completely automated processing are essential. This means also that range and velocity ambiguities must be removed completely.

While a dual-Doppler configuration may give a clear picture as in Fig. 3.13 it is unsuitable for an operational context. One possibility is to have a Doppler radar (of short range and low power and a frequency of 10 GHz, to get high power return) at the end of each runway to monitor the wind changes along the runway at low levels. This will be a compact equipment. Such systems are commercially available. But if the objective is to cover hazards other than LLWS also, a more versatile radar like the TDWR is necessary. The use of a radar sited at the airport has the disadvantage of clutter at near ranges. It also needs a 360° scan which if done up to very high elevations will take an unduly long acquisition time. An alternative is a radar situated several kilometres from the airport with its data transmitted to the airport. It can scan a sector in the direction of the runway even at low elevation without clutter contamination, assuming that an unobstructed view is available.. However if there is more than one runway, the true shear along the runway which is not in the direction of scanning cannot be obtained. Since microburst outflow is asymmetric what is obtained at one azimuth may not represent the true shear along the runway.

9.1.13 Use of Wind Profiler Radars

Clear air radars or wind profilers are discussed in Chapter 10. Profilers can give the following products of interest to aviation [Carlson and Sundararaman (1982), Serafin and Dabberdt (1990)].

1. Winds at various levels and vertical wind shear
2. Wind variability. The high temporal resolution of wind profilers enables keeping track of wind changes with time.
3. Turbulence parameters including Clear Air Turbulence (CAT).
4. Detection of inversions, fronts, tropopause and jet stream.

Chapter 10

OBSERVATIONS WITH "CLEAR AIR" RADARS

In the previous Chapters we have largely dealt with microwave weather radars of the type which are used in operational meteorology to observe weather systems associated with precipitation. We referred briefly (Chapter 2) to scattering and reflection in the clear air due to discontinuities or fluctuations in radio refractive index and mentioned that echoes due to such phenomena are better observed at longer wavelengths i.e. in the UHF and VHF frequencies. We also discussed briefly (Chapter 4) several clear air phenomena, besides hydrometeors, which can be detected by microwave radar. In the present Chapter, we shall deal with VHF and UHF radar systems which are particularly suited for clear air observations. and are called "clear air" radars. It has been recognised in recent years that, despite their name, these radars are capable of detecting echoes from hydrometeors also. This capability, which will also be discussed in this chapter, is the reason for the use of the inverted commas with the name "clear air". Some authors prefer to call them simply VHF/UHF radars.

In section 10.1 we introduce the basic principles of these radars and discuss the echoing mechanisms. In section 10.2 a description is given of the Indian MST radar as a typical example of a research radar. Section 10.3 deals with wind profilers, operational networks of profilers and their comparison with radiosondes. The direct measurement of vertical velocity which is an important capability of profilers is also discussed. In section 10.4 we discuss the derivation of products such as divergence, momentum flux and turbulence and vertical profiles of refractive index, temperature and humidity. Section 10.5 deals with the detection and study of precipitation using VHF/UHF radars. Finally in section 10.6 we give a brief account of the Radio Acoustic Sounding System (RASS).

This type of radar has been extensively used for studies of the upper atmosphere; but such studies are beyond the scope of this book.

10.1 MST/ST Radars

10.1.1 Development of MST/ST Radars

Large VHF radars were originally built for studies of scattering from the ionosphere. They obtain returns because the ionisation contributes a term due to electron density to the refractive index equation (2.10). The paper of Woodman and Guillen (1974) was a landmark paper demonstrating the capability of such a radar to monitor winds in the (non-ionised) stratosphere and mesosphere. The implicit assumption is that the refractive index irregularities, discussed in Chapter 2, are tracers of air motion. Because of their capability to observe the Mesosphere, Stratosphere and Troposphere, they are now called MST Radars. Echoes from clear air have also been observed in the troposphere and lower stratosphere with microwave and UHF radars. Radars used for clear air observations in the Stratosphere and Troposphere are called ST radars.

While a scanning microwave weather radar, which we have so far discussed, normally uses a steerable dish antenna which scans over a wide range of azimuth/elevation, the "clear air" radar usually uses an array of discrete antenna elements, which generates a vertically pointing beam that can be switched electronically to off-vertical angles. This enables the detection of vertical and horizontal motions of the clear air scatterers. Hence, these radars are used to determine the three components of air motion. The wind can be determined in clear air conditions without depending on the presence of hydrometeors. More important, vertical air motion can be directly and almost continuously monitored instead of being indirectly derived from radiosonde observations. Turbulence, wind shear and divergence can also be determined. Equipment of this type, known as wind profiler, operating in the UHF, has come into operational use. Variations of this can be used for studies of the atmospheric boundary layer or of stratified layers such as the tropopause. Though they are called "clear air" radars, VHF/UHF radars also get echo returns from precipitation.

This category of radars has been rated as the most exciting development in radar meteorology in the past two decades. In this book with its emphasis on operational meteorology we shall confine our attention to the application of these radars for observation of meteorological phenomena in the troposphere and lower stratosphere especially in the operational context. A brief description of the Indian MST radar, in particular, is also given. An overview of atmospheric measurements with these radars may be seen in Green et al. (1979). Later reviews of clear air radars may be seen in Gossard (1990), Hardy and Gage (1990), Jain et al. (2000), Rao(1990) and Röttger and Larsen (1990).

10.1.2 Signal Processing

Signal processing is, by and large, similar to that for Doppler weather radars. Details may be seen in Clothiaux et al. (1994), Keeler and Passarelli (1990) and May and Strauch (1989).

The VHF and UHF returns from turbulent scatter are quite weak. The correlation times of the scattering process are large in comparison with the inter-pulse period of the radar. Hence a number of samples of the I and Q signals can be separately integrated. This is called coherent integration because the integration is done in the same phase. Coherent integration of N samples improves signal-to-noise ratio by a factor N. It also reduces the volume of data to be processed. The power spectrum is computed using the Fast Fourier Transform (FFT) algorithm. The power spectrum exhibits a statistical fluctuation due to the random nature of the turbulence. The receiver output is a time series of data of a random process with Gaussian distribution. FFT being a linear transformation, the Power Spectral Density (PSD) is the sum of the squares of the real and imaginary parts of the spectral components and the standard deviation of the power spectrum is proportional to the spectral density. The amplitude of statistical fluctuation can be reduced by incoherent integration of successive samples as discussed in Chapter 3 for the case of scanning radars. In this case, integration of N samples reduces the standard deviation of the power spectrum by a factor $N^{1/2}$. Assuming that the power spectrum is Gaussian, the zeroth, first and second moments of the spectral density are computed[1]. These correspond respectively to the signal power, mean Doppler velocity and the spectral width.

10.1.3 Wind Measurement - the Doppler Beam Swinging Method

The most common mode of operation of a VHF or UHF radar is the Doppler

[1] Signals are often contaminated because of aliasing of higher frequency signals into the clear-air portion of the spectrum and a broadening of the spectral contaminants caused by the relatively short time series used to generate Doppler spectra. By processing longer time series of data at greatly increased speeds it is possible to reduce contamination in the spectra (Wilfong et al. 1999).

beam swinging method illustrated in Fig. 10.1. The antenna consisting of a square array has a narrow beam pointing vertically up. The beam is switchable to off-vertical directions at a zenith angle $\psi = (90 - \theta)$, θ being the angle of elevation), say, to the north, east, south or west[2] of the vertical, by controlling the relative phases of signal fed to different antenna elements. A cycle of switching may typically consist of the four oblique positions and two vertical beams in orthogonal polarisations. The Doppler spectra for each beam can be displayed and the Doppler moments computed, the first moment giving the mean Doppler velocity. If we assume that the wind is uniform over the area sampled by all the beams and if the Doppler velocities (designated positive when away from the radar) in the various positions are V_z (vertical beam), V_n (north beam), V_e (east), V_s (south), V_w (west) and u, v, w, are the three orthogonal components of wind, we have the relations,

$$V_z = w \tag{10.1}$$

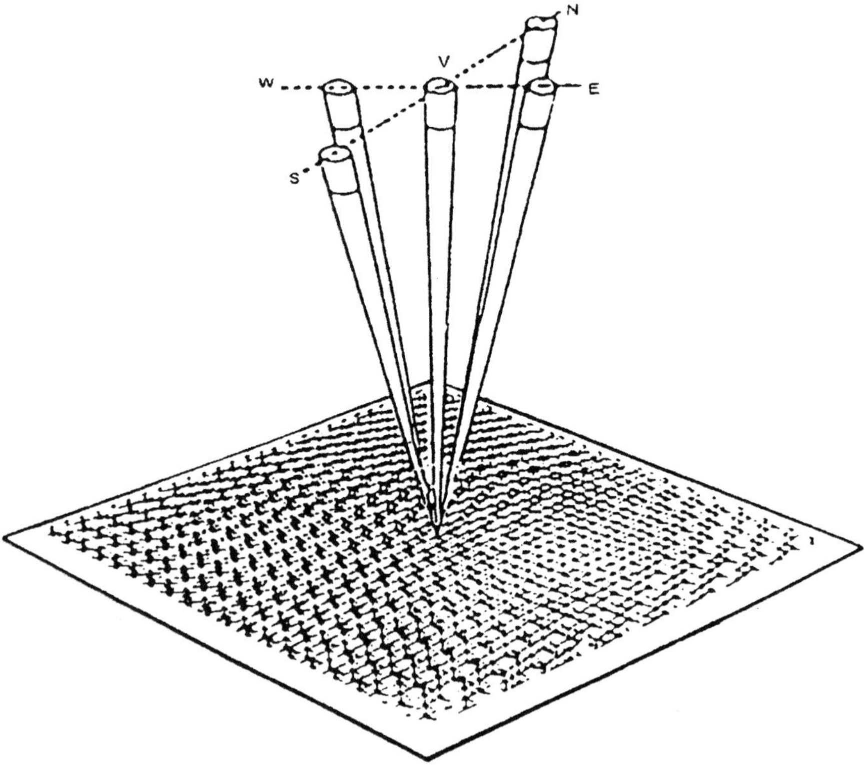

Fig. 10.1. Doppler beam swinging

[2] The four cardinal directions mentioned here are convenient, but the tilt can be to any azimuth and any number of azimuths.

$$V_e = +u\cos\theta + w\sin\theta \tag{10.2}$$

$$V_n = +v\cos\theta + w\sin\theta \tag{10.3}$$

There will be similar equations for V_s and V_w. Substituting the first equation in the other two, u and v can be determined. Or, using the equations for V_e and V_w, u and w can be determined and similarly using V_n and V_s, v and w can be determined. The value of w from the first equation can be used as a check on that obtained from the oblique beams. Three beams are enough to determine three components of wind, but the availability of the two extra beams provides redundant data which can be used to get a least square estimate and improve the accuracy.

We have omitted here the fall velocity of precipitation particles, if present and detected, which may be mixed up with w. We have also omitted possible errors in the radial velocity measurements. These will be considered in section 10.3.

In the general case where there is a number of beams in three or more different directions, the magnitude of the radial velocity in each beam direction can be expressed as the scalar product of the wind velocity vector and the unit vector along the beam direction. The equations can be solved by least square methods to compute the three components of the velocity vector. If frequent observations are made, a "consensus average"[3] is taken to obtain an hourly value.

It will be seen that the above is a simplified version of the VAD technique of Chapter 3.

The choice of the angle ψ for the oblique beams is based on a trade-off between the following considerations:

1. The higher the elevation the closer is the effective aperture to the physical area of the antenna.
2. With higher elevation the slant range for a given height is reduced and signal strength increases.
3. At high elevation the sampling volume is nearly at the same altitude for the vertical and oblique beams.
4. Ground clutter contamination is reduced at high elevation.
5. The area occupied by all the beams at a given height is reduced at high elevation angle. Hence the homogeneity of wind field is more likely to be realised.,
6. Error in antenna pointing direction causes greater error in computed wind vector if beam elevation is high.

---

[3] A consensus average is suitable for synoptic analysis. But in situations involving rapid temporal changes this cannot be applied.

7. The reflection from stratified layers increases steeply at angles close to zenith. This will result in bias of the Doppler values to lower velocities.
8. At low elevation the height resolution is degraded.
9. For low elevation the relative error in horizontal velocity is reduced.

10.1.4 The Spaced Antenna Method

In this method (originally used for ionospheric studies) the transmitting antenna is a vertically pointing array as before. There are three or more receiving antenna arrays spaced well apart (typically 30 to 50 metres for a VHF radar). The echo return from any clear air scattering feature will arrive at these antennae with some time difference depending on the velocity of movement of the feature over the antennae. The time delays can be deduced from the cross-correlation functions of these signals and the horizontal wind causing movement of the feature can be computed (Larsen and Röttger 1989). Doppler parameters are not involved and no beam switching is involved. However, Liu et al. (1991b) have shown that it is possible to use the spaced antenna method with oblique beams also. According to them, this can give winds at specific points without assumptions about homogeneity of wind field. They also compared winds from this technique with winds from vertical beam spaced antenna observations and with Doppler beam swinging method and found general agreement.

10.1.5 Echoing Mechanisms and Wavelength Dependence

Three mechanisms causing the echo return from clear air were identified in Chapter 2. The dominant one is the Bragg scatter from irregularities having a dimension of half the wavelength of the radar. The inertial subrange extends over a scale length from about 10^{-2} to 10^{2} metres at the surface and narrows down at higher altitudes. The inner scale of turbulence (i.e. the boundary between the inertial and viscous ranges) l_0 is related to the eddy dissipation rate ε and kinematic viscosity ν (m^2 s^{-2}) by the relation (Gage 1990, Hocking 1985),

$$l_0 = 7.4\eta_0 = 7.4\left(\frac{\nu^3}{\varepsilon}\right)^{1/4} \tag{10.4}$$

η_0 expressed in metres is called the Kolmogorov microscale. To detect turbulence, the radar wavelength must be greater than a critical value of $4.2l_0$.

Shorter wavelengths are damped. Thus a 50 MHz MST radar is sensitive to 3 m scale irregularities while a 400 MHz wind profiler is sensitive to 38 cm scale irregularities and 3 GHz microwave radar senses 5 cm irregularities. Hence, the longer wavelength radars can get better echo returns from higher altitudes from these irregularities (Strauch et al. 1986). The short wavelength cutoff of the inertial subrange is shown in Fig. 10.2. and provides the upper limit up to which observations are possible at a given radar wavelength. This limit is not sharp and may vary with atmospheric conditions.

The detectability of a given level of refractive index turbulence (C_n^2, defined in Chapter 2, being its measure) will, depend on the sensitivity of the radar. The signal-to-noise ratio (SNR) for a clear air radar is taken as (Strauch et al. 1986),

$$\frac{P_{tav} A_e \lambda^{1/6} \Delta r C_n^2 (T_0 \Delta v)^{1/2}}{r^2 \sigma_v T_{op} L} \tag{10.5}$$

where P_{tav} = average transmitted power, A_e = effective antenna area, λ = radar wavelength, r = range, Δr = range resolution, σ_v = width of the Doppler velocity spectrum, L = system loss not included in P_{tav} or A_e, T_{op} = system noise temperature, T_0 = observation time and Δv the velocity resolution of the data processor. The portion of the above expression which is dependent on the radar parameters is,

$$\frac{P_{tav} A_e \lambda^{1/6}}{T_{op} L} \tag{10.6}$$

This expression can be used to compare different radars. The product $P_{tav} \times A_e$ is referred to as the power-aperture product. The SNR can be obtained from the zeroth moment of the spectrum and therefore C_n^2 can be computed from equation (10.5).

The wavelength and design parameters of the radar are chosen depending on the height range over which observations are required. A knowledge of the C_n^2 values likely at various heights in all seasons will be needed for this purpose. C_n^2, of course, decreases with height. As is to be expected it is strongly correlated with vertical and horizontal wind speed. In convective activity it increases[4] [see e.g. Green et al. (1978), Knight and Miller (1993)]. There is also a seasonal variation. Warnock et al. (1986) compared these variations with the model of VanZandt et al. (1978) referred to in Chapter 2. They found that the model understimates C_n^2 in the troposphere.

A climatological study of C_n^2 variations in the boundary layer by

[4] Sidelobe returns from strong convective cells can also increase the reflectivity.

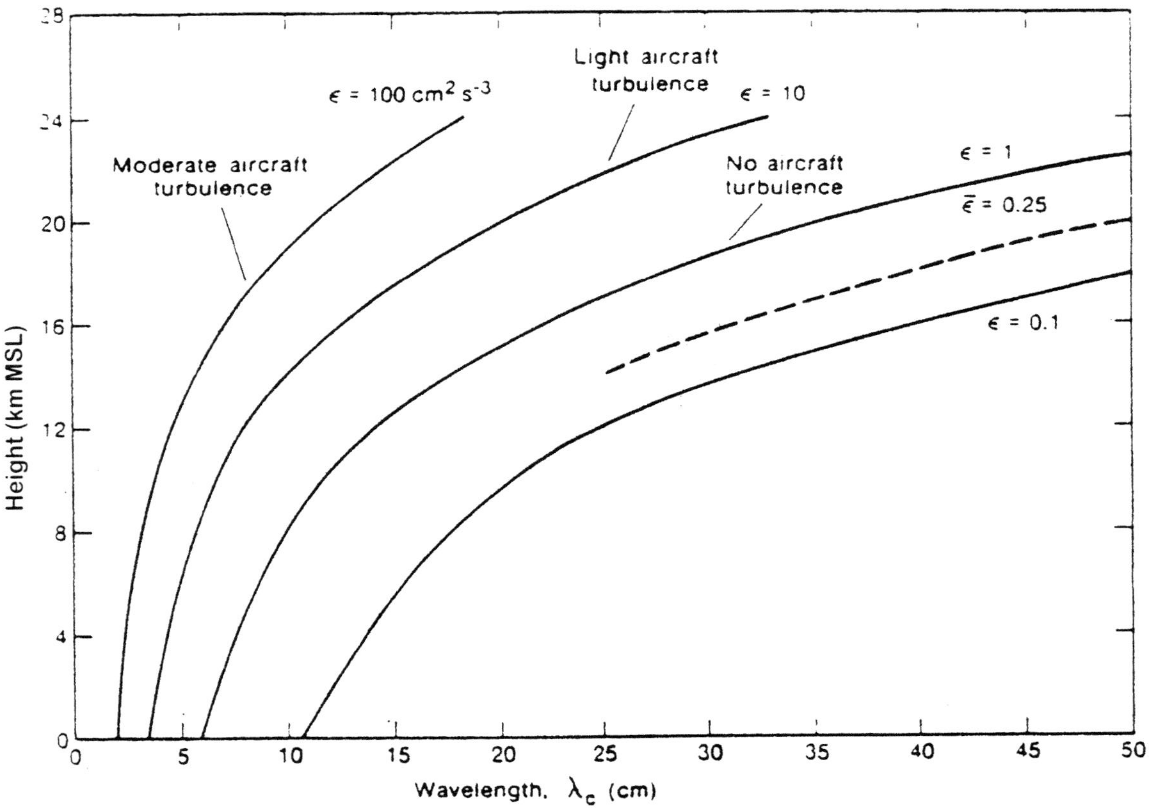

Fig. 10.2 Cutoff wavelength versus height for various levels of mechanical turbulence. Routine clear-air wind profiling should be possible with radar wavelengths greater than the cutoff wavelength, that is, to the right of the curves. (from Strauch et al. 1984; courtesy AMS).

Chadwick and Moran (1980) was referred to in Chapter 2. Measurements at various tropospheric heights have been presented by Doviak et al. (1983) and show ($\log C_n^2$) values between about − 14.5 near the surface varying exponentially with height up to about −17 at 12 km altitude. Balsley (1978) and Doviak et al. (1983) have discussed the criteria for successful clear air wind measurements. Jain et al. (1995) computed ($\log C_n^2$) from signal-to-noise ratio measurements with the Indian MST radar (50 MHz; see section 10.2) and obtained values of about −16 near the surface decreasing to about −19 at 12 km altitude. The values showed a large variation within a day and from day to day. Narayana Rao et al. (1996, 1997) extended the computation to 20 km and for the whole year. They obtained values from −16 to −20. Narayana Rao et al. (2001c) and Ghosh et al. (2001) studied the seasonal variations of ($\log C_n^2$) in the 4 to 20 km altitude range with the Indian MST radar. They also found values of −17 to −19. These are all rather lower than the values indicated by Doviak et al. (1983). Frisch et al. (1990) found that in the troposphere there is a year to year increase also! They suggest that profilers should be designed with an additional margin of sensitivity of 5 to 10 dB to provide for any long term decreases of C_n^2 in future. Frisch and Weber (1992) compared C_n^2 distribution determined with 50, 405 and 915 MHz profilers. They found that they were similar in the lower troposphere but differed near and above the tropopause.

The aperture and wavelength determine the limit of the near field in which observations are difficult to interpret as the far field equations do not apply. This limit is given by $2D^2/\lambda$ where D is the antenna dimension (see Chapter 1) and represents the lowest altitude at which observations are available. MST radars have a large aperture and therefore have a minimum altitude of 3 to 4 km (the maximum altitude of observation being about 100 km) despite using a long wavelength (typically 6 metres). UHF wind profilers have smaller apertures and may typically operate between a lower limit of 100 to 500 m and an upper limit of 5 to 18 km. In the case of such low altitudes the T-R switch recovery time may set the lower limit.

Besides Bragg scatter, quasi-specular reflections from discontinuities of refractive index associated with inversion layers or the tropopause cause an echo return which is prominent in the vertical beam in VHF radars. The power return is thus much stronger at vertical beam than at off-vertical angles. This is called aspect sensitivity. Whether the echoing mechanism is scatter or reflection can be decided by seeing the dependence of the signal strength on the zenith angle of the beam. If the signal diminishes rapidly as zenith angle increases, the mechanism is reflection. If the width of the dependence on zenith angle is much larger than the transmitting antenna beam width, the mechanism is scattering.

The Doppler spectra are also wider at the off-vertical angles. The occurrence of an intense thin layer of turbulence not filling the beam may cause the effective centre of the scattering region to be shifted. The tilt angle of the beam therefore shifts and a false indication of large wind shear can appear. This is called finite range volume effect [Fukao et al. (1988, 1988b), May et al. (1988)].

Another important wavelength-dependent phenomenon to be considered is the relative echo strength from clear air and hydrometeors. Figure 10.3 reproduced from Larsen and Röttger (1987) compares the reflectivity from turbulent scatter (with two assumed C_n^2 values) with that from precipitation (with two assumed dBZ values). It is seen that while turbulent reflectivity is comparable with that from moderate precipitation at VHF, precipitation dominates at UHF. The result is that the Doppler spectra will exhibit downward vertical velocities due to hydrometeors. Hence in precipitation, the wind velocity derived from equations (10.1) to (10.3) can be in error during precipitation. The precipitation echo has been used for studying precipitation phenomena (section 10.5).

For MST radars, VHF is advantageous (Green et al. 1979) because, apart from the altitudes attainable,

(1) large array antennas can be constructed at relatively less cost.,

(2) stable horizontal laminae can be better observed at vertical incidence by Fresnel reflection,

(3) more effective turbulence measurement is possible in the mesosphere where both the increased kinematic vicosity and the dispersive refractive index favour longer wavelengths.,

(4) VHF radars are less sensitive to hydrometeors.

On the other hand, UHF (typically 400 MHz) with smaller aperture arrays is

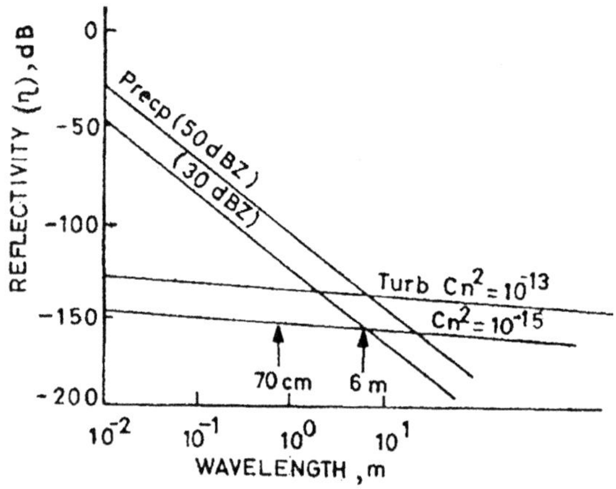

Fig. 10.3 Comparison of reflectivity for turbulent scatter and precipitation as a function of wavelength (from Larsen and Röttger 1987, courtesy AMS).

preferred if wind determination in the troposphere and lower stratosphere (say 0.5 to 17 km) is the principal requirement. For study of the atmospheric boundary layer (0.1 to 3 km say), higher UHF (say 900 to 1500 MHz) is preferred. For study of precipitation, UHF or a combination of UHF and VHF is advantageous. UHF radars are compact and can be made portable.

Frisch et al. (1986) compared profilers operating at 50, 405 and 915 MHz. and found that the 50 MHz profiler has the greatest change in height (i.e. maximum height to which observations are available) for the same change in sensitivity, the 405 MHz the next greatest and 915 MHz the least. For a 7 dB improvement in sensitivity the height performance increased by 7 km at 50 MHz, 5 km at 405 MHz and 2 Km at 915 MHz.

10.2 The Indian MST Radar

10.2.1 MST/ST Radars for Research

There is a number of MST/ST radars around the world devoted to research in meteorology, upper atmosphere and radio astronomy. The Jicamarca radar in Peru, the SOUSY[5] VHF radar (Germany), the PROUST[6] radar (France), the MU[7] radar (Japan), the Poker Flat radar (Alaska, USA), the Arecibo radar (Puerto Rico, USA), the Chung-Li radar (Taiwan), the Sunset radar (Colorado, USA), the Pohnpei radar (East Caroline Islands) and the CLOVAR[8] ST radar (Canada) may be mentioned among these. Their specifications may be seen in Röttger and Larsen (1990) and Hocking (1997b).

Some of them use more than one frequency but the most important frequency of operation is the 50 MHz band. Here, we give a description of the Indian MST radar as representative of this class of radars.

10.2.2 The Radar System

The radar is located at Gadanki (13°.47 N, 79°.18 E) in Andhra Pradesh, India. Its specifications are shown in Table 10.1, a picture of the antenna array in a bowl-shaped landscape in Fig. 10.4 and a block diagram of the radar

[5] SOUnding System.

[6] Prototype de Radar pour l'Observation en Uhf de la Stratosphère et Troposphère.

[7] Middle and Upper atmosphere.

[8] Canadian (London, Ontario) Vhf Atmospheric Radar, described in detail in Hocking (1997).

Fig. 10.4. MST Radar antenna array at Gadanki, Andhra Pradesh, India. [courtesy Dr. A.R. Jain, Director, National MST Radar Facility (NMRF)]

in Fig. 10.5. The antenna consists of 1024 crossed three-element Yagi antenna elements arranged in a 32×32 matrix over an area of about 15,000 m^2. Thirty-two modular transmitter units (Master Oscillator Power Amplifier type) energise the 32 sub-arrays in each polarisation (EW & NS) using a centre-fed excitation. The antenna has an illumination taper to ensure low side lobe level. The transmitter waveform is programmable with complementary biphase coding. When a long pulse width is used to get high signal return, the range resolution suffers. To improve the resolution, the pulse (of duration T) is phase modulated at subintervals T_b called the band length. The simplest case is that phase of each sub-pulse is chosen to be 0° or 180°. The sub-pulses are recognised by their phase. Regarding pulse coding techniques see Schmidt et al. (1979).

The pulse width, coding, p.r.f.[9] the number of beam positions, the beam sequence, the number of pulses to be coherently averaged, the number of FFT points to provide the needed velocity resolution and the number of pulses for incoherent averaging to provide the necessary time resolution and spectral averaging can all be selected through a radar controller.

Coherence is achieved by using a highly stable oven-controlled crystal oscillator as the master reference oscillator. A two-channel synthesiser

[9] Note that compared to scanning weather radars, the maximum unambiguous range needed is relatively small, permitting high p.r.f. to be used. Also, because of high elevation operation the component of horizontal velocity in the beam direction is small; the wavelength is also longer. Hence maximum unambiguous velocity requirement is not stringent. However, see section 10.3. regarding unfolding of aliased velocities.

Table 10.1. Indian MST radar system specifications.

Parameter	Value	Parameter	Value
SYSTEM		**Receiver**	2 channel (I & Q) coherent
Operating Frequency	53 MHz.	Overall gain	110 dB
Peak Power Aperture Product	3×10^{10} W m^2	Dynamic range	70 dB
Height range	5 to 100 km	Coho stability	1×10^{10} (short term)
Spatial Resolution			
Range	150 m (pulse width)	Data acquisition and signal processing	Real time computer-controlled
Angle	3° (beam width)		
Velocity resolution	0.1 m s^{-1}	Data resolution	12 bits (bipolar)
Time resolution	0.5 minute	Sampling rate	1 MHz per channel
Wave form	Selectable pulse widths and p.r.f.'s including pulse compression	No. of range gates	up to 256
		No. of points for spectral estimation	64 to 512
Pulse compression	Pseudo random coding	Velocity resolution	0.1 m s^{-1}
Signal Processing	Real time digital (FFT based)	Signal enhancement by coherent integration	20 dB (nominal)
SUBSYSTEMS		Spectrum integration period	Selectable 5 s to 10 min. in steps
Antenna	Phased array 1024 crossed Yagi elements	System computer 32 bit (MASSCOMP 5600)	
Gain	36 dB (nominal)	Operating system	Real time UNIX (RTU)
Beam width	3°	Online memory	8 M Bytes (expandable)
Beam positions	Zenith, up to 20° off-zenith E, W, N and S	Storage	Hard disk, floppy & mag tape
Sidelobe	– 20 dB	Display	CRT with colour graphics
Size	130 × 130 m	Hard copy	Printer, Plotter
Transmitter	Coherent, modular with variable pulse width and p.r.f.		
Peak Power	2.5 MW		
Duty ratio	2.5%		
Pulse width	Selectable 1 to 32 μs		
p.r.f.	62.5 Hz to 8 kHz		

connected to the master reference oscillator generates the coded waveform at intermediate frequency to be used as transmitter input or as a simulated signal

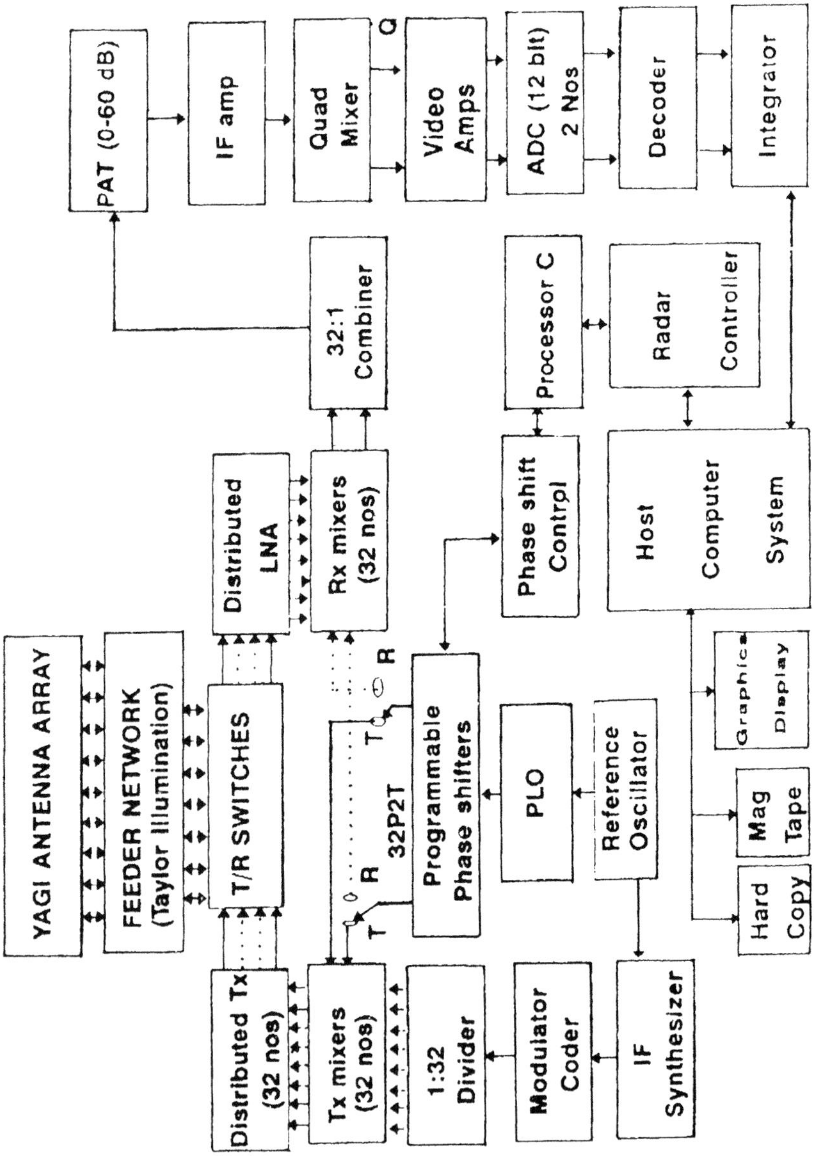

Fig. 10.5 Functional block diagram of Indian MST radar. [courtesy NMRF and Editor, IJRSP].

for performance verification after conversion to 53 MHz. A two-channel I and Q receiver with a 70 dB dynamic range is employed. The interpulse period is gated with a programmable gating scheme with a maximum of 256 gates, the height resolution being 150 m. Quadrature outputs of the receiver are digitised using high speed 12 bit bipolar A/D converters. An on-line digital signal processor is used for decoding and coherent integration generating time series data. A dedicated on-line computer derives the spectral information. Moments are extracted off-line. The wind components *u*, *v*, *w* are extracted from the moments. In addition to the on-line data acquisition and processing system an off-line data processing facility is provided. Rao et al. (1994) have given an overview of the radar in the ST mode. Data processing is discussed by Raghava Reddi and Muralidharan Nair (1994). System description and reviews of results may be seen in Jain et al. (1995, 2000) and Rao et al. (1995).

10.2.3 Outputs

A typical five-beam spectral scan in the ST mode is shown in Fig. 10.6. Signal amplitude and Doppler shift are shown in each range gate of 150 metres width corresponding to a band length of 1 μs. The horizontal scale is Doppler shift in Hz, one Hz corresponding to a velocity of 2.83 m s^{-1}. The spectra (a), (b), (c), (d), (e) are respectively for the east, west, vertical, north and south beams.. Note that in the vertical beam the signal is near zero Doppler velocity as vertical velocities are very small. The tilted beams give components of wind velocity in their directions. The east and west beams show equal and opposite components and so do the north and south beams. Beyond 10 km altitude the signal is swamped by noise. An "isometric" display with Doppler shift on the X-axis, range on Y-axis and intensity on a Z-axis is also possible. A Range-Time-Intensity (RTI) plot can also be displayed using the vertical beam. This gives time on the X-axis, range (i.e. height) on the Y-axis and intensity in terms of colours. Fig. 10.7 shows the computed height profiles of *u*, *v* and *w* components and of the horizontal wind speed. In this figure *u* is positive if wind is westerly and it can be seen that wind is westerly up to about 7 km and easterly at higher levels, typical of the summer monsoon season in India.

10.2.4 Calibration

Calibration of the receiver is done by a simulated echo with a known and precise frequency offset (say ± 5 or 10 Hz) and the signal is processed in the

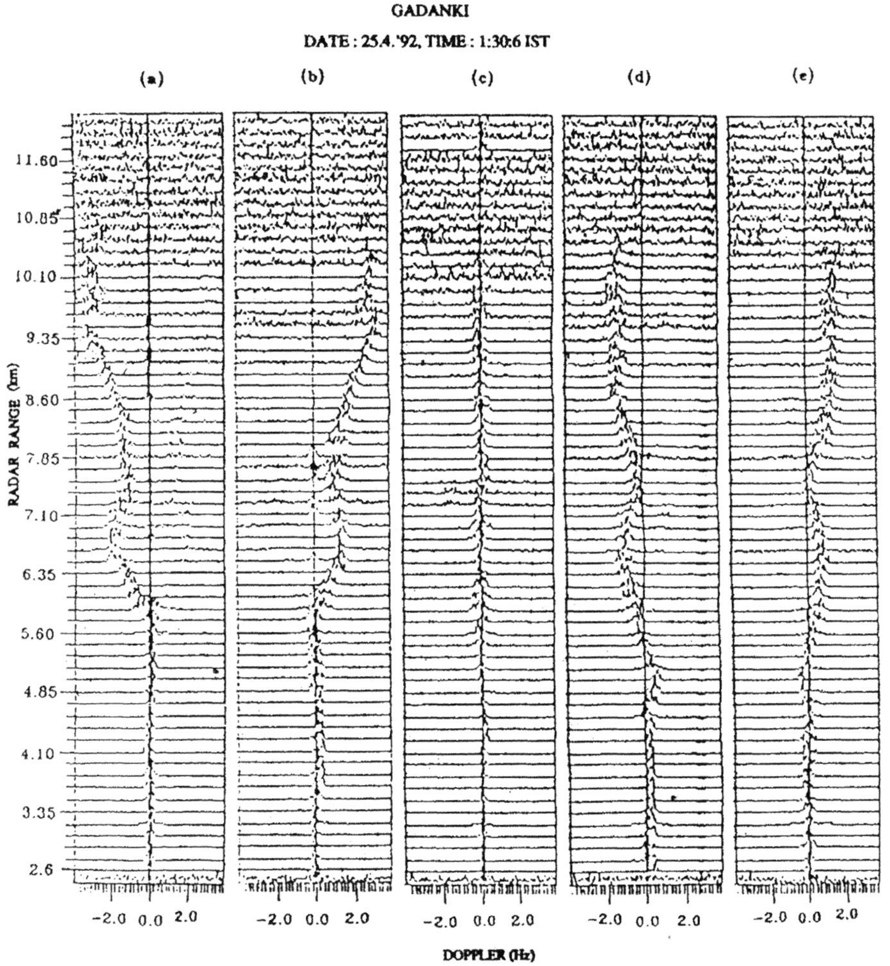

Fig. 10.6. A typical five beam spectral scan of Indian MST radar (from Kishore et al. 1994; courtesy NMRF and Editor, IJRSP). See text for details.

usual manner. This signal can be inserted at a range gate of, say, 40 km where atmospheric echo is not expected, so that the calibration data are available along with any experimental data. Absolute calibration of MST/ST radars with a noise source has been discussed by Green et al. (1983). However, for operation purely for wind measurements, only Doppler velocities are needed and absolute calibration of signal intensity is not important. Calibration for precipitation studies is discussed in section 10.5.

The antenna beam pattern can be mapped by observing the sun whenever it passes over the antenna but this is not convenient because unlike the case of scanning radars discussed in Chapter 5, solar passes will occur over the vertical

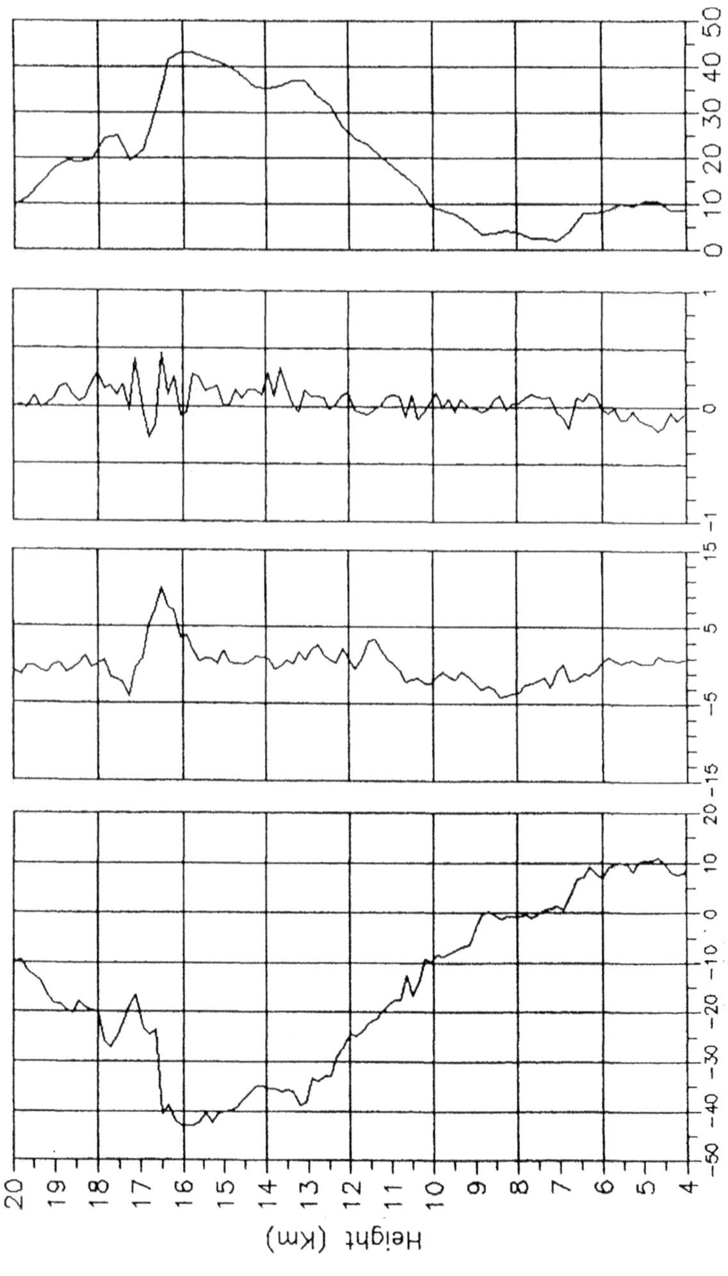

Fig. 10.7 Height profiles of u, v and w components and of the horizontal wind speed from MST radar (courtesy NMRF).

pointing antenna[10] only on some days of the year. A radio-star has therefore been used as a point source (Chakravarty et al. 1993). Viswanathan et al. (1993) plotted the vertical velocities from orthogonally polarised vertical beams against each other on Cartesian axes and found all points falling close to the 45° line. According to them this shows that the phasing is correct.

10.3 Wind Profilers

10.3.1 Measurement of Horizontal Wind

The measurement of horizontal wind by the beam swinging method is the most widely used operational application of the ST radar or wind profiler. The profiler enables near-continuous monitoring of upper winds over a site and a network of profilers can therefore be considered as an alternative to a radiosonde network. What are the relative merits of the two methods of measurement?

10.3.2 The Radiosonde

The radiosonde has been in use in various forms for over 50 years. It enabled, for the first time and as a routine, measurement of atmospheric parameters (not only the wind) at various levels up to an altitude of about 30 km and is the mainstay of meteorological services the world over. The radiosonde technique however is an expensive one, as it involves one-time use of a sophisticated instrument and facilities for launching balloons. Hence the frequency of observation is restricted to twice (or in some countries 4 times) a day. Information on short period changes in winds (required for aviation, for example) is, therefore, not available. The balloon takes over an hour to reach its maximum height and so that is the minimum time in which the data can be gathered. The balloon traverses a path which may take it 100 km or more from the place of launch. Therefore, the profile obtained by the radiosonde does not represent the vertical profile over a single point on the earth and may contain the effect of changes in the horizontal plane. The radiosonde usually yields data with a vertical resolution of about 300 m which is adequate for general weather forecasting, but not for specialised

[10] They will not occur at all at locations outside the tropics.

applications such as pollution studies and radio communication studies for which special instrumentation would have to be used. The radiosonde makes an in situ spot measurement while travelling rapidly through the air and it is not possible to average a large number of samples to minimise the effect of short period fluctuations and measurement errors. The measurement by the radiosonde represents only the horizontal component of motion of the wind. The vertical motion is important for certain applications (e.g. for studying or modelling convection, or subsidence or the rise or spread of pollutants) but cannot be directly obtained by the radiosonde. Besides the average air motion i.e. the wind, knowledge of turbulence is also important in a variety of applications but cannot be measured by the radiosonde.

10.3.3 Merits and Limitations of Wind Profilers

The principal merit of the profiler is the high temporal resolution which can be achieved. Wind profiles can be obtained as often as necessary at little additional cost without expendable instrumentation; real time computation and display is possible. Frequent observations with a single profiler enable a time section to be prepared for study of movement of weather systems. Unattended operation except for occasional maintenance, is possible. Besides wind, vertical motion, divergence and turbulence can be monitored. The data represent a profile over a limited area[11] of a few km across, over which the wind can be assumed to be constant in many situations. The height resolution is better than the radiosonde and a number of samples can be averaged to obtain greater accuracy. Because of the time integration the profiler effectively samples a much larger volume than the radiosonde.

Limitations in operational use are:

1. Data are available with current operational VHF/UHF profilers only up to 16 or 18 km as C_n^2 values are too low at higher altitudes to get reliable data. This maximum height can vary considerably with variations of C_n^2. There is also a lower limit of altitude.

2. Profilers cannot yield temperature and humidity data except in an indirect manner (section 10.4). Even when supplemented by RASS (section 10.6.) temperatures can usually be obtained only in the lower troposphere.,

3. The assumption of homogeneity of wind in the observing volume is vitiated, if there is convection or precipitation causing variations in the vertical or

[11] For example, if the oblique beam is at 15° inclination its offset from the vertical beam at 10 km altitude is 2.7 km.

horizontal air motion within this volume. In convective storms there can also be variations of wind field with time during the time taken to complete one set of observations, either due to advection of the storm across the beam or the rapid temporal evolution of the storm. Vertical motion of precipitation particles may vary within this volume. Similar problems can be caused by the occurrence of gravity waves.

4. Birds, insect swarms, aircraft and lightning can give echoes, besides radio interference from other sources of radiation. These echoes may be stronger than the clear air echo and need to be recognised and eliminated.[12]

10.3.4 Precision and Accuracy of Winds from Profilers

Strauch et al. (1987) made a study of the basic accuracy of measurements using a five-beam profiler (405 MHz) in Colorado, USA. If we take the algebraic sum of the vertical motion and that of falling hydrometeors as W, and errors in measured Doppler velocities V_e, V_n etc., in the east, north etc. beams as δV_e, δV_n etc., equations (10.1) to (10.3) can be rewritten as,

$$V_z = +W + \delta V_z \tag{10.7}$$

$$V_e = +u\cos\theta + W\sin\theta + \delta V_e \tag{10.8}$$

$$V_n = +v\cos\theta + W\sin\theta + \delta V_n \tag{10.9}$$

and similarly

$$V_s = -v\cos\theta + W\sin\theta + \delta V_s \tag{10.10}$$

$$V_w = -u\cos\theta + W\sin\theta + \delta V_w \tag{10.11}$$

There are two independent measurements of the horizontal wind components which may be designated v_n, u_e, v_s, u_w. In clear air, these are given by,

$$v_n = +V_n\sec\theta - V_z\tan\theta = v + \delta v_n \tag{10.12}$$

$$v_s = -V_s\sec\theta + V_z\tan\theta = v + \delta v_s \tag{10.13}$$

$$u_e = +V_e\sec\theta - V_z\tan\theta = u + \delta u_e \tag{10.14}$$

$$u_w = -V_w\sec\theta + V_z\tan\theta = u + \delta u_w \tag{10.15}$$

[12] Methods for recognising such extraneous echoes may be seen in Merritt (1995) and Wilczak et al. (1995). A quality control (QC) algorithm designed to flag winds that are suspected of being biased in wind velocity or direction by migrating birds, has been implemented on the hourly winds from the NOAA Profiler Network, USA (Miller et al. 1997). On the other hand, profilers can be used to study bird migration.

The errors in these are given by,

$$\delta v_n = -\delta V_z \tan\theta + \delta V_n \sec\theta \quad (10.16)$$

$$\delta v_s = +\delta V_z \tan\theta - \delta V_s \sec\theta \quad (10.17)$$

$$\delta u_e = -\delta V_z \tan\theta + \delta V_e \sec\theta \quad (10.18)$$

$$\delta u_w = +\delta V_z \tan\theta - \delta V_w \sec\theta \quad (10.19)$$

The differences between the independent measurements are given by

$$Du = u_e - u_w = \delta u_e - \delta u_w = -2\delta V_z \tan\theta + (\delta V_e - \delta V_w)\sec\theta \quad (10.20)$$

$$Dv = v_n - v_s = \delta v_n - \delta v_s = -2\delta V_z \tan\theta + (\delta V_n - \delta V_s)\sec\theta \quad (10.21)$$

The term $2\delta V_z \tan\theta$ is contributed by the vertical motion and the other term by random errors in measurement. If these differences are plotted with *Du* on the X-axis and *Dv* on the Y-axis a scatter diagram will result in which the points tend to get spread along the diagonal from lower left to upper right owing the vertical motion term. The random error term will spread the points all around. Strauch et al., found an elliptical spread of points along this diagonal showing that the vertical motion term was more important than the random error term.,

The error terms in the radial velocity measurements may be due to instrumental errors or due to inhomogeneities in the wind field. May et al. (1989) examined the errors due to data processing methods and concluded that they are small compared to the error due to wind variability. Hence, according to them, it is not worthwhile employing complicated algorithms to make small improvements in the processing errors. See May and Strauch (1989) and Clothiaux et al. (1994) for discussions of signal processing methods.

Since the zenith angle of the oblique beam is small, the vertical component of air motion has a large effect on the derived horizontal wind. Using observations in clear air conditions, Strauch et al. (1987) estimated the errors in the derived horizontal components of wind from "consensus-averaged"[13] hourly values assuming several cases.

Case (i) is a "two beam system" i.e. n and e beams only are used and no correction for vertical motion is made using the zenith beam measurement. In this case the standard deviation of the error in horizontal wind was 1.7 m s^{-1}.

[13] The consensus consists of 12 values in an hour of the mean Doppler velocities for each height level in each beam direction, but often fewer than 12 values at higher levels because weak signals are excluded. A minimum of 4 values out of 12, are considered necessary to form a consensus. They must lie within a window that is 1/16 of the Nyquist velocity interval (Frisch et al. 1986). Otherwise, the set of observations is rejected. If a velocity consensus is found, the signal power and spectral width are averaged for that set. For a wind vector to be calculated, both horizontal wind components from the oblique antennas must pass the consensus test.

(ii) two three beam systems with a common zenith beam to correct for vertical velocity in all the four off-zenith beams for which the error analysis was given above.
(iii) two three beam systems with two zenith beam observations to correct the n-e and s-w pairs independently for vertical velocity.

In cases (ii) and (iii) the standard deviation of the error was 1.3 m s^{-1} and 0.9 m s^{-1} respectively, if, only data with high signal-to-noise ratio were used. They stress the importance of making correction for vertical motion.

This study was followed with one (Wuertz et al. 1988) in conditions of precipitation where the fall velocities of drops were several metres per second. These large vertical velocities can make the radial velocity go beyond the maximum measurable velocity causing aliasing unless the maximum unambiguous velocity is set high. This can cause very large errors in the computed horizontal wind. The fall speeds of hydrometeors can also effectively reduce the maximum horizontal velocity that can be measured.

Wuertz et al. (1988) considered three cases: (i) uniform precipitation over the observing volume (ii) time-varying precipitation (within the hour over which the data are averaged) and (iii) space-varying precipitation. In the first case, a simple correction for vertical motion is adequate and the precision (2 m s^{-1}) is not appreciably worse than in clear air. But uniform precipitation in time and space will rarely occur. In case (ii) the correction for vertical motion should be carried out on each set of observations before the consensus averaging. Also, since all the beam measurements are not simultaneous, a time adjustment is necessary. Case (iii) is more complicated as the basic assumption of homogeneity is violated. This condition can be detected and the erroneous wind observations "flagged". Yoe et al. (1992) have also examined errors involved in horizontal wind estimates in the presence of precipitation and lightning. Since the spectra in such cases show large variations in Doppler velocities, they eliminated radial-velocity outliers and spectra with large spectral widths before consensus averaging.

Weber et al. (1992, 1993) proposed a "continuity" method instead of the consensus method of averaging. First, velocity-aliased data are unfolded[14] and then the algorithm looks for consistency in data in height and time by

---

[14] Though it was stated that velocity ambiguities are less common than in scanning radars, folding of velocities does occur. To unfold these in the 405 MHz profilers in the USA, an algorithm has been implemented (Miller et al. 1994). This compares the measured velocity in a range gate as well as the two possible aliased velocities on either side (i.e. measured velocity ± twice the maximum unambiguous velocity) with the median of the velocities measured in the adjacent seven range gates below and above. Whichever of these is nearest to the median is taken as correct. It is assumed that more than one fold will not occur.

comparison with neighbouring points. Data which do not fit are replaced by interpolations from their neighbours.

10.3.5 Comparison with Radiosondes

Another method of evaluating wind profiler accuracy is to compare with radiosonde data which are the standard data used by the meteorologist.[15] Kishore et al. (2000) compared horizontal winds from MST radar up to an altitude of 18 km with near-simultaneous radiosonde data from a station 90 km away. They found an underestimate by the radar, the standard deviation of the differences being 2 to 3 m s^{-1}. They attributed this to aspect sensitivity[16] of the radar. Narayana Rao et al. (1998) applied quality control algorithms for consensus and continuity and found that the agreement between radar and rawinsonde data improved as in Fig. 10.8 which shows scatter diagrams of the meridional wind component from the two types of measurement, before and after application of the algorithm. One source of difference at low altitudes could be the physical separation of the two sets of observation as well as time differences. Differences can also depend on wind variability at particular sites being greater due to topography and other local factors at low levels.

Comparisons have been made in a wide variety of conditions, in various parts of the world, in some cases with radiosonde and radar stations collocated, but differences of the same order have been documented. Fukao et al. (1982) compared vertical profiles of horizontal winds from the UHF radar at Arecibo (Puerto Rico) with those from a radiosonde station 80 km away. The standard deviation of the differences was 4.9 m s^{-1} in the upper troposphere and was mainly due to spatial and temporal variations in the wind field. The standard deviation of 3.3. m s^{-1} in the lower stratosphere was attributed to experimental error of the radiosonde. Larsen (1983) compared the Poker Flat (Alaska, USA) radar with the twice daily observations of the nearest five radiosonde stations and found differences of 3 to 4 m s^{-1}. He also compared both with the geostrophic wind derived from the geopotential height field generated by objective analysis of the same radiosonde data and found a difference of 5 to 6 m s^{-1}. Weber and Wuertz (1990) made a comparison of twice daily radiosonde and radar measurements over nearly two years covering various seasonal conditions.

---

[15] Comparisons have been made with other techniques of measurement also, e.g. lidars.

[16] Jain et al. (1997) found that the aspect sensitivity of the Indian MST radar is greatest between 15 and 21 km altitude and is different for the east-west and north-south planes i.e. anisotropic. According to them, the underestimate of horizontal winds will be about 5% for beam zenith angles greater than 8° and up to 30% for angles of about 4°.

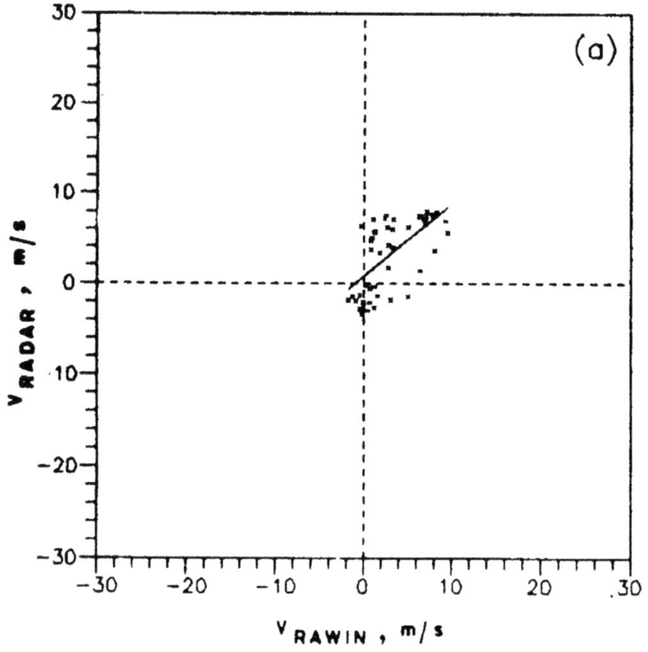

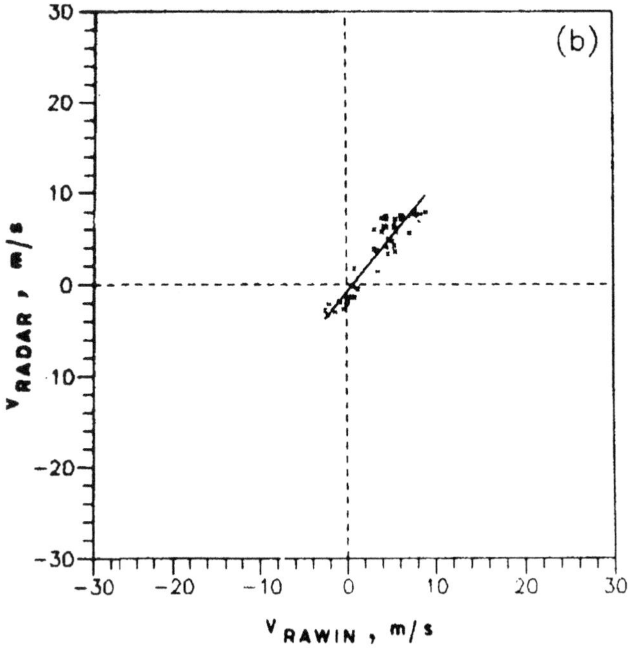

Fig. 10.8. Scatter plots between meridional components of wind from radar (V_{RADAR}) and rawinsonde (V_{RAWIN}), (a) before and (b) after application of quality control algorithm (from Narayana Rao et al. 1998; courtesy IJRSP).

They found that the standard deviation of the differences was about 2.5 m s^{-1}, which they attributed mainly to meteorological variability. Weber et al. (1990) compared one of the 405 MHz wind profilers of the U.S. demonstration network (see Paragraph 10.3.6.) with a 50 MHz radar nearby and also the nearest radiosonde. The standard deviations of the differences were a little over 2 m s^{-1} between the profilers and 3 m s^{-1} with respect to the radiosonde. These comparisons have also served to test the reliability of the profilers. Data were reported consistently up to 18 km with few outages. An outage may be due to equipment failure or more commonly to weak signal. May (1993) compared VHF wind profiler data at a tropical station with radiosonde ascents and found r.m.s. differences of 1.5 m s^{-1} at low altitudes increasing to 2.5 m s^{-1} at higher levels. In Canada with the CLOVAR radar (Hocking, 1997), in comparisons with collocated radiosonde flights and with Toronto radiosonde station 200 km to the east, an underestimation of wind speeds by radar by about 10% was noticed, while wind directions agreed well. Hocking attributes this to anisotropy of the scatterers..

Luce et al. (2001) compared the winds from MU radar in Japan with GPS[17] radiosonde observations launched 30 km west of the radar. The standard deviation of the differences was less than 2.6 m s^{-1} using radar data up to an altitude of 20 km and less than 1.5 m s^{-1} for a radar signal-to-noise ratio larger than 0 dB in oblique directions and a horizontal radar-balloon distance smaller than 50 km.

Nehrkorn (2000) has described an Optimum Interpolation (OI) technique for quality control and combination of wind profiler data with other observations of wind profiles.

10.3.6 Operational Networks of Profilers

While a number of research profilers have been in use in many parts of the world, operational profilers in the observational networks have been inducted in several countries in recent years. From about 1989, a Wind Profiler Demonstration Network (WPDN) was established in the USA consisting of 405 MHz wind profilers[18] with a view to eventual operational use (specifications in Table 10.2). There are 32 stations (listed in Ralph et al. 1995). These profilers have two modes of data collection: (1) low-altitude [0.5 to 9.25 km

---

[17] GPS = Global Positioning System.

[18] Renamed now as NOAA Profiler Network, the frequency has been changed to 449 MHz recently, because of radio-frequency allocation considerations.

above ground level (a.g.l.)] and (2) high-altitude (7.5 to 16.25 km a.g.l.). Three beams (beam width 4°), one vertical and two tilted at 16.3° in two orthogonal directions, are used. It takes about a minute to collect data in each beam in each mode. Hence a complete set of data is collected in 6 minutes. Data from each profiler is sent every 6 minutes to a centralised hub computer. Here the 6-minute data are quality controlled and averaged to produce hourly profiles of vertical and horizontal winds. Data from this network go into the Global (Meteorological) Telecommunication System (GTS) for real time use in weather analysis.

A review of developments in profiler networks and their capabilities may be seen in Serafin and Dabberdt (1990).

Table 10.2. NOAA demonstration wind profiler specifications (USA)

Parameter	value
Max data height	16.25 km
Min data height	0.5 km
Vertical cell spacing	250 m
Max horizontal wind	200 m s^{-1}
Max vertical wind	22.4 m s^{-1}
Frequency	404.37 MHz (now 449 MHz)
Peak power	16 kW
Average power	2200 W
Pulse width (compressed)	1.67/6.67 μs
Pulse repetition period	100/153 μs nominal
Antenna type	coaxial collinear array
Gain	> 32 dB
Sequential beams	east, north, vertical
Sidelobe level	– 20 dB to – 40 dB
Receiver noise figure	0.5 dB
MTBF	> 4400 h
MTTR	60 min
Operational conditions	
Temperature	– 40 to +50 deg C
Humidity	0 to 100%
Wind	63 m s^{-1} (83 m s^{-1} gust)
Rain	7.6 cm h^{-1}
snow	1.22 m
ice	7.6 cm radial

In Europe, several countries co-operated in the COST-74 (Co-operation for Science and Technology 74) programme (Lafaysse 1994) which was meant to study the feasibility of establishing an operational network. COST-74 has drawn up standard specifications (Gilet 1994) which are shown in Table 10.3.

The types are recommended according to the use proposed. Besides the meteorological considerations, the choice of frequency is constrained by the availability of frequency allocation.

A critical study of operational problems in a network has been made by Pilon (1992) in France. The first operational profiler in France was established in 1995 (Pilon et al. 1995). This is a VHF radar with a 3500 m array of 156 Yagi antennas. A correction for Fresnel reflection has been applied (Klaus et al. 2000). It operates with one vertical and four oblique beams (11° off-zenith). It is connected by a high-speed communication link to various Météo-France centres.

In the tropical Pacific areas, the USA has set up a number of VHF and UHF wind profilers [Balsley et al. (1991), Gage et al. (1994)]. These have given considerable experience of the use of profilers in tropical regions[20]. A merging procedure for combining radiosonde and profiler data to obtain an integrated, high quality upper air data set has been developed by Ciesielski et al. (1997).

In India, Raghavan and Viswanathan (1995) proposed a hybrid national network of UHF profilers and radiosonde stations to exploit the advantages

Table 10.3. European Specifications for Wind Profilers[19].

Parameter	Type 1	Type 2	Type 3
1. Frequency of operation MHz.	40 to 70	420-510	900-1300
2. Minimum altitude m	1500	300	100
3. Required upper altitude km	25	15	3
4. Vertical resolution m	500	300	150
5. Period of measurement minutes	30	20	10
6. Altitude accuracy m	50	50	50
7. Maximum instrumented altitude km	30	20	10

[19] The altitude parameters 2, 3 and 7 in table 10.3 are defined as follows:

2. Minimum altitude of measurement is the lowest vertical position above ground level where wind measurements are available.

3. Required upper altitude is defined as the height above ground level to which at least one complete wind profile can be measured in any given three hour period, 95% of the time.

7. Maximum instrumental altitude which the profiler should reach in cases where the atmospheric backscattering properties are strong enough so that there is no difficulty in obtaining a signal up to an unlimited altitude.

[20] Data from some of these go into the Global Telecommunication System.

of both. They have argued that in the tropics where horizontal gradients of upper air temperatures are small, the spatial density of wind observations should be high and that of temperature observations can be lower and this configuration can be met by a hybrid network, reducing the recurring expenditure on radiosonde observations. Designing of optimal configuration of observing networks can be guided by Observing System Simulation Experiments (OSSE). In these experiments simulated observations from the locations where a new equipment is proposed to be installed, are extracted from a reference model atmosphere and a test run of a model is made by assimilating the simulated as well as real observations[21].

Evaluation of the performance of wind profilers and allied equipment has been made in the US [Martner et al. (1993), Barth et al. (1994)] and types of failures have been identified. Specifically, the profiler data from the US network was found useful (Decker and Ewald 1994) in,

(1) identifying, tracking and monitoring (a) short wave troughs and ridges and low, middle and high level jets (especially short-lived ones),

(2) examining the vertical structure of air masses.

(3) identifying and monitoring the strength of thermal and moisture advection.

(4) identifying the beginning and ending of orographically induced precipitation and stratiform precipitation.

(5) determining the probability that thunderstorms will be severe and produce tornadoes.

(6) determining favourable conditions (moisture inflow, shear and storm movement) for flash flooding.

A number of case studies of application of profiler data in aviation in the USA (NOAA, 1992) have demonstrated their utility in short-period forecasts of severe weather, low level jets, turbulence and mountain winds.

The use of profiler data in Numerical Weather Prediction (NWP) has been examined by Kuo et al. (1987, 1987b) and Kuo and Guo, (1989). For global and regional NWP models, the World Meteorological Organization (WMO, 1986) laid down minimum requirements for upper wind observations as in Table 10.4. American wind profiler data have been successfully used since 1999 at the ECMWF; American and European wind profiler data assimilated in their "4-D Var algorithm" are at least as good as radiosonde data in the mid-troposphere (F. Bouttier, 2000, personal communication).

[21] Details are beyond the scope of this book. See e.g. Schlatter et al. (1999), Tijin and Wu (2000).

Table 10.4. WMO requirements for wind profilers

Parameter	Value
Horizontal resolution km.	50 to 100
Vertical Interval and Resolution	
upto 2 km altitude	500 m
upto 15 km	1 km
upto 30 km	3 km
Observational Error (r.m.s)	
Troposphere	$\pm$ 1 to 2 m s^{-1}
Stratosphere	$\pm$ 2 to 3 m s^{-1}
Minimum Frequency of Observation	4 times per day

All these requirements can be fulfilled by the operational profiler Type 2 in Table 10.3 except that data will be available only up to an altitude of 15 to 20 km. Data for the higher levels are obtainable from the radiosonde network and since the temperature and wind fields are not independent of each other, the poorer horizontal resolution for higher levels is not a serious impediment to numerical analysis unless very small scales are involved.

In setting up operational networks, the allocation of radio frequencies with adequate bandwidth is a major problem. UHF bands are very crowded. The "occupied bandwidth" (i.e. 99% of power is in this width) of a profiler is several MHz. It may be necessary to locate the profiler, if possible, in a bowl-shaped terrain to minimise interference with other users of the electromagnetic spectrum. Other technical considerations are a high degree of reliability for unattended operation and easy maintainability (see also Chapter 1 regarding siting of weather radars).

A flagging system is necessary to evaluate the data reliability in real time. A central computer can monitor the operational status of each profiler, including the signal-to-noise ratio, average noise level and ground clutter, the spectral width and multiple peaks (to identify and exclude signals due to specular reflection in vertical beam, returns from hydrometeors, lightning, aircraft etc.). It can also check data against numerically forecast fields to eliminate incompatible values.

Operational wind profilers are usually equipped to yield only the Doppler moments. The complete Doppler spectra are not recorded (because of the huge volume of data). While signal power may be recorded, there is no calibrated output of reflectivity or C_n^2. Hence, if any study is planned using spectral data, special arrangements should be made to collect and archive data in the required form. If needed, quantitative calibration of reflectivity should also be carried out.

Operational profilers provide an excellent opportunity to build up climatologies of C_n^2, ε and other parameters besides the wind, but only if data are archived in an appropriate manner.

Besides networks of profilers for synoptic meteorology, stand-alone profilers are used at individual sites for specific applications[22]. e.g. for boundary layer studies, pollution dispersal studies and support facility for space craft launch[23]. Specifications in these cases are tailored to meet the specific requirements.

10.3.7 Vertical Velocity Measurements

ST radars/wind profilers have provided for the first time a method of direct measurement of vertical air velocity as a routine. When it is not contaminated by returns from precipitation particles, vertical velocity is simply obtained from the vertical beam. The assumption is implicit that the scatterers are true tracers of the air motion. In the case of horizontal wind components, successful comparison with radiosondes may be taken as a verification of the assumption. Radiosondes do not directly yield vertical motion, but analyses of radiosonde network data do so and can be compared with vertical velocity data of profilers. This was done by Larsen et al. (1988) using ECMWF (European Centre for Medium range Weather Forecasting) analyses considering six grid points nearest to the radar site. They found that the magnitude of w variations were comparable in the troposphere but the radar showed large fluctuations in the stratosphere.,

Nastrom and Gage (1983) who studied the variability of vertical velocity using the MST radar at Poker Flat (Alaska, USA), found that while the velocity was usually -40 to $+40$ cm s^{-1} with a standard deviation of about 20 cm s^{-1}, there were episodes of much larger standard deviations. They attributed these to the presence of gravity waves. Individual observations may not be representative of larger scale features. Nastrom et al. (1985) pointed out that local effects such as standing lee waves near mountains may dominate the observation. They compared radar-determined vertical velocities averaged over several hours with large-scale vertical velocities indirectly estimated by adiabatic, kinematic and quasi-geostrophic omega equation methods. The radar-

---

[22] While frequencies of observation more than once in three hours may not be important for large scale forecasting, higher frequencies made possible by profilers will be particularly valuable for local forecasting and for various specialised applications.

[23] For example in France, UHF profilers are operated at Nice Airport for wind shear detection and at Marseille Airport for air pollution measurements (V. Klaus, 2002, personal communication).

measured values were found to be larger by a factor of two. They attributed this to sub-synoptic influences which will not figure in the synoptic-scale estimates.

Vertical velocity averages over various time periods as determined by two VHF radars spaced 23 km apart in flat terrain in the USA were compared by Warnock et al. (1994). They found excellent agreement between the two in 5.25 day averages, good to excellent agreement in 1 day averages. However in shorter period (6 hours) averages there were, at times, large discrepancies. Measurements below 5 km altitude were not reliable probably due to precipitation contamination. They concluded that synoptic-scale variations of vertical velocity can be reliably measured by wind profilers.

Balsley et al. (1988) computed averages of observed vertical velocities with a VHF profiler on Pohnpei (a tropical Pacific Island). In clear air conditions the average vertical profile of vertical velocity from about 3.5 km to 20 km altitude showed subsidence throughout (mostly within 1 cm s^{-1}) with a maximum at about 14 km. In the troposphere, this compared well with profiles obtained from conventional techniques (i.e. derived from radiosonde networks or aircraft observations) in the tropics (not necessarily in the same location). In predominantly stratiform rainfall conditions their profile showed positive values with a maximum of about 15 cm s^{-1} near the tropopause. This also broadly agreed with conventional profiles in tropical stratiform anvils (see discussion in Chapter 6) except that the conventional techniques indicated a maximum at a lower altitude[24]. In the case of convective precipitation, however, they obtained increasing positive values up to 180 cm s^{-1} while conventional techniques in the Atlantic had indicated a smaller maximum at a lower altitude. (They attributed the difference to differences between Atlantic and Pacific convection). The value of 180 cm s^{-1} is much smaller than the upward velocities common in convective storms. It is apparently due to averaging over one hour periods; high vertical velocities do not persist for the entire hour. Vertical velocities in the convective, stratiform and transition regions[25] of tropical mesoscale convective systems in the Australian region were analysed using data from a 50 MHz profiler by Cifelli and Rutledge (1994). Jain et al. (2000b) and Narayana Rao et al. (2001b) have reported vertical velocities of several m s^{-1} with peaks in the upper troposphere in convective situations in India. During convection, vertical velocity may vary within the beam and the radar may be smoothing out the variations.

---

[24] It should be noted that in the presence of precipitation the clear air echo should be distinguished from the precipitation echo (see section 10.5) before computing vertical velocity.

[25] See Chapter 6 for an explanation of the convective, stratiform and transition regions.

A small error in the pointing of the vertical beam can result in a component of the horizontal wind appearing in the vertical beam. Since the u and v components are usually far larger than the w, the error in measurement of the latter will be large. This is especially the case with long-term averages (Huaman and Balsley 1996). The pointing of the vertical beam can be checked with solar observations (see Chapter 5) or better with a radio star (Chakravarty et al. 1993). Another source of error is due to echo return by Fresnel reflection from a stratified layer tilted with reference to the horizontal. A strong echo of this type can come from within the beam width and make the effective angle of the beam tilt to one side of vertical. This was found to be the main source of error in vertical beam measurements of vertical motion by Larsen et al. (1991) who compared them with data obtained by VAD analysis of 15 oblique beams using the MU radar in Japan. Larsen and Röttger (1991) devised a method of correcting the vertical velocity for this error by measuring the incidence angles with a spaced antenna mode of observation. Palmer et al. (1991) (see also Larsen et al. 1992) verified the bias using a Radar Interferometry (RI) technique i.e. comparing the vertical velocity obtained from the vertical beam with that calculated from measurements in two oblique beams 180° apart in azimuth. Differences between the two are attributed to aspect sensitivity effects.

10.3.8 915 MHz and 1357 MHz Profilers

A low-cost 915 MHz profiler was designed in the USA for lower tropospheric profiling (Ecklund et al. 1988). At this frequency, the far field can be formed at a range of about 100 m and the system recovery time also can be kept low. Short pulse necessary for high resolution can be used, as the required receiver bandwidth is easily achievable at this frequency. Ecklund et al. used a small printed "microstrip" antenna enclosed in a fence to reduce ground clutter. The transmitter-receiver is mounted under the antenna and connected to the IF receiver and processor units indoors. This is a low power system and is also portable. It is relatively inexpensive and can be operated automatically even at remote locations with minimal maintenance. Profilers of this type have been installed at a number of sites in the tropical Pacific as part of various research projects. Typical specifications are given in Table 10.5.

The profiler was mounted on board a ship on a gyro-stabilised platform (Carter et al. 1992). The observations in this case were corrected for ship movement, using a satellite Global Positioning System (GPS). The wind data

Table 10.5. Manus Island 915 MHz wind profiler parameters (from Williams et al.1995; courtesy AMS).

Parameter	Value
Wavelength	32.8 cm,
Peak Power	500 W
Antenna type	Microstrip phased array
Antenna beams	3 beams (One vertical; two 21° from zenith)
Antenna steering	Electrical,
Antenna size	1.8 × 1.8 m,
Beam width (two-way)	6.4°
Range gate separation	105 m (low mode), 255 m (high mode)
Maximum height sampled	5.2 km (low mode), 12.6 km (high mode)
Range resolution	105 m (low mode), 495 m (high mode)
Max. unambiguous velocity	11 m s^{-1}
Spectral points	64
Spectral averages coherent averages	60
Dwell time	~ 30 s
Data recording	900 Mb optical disk.

were compared with balloon soundings taken at some points on the ship's route. This profiler may be suitable for operational use both over land and sea.

In Japan a Lower Atmospheric Wind Profiler (LAWP) operating at 1357.5 MHz has been developed for studies of the boundary layer. It can also be used for precipitation studies. The transmitted peak power is 1 kW distributed to a phased array antenna consisting of 4 blocks of 12×12 element arrays. Using phase shifters the beam of 4° width is tilted 15° to the north and using phasing cables it is tilted 15° east. The three beam cycle is repeated continuously. Vertical profiles of horizontal wind are displayed in near real time up to a maximum height of about 6 km. There is a choice of three pulse widths. There is also an associated RASS system (see section 10.6) for temperature measurements. Using collocated LAWP and MST radar, Krishna Reddy et al. (2002) found that r.m.s. differences in wind speed measurements by the two instruments were within 1.6 m s^{-1}.

10.4 Derived Parameters from ST Radars

10.4.1 Divergence

We may recall the equation of continuity (3.35) renumbered here as (10.22),

$$\frac{\partial u}{\partial x}+\frac{\partial v}{\partial y}+\frac{\partial w}{\partial z}=-\frac{1}{\rho_a}\frac{d\rho_a}{dt} \tag{10.22}$$

where u, v, w, are the wind components, ρ_a is air density and t represents time. In conditions in which the r.h.s. of equation (10.22) can be neglected, we have,

$$\frac{\partial w}{\partial z} = -\left(\frac{\partial u}{\partial x} + \frac{\partial v}{\partial y}\right) \tag{10.23}$$

The l.h.s. is the vertical gradient of vertical velocity and it is equated to the horizontal divergence on the r.h.s. From a vertical profile of vertical velocity measured with the vertical beam of a wind profiler, $\partial w/\partial z$ is available and the divergence is determined. This is the reverse of the usual determination of vertical velocity from measurements of horizontal wind. Clark et al. (1986) obtained from this, a vertical time section of divergence over their radar and could detect the passage of a cold front. In the case of a network of profilers, the divergence and vorticity can be computed in the conventional manner (i.e. as from radiowind observations) from measurements of the horizontal wind. Using a network of three VHF radars in Colorado, USA, Zamora et al. (1987)[26] computed divergence as well as the area-averaged geostrophic and ageostrophic wind. Narayana Rao et al. (2001b) have computed convergence and divergence of up to 50×10^{-4} s^{-1} in the lower and upper troposphere respectively.

10.4.2 Momentum Flux, Turbulence and Gravity Waves

The deduction of turbulent air motion in a snowstorm from the fluctuations of the VAD sine pattern was mentioned in Chapter 6. Similar observations with particle echoes (not precipitation) were made by Kropfli (1986) with microwave radar. These fluctuations attributed to turbulence were computed from an ensemble of scans and the vertical flux of horizontal momentum was derived.

With VHF MST radar, Fukao et al. (1988c) determined the vertical flux of horizontal momentum in the upper troposphere and lower stratosphere, by a method originally proposed by Vincent and Reid (1983). The instantaneous radial Doppler velocities V_1 and V_2 in a fluctuating wind field in two opposite oblique beams (say, east and west), at zenith angle $\psi\,(=90-\theta)$ are, following equation (10.2.),

$$V_1 = u_1 \sin\psi + w_1 \cos\psi \tag{10.24}$$

$$V_2 = -u_2 \sin\psi + w_2 \cos\psi \tag{10.25}$$

where u_1, u_2, w_1, w_2 are the instantaneous horizontal and vertical wind

[26] See also Lemaître et al. 1994.

components. Denoting the fluctuating part of each velocity (i.e. instantaneous minus mean value) by a prime, they give the vertical flux of horizontal (zonal in this case) momentum per unit mass as,

$$\overline{u'w'} = \frac{\overline{V_1'^2} - \overline{V_2'^2}}{2\sin 2\psi} \tag{10.26}$$

the overbars indicating averaging. The meridional momentum flux $\overline{v'w'}$ can be determined by using the north and south beams. Fukao et al., give an expression for momentum flux for the case where a vertical and two oblique beams are used. Using this method with the Indian MST Radar, Narayana Rao et al., (1997) determined the vertical momentum fluxes up to an altitude of 20 km. They obtained values in the range 0.7 to − 0.6 m^2 s^{-2} for the zonal momentum flux and 0.7 to − 0.3 m^2 s^{-2} for the meridional. Peters and Kirtzel (1994) have used a similar method of deriving momentum flux but by using a Radio Acoustic Sounding System (RASS; see section 10.6.) signal instead of the natural clear air signal.

The turbulent energy dissipation rate ε was discussed in Chapter 6 in the context of convective storms. In clear air, it can be computed from the measured $C_n^2(r)$ [see Chapter 2 for definition of $C_n^2(r)$]. Hocking (1985) gives, for a dry atmosphere,

$$\varepsilon = \gamma \left[C_n^2(r) \right]^{3/2} (T/p)^3 \tag{10.27}$$

where $\gamma = 1.08 \times 10^{22}$ for a dry troposphere, $\gamma = 3.25 \times 10^{21}$ for the stratosphere, p is pressure in hPa; T is temperature K (climatological values can be used), $C_n^2(r)$ is in $m^{-2/3}$ and ε is in W kg^{-1} or m^2 s^{-3}. Contribution of quasi-specular reflection to the received power must be removed before computing $C_n^2(r)$. This determination gives the mass energy dissipation rate over the radar sampling volume.

Alternatively ε can be determined from the spectrum width, in which case the value represents dissipation rate within a scattering layer. Using the vertical beam the mean square fluctuating velocity $\overline{v^2}$ can be obtained from the spectrum width (after subtracting the contribution due to beam broadening and shear broadening). Then ε is given by (Weinstock 1981),

$$\varepsilon = c\overline{v^2}N \tag{10.28}$$

where c is a numerical constant (about 0.4), and N is the Brunt-Väisälä Frequency (BVF).

The Brunt-Väisälä frequency N (units: radians s^{-1} or $N/2\pi$ Hz) is defined (in unsaturated air) by,

$$N^2 = \frac{g}{\theta}\frac{\partial \theta}{\partial z} \tag{10.29}$$

where θ is the potential temperature, z is the altitude and g the acceleration due to gravity. It is the frequency at which an air parcel will oscillate when subjected to an infinitesimal perturbation in a stably stratified atmosphere. It can be calculated from a radiosonde observation. Cohn (1995) compared the ε values obtained by the backscattered power method and spectrum width method and found good agreement in the middle troposphere despite the fact that the two methods measure rather different quantities. Narayana Rao et al. (1997, 2001c) computed the eddy dissipation rates by both these methods with the Indian MST radar. The radiosonde data of the nearest station (Chennai, 130 km away) were used for the C_n^2 method. Their values are in the range 10^{-6} to 10^{-3} m^2 s^{-3}. They claimed a good agreement between the values up to 20 km altitude.

Using the high resolution PROUST radar in France, Delage et al. (1997) determined the eddy coefficient by these two methods and also by what they call a 'thickness' method in which the statistical characteristics of the observed wind field are used. According to them the results from the three methods are reasonably similar.

Equation (10.29) can also be rewritten as ,

$$N^2 = \frac{g}{T}\left(\frac{\partial T}{\partial z} + \Gamma\right) \tag{10.30}$$

where T is the air temperature and Γ the dry adiabatic lapse rate.

Richardson's number R_i is a quantity which is used as a criterion of stability or instability in the clear air. A value less than 0.25 is taken to indicate instability. R_i is defined by,

$$R_i = \left[\frac{g}{\theta}\frac{\partial \theta}{\partial z}\right]\left(\frac{\partial v}{\partial z}\right)^{-2} \tag{10.31}$$

where v is the horizontal wind. The quantity in square brackets is N^2 as defined earlier (VanZandt et al. 1980). The quantity in the other brackets is the vertical shear of the horizontal wind. The latter can be determined by a profiler. The former can be obtained from radiosonde observation or from the profiler itself by the method mentioned in Paragraph 10.4.4. Hence it should be possible to detect Clear Air Turbulence (CAT) which is of practical importance as an aviation hazard.

In a stably stratified atmosphere, buoyancy acts as a restoring force for small perturbations. As a result, the atmosphere sustains a class of propagating waves known as internal gravity waves. The maximum frequency of these waves is the Brunt-Väisälä frequency already mentioned. The cause of initiation of the perturbations may be strong vertical shear, jet streams, fronts or interaction of winds with mountainous terrain. A measure of the dynamic stability is the Richardson number referred to above. These waves are detectable by a profiler radar in the form of perturbations of the *u*, *v*, *w* components.

Rüster and Czechowsky (1980) observed the shear instability generated by the strong vertical gradient of the horizontal velocity at the bottom of the jet stream, with a VHF radar. Spatial scales involved are of the order of 100 m; hence a short pulse length is necessary. Yoe and Rüster (1992) observed with a VHF radar, a downward vertical velocity of up to 0.5 m s^{-1}, below the jet stream wind maximum and upward motion above it. The wind maximum itself is detectable from a vertical time section from a single profiler or a cross section from a network of profilers. Application to aviation is obvious.

In a study with the Indian MST radar, Nagpal et al. (1994) subjected these perturbations to spectral analysis and found wave periods of several minutes in the upper troposphere in the region of a wind maximum (which they call a tropospheric jet stream). The waves appeared to propagate from a height of 9 km both above and below. Similar studies have been reported by Revathy et al. (1994) and Keshav Kumar et al. (1994).

10.4.3 Detection of Stable Layers and Tropopause

Fresnel reflection and scattering produce an echo return from horizontal layers of discontinuity in refractive index, especially at VHF. Hence, the tropopause and other inversion layers can be detected on the vertical beam. The tropopause can be identified by the height at which the echo intensity from the vertical beam increases abruptly (Gage and Green 1979). This height agreed well with the tropopause height determined by radiosonde. Gage and Green (1982b) presented an objective method for doing this, based on the Fresnel scattering model of Gage et al. (1981). This involves a comparison of the observed height variation of backscattered power with the prediction of the model for a lapse rate[27] of 2 K km^{-1}. It yields values within 300 metres of radiosonde-determined tropopause heights. Gage et al. (1986) determined the tropopause height with a low resolution radar, to make the technique suitable for routine use. Multiple thin layers of turbulence were observed in the upper troposphere and lower stratosphere with the Arecibo (UHF) radar by Sato and Woodman (1982, 1982b).

In the tropics, from vertical beam observations with the Indian MST radar,

---

[27] 2 K km^{-1} is the limit normally used to distinguish between tropospheric and stratospheric lapse rates.

Jaya Rao et al. (1994) found the tropopause at Gadanki to have a multilayered structure. On these occasions the nearest radiosonde station (Chennai) showed a single tropopause[28]. They also observed stable layers at lower heights in the troposphere which persisted for up to 6 hours. Keshav Kumar et al. (1994b) have reported similar observations at the same site. Fronts at the surface and upper levels have been located by several authors from vertical time sections or from cross sections using a network of profilers (e.g. Shapiro et al. 1984).

Stable layers and tropopause are identified from the reflectivity in the vertical beam due to Fresnel reflection or scattering. . The oblique beam gets returns from Bragg scattering. Jain et al. (2000b) found that during a period of strong convection the oblique and vertical beam reflectivities are close to each other indicating absence or weakening of the stable layer structure at the tropopause. After the convection event, the difference between the reflectivities increases indicating restoration of the stable layer.

Rama Verma Raja et al. (1999) using the Indian MST radar, observed three layers of high reflectivities, at altitudes of 5 to 6 km, 10 km and 17 km. Correspondingly they found alternate layers of upward and downward motion the depth of each layer being about 2.5 km. They attributed them to inertio-gravity waves of wavelength 5 km.

10.4.4 Vertical Gradients of Refractive Index, Temperature and Humidity

Gage and Green (1982) used the determination of tropopause height from the vertical beam as above, to assign a most probable tropopause temperature for a given station and season. Measurements of atmospheric stability permit the temperature profile to be continued from tropopause upwards. The profile is continued downward by assuming a uniform lapse rate in the troposphere.

It has been shown that the height gradient of the potential refractive index (see Chapter 2 for its definition) ϕ can be expressed in terms of radar-observed parameters as (Gossard et al.1998),

$$\left(\frac{d\phi}{dz}\right)^2 = C\frac{C_\phi^2}{C_w^2} \qquad (10.32)$$

where z is the vertical coordinate, C_ϕ^2 is the structure parameter of potential refractive index[29], C_w^2 is the structure parameter of vertical velocity w and the

[28] Multiple tropopauses are often observed by radiosonde also.

[29] Regarding the conversion of structure parameter of refractive index to that of potential refractive index, see Gossard and Strauch (1983) Appendix B, pp 251-253.

coefficient C includes quantities relating to the length scale etc. [see Gossard et al. (1998) for details]. The structure parameters C_ϕ^2 and C_w^2 are defined in the same way as C_n^2 was defined in Chapter 2, i.e. if parameter x is $x(r_0)$ at position r_0 and $x(r_0+r)$ at position r_0+r,

$$\overline{|x(r_0+r)-x(r_0)|^2} = C_x^2 r^{2/3}, \; x = n, \phi \text{ or } w \tag{10.33}$$

C_ϕ^2 can be used in equation (2.94) in place of C_n^2 and can be determined from measured reflectivity (i.e. the zeroth moment of the spectrum). C_w^2 can be determined from the spectral width (i.e. the second moment of the Doppler spectrum). Since the vertical variation of temperature and humidity both contribute to the vertical gradient of refractive index, the latter contains information on the vertical profile of temperature and humidity. If the humidity can be neglected at higher tropospheric levels and in the stratosphere, the vertical gradient of refractive index yields the vertical gradient of temperature. If a reference temperature at one altitude is available, the temperature profile can be determined.

The power spectrum of the vertical velocity at any altitude from the radar shows a peak at the Brunt-Väisälä Frequency (BVF; see Paragraph 10.4.2. for definition of BVF) (Röttger 1986). The spectrum shows a sharp cutoff on the higher frequency side and there are smaller peaks on the lower frequency side. Revathy et al. (1996) used this to identify the BVF. Using the Indian MST radar they obtained a time series of vertical velocity data at intervals of 46 seconds in the height range 3.75 to 30.6 km with 150 metres resolution. These were subjected to a Fast Fourier Transform to obtain the spectra. Fig. 10.9 shows spectra for altitudes of 3.75 km (lower troposphere) and 15 km (near the tropopause). The spectral peak indicated the BVF[30]. From the altitude profile of the BVF they determined the vertical temperature profile in the troposphere and lower stratosphere. The temperature gradient is obtained from equation (10.30). A temperature value for a reference altitude is needed to get the actual temperature profile. The reference temperature can be taken from a radiosonde observation or by using the surface temperature with an assumed lapse rate up to the reference level. The technique requires vertical velocity data over a period of about one hour; also the technique is not suitable for a convectively unstable atmosphere.

In the troposphere, if the temperature profile is known (e.g. by a radiosonde or RASS measurement; see section 10.6 for RASS), the humidity gradient profile can be determined from the refractive index gradient profile. Humidity

[30] The observed BVF is to be corrected for the effect of Doppler shifting due to atmospheric gravity waves. The procedure for this is discussed by Mohan et al. (2001).

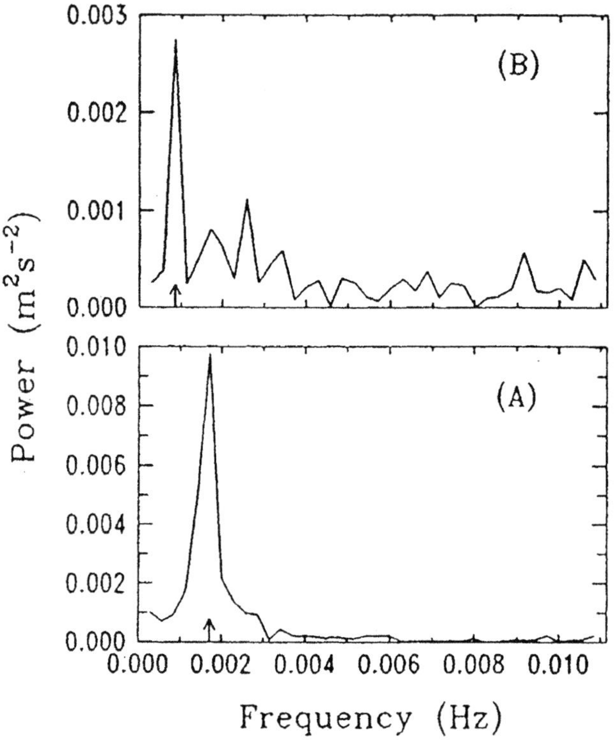

Fig.10.9. Temporal spectra of vertical wind velocity at (A) 3.75 km and (B) 15.0 km [from Revathy et al. 1996. Copyright by the American Geophysical Union (AGU). Reproduced courtesy AGU]. The arrow marks indicate the BVF.

at a reference level should be known in order to get the actual humidity profile. Also, since equation (10.32) gives the square of the vertical gradient of ϕ there will be a sign ambiguity in the humidity gradient profile. Vertical gradient of humidity is usually negative but there can be exceptions. Gossard et al. (1999, 1999b) used GPS measurements of total precipitable water to constrain the integration of the ϕ gradient[31].

Tsuda et al. (2001) using RASS to determine the temperature gradient, derived the humidity profiles from MU radar observations[32]. They expressed the observed reflectivity η in terms of the eddy dissipation rate ε, the BVF N and vertical gradient of potential refractive index M,

$$\eta = C\varepsilon^{2/3}M^2N^{-2} \qquad (10.34)$$

[31] See Chapter 11 for combined ground- and space-based techniques for temperature and humidity retrievals.

[32] Oblique beams were used in order to avoid Fresnel reflections.

where C is a constant to take account of radar parameters. ε is determined from spectral width and N^2 from a RASS temperature measurement[33]. They express M as,

$$M = -\frac{77.6\times10^{-6}}{T}\left(\frac{N^2}{g}+\frac{15600q}{T}\frac{N^2}{g}+\frac{7800}{T}\frac{dq}{dz}\right) \tag{10.35}$$

It may be noted that this expression is the same as that in equation (2.93) except that the values of the constants are slightly different from those of Tatarski (1971). Tsuda et al. omit the second term in equation (10.35) as relatively small and rewrite the equation as,

$$M = -\frac{77.6\times10^{-6}p}{T}\left(\frac{N^2}{g}+\frac{7800}{T}\frac{dq}{dz}\right) \tag{10.36}$$

From this dq/dz is computed. A boundary condition is necessary. Surface humidity measurement can be used, but since MU radar has a minimum height of 1 km for measurements, they assumed that $q = 0$ at an altitude of 10 km and integrated from 10 km downwards. They obtained a fine structure in the humidity which the radiosonde did not detect. Since equation (10.34) contains M^2, there is ambiguity about the sign of M and of dq/dz They are usually negative. Positive values occurring at cloud boundaries can be identified from radiosonde or ceilometer. Furumoto et al. (2003) added the constraint for integrated water vapor from simultaneous GPS measurement and also time-interpolated radiosonde results. They also verified that clouds passing over the radar as well as cloud top height data from satellite measurements coincided with increases in humidity shown by the radar.

Mohan et al. (2001) used the Revathy et al. (1996) method to derive the temperature profile and used it to determine the humidity profile by the Tsuda et al. (2001) technique. They obtained the sign of M from simultaneous radiosonde ascents. The radar generally yielded higher values of humidity than the radiosonde. They have stressed the need for good calibration of the radar as the constant C is involved in equation (10.34). They also point out that the neglect of the second term in equation (10.35) may not be valid at low altitudes. Clearly, the technique needs further refinement for reliable operational use.

10.4.5 Land and Sea Breeze Circulations

A 1357 MHz profiler has been developed at the Communications

[33] RASS gives the virtual temperature, which has been used in this case instead of the dry bulb temperature.

Research Laboratory in Koganei, Japan. This has been used to observe the land and sea breeze circulation. The diurnal variation of horizontal wind and that of the vertical wind observed over 2 years portrays the sea breeze circulation.

10.5 Observation of Precipitating Systems with ST Radars

10.5.1 Echoes from Precipitating Weather Systems

Green et al. (1978) observed considerable enhancement of the clear air reflectivity in a VHF radar when a convective cloud passed overhead. They attributed this to increase in turbulence and humidity within the cloud. The enhancement of C_n^2 extended in some cases well above the cloud, indicating turbulence in the lower stratosphere. Besides this, they also observed, during thunderstorms, a separate echo which they attributed to scattering from precipitation particles.

The reflectivity of precipitation echo is the same at both VHF and UHF but the clear air reflectivity depends on the wavelength. From Fig. 10.3 it can be seen that besides echo from clear air, an ST radar/wind profiler, especially one operating at UHF can get appreciable signal return from hydrometeors in precipitation. In the Doppler spectra, the fall speeds of the hydrometeors will be exhibited as a range of downward velocities depending on the distribution of hydrometeor sizes. The spectral width will be broadened depending on the drop size distribution[34]. A schematic of Doppler power spectra from the vertical beam of a UHF profiler is shown in Fig. 10.10 from Ralph et al. (1995). Fig.(a) represents a case when only clear air echo is seen, while (b) shows a precipitation echo also. Note that in this figure, downward velocities are positive. In case (a), the measured radial Doppler velocity V_r is equal to V_a, the reflectivity-weighted mean air velocity from the clear air echo. In (b) the precipitation echo is stronger than the clear air echo V_r in this case is a combination of hydrometeor terminal velocity V_t and air velocity. If the latter is determined from the clear air echo V_t can be calculated. In the figure, S is measured signal power, σ_w the spectrum

[34] Besides the spectral broadening due to the range of fall velocities, there will be additional broadening due to turbulent motion of the hydrometeors, wind shear, and beam broadening. The last of these is explained as follows. If the beam width of the radar antenna is substantial (as is the case with VHF radars) a horizontal wind across the vertical beam will produce a velocity component towards the radar at the leading edge of the beam and away from the radar at the trailing edge. This increases the observed spectrum width (Sloss and Atlas 1968).

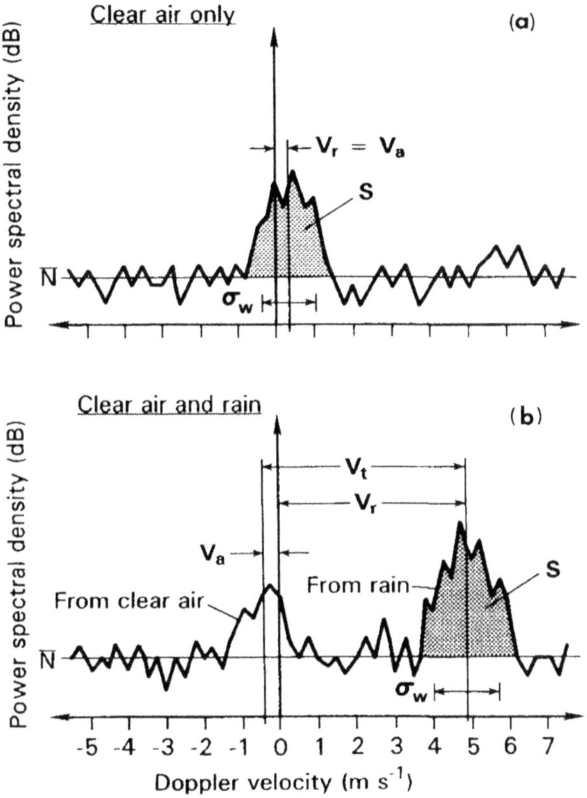

Fig. 10.10. Schematic of power spectra from a UHF profiler for (a) case with clear air echo only and (b) case with precipitation echo in addition to clear air echo (from Ralph et al. 1995; courtesy AMS). (See text Paragraph 10.5.1 for details).

width and N the background noise level. Compare this with the techniques mentioned in Chapters 2, 3 and 7 with microwave Doppler radars where several assumptions had to be made to separate V_t and the vertical velocity w.

In stratiform precipitation when the vertical air motion is small, the clear air echo will be close to the zero velocity while the particle return has a downward velocity. In a VHF radar, only the larger particles give an echo and since they have large fall velocities the precipitation echo will show up as a separate peak in the spectra and can usually be distinguished easily. At UHF however, small particles with low fall speeds will also return appreciable echo and, therefore, the distinction may be difficult. In a convective storm the up- and down-draft speeds may produce clear air echoes with various Doppler velocities. The downdrafts may therefore get mixed up with hydrometeor echoes.

S, the area under the more prominent peak in Figure 10.10 gives the signal power. For determination of moments, if the processing algorithm picks out the strongest echo and ignores the rest as is often the case, the precipitation echo being stronger may get into the calculation of moments and introduce large error in wind determination. Also, in convective storms the vertical air motion as well as particle fall speeds may vary within the volume sampled by all the beams, leading to large errors in computed moments. The consequences for horizontal wind determination have been discussed in section 10.3.

The capability of ST radars to detect precipitation particles can be exploited as a means of studying precipitating systems. Fukao et al. (1985) showed that VHF radar provides the unique facility of detecting precipitation particles and measuring their fall speeds as well as the air motions simultaneously. The profiler is thus an ideal multipurpose tool for observing weather systems passing overhead.

10.5.2 Doppler Spectra in VHF and UHF Radars

Fig. 10.11 from Larsen and Röttger (1987), shows the vertical velocity spectra in a number of range gates (each gate is 150 m) obtained from a VHF radar. In gate nos. 16 to 24 (3.5 to 4.5 km altitude) two velocity peaks are seen, the one on the right (downward; here downward is negative) being due to fall speeds of drops and the other due to air motion. Thus, by viewing the complete Doppler spectra the particle velocity can be distinguished from the air motion. However, the peaks due to the precipitation tend to merge with the turbulent scatter peak at higher altitudes (because of lower fall velocity)[35]. All the spectra show broadening beyond usual clear-air widths. The enhancement of signal in the lower troposphere can be seen from the curve on the right hand side. For comparison, Fig. 10.12 shows the case of a UHF radar where the hydrometeor signal dominates. There is a much larger enhancement of signal and a considerable broadening of the spectrum but no evidence of a separate clear air signal. Figs. 10.13, from Narayana Rao et al. (2001) shows the Doppler spectra in convective, transition and stratiform precipitation spectra observed by a Lower Atmospheric Wind Profiler (LAWP) operating at 1357 MHz. Fig. 10.14 shows similar spectra[36] with the 53 MHz MST radar, one Hz of Doppler shift corresponding to 2.83 m s^{-1}.

---

[35] Note also that the turbulent scatter echo (at VHF) in the figure is stronger than the precipitation echo at the same level. This is generally true except at the melting band where the precipitation echo is stronger. However Chu and Lin (1994) report a case in Taiwan where the turbulent scatter echo was about 15 dB weaker than the precipitation echo even below the bright band. They attribute this to entrainment of cool and dry air in the presence of intense updrafts, reducing the variance of atmospheric refractive index.

[36] The two figures are not for the same occasion.

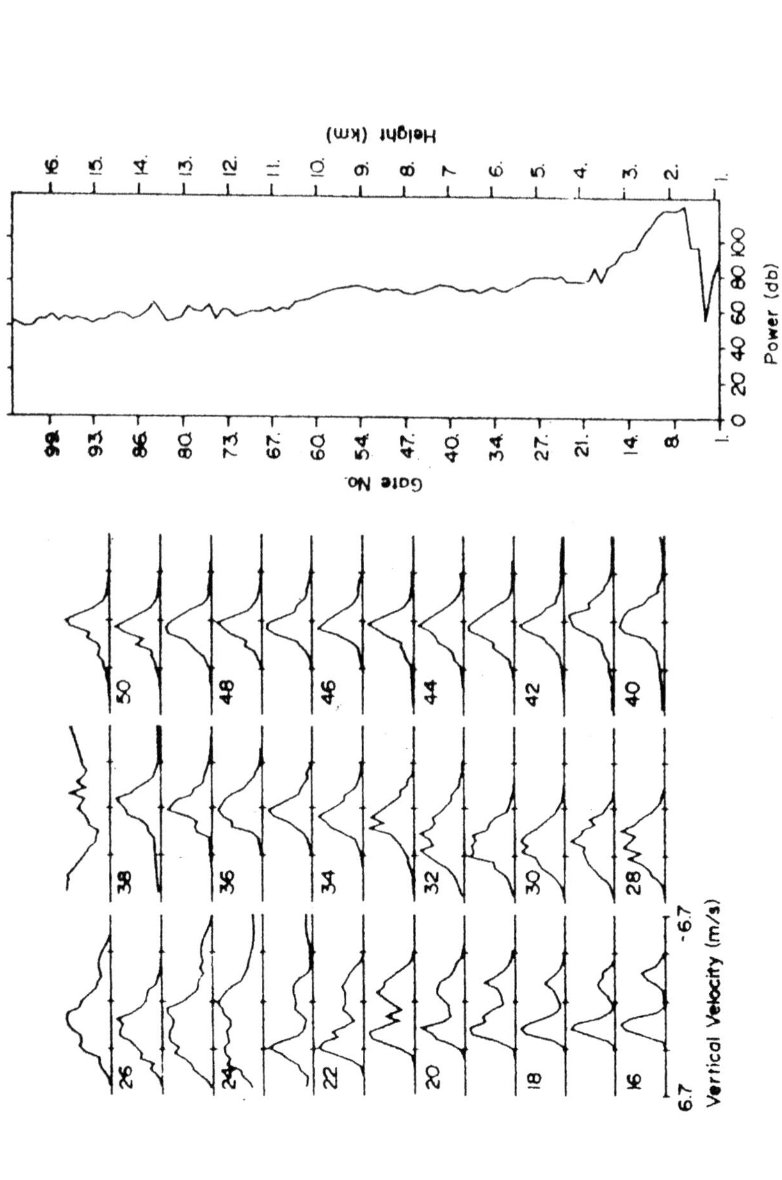

Fig. 10.11. Doppler spectra from a VHF radar exhibiting precipitation echo (from Larsen and Röttger 1987; courtesy AMS). At left: vertical velocity profile in range gate numbers 16 to 50 (each gate 150 m). Gates 16 to 24 show double-peaked spectra. The one at right is due to falling particles, the other is due to air motion. At right: Signal power profile.

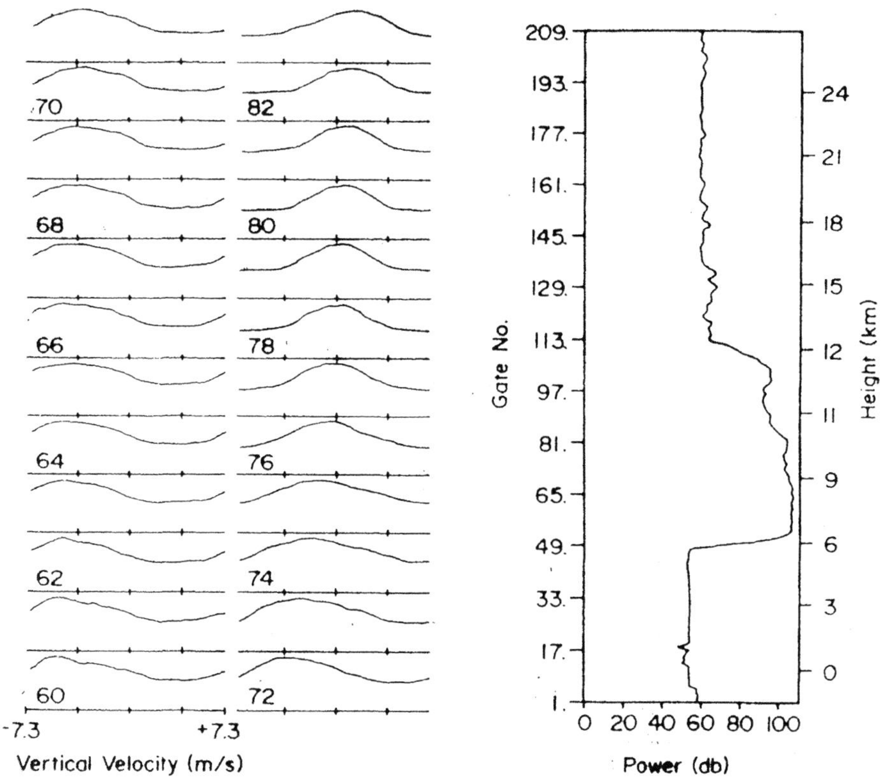

Fig.10.12. Doppler spectra from a UHF radar exhibiting precipitation echo (from Larsen and Röttger 1987; courtesy AMS). At left: vertical velocity profile in range gate numbers 60 upwards. Note that negative (downward) velocities are on the left. At right: Signal power profile.

Profilers thus have the advantage over scanning microwave radars that the vertical air motion and hydrometeor motion can be determined separately The assumptions made in Chapter 3 in order to separate vertical air velocity and hydrometeor fall speed are not necessary. The Drop Size Disribution (DSD) in precipitation can also be deduced by VHF radar by a method similar to that of Paragraph 7.5.4. Various other properties of the precipitation such as melting band structure and *Z-R* relationships have been studied using VHF or UHF radars or a combination of the two.

UHF profilers are more useful for studies of precipitating clouds because Rayleigh scattering from hydrometeors dominates at the higher frequencies. Ecklund et al. (1999) have developed a 3-GHz profiler which is stated to give a 20-dB improvement in sensitivity over the 915 MHz profiler.

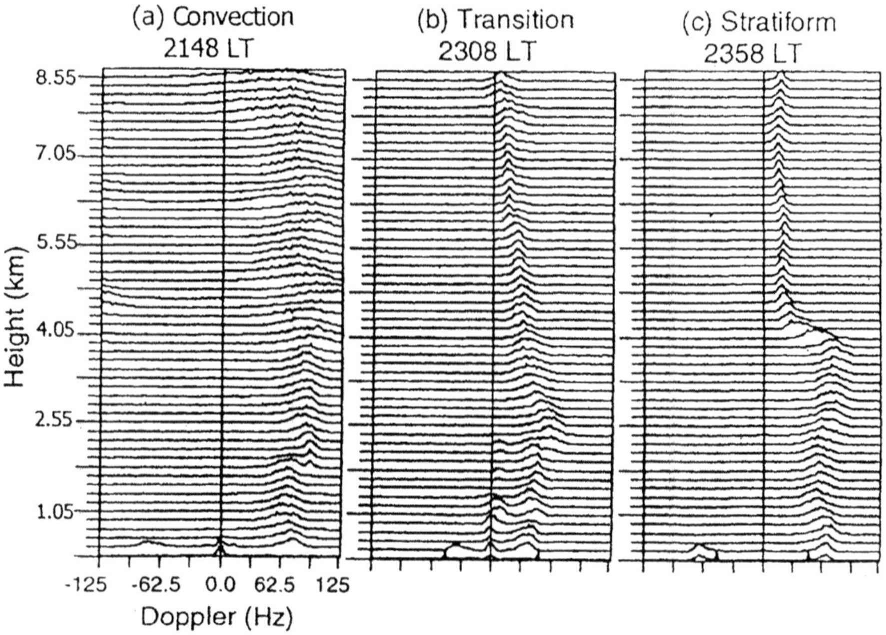

Fig. 10.13. Doppler spectra from 1357 MHz wind profiler in (a) convective (b) transition and (c) stratiform regions of precipitation [from Narayana Rao et al. 2001. Copyright by the American Geophysical Union (AGU). Reproduced courtesy AGU]. The X-axis shows Doppler shifts in Hz. One Hz corresponds to a Doppler velocity of 0.11 m s^{-1}.

10.5.3 Identification of Precipitation Echo and Determination of Drop Size Distribution

But as a first step it is important to identify and separate the clear air and hydrometeor echoes. Wakasugi et al. (1986, 1987) considered the Doppler spectrum from the vertical beam of a VHF radar to be the sum of the spectra $S1(w)$ and $S2(w)$ due to hydrometeors and refractive index irregularities respectively. The former is influenced by the response of the hydrometeors to turbulence. If there is a mean vertical air velocity, the spectrum is shifted by that amount (see Fig. 10.10 and also Paragraph 7.5.4). They fitted a spectrum based on the Marshall-Palmer distribution to the observed precipitation component of the spectra and a Gaussian curve to the turbulence component. They deduced $S1(w)$ from the observed spectra and derived the DSD from an equation similar to (7.64). Gossard (1988) used a similar method with UHF radar. Sato et al. (1990) automated this procedure.

Narayana Rao et al. (1999) attempted the separation of the clear air and precipitation echoes by two methods (1) assuming that the point of minimum

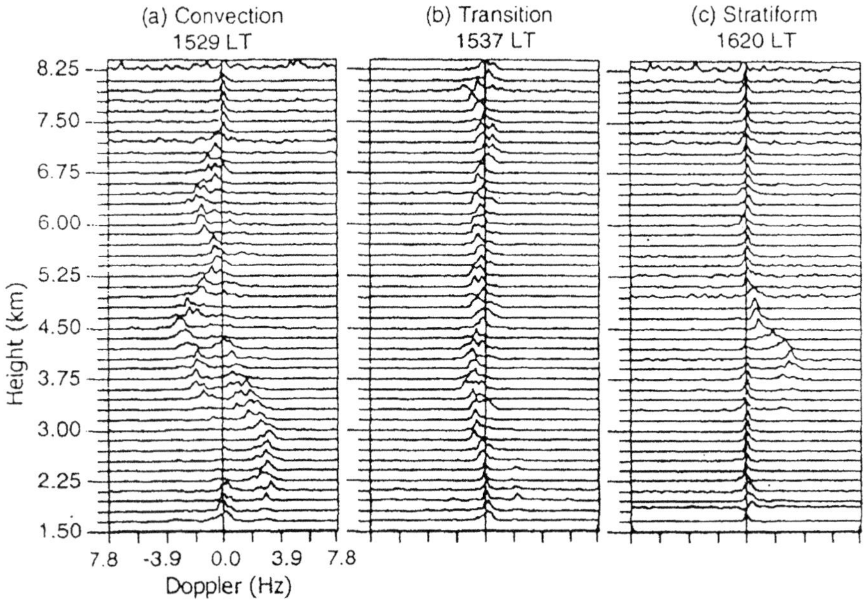

Fig. 10.14. Doppler spectra from 53 MHz MST radar in (a) convective (b) transition and (c) stratiform regions of precipitation [from Narayana Rao et al. 2001. Copyright by the American Geophysical Union (AGU). Reproduced courtesy AGU]. The X-axis shows Doppler shifts in Hz. One Hz corresponds to a Doppler velocity of 2.83 m s^{-1}.

between the two peaks represents the boundary between the spectra and (2) fitting a Gaussian distribution to both the echoes. Fig. 10.15 (a) shows the first method while Fig. 10.15(b) and (c) show Gaussian fits to the clear air echo and precipitation echo respectively. Rajopadhyaya et al. (1993) proposed a method not involving any particular assumption about the shape of the DSD. It was noted in Fig. 10.10, that the signal peaks due to precipitation and turbulence tend to merge at higher altitudes. This may result in the algorithm merging the two signals and fitting a single Gaussian curve with its centre well to the negative (downward) velocity side of the turbulence peak. Cifelli and Rutledge (1994), found that the merging of the two peaks occurred most in the lower troposphere in the convective part of tropical mesoscale convective systems and in the melting layer in the stratiform anvil. In such cases the peaks could be distinguished by manual intervention.

To overcome the difficulty, especially in the case of UHF radars, of separating the precipitation and turbulence echoes, some authors have used dual-frequency observations at VHF and UHF. The former yields the vertical wind in the sampling volume while the latter gives the precipitation signal [Chilson et al. (1993), Currier et al. (1991), Maguire and Avery (1994)]. Frequencies of 915 MHz and 3 GHz have also been used for this purpose

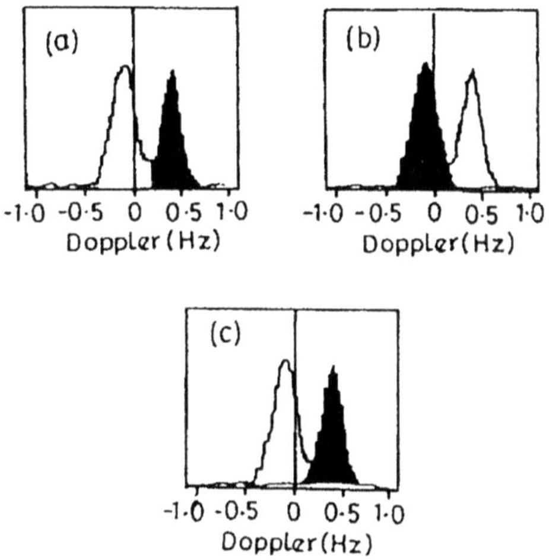

Fig. 10.15. Separation of clear air and precipitation echoes. (a) minimum point in between is taken as boundary between the two (b) fitting Gaussian spectrum to clear air echo (c) fitting Gaussian spectrum to precipitation echo [from Narayana Rao et al. 1999. Copyright by the American Geophysical Union (AGU). Reproduced courtesy AGU]. The shaded area represents the precipitation echo and the open area the clear air echo.

[Ecklund et al. (1999), Gage et al. (1999), Williams et al. (2000)]. Since the equivalent reflectivity factors (Z_e) for Bragg scatter[37] are different at the two frequencies and the Rayleigh reflectivity factors are nearly the same, thresholds of difference in the equivalent reflectivity factors at the two frequencies can be set up. For example, for clear air, the Z_e difference between 915 MHz. and 3 GHz is of the order of 18 dB and for hydrometeors the difference is within about 2 dB. The use of two profilers has another advantage. The VHF profiler does not resolve small drops from the combined clear air and precipitation return. Hence the median volume diameter derived from it tends to be biased high (Rajopadhyaya et al. 1999). The VHF/UHF combination gives an unbiased DSD. However a dual-frequency profiler, while feasible for research purposes, is not convenient in operational work.

Rajopadhyaya et al. (1994) have described a technique to retrieve size information on ice particles from 50 MHz profiler data from the trailing stratiform region of a squall line in Australia, assuming an exponential distribution. The method is basically similar to that of Wakasugi et al. (1986) discussed above.

Richter (1993) used the fluctuations of vertical sound velocity obtained from a Radio Acoustic Sounding System (RASS; see section 10.6) in conjunction with a 1.235 GHz FM-CW radar to estimate vertical wind

[37] Equivalent reflectivity factor (Z_e) for Bragg scatter is defined by equation (2.97).

fluctuations. These were used to correct measured Doppler radar spectra in precipitation. The DSD, rainfall rate and liquid water content were then obtained from the spectra.

Operational wind profilers usually do not provide the complete Doppler spectrum but only the three Doppler moments (signal power, mean radial Doppler velocity and spectral width). Even the signal power is not usually calibrated to yield reflectivity factor as the latter is not needed for routine wind computation. Gossard et al. (1992) used the spectral moment data from a 915-MHz profiler but had access to occasional complete spectra. They used the latter to identify and remove the up- and downdrafts thus getting the fall velocity. They could compute Z from signal power. Assuming a shape for the DSD (not necessarily exponential) they devised what they call a Z-R-V_f diagram (V_f standing for fall velocity) on which the observed Z and V_f could be plotted.

The diagram can be used, according to them, to monitor (1) cloud-layer structure above lower overcasts that preclude optical observation, (2) height profiles of cloud population drop size continuously, (3) the transition from ice to water (and with RASS, comparison with the 0° isotherm) and (4) height profiles of liquid water flux (rain). Identification of hydrometeors and their evolution in time and height will probably be possible.

Ralph and co-authors in NOAA, USA, have studied (Ralph 1995, Ralph et al. 1995, 1996) the identification of precipitation echoes and derivation of precipitation parameters from the routine spectral moment data available from the US Wind Profiler Demonstration Network. This is important for operational use of the data, as the inspection of spectra relied upon by Gossard et al. (1992) is not possible operationally. Vertical profiles of range-corrected signal-to-noise ratio (SNR), which is proportional to reflectivity, in convective, transition and stratiform regions, from a 1357 MHz profiler are shown in Fig. 10.16 (a) from Narayana Rao et al. (2001). Doppler velocity and spectrum width profiles are shown Figs. 10.16 (b) and (c) respectively. The reflectivity peak at the bright band, the low fall velocities in the solid precipitation above the melting layer and the rapid increase in the vertical velocity at the melting layer are all clearly seen in the case of stratiform precipitation[38]. The convective region has higher SNR's extending to greater heights, higher fall velocities and higher spectrum width. The transition region shows intermediate values. Such profiles can be used to distinguish between stratiform and convective precipitation. Vertical time sections of these parameters help to watch their evolution.

Ralph (1995) and Ralph et al. (1996) established thresholds of vertical

[38] Compare with Figures 3.16(a) and 4.5.

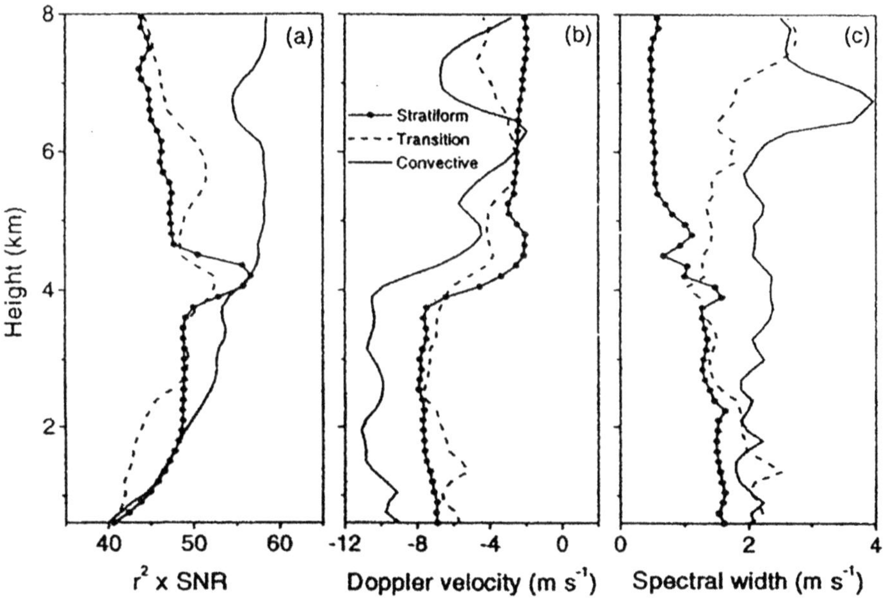

Fig. 10.16. Vertical profiles of range-corrected signal-to-noise ratio ($r^2 \times$ SNR), Doppler velocity and spectral width observed with LAWP [from Narayana Rao et al. 2001. Copyright by the American Geophysical Union (AGU). Reproduced courtesy AGU]. The solid line, dashed line and the line with circles represent respectively the convective, transition and stratiform regions.

velocity[39] for hydrometeor echoes. Correlations between reflectivity (represented by range-corrected SNR) and vertical radial velocities were used by them, to separate air and precipitation motions and to identify the type of precipitation.

10.5.4 Melting Band Structure

Chu et al. (1991) studied the melting band structure during the passage of a typhoon over Taiwan using the Chung-Li VHF radar. They found that
(1) while the precipitation echo was generally weaker than that due to refractivity turbulence it is 5 dB more than the latter at the melting level.
(2) the melting band is not detectable in the refractivity echo power profile.
(3) the fall speeds of rain drops were about 6 to 8 m s^{-1} while those of ice particles were about 1 m s^{-1} consistent with other observations (see Chapter 4).
(4) echo power profiles observed in vertical and oblique beams showed aspect

[39] Chu et al. (1999) studied the fall velocity versus range-corrected SNR relationship (see fall velocity-reflectivity relationship in Chapter 2) and found that they break down within the melting layer.

sensitivity in the case of precipitation echoes but not in the case of refractivity echoes.
(5) there was a high correlation between the horizontal wind speed and the Doppler spectral width of precipitation signal.

The last result was attributed to the beam-broadening effect of horizontal wind. Chu et al. (1991) point out that this contribution to spectral width should be considered while analysing precipitation data.

Chu and Song (1998) found that the peak reflectivity at the melting level was absent at times of convective rainfall with appreciable vertical motion (an expected result which is shown in Fig. 10.16). Chu and Lin (1994) and Chu and Song (1998) found a depletion of the clear-echo power as well as the intensity of the bright band when there is a large updraft. They attributed it to turbulent mixing between warm and humid in-cloud air and cool and dry ambient air entrained into the cloud following a strong updraft.

The work of Fabry and Zawadzki (1995) using vertically pointing X-band radar and UHF profiler has been referred to in Chapter 4. Narayana Rao et al. (1999) found the thickness of the bright band to be positively correlated with reflectivity; the average thickness was 900 metres.

The Doppler spectrum of stratiform precipitation in UHF/microwave radars is sometimes bimodal i.e. there are peaks at two different velocities. Ekpenyong and Srivastava (1970) explained the peak at the lower Doppler velocity as due to small particles which have completely melted and large ones which have melted little and have not speeded up substantially. The peak at higher velocity is due to medium size particles which contribute much of the reflectivity and have acquired high speed because of their size and melting[40]. At VHF, the bimodal spectrum is usually not seen perhaps because the small particles are not detected by the radar.

White et al. (2002) using a Doppler wind profiling radar, found that the bright band height is, on average, 192 m lower than the melting level measured with rawinsondes. They developed an objective algorithm to detect the bright band height from profiles of radar reflectivity and Doppler vertical velocity in near real time. The information was used in river forecasting.

10.5.5 Z-R Relationship

Knowing the DSD and terminal velocity, the rain rate can be computed. Rain rates determined by radar have been compared with rain gauge

[40] This seems to have been qualitatively anticipated by Dyer (1968).

measurements at the same site (e.g. Sato et al. 1990). Hence Z-R relationships have also been derived. Chu and Song (1998) have attempted a simpler method. The received power P_s from the precipitation echo at various heights was measured. This is related to Z by equation (2.62).

$$P_s = \frac{C \cdot |K|^2 \cdot Z}{r^2} \tag{10.37}$$

If we put $P_s r^2 = P_r$ and since $Z = AR^b$ we get

$$P_r = aR^b \text{ where } a = C|K|^2 A \tag{10.38}$$

(All quantities above except P_s, P_r and a are as defined in Chapter 2). Therefore

$$\log P_r = \log a + b \log R \tag{10.39}$$

The rain rate R at the ground was measured by an optical rain gauge[41]. The coefficient A and exponent b could then be determined from the observed P_s at various levels and R at the ground. They found that the value of b decreased from about 1.3 at 2.4 km altitude to about 1.0 at 3.9 km. Value of a did not vary appreciably with height. Using these values, rain rates can be derived from subsequent radar observations, but probably more data are needed for establishing *Z-R* relationships for various situations.

Rainfall estimate is possible, of course, only for the radar site, but with a network of profilers a number of useful observations can be collected over an area.

The DSD and rain rate at the ground can be determined, independently of the radar, by a disdrometer and values of Z can be computed. Using the disdrometer-derived Z and R values and using the radar only to identify the type of precipitation, Narayana Rao et al. (2001) plotted *Z-R* regression lines for convective, transition and stratiform precipitation and determined the coefficient and exponent of the relationship for each region. Their values are given in Table 7.2.

10.5.6 Classification of Precipitating Systems

Using the tropical Pacific 915 MHz profilers (see section 10.3) a method of classification of tropical precipitating systems has been devised

[41] Chu and Song also monitored the time difference between the precipitation echo at various levels and surface precipitation from the gauge and took these into account (assuming that the same precipitation reaches the gauge without horizontal advection).

(Gage et al. 1994b, Williams et al., 1995). The profiler-based study identifies four regions called, (1) deep convective, (2) stratiform, these two corresponding to the convective and stratiform regions respectively in the Leary-Houze model, (3) a mixed stratiform/convective category and (4) a shallow convective category.

If a melting band signature, as discussed earlier, is observed by the profiler and there is also enhanced turbulence above the melting layer, the system is identified as category 3. If the latter is not present, the category is (2) i.e. stratiform. If there is no melting band and there are hydrometeors extending above the melting level it is termed deep convection (category 1). If there are no hydrometeors above the freezing level, it is assigned category (4) shallow convection (warm rain). The identification of the melting band is itself based on the Doppler velocity gradient and maximum spectral width being above or below certain preset thresholds. Only the spectral moment data need be used and not the complete spectra. In a similar way, Cifelli et al. (2000) in their study of tropical mesoscale convective systems (MCS) near Darwin, Australia, using dual-frequency (UHF and VHF) wind profiler technique, partitioned the data into a three-tier classification scheme (i.e., convective, mixed convective-stratiform, and stratiform). They determined DSD characteristics and developed Z-R relations for each precipitation category as a function of height using linear regressions to the profiler retrievals of R and Z. The rainfall decreased for a given reflectivity as the precipitation type changed from convective to stratiform. According to them, the profilers which can observe the evolution of the drop size distribution with relatively high vertical resolution, have the potential to serve as calibration tools for scanning radar estimates of surface rainfall using empirical Z-R relations.

Also using a 915 MHz profiler, Ecklund et al. (1995) studied tropical convective systems. The moments were calculated using the strongest spectral peak. At this frequency, the precipitation echo was the stronger one except at very low levels. When precipitation was present, there was an appreciable increase in the maximum height range of the profiler. Winds could be computed using the precipitation signal. Precipitation could be identified from the vertical velocities and also from the reflectivity profile. They found many cases of precipitation echo well above the freezing level without rain at the surface. They have discussed the implications of the high level anvils for heating rates in the troposphere (see also Chapter 6).

Comparing a 404 MHz profiler with a Ka band radar, Orr and Martner (1996) have shown that the former can detect non-precipitating or weakly precipitating clouds under certain circumstances that can be approximately delineated by profiler reflectivity and vertical velocity thresholds.

10.5.7 Calibration for Precipitation Studies

As mentioned in Paragraph 10.2.4, absolute calibration of the receiver is not important for wind profiling. Wind profiler outputs usually give the range-corrected signal-to-noise ratio, which is a measure of reflectivity. Hence reflectivity is given in arbitrary units, which is useful for comparisons of values. However, for precipitation studies it is necessary to have the absolute value of received power. A known signal (typically – 110 dBm) from a standard signal generator can be injected at the input to the receiver [usually a Low Noise Amplifier (LNA)] during the entire observation period. The signal is placed at a range and a Doppler bin where other echoes are not expected. The amplitude of this signal can be monitored or printed out in arbitrary units as in the case of the clear air and precipitation echoes. The absolute signal power of the precipitation echo at the receiver input can therefore be computed[42], assuming that the receiver response is linear[43] throughout its dynamic range. This needs to be corrected for the path losses from antenna to receiver. For calculating the reflectivity factor Z it is necessary to have accurate values of all the radar constants and the path loss from transmitter to antenna. If there are multiple transmitters it is possible that some transmitters may not be working at a given time. Hence the actual transmitted power has to be used in the calculation.

Reflectivity factor obtained from the profiler at the lower range gates can be compared with that computed from DSD measured by a collocated disdrometer. The lowest one or two gates cannot be used because of ground clutter and receiver recovery time considerations. If data at an altitude of say 500 m from a UHF profiler are used, a time lag has to be applied to the profiler data to allow for the time of fall of the drops to the disdrometer at the ground. The sampling volume of the disdrometer depends on the fall speed of the drops and may be of the order of 1 m^3 for a 30-second sample while the profiler sampling volume may be up to 10^6 times that value depending on the altitude of the range gate used. However, Gage et al. (2000) showed that the two reflectivities agree reasonably well. Although, disdrometers have some limitations such as insensitivity to very small drops, the D^6 dependence of Z mitigates the error involved. Gage et al. (2000) used the difference between the two to calibrate the profiler. Note that only reflectivities are being compared and not rain rates and therefore no Z-R relations are involved.

Scanning radars have still larger sampling volumes which are difficult to compare with a single rain gauge or disdrometer. As seen in Chapter 7, a network

[42] If any STC or other attenuation is applied at near ranges, that is to be taken into account.

[43] Linearity is to be verified as in the case of scanning radars (see Chapter 5).

of gauges is used for that purpose. Gage et al. (2000) proposed that the profiler, which has an intermediate sampling volume can be used as a transfer standard to calibrate reflectivity measurements from scanning radars, i.e. the profiler is calibrated with reference to the disdrometer and in turn calibrates the scanning radar if it is not too far away from the profiler.

10.6 Radio Acoustic Sounding System (RASS)

10.6.1 Sound Velocity and Temperature

The measurement of the velocity of sound has been used as a means of determining upper air temperatures long before the advent of radar. The relationship between the virtual temperature T_v (K) of air at any point and the sound velocity C_a (m s^{-1}) in air at that point is,

$$T_v = \left(\frac{C_a}{20.047} \right)^2 \tag{10.40}$$

Equation (10.35) is obtained by substituting suitable values of the constants in the theoretical relationship

$$C_a = \left(\frac{\Gamma RT}{M} \right)^{0.5} \tag{10.41}$$

where Γ is the ratio of the specific heats of air, R the gas constant, T the temperature and M the molecular weight of air.

The principle of the Radio Acoustic Sounding System (RASS) is to track a propagating acoustic wavefront by Doppler radar to determine C_a (May et al. 1990). The technique has been studied from the sixties but its implementation has become practicable with the advent of the profiler radar and modern digital processing methods since the seventies [e.g. North et al. (1973)[44], Frankel et al. (1977)]. While only a brief treatment is given here, a more detailed discussion may be seen in Doviak and Zrnić (1993) and the other references cited in this section. A review of the application of RASS for study of the lower atmosphere may be seen in Singal and Goel (1997).

---

[44] The first successful attempt by North et al. (1973) used a 8.15 m radar of 5 watts output with an acoustic system operating at a wavelength of 4.07 m with 70 watts output.

10.6.2 Implementation of RASS

A short pulse of sound at a wavelength approximately half of the radar wavelength is transmitted upwards from the radar site by an acoustic source. For example, if radar frequency is 405 MHz the acoustic frequency used will be 880 Hz. The pressure fluctuations due to the sound wave generate a variation of radio refractive index along the path of the sound beam. These perturbations can be detected by the vertical beam of the ST radar in the form of a Doppler velocity spectrum from which the sound velocity can be determined. The vertical air velocity is neglected or is determined in a separate observation with the radar and subtracted. To get an efficient backscatter the sound wavelength should be half the radar wavelength and the directions of propagation should be the same. The sound wavelength for a given frequency will vary with altitude because of the temperature variation. It is, therefore, necessary to transmit a range of frequencies to satisfy the wavelength condition at various altitudes. The sound direction may change because of refraction and also because of air motion (horizontal winds and turbulence). The acoustic transmitter can be located on the windward side of the radar antenna so that the sound wave front comes over the radar beam at higher altitudes.

Instead of pulsed acoustic transmission, continuous white noise can be transmitted. Spectral analysis of the radar signal retrieves the acoustic frequency which matches with half the radar wavelength (Bragg matching). Since the latter is known, C_a can be determined.

RASS was developed initially for boundary layer measurements. Because of strong attenuation of sound at the frequencies used, and the spread of the backscatter region due to horizontal wind it was not possible to make measurements at altitudes of more than a km or two. However Matuura et al. (1986) used an acoustic frequency of around 100 Hz at which attenuation is low, along with a radar of matching wavelength (50 MHz; the MU radar in Japan) and succeeded in getting temperature data up to 20 km altitude. A wide acoustic beam (45°) was used and to keep the acoustic wavefront in the radar beam, the acoustic transmitter was moved to three appropriate positions within a few hundred metres of the radar on the windward side of the radar antenna. The radar beam direction was swept over a zenith angle range of 10° on either side.

Masuda (1988) studied the effect of wind and temperature on the maximum height attainable by RASS. It was found that with fixed beams the maximum sounding height is inversely proportional to the square root of the wind velocity gradient and directly to the square root of radar antenna diameter. A moveable acoustic source and a steerable radar beam enable much greater heights to be attained. Maximum sounding height is lower in windy conditions than in calm conditions.

Early RASS measurements neglected the vertical velocity (not more than 1 m s^{-1} in most cases) in comparison with the sound velocity (about 350 m s^{-1} near the surface). Alternatively, the vertical velocity was measured in a separate observation and used to correct the sound velocity. Weber et al. (1992) stressed the importance of the two measurements to be made simultaneously. This has been achieved by Angevine et al. (1994) using a 2048-point FFT that could encompass both velocities while maintaining full resolution of the spectrum. Thus, both could be measured simultaneously and the sound velocity corrected. While this is possible in fair weather, it may be difficult during precipitation in the case of 915 MHz profilers in which the precipitation echo dominates over the clear air echo. The technique is particularly valuable if oblique radar beams are used instead of the vertical as in these cases the radial air velocity will be large and cannot be neglected. The use of the oblique radar beam is advantageous to select a position such that the acoustic wavefront will come within the beam in the presence of horizontal wind.

Angevine and Ecklund (1994) examined the errors in RASS measurements. For a 915 MHz wind profiler the corresponding RASS frequency is about 2 kHz. This frequency is subject to heavy attenuation in the atmosphere. The signal strength therefore varies within a range resolution cell. Since the temperature measured by RASS is a weighted average over the cell, there will be an error in temperature introduced by the signal variation. Other sources of error are wind (already discussed) and turbulence. May et al. (1989b) compared RASS-profiler data (50 MHz and 915 MHz profilers with corresponding sound frequencies of about 110 and 2000 Hz.) with radiosonde data and found an r.m.s. difference of 0.7°C. Tsuda et al. (1994) compared RASS temperatures averaged over 40 hours with virtual temperatures measured by radiosonde (mean of six soundings) and found agreement within about 1°C up to an altitude of 7 km. Moran and Strauch (1994) compared RASS (with 50 MHz profiler) virtual temperatures (up to 7 km altitude) without and with vertical air velocity correction, with radiosonde observations and found closer agreement (mean improvement of 0.7°C) in the latter case.

Performance of RASS varies with frequency. As already mentioned, RASS with 50 MHz profiler gives temperatures up to greater heights because of the lower attenuation of the matching sound frequency, but the lowest sounding level will be high (typically 2 to 3 km) because of the radar limitation. On the other hand RASS with a 915 MHz profiler can give data from about 0.15 km to 0.7 km (Martner et al. 1993) most of the time.

The utility of RASS temperatures with high spatial and temporal resolution in conjunction with wind data from wind profilers in tracking mesoscale weather systems in the lower and middle troposphere has been demonstrated by Neiman et al. (1992). Some of the US WPDN profilers and some of the 915 MHz profilers in the tropical Pacific are equipped with RASS. Although there are height limitations RASS has two advantages (1) it can operate through clouds and (2) like profilers it can yield almost continuous measurements. It may be expected that RASS may become a common operational technique before long.

RASS data at low levels were combined with TOVS[45] temperature profiles by Schroeder et al. (1991). The two are complementary because accuracy of TOVS profiles is poor near the surface while the TOVS extends the profile to heights which RASS does not give.

Rubin (2000) has demonstrated that a RASS can detect the wake vortices behind an aircraft (which can pose danger to a following aircraft). When the acoustic wave passes through the rotational flow, some sections of the wavefront speed up and others slow down. This results in a characteristic Doppler spectrum from which the spatial velocity distribution in the vortex can be evaluated. The Doppler spectrum is centred at the speed of sound and can be distinguished easily from the Doppler spectra of ground clutter and aircraft echoes, since wind and aircraft speeds near an airport are much smaller than the speed of sound.

---

[45] TOVS = TIROS-N Operational Vertical Sounder

Chapter 11

SPECIAL APPLICATIONS, NEW RADARS AND THE FUTURE

In the previous chapters the focus has been on operational aspects of weather radar. In this concluding Chapter we shall look briefly at some special applications, new types of radars and finally, speculate on the future of radar meteorology. Section 11.1 deals with applications to weather modification and space launch support. In section 11.2 we discuss the use of radars at millimetre wavelengths for cloud as well as precipitation studies. We also deal with new developments in the form of bistatic networks, rapid scan systems and the like which may eventually become operational. Finally in section 11.3 we review the progress of radar meteorology and attempt a peep into the future.

11.1 Some Special Applications of Radar

11.1.1 Weather Modification

Artificial modification of weather has been attempted in various ways. By a coincidence, these efforts started around the time when radar was invented and radar has played an important part in many weather modification programmes.

Weather modification may be designed to

1. Enhance precipitation by seeding individual clouds with silver iodide (or other substances such as common salt or lead iodide) from ground or aircraft or rockets.
2. Mitigate severity of weather phenomena such as thunderstorms, hail, and tropical cyclones.

3. Enhance precipitation over larger spatial scales by influencing radiation absorption.
4. Disperse fog which is an aviation hazard.

A treatment of various weather modification techniques may be seen in WMO (1974). We shall discuss only the first two in which radar has been a significant aid. Apart from the role of radar in studying and understanding the physics of clouds and precipitation which, of course, is essential for formulating weather modification techniques, radar has been applied in weather modification programmes in two ways.

1. To identify the weather system or the individual cloud or the part of it which is to be seeded.
2. To evaluate the effect of seeding on cloud parameters and result of the modification attempt (Nania 2002).

Radar has also been used to track the path of seeding material (see Para 11.2.6).

Weather modification efforts have immediate practical benefits in view. Precipitation enhancement is usually attempted in order to relieve drought and augment water resources. Sometimes it may be to quell forest fires. Hail or other severe weather suppression is designed to reduce damage. While there have been many well-designed scientific experiments to study the feasibility of significant modification, there are also many programmes of "operational" modification undertaken by governments or commercial interests at the instance of those who want more water or those who want to save their crops from damage. In the latter kind of projects science takes a backseat. The effort is undertaken without going deep into the scientific basis, in the hope that some benefit would accrue. But even in these cases identification of the suitable clouds and evaluation of the results are necessary.

11.1.2 Precipitation Enhancement

In the case of precipitation enhancement programmes, the seeding introduces nuclei to promote condensation and growth of precipitation particles. It is also often designed to promote the further growth of the cloud itself and thereby increase precipitation. Radar (ground-based or airborne) is used along with other instrumentation, to observe the height, rate of growth, reflectivity distribution, vertical motion within the cloud, wind field around the cloud, structure of the cloud (e.g. presence of a bright band) and temperatures in the cloud. A decision to seed or not to seed is taken on this basis.

In the case of scientific experiments, there is usually a control area and a target area, so that the resulting rainfall over the target area can be compared with that from the control area to verify the success of the seeding. Alternatively, clouds may be chosen at random for seeding and the unseeded clouds may be used as controls. The spatial and temporal distribution of precipitation in both areas are determined by radar in conjunction with rain gauges.

Precipitation enhancement programmes - research or operational - have been carried out in several countries and in all these, radar (ground-based or airborne) has played an important role.

11.1.3 Mitigation of Severe Weather - Hail Suppression

Detection of hail by conventional, Doppler, polarimetric and dual wavelength radars has been discussed at length in Chapter 6.

Both scientific and operational programmes of hail suppression have been carried out extensively in the erstwhile USSR where a high degree of success has been claimed. (Verification has been by ground measurements with "hailpads" and from insurance statistics). The USSR methods have also been followed in several European countries. The method consists of identifying, by radar, storm clouds containing hail or likely to develop into hailstorms and seeding them at an appropriate altitude (typically −5°C level) by material delivered by rockets or aircraft or ground generators. In the USSR model, an "accumulation zone" of large supercooled water drops is identified by radar. The hail grows rapidly in this zone. Seeding is expected to freeze many of the large drops which could serve as hail embryos. The large number of embryos will compete for available supercooled water resulting in reduction of mean stone size and perhaps their melting before they reach the ground. Radar methods and equipment have been described by Abshaev and Kartsivadze (1974). Dual wavelength radar has also been used. Cells have been tracked from tens of kilometres away but Battan (1975) pointed out that that the ice nuclei have to be introduced into the updraft between −5° and −10°C levels. These regions do not have high reflectivity and cannot be easily identified by conventional radar.

An important experiment to verify the USSR method was the "Grossversuch IV" programme in Switzerland, France and Italy. The National Hail Research Experiment in the USA and the South African experiment have been mentioned in Chapter 6. Federer (1977), Foote and Knight (1979) and Sulakvelidze et al. (1974), may be seen for further details.

The use of radar in attempts to modify tropical cyclones has been discussed in Chapter 8.

11.1.4 Meteorological Support for Spacecraft Launch

Meteorological conditions at the surface and in the upper air are critical for operations relating to launch (including transport of the vehicle to the launch site and loading of fuel) and reentry of spacecraft (NTIS 1988, Bauman and Businger 1996). A knowledge of the wind structure in the troposphere and lower stratosphere is essential. Climatology of the wind profile is needed in mission planning to decide the payload limits, optimum trajectories and fuel requirements as well as the favourable season or time of day for the launch. Possible hazards to be monitored include thunderstorms, lightning, wind shear and likelihood of icing. Lightning can be triggered by the launch vehicle itself as the vehicle and its plume act as conductors and decrease the electric field necessary to generate a lightning flash. Radars and lightning detection systems have, therefore, a major role to play in rendering meteorological support for this activity.

A wind profiler at the launch site can be used for:
(1) the observation of vertical wind profile just before launch, as often as necessary, (conventional upper air observations have to be terminated a couple of hours before launch),
(2) building up a climatology of the wind profile at the site, including a climatology of frequency of wind shears and significant wind changes in short periods,
(3) prediction of the wind profile reasonably in advance.

A network of profilers in the region can be used for weather analysis and prediction of the wind field using appropriate numerical models.

Scanning weather radar can detect thunderstorms as well as precursor conditions for their formation. The relationship between radar-observed parameters and lightning has been discussed in Chapter 4. Launch Commit Criteria (LCC) relating to lightning have been drawn up for deciding when a launch can be allowed or when it should be aborted. Details of LCC's and radar observational strategies may be seen in Boyd et al. (1999), Garner and Oram (2000) Gremillion and Orville (1999), Roeder and Pinder (1998), Roeder et al. (1999), and Short et al. (2000).

Polarimetric radar is expected to be useful in detecting icing conditions and determining drop size distribution.

11.2 New Types of Radar

11.2.1 Cloud Studies with Millimetre Radars

In the previous chapters the emphasis has been on the radar observation of precipitating clouds. Besides operational products, many cloud physical phenomena of research interest have been discussed. E.g. melting band structure, drop size distribution studies, phase of hydrometeors, vertical motion in clouds, latent heat release, and thermodynamic and microphysical retrievals. Non-precipitating clouds, however, are usually poorly detected at the wavelengths used in operational radar networks. These can be better studied at higher frequencies at which cloud size hydrometeors give large return. Light precipitation can also be observed at higher frequencies. Even in phenomena involving heavy precipitation, high frequencies are suitable for observation at close range, as light and portable radars can be designed at these frequencies. Not all frequencies can be used as there are absorption bands of atmospheric gases at some frequencies. While these frequencies suffer from heavy attenuation in precipitation they have several advantages for cloud observations at short range. The radars emit low power (which is adequate because of the high reflectivity and short range), are light and small and are portable. Their sensitivity can be as good as −30 dBZ at a range of 10 km. They can have quite narrow beams and spatial resolution as high as a few metres with a small antenna and are not affected much by ground clutter. One disadvantage for Doppler measurements is the small value of the $r_{max} \times v_{max}$ product [(see equation (1.10)]. Even at short ranges v_{max} can be small.

The Ka band (8.6 mm wavelength; 35 GHz) has been extensively used. Pasqualucci et al. (1983) developed a radar in this band with linear and circular polarisation capabilities for cloud and precipitation studies. Dual wavelength radars (e.g. X and Ka bands) have been discussed in Chapter 7 in connection with airborne and satellite-borne observations. Dual wavelength systems have been proposed for discrimination between supercooled water and ice and to measure the mass content in each (Gosset and Sauvageot 1992). Their method consists of measurement of the differential attenuation between the X and Ka bands and the reflectivity factor (see Chapter 7). The former is a linear function of the liquid water content. The two measurements provide two equations from which liquid and ice water content can be determined separately.

In the case of cloud drops the attenuation at X band can be neglected. Hence the reflectivity can be determined with X band radar. The attenuation over a short range interval is obtained from the Ka band radar. This is a linear function of the LWC which can therefore be determined, as the contribution of ice hydrometeors is negligible. Vivekanandan et al. (2001) proposed that dual wavelength radar can be used for detection of aircraft icing conditions, provided a method is available for removal of Mie scatterers and ice crystals. Icing is mainly contributed by supercooled[1] large liquid drops. They proposed determination of LWC and a new parameter called Radar-Estimated Size (RES) and use of these to estimate the ice accretion. RES is defined by them as

---

[1] Rosenfeld and Woodley (2000) have found (from aircraft measurements) that supercooled small drops can exist even at temperatures down to – 37°.5C. Larger drops will be confined to lower altitudes.

$$\text{RES}=\left(\frac{\langle D^6\rangle}{\langle D^3\rangle}\right)^{1/3} \tag{11.1}$$

where $<D^6>$ and $<D^3>$ respectively are the sixth and third moments of the DSD. The former can be recognised as equal to the reflectivity factor Z and the latter is proportional to the attenuation A. Using the units mm for RES, mm^6 m^{-3} for Z and dB km^{-1} for A they get,

$$\text{RES}=\left(7.12\times10^{-4}\,\frac{Z}{A}\right)^{1/3} \tag{11.2}$$

According to them RES discriminates more effectively between cloud, drizzle and rain spectra than other parameters.

Gossard (1994) and Gossard et al. (1997) developed a technique for determining cloud droplet size spectra. Since the droplets are small, their fall or settling velocities are small and their motion is determined by the turbulent fluctuations in the medium. Hence the backscatter from the cloud droplets cannot be easily separated from the turbulent scatter in the manner discussed in Chapter 10 for rain drops. The measured spectra have to be deconvolved to recover the cloud spectrum and separate it from precipitation. Frisch et al. (1995, 1995b) used a vertically pointing Ka band Doppler radar for measuring vertical motion in clouds. The droplet fall velocity contamination could be made negligible by using only measurements associated with a reflectivity factor less than −17 dBZ. Another technique using Ka band radar is for study of cirrus clouds with elliptical polarisation (Matrosov and Kropfli 1993). Moran et al. (1998) have described a vertically pointing Ka band Doppler radar, which can operate unattended and can observe non-precipitating clouds, including cirrus, continuously, providing a cloud climatology[2] over the site.

With the advancement of microwave technology for still higher frequencies, hardware to implement radar systems has become available. Lhermitte (1987) developed a 94 GHz (W band) Doppler radar for cloud observations. Lhermitte's radar transmitter uses a power oscillator called "Extended Interaction Oscillator (EIO)" delivering 1 kW peak power. Very short pulse length with very low rise and fall times are possible. This gives high range resolution. IF has to be very high in order to get sufficient receiver bandwidth. Lhermitte observed cirrus clouds with his radar and also the melting band in light stratiform precipitation. The signature of the melting band is

[2] Uttal and Kropfli (2001) have pointed out that climatological statistics can be affected by the pulse length of the radar.

different from that at lower frequencies as Mie scattering occurs even with small precipitation particles. Since Rayleigh scattering cannot be assumed, simple Z-R relationships cannot be used.

An evaluation of the 94 GHz radar for remote sensing of cloud properties has been made by Clothiaux et al. (1995) and Klugmann (2001). Pazmany et al. (1993, 1994), Galloway et al. (1997) and Vali et al. (1993) have described airborne 95 GHz polarimetric radars for cloud studies.

A Cloud Profiling Radar System (CPRS) has been developed at the University of Massachusetts, USA. This consists of a transportable dual frequency 33 and 95 GHz polarimetric Doppler radar which simultaneously transmits and receives H and V polarised radiation at both frequencies through a single horn-fed lens antenna (Lohmeier et al. 1997). This system has been used to identify regions of ice, cloud droplets, rain, mixed-phase hydrometeors and insects in stratiform clouds using the measured reflectivities at the two wavelengths and their ratio (DFR), LDR, and velocities. After correcting for attenuation, the data are used to train a neural net (see Paragraph 7.2.7). A scheme for classification has been worked out on this basis.

Attenuation is a serious problem at the 94 GHz frequency. Li et al. (2001) used a dual radar arrangement, one from the ground looking upwards and the other a nadir-pointing airborne radar, looking at the same cloud from above. This arrangement enables determination of attenuation as well as the true reflectivity.

11.2.2 Precipitation Studies with Millimetre Radar

Lhermitte (1990) has discussed the attenuation and scattering at frequencies of 35, 94, 140 and 240 GHz by clouds and precipitation. Since Mie scattering occurs in the case of most hydrometeors at these frequencies the radar cross section oscillates as particle diameter increases (Fig. 2.5). Hence the conventional Z-R relationships cannot be formulated. However, consider a vertically pointing Doppler radar as in Paragraph 3.3.7. If there is no vertical air motion ($w = 0$), the Doppler velocity spectrum represents the distribution of fall velocities of the drops. If the radar operates on the X, C or S-band, the Doppler spectrum will be as shown in the topmost diagram of Fig. 11.1 (from Kollias et al. 2002). There is a single maximum in power received at the mean fall velocity V_t. (The fluctuations are due to turbulence, shear etc.). If a VHF or UHF profiler is used (middle diagram in Fig. 11.1) there are two maxima, one at zero velocity due to clear air and the other due to the mean fall speed of hydrometeors. In the case of a W band (94 GHz) radar (bottom diagram in Fig. 11.1), there are maxima and minima at various velocity values because of the Mie oscillations. Each of

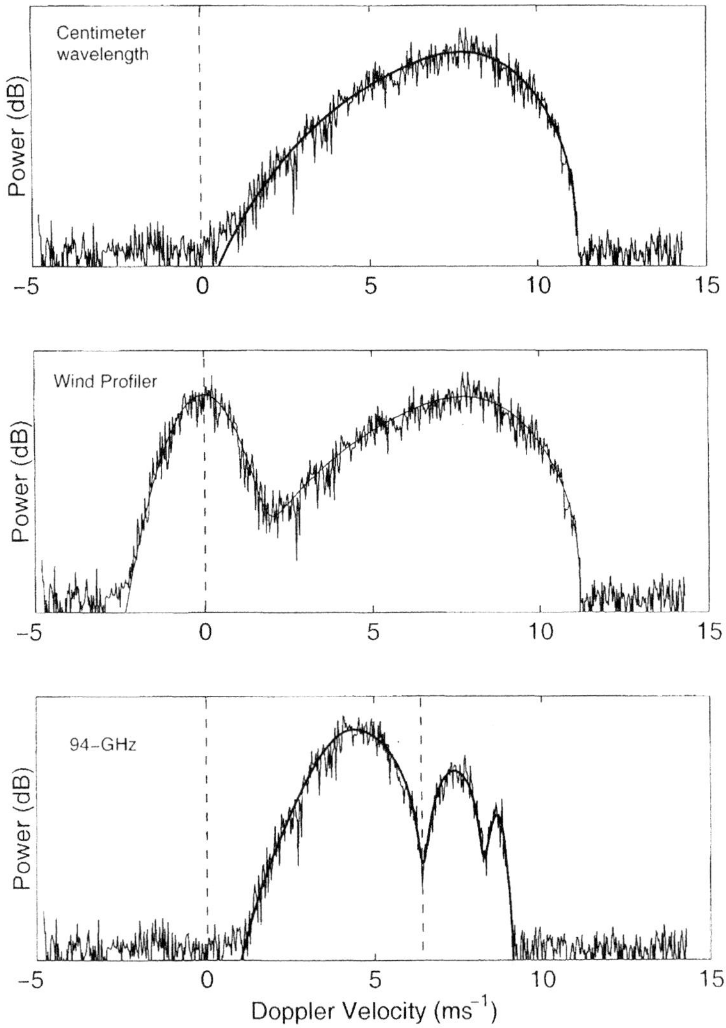

Fig. 11.1. Simulated Doppler spectra of rain drops using a vertically pointing radar for (top) cm wavelength radar (X,C,S, band), (middle) wind profiler (VHF-UHF) and (bottom) a W-band cloud radar (from Kollias et al. 2002; courtesy Dr. P. Kollias and AMS).

these maxima and minima corresponds to a drop size at which a maximum or minimum occurs in the radar cross section in Fig. 2.5[3]. The first Mie minimum[4] in the back-scattering cross section at this frequency, for an oblate spheroid, occurs at a drop size of 1.7 mm. Since the drop size can be related to terminal

[3] Fig.2.5 gives back-scattering cross sections for spheres as a function of $2\pi a/\lambda$. Curves of normalised back-scattering cross sections for spheres and spheroids as a function of rain drop diameter for this frequency can be seen in Kollias et al. (2002).

[4] The position will depend slightly on temperature.

velocity by an equation such as (2.45) or (2.46), the terminal velocity corresponding to 1.7 mm can be deduced as 5.95 m s^{-1}. Hence the minimum shown in the bottom diagram of Fig. 11.1 at 5.95 m s^{-1} corresponds to a drop diameter of 1.7 mm. If there is a vertical velocity, the maxima and minima in the Doppler spectrum will be shifted as shown in Fig. 11.2. Hence the vertical air velocity can be determined accurately as the difference between the two positions. Once this is done, the spectrum can be shifted to correspond to zero vertical air motion. The spectrum can be fitted to correspond to a model DSD (say gamma) and the DSD derived. Lhermitte (1988) had the insight to recognise this and proposed the use of the Mie oscillations, overturning the conventional thinking that the W band is not suitable for precipitation studies. The method has been developed further by Firda et al. (1999)[5] and Kollias et al. (1999, 2001, 2002) which may be seen for details.

A van-mounted mobile version of a 95 GHz Doppler radar has been used by

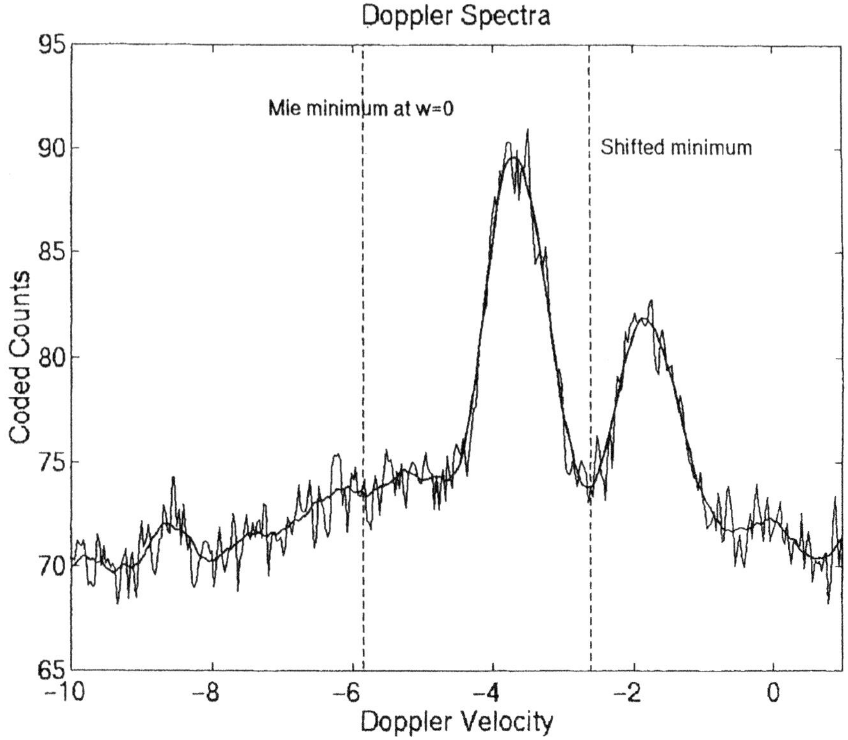

Fig. 11.2. Mie minima in Doppler spectra from a 94-GHz radar (from Kollias et al. 2002; courtesy Dr. P. Kollias and AMS).

[5] Firda et al. (1999), used the CPRS mentioned above and therefore their algorithm is based on the dual frequency ratio (see Chapter 7).

Bluestein et al. (1995) to observe small scale phenomena such as mesocyclones and hook echoes at close range (up to 4 km). Their radar uses an antenna which can be extended above the van. The rest of the system is inside the van. A generator is used for power supply. To obtain adequate unambiguous range and velocity they use the polarisation diversity method proposed by Doviak and Sirmans (1973) (see Paragraph 3.3.5), which they call DPPP (Dual Polarisation Pulse Pair)[6] (Pazmany et al. 1999, 1999b). A horizontally polarised pulse and a vertically polarised one are transmitted as a pair with an interval of 10 μs between them. The effective p.r.f. between the H and V pulses is therefore 100,000 giving a v_{max} of 80 m s^{-1}. The H and V signals are processed by two receiving channels with a phase correction for the difference in time of transmission and the depolarisation in precipitation. The p.r.f. with respect to successive pulse pairs is 15 kHz (r_{max} = 10 km). Since the DPPP method can be affected by depolarisation in the target and isolation problems in the radar system, a lower single polarisation p.r.f. was also used. The radar has been upgraded (specifications in Table 11.1). The reflectivity and velocity fields in Fig. 11.3 show a tornado with a spatial resolution of 15 metres. The eye structure seen in the figure is discussed in the context of centrifuging of scatterers, by Dowell et al. (2001).

A portable radar operating at 215 GHz (1.4 mm wavelength) for detection of non-precipitating clouds and fog has been described by Mead et al. (1989). This uses an EIO with 60 W peak power output and separate transmitting and receiving antennas (the latter is necessary because the performance of circulators and T-R switches is not satisfactory at this frequency). Attenuation due to water vapour and liquid water are very high at this frequency. Hence, measurements of Z_e are corrected for water vapour attenuation by using surface measurements and a model of variation of water vapour with height. Correction for attenuation by liquid water is made using the measured reflectivity and an assumed drop size distribution.

A 94 GHz polarimetric Doppler radar has been designed in the USA (Racette et al. 1993) for use on a high-altitude aircraft. The radar can also be adapted for ground use. This will observe clouds. Since the aircraft has also a 10 GHz polarimetric Doppler radar, dual wavelength observations are possible.

11.2.3 Bistatic Radar Networks

In many field experiments multiple-Doppler radars have been used. This

[6] Also called Polarisation Diversity Pulse Pairs (PDPP) (Pazmany et al. 1999).

Table 11.1. Specifications of W band radar of University of Massachusetts (from Pazmany et al. 1999).

Parameter	Value
Transmitter	
Operating frequency	95.04 GHz
power	1.2 kW, 1 % duty
pulse length	200 ns - 2 ps
Polarisation	V or H pol.
Receiver	
Noise figure	10 dB
Bandwidth	2, 5 or 10 MHz
Dynamic range	75 dB
Sampling Resolution	15 m
Maximum Range	15 km
Radar Parameters	Reflectivity and Pulse Pair Velocity
Unambiguous Velocity	$\pm$ 80 m s $^{-1}$
Antenna	
Type	Cassegrain dish
Diameter	1.2 m
Beamwidth (3 dB)	0.18°
Gain	59 dB
Sensitivity	
Single Pulse, 0 dB SNR	–27 dBZe at 1 km

is very expensive and the radars are not easily movable from place to place. Wurman (1994) has suggested use of a bistatic network consisting of an existing radar transmitter in conjunction with a number of non-scanning non-transmitting receive-only units. The receivers placed at a number of points will have simple stationary broad beam antennas. The signal from any sampling volume illuminated by the single transmitter is received in various directions by these receivers. The data can be analysed in a multiple-Doppler mode. The cost of additional complete radar systems is saved. Further refinement of this technique may be seen in Wurman and Randall (2001). Sato and Wurman (2001) have examined the accuracy of wind fields derived from such systems. Sidelobe contamination can be a serious problem (de Elía and Zawadzki 2000). De Elía and Zawadzki (2001) have formulated criteria for optimal layout of a bistatic network.

11.2.4 Portable Radars

Portable low-power equipment is of advantage for small scale or short-lived

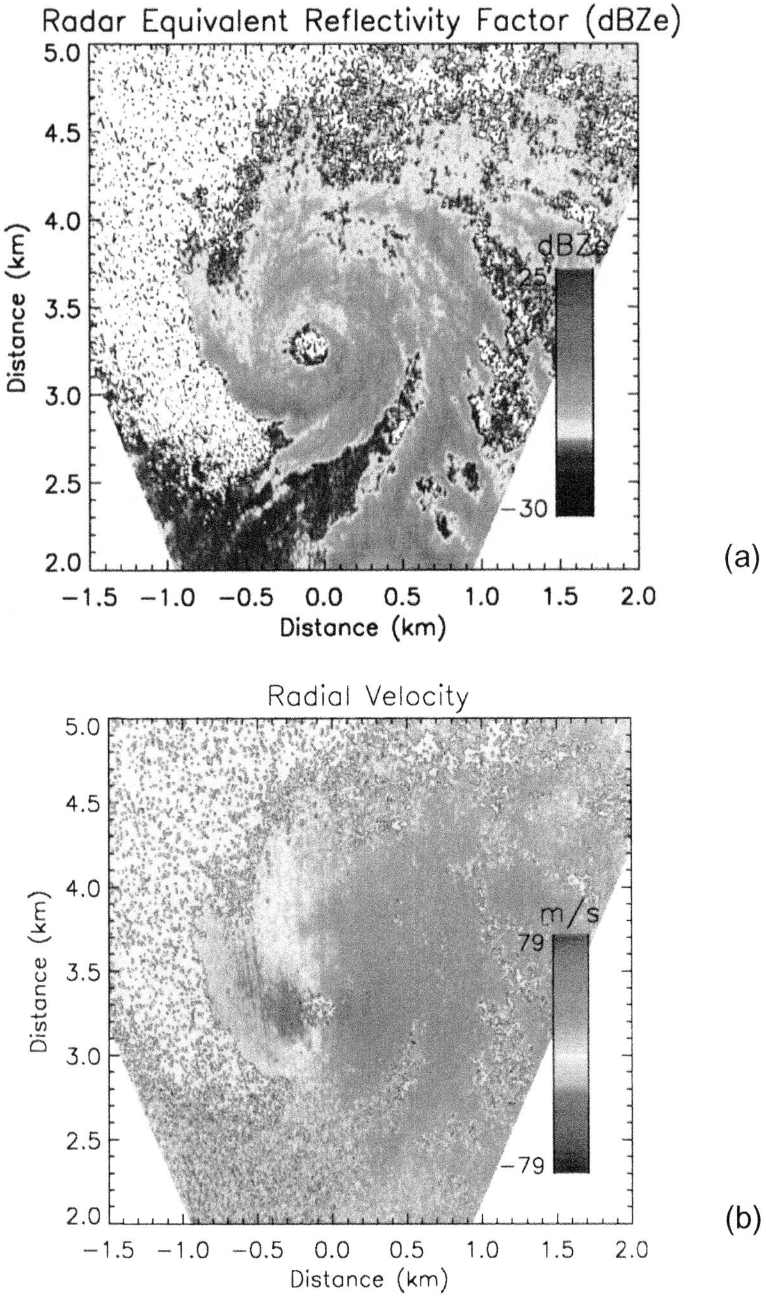

Fig. 11.3. Tornado in Oklahoma, USA, on 3 May 1999, seen by W band mobile radar. (a) reflectivity (b) velocity. Note the very high resolution and the clear view of the "eye" of the tornado (figure courtesy Dr. A.L. Pazmany, University of Massachusetts).

phenomena e.g. tornadoes, at close range. Continuous Wave (CW) Doppler radar of 1 W power output has been used (e.g. Bluestein and Unruh 1989; 3 cm wavelength in this case). This cannot give range information but the Doppler velocities can be measured up to a maximum range of about 10 km. Bluestein and Unruh modified their radar subsequently into a FM-CW radar. With, this, the range as well as Doppler velocity can be measured. A mobile vertical polarisation X-band radar called Doppler On Wheels (DOW) has been constructed (Wurman et al. 1997) in the USA. This has very narrow beam width at short ranges (typically 3 km) and can observe small scale phenomena such as tornadoes and dust devils. It can be used as a pseudo-dual-Doppler system by moving it around a weather phenomenon. A dual polarisation version called DOW4 has also been developed (Wurman 2001).

11.2.5 Surface Precipitation Sensor

Efforts have been made in the past several years to automate those surface meteorological observations which are traditionally carried out visually by human observers [see e.g. Pilon (1989), Gallagher (1989)]. Examples are present weather, past weather and times of commencement and cessation of precipitation. To facilitate this, a low power X-band continuous wave Doppler radar was developed by the Atmospheric Environment Service of Canada by modifying a traffic control radar (Sheppard et al. 1983). This is called "Precipitation Occurrence Sensor System" (POSS). It observes in the vertically pointing mode at a range up to 2 metres. A fast Fourier transform produces, from the signal, a power spectrum of fall velocities. The total power in the spectrum is used to identify times of commencement and cessation of precipitation. The peak fall velocity is used to classify precipitation as snow (< 1.5 m s^{-1}), rain or drizzle (1.5 m s^{-1} to 9 m s^{-1}) or hail (> 9 m s^{-1}). The total rainfall during the spell of rain is also estimated. Impact of precipitation particles on the radome gave spurious signals. To overcome this, the system was converted into a bistatic radar with tilted transmitting and receiving beams meeting over a sampling volume above the radar. This system has been used by Sheppard (1990) for continuous real time measurement of rain Drop Size Distribution (DSD). The DSD is retrieved from the Doppler power spectrum. Comparisons with more conventional instruments (disdrometer, particle measuring system) were satisfactory (Sheppard and Joe 1994).

A portable CW system for the same purpose but a monostatic one operating at 24.5 GHz has been developed in France (Duvernoy et al. 1991) and uses the 3P and 3V methods discussed in Chapter 7 for precipitation identification. A bistatic system working at 35 GHz has also been developed (Duvernoy and Gaumet 1996).

11.2.6 Chaff Tracking

Chaff released from rockets or aircraft has been used for a long time as a radar target for determining upper winds. Chaff consists of numerous half-wave dipoles in the form of very thin metal-coated fibres. It can be dropped from aircraft or rockets and can trace the horizontal air motion accurately. In the vertical, it has a fall velocity but this is small compared to vertical air motions in convective storms. Hence chaff has been used to study air motions within storms as mentioned in Chapter 6. Since the dipoles disperse as they fall, the diffusion in the air can also be determined. While the method is fairly straightforward[7] in clear air, the chaff echo may be masked by the hydrometeor echo in clouds and precipitation. Moninger and Kropfli (1987) proposed that dual polarisation measurement be used to overcome this problem. If the chaff dipoles are randomly oriented, Z_{DR} of chaff will be 0 dB; if they fall with preferred orientation there will be a finite Z_{DR}. In either case the chaff will not be readily distinguishable in precipitation. However, if circular polarisation is transmitted, the dipoles will return linear polarisation which can be resolved into left and right circular polarised components of equal amplitude. Hence the circular depolarisation ratio (CDR) will be 0 dB irrespective of orientation. Hydrometeors, on the other hand, have CDR several dB lower. Hence the chaff is easily traceable. This technique called TRACIR (TRacking Air with CIrcular-polarisation Radar) has been used by Martner et al. (1992) to track air motion and entrainment in clouds. Boe et al. (1992) have used the technique to trace silver iodide aerosols in cloud-seeding. These measurements used X-band radars.

11.2.7 Over-The-Horizon (OTH) Radar

Meteorologists never get resource allocations comparable to other activities e.g. aviation or defence. But though short of funds they are not short of ingenuity. Several non-meteorological radars have been put to meteorological use right from the early days of radar meteorology (see section 1.1). One example is the use of airport surveillance radar (ASR) discussed in Chapter 9. Another example is the use of Over-The-Horizon (OTH) radars. These are military

[7] An account of measurements with chaff by dual-Doppler radar in clear air and close to precipitating clouds may be seen in Johnson et al. (1993).

installations for surveillance of surface or airborne targets over very long distances. They operate with large antenna arrays in HF bands (less than 30 MHz). The radiation is reflected by the ionosphere and the resulting sky wave can be incident on the ocean surface up to a few thousand km away from the radar and give a signal return through reflection again from the ionosphere. The ocean surface wind field can be mapped using the scatterometer principle (see Chapter 4).

Echoes from ocean surface waves travelling towards and away from the radar are Doppler shifted by the phase velocity of the waves. The advancing and receding waves can be recognised as two peaks on the Doppler spectrum. The relative amplitudes of the two waves can be found from the ratio of the echo amplitudes. Assuming that the wind direction coincides with that of the strongest waves, the direction can be found with a 180° ambiguity which may be resolved using independent meteorological information (Georges and Thome 1990). Using a HF sky wave radar in California with 0.5° beam at 15 MHz, Maresca and Barnum (1979) mapped the surface wind field from the direction of the ocean gravity waves of 8 m wavelength. The wind directions agreed with available ocean buoy winds within 10%. A hurricane in the Gulf of Mexico could be detected. The direction ambiguity was resolved from the general circulation of the hurricane. In a similar analysis using a sky wave radar in Australia (Keenan and Anderson 1987) the vortex of a developing tropical cyclone could be identified; cyclone positions determined were within 30 km of the official positions of the Bureau of Meteorology. A US air force OTH radar in the northeastern USA was similarly used to track a hurricane 2400 km away (Georges et al.1993).

11.2.8 Rapid Scan Radar

We have seen that a volume scan of several elevations, takes 5 to 10 minutes with most radars. In rapidly changing meteorological phenomena there is a need to reduce this time to about a minute. Applications such as aviation demand rapid updating of data. One possibility that has been considered is the use of electronic scanning. While electronic beam switching is standard practice in MST/ST radars, its implementation in scanning weather radars is a more complicated problem. To cover a whole hemisphere with a 1° beam without gaps requires over 20,000 beam positions. If these are to be scanned in a minute, the dwell time per beam is reduced to about 3 ms. To get adequate number of samples for averaging, the p.r.f. should be increased and the maximum range reduced. The rapid continuous scan will induce a spectral broadening, reducing the accuracy of Doppler velocity measurement. Electronic scanning involves not just a change in antenna design but major changes in the transmitter, receiver and signal processing areas. Also, to produce electronic scanning in two dimensions will be extremely expensive.

The problems have been examined by several authors [Carbone and Carpenter (1983), Holloway and Keeler (1993), Josefsson (1991), Joss and Collier (1991), Keeler and Frush (1983), Keeler et al. (1993), Meischner et al. (1997), Doviak et al. (2001)] and solutions suggested. A possibility is to use a mechanical rotation for the azimuth and electronic scanning for elevation only. One option is a multiple beam antenna observing several elevation directions simultaneously[8]. This reduces the scan rate per beam and spectral spread and increases the dwell time. The number of steps in elevation can be limited as very high elevations are not needed for many purposes. Instead of a dish, a stack of slotted waveguides can be used with elevation scanned electronically in discrete steps to reduce spectral broadening. A wide bandwidth gives more independent samples per given range interval. This can be used with pulse compression techniques, maintaining a high average power. While the scatterers in one resolution volume are reshuffling into an independent configuration, the phase array radar can switch its beam to other directions and come back to the original direction after the signal is no longer correlated. Adaptive scanning consisting of selective scanning of sectors having echoes, or specific beams can be made, with one preliminary scan to identify those sectors. Adaptability according to the type of weather phenomenon to be observed is also possible, in terms of dwell time, pulse width and p.r.f.

A rapid-scanning mobile radar called "Rapid-DOW" using hybrid electronic/mechanical scanning with a multiple beam system to collect partial volumetric data in about 10 seconds, is being developed in the USA (Wurman and Randall 2001b). The Rapid-DOW has a slotted waveguide array antenna producing frequency-dependent beam steering. Energy from a single transmitter, producing 6 to 10 different frequencies nearly simultaneously, is emitted in 6 to 10 pencil-beams, each with a different elevation angle. Fast mechanical azimuthal scanning will result in a sweep of the sky over 6 to 10 beams in six seconds

11.2.9 Integrated Observing Systems

With the recognition that various observational techniques are complementary to each other, the trend in recent years is to combine them into

[8] Frequency offsets are required between the beams; this implies availability of a large RF bandwidth.

an integrated observing system. We shall confine ourselves to a discussion of the role of radars in such systems. Outputs from scanning radars or ST radar-RASS systems can be combined with those from other surface-based systems such as lightning detectors, passive radiometers, radiosondes and lidars or space-based observations such as brightness temperatures, cloud (or water vapour) motion vector winds, passive microwave radiometers, GPS total water vapour measurements and GPS measurements of refractivity gradients[9].

Data of radar and other sensors (lightning sensor and passive microwave imager) on the TRMM satellite have been combined for studies of convective precipitation (e.g. Petersen et al. 2002). Hardaker et al. (1997) suggested the use of attenuation differences between 19.77 and 12.502 GHz line-of-sight communication links to estimate rainfall and use the estimate to adjust radar measurements where the latter are contaminated by ground clutter or anomalous propagation.

An example of a multisensor system is the combination of temperature profiles in the lower atmosphere from ground-based RASS/wind profilers or from atmospheric radiation sensed by ground-based multi-channel microwave radiometers with data at higher altitudes from satellite-based (TOVS) profiles (Schroeder et al. 1991). Multisensor systems for deriving temperature and humidity profiles and liquid water monitoring, which include Ka-band radar to determine heights of cloud layers (where saturation conditions are presumed) have been developed [Stankov (1996, 1998), Stankov et al. (1995, 1996)]. Temperature and humidity profiles can be derived from profilers as discussed in Chapter 10, but to resolve ambiguities, the data of one sensor have to be constrained by data from others.

Assimilation of measurements into numerical models, is an important part of an integrated observing system. Assimilation consists, on the one hand, of the incorporation of measurements into computer prediction models and, on the other hand, the accommodation of such models to a set of observations. A regular four-dimensional representation of the atmosphere is sought to be produced from a heterogeneous array of observations, which sample imperfectly and irregularly in space and time. Many of the new observational techniques mentioned above do not directly yield the conventionally measured parameters such as temperature and humidity (Raghavan 1999b). Thus, Doppler radars give radial components of winds, satellites give radiance in specific wavelength regions, and GPS techniques yield atmospheric refractivity gradients. Instead of converting these into the conventional parameters, it is advantageous to develop models in which the measured quantities can be input directly (see e.g. Dabberdt and Schlatter 1996).

---

[9] For a discussion on the technique, see Rocken et al. 1997.

thrust needs to be on development of reliable algorithms to detect severe weather. Plans for upgradation of existing operational systems taking advantage of advances in computer hardware and software are under way (e.g. Crum et al. 1998, Meischner et al. 1997). Polarimetric radar may very soon be implemented operationally at least in limited areas. It should improve accuracy of precipitation estimates and also contribute to severe weather recognition.

Radar measurement of precipitation is not satisfactory even after 50 years of research. Estimation of precipitation over large areas where implemented is still beset with problems of accuracy. The problem is acute in hilly terrain. But advances are being made. We should expect to see radar networking on a larger scale for this purpose combined with satellite data, with real time availability of data to all users integrated with other hydrological data e.g. river levels.

In many countries radar coverage is poor. The cost of radars is a deterring factor. It is not clear whether the new technologies will bring down costs substantially. It is likely that for some purposes space-based radar and other observations will be a substitute for high cost ground-based radars. Perhaps, in a decade from now space-borne radars may become operational and be able to give area-integrated precipitation data over the oceans as a routine. But other applications such as local severe weather monitoring, aviation and hydrometeorology will continue to deploy ground-based equipment.

Profiler networks will become operational in most countries and yield high resolution wind data for use in global and local analyses. These will probably be used in conjunction with RASS, radiometers, improved radiosonde systems and satellites (sounding as well as GPS) to provide an integrated atmospheric sounding system [see e.g. Douglas and Stensrud (1996) regarding the North American upper air network].

In all these advancements, it is to be expected that radar data are integrated with other observations e.g. satellites, lightning detection systems, surface and upper air observations through models.

11.3.5 The Human Element

Over and above the advances in technology and automation, the most important element in past progress is perhaps the initiative of the innovative research meteorologist and the resourceful operational meteorologist. It is important that the human meteorologist continues to interpret the observations and overrides the automated systems whenever necessary.

With these prospects, education and training in radar meteorology will be very important. It is hoped that this book will stimulate interest in this area.

REFERENCES

Abbi S.D.S., 1972, "Hydrometeorological analysis", pp 45-68, in *Manual of Hydrometeorology*, Part I, India Meteorological Department (IMD), New Delhi, 175 pp.

Abdullah A.J., 1966, "The spiral bands of a hurricane: a possible dynamic explanation", *J. Atmos. Sci.*, **23**, 367-375.

Abshaev M.T. and A.I. Kartsivadze, 1974, "Radar methods and equipment for hail suppression", pp 343-351 in *Proc. WMO/IAMAP Scientific Conf. on Weather Modification, Tashkent, 1-7 October 1973*, WMO-No. 399, 538 pp.

Achtemeier G.L., 1991, "The use of insects as tracers for 'clear air' boundary-layer studies by Doppler radar", *J. Atmos. Ocean. Technol.*, **8**, 746-765.

Aden A.L. and M. Kerker, 1951, "Scattering of electromagnetic waves by two concentric ice spheres", *J.Appl.Phys.*, **22**, 1242–1246.

Adler R.F., 1981, "Severe storms requirements for precipitation information", *Workshop Report, Precipitation Measurements from Space*. Eds. D. Atlas and O.W. Thiele, Goddard Space Flight Center, USA., pp D31-33.

Ahnert P.R., M.D. Hudlow, E.R. Johnson, D.R. Greene and M.P.R. Dias, 1983, "Proposed 'on-site' precipitation processing system for NEXRAD", *Preprints, 21st Conf. Radar Meteor.*, Amer. Meteor. Soc., 378-385.

Ajayi G.O. and E.B.C. Ofoche, 1984, "Some tropical rainfall rate characteristics at Ile-Ife for microwave and millimeter wave applications", *J. Climate Appl. Meteor.*, **23**, 562-567.

Albers S.C., 1989, "Two-dimensional velocity dealiasing in highly sheared environments", *Proc. 24th Conf. Radar Meteor.*, Amer. Meteor. Soc., 411-414.

Albers S.C., 1991, "The detection of isolated tornadic vortex signatures with NEXRAD algorithms", *Preprints, 25th International Conf. Radar Meteor.*, Amer. Meteor. Soc., 123-126.

Amayenc P., J.P. Diguet, M. Marzoug and T. Tani, 1996, "A class of single- and dual-frequency algorithms for rain-rate profiling from a spaceborne radar. Part II: Tests from airborne radar measurements ", *J. Atmos. Ocean. Technol.*, **13**, 142-164.

Amayenc P., J. Testud, and M. Marzoug, 1993, " Proposal for a spaceborne dual-beam rain radar with Doppler capability", *J. Atmos. Ocean. Technol.*, **10**, 262-276.

Anagnostou E.N. and W.F. Krajewski , 1998, "Calibration of the WSR-88D Precipitation Processing Subsystem", *Wea. Forecasting*, **13**, 396–406.

Anagnostou E.N. and W.F. Krajewski , 1999, "Real-Time Radar Rainfall Estimation. Part I: Algorithm Formulation", *J. Atmos. Ocean. Technol.*, **16**, 189–197.

Anagnostou E.N., C. A. Morales and T. Dinku, 2001, "The Use of TRMM Precipitation Radar observations in determining ground radar calibration biases", *J. Atmos. Ocean. Technol.*, **18**, 616–628.

Anderson J.R., 1987, "The measurement of Doppler wind fields with fast scanning radars: Signal processing techniques", *J. Atmos. Ocean. Technol.*, **4**, 627-633.

Anderson L.C. and G.D. Green, 1991, "The WSR-88D combined moment product", *Preprints, 25th International Conf. Radar Meteor.*, Amer. Meteor. Soc., 233-236.

Anderson L.C. and G.D. Green, 1992, "Application of the combined moment product in aviation forecasting", pp194-198 in *Postprint Volume National Weather Service Aviation Workshop*, December 1991, NOAA Tech. Memo. NWS CR-102.

Anderson T. and B.Lindgren, 1992, "A sea breeze front seen by radar", *Meteor. Mag.*, **121**, 239-241.

Andrieu H. and J.D. Creutin, 1995, "Identification of vertical profiles of radar reflectivity for hydrological applications using an inverse method. Part I: Formulation", *J. Appl. Meteor.*, **34**, 225-239.

Angevine W.M. and W.L. Ecklund, 1994, "Errors in radio acoustic sounding of temperature", *J. Atmos. Ocean. Technol.*, **11**, 837-842.

Angevine W.M., W.L. Ecklund, D.A. Carter, K.S. Gage and K.P. Moran, 1994, "Improved radio acoustic sounding techniques", *J. Atmos. Ocean. Technol.*, **11**, 42-49.

Antipovsky S.V., A.V. Kochin, V.R. Megalinsky and A.B. Shupiatsky, 1993, "Application aircraft meteorological polarization radar for studing meso- and micro- structure of clouds and precipitation", *Preprints, 26th International Conf. Radar Meteor.*, Amer. Meteor. Soc., 726-729.

Aoyagi J., 1969, "Mean Doppler velocities of precipitation near the ground", *Proc., 13th Wea. Radar Conf.* Amer. Meteor. Soc., 22-25.

Aoyagi J., 1983, "A study on the MTI weather radar system for rejecting ground clutter", *Pap. Meteor. Geophys.*, **33**, 187-243.

Arkin P.A., 1979, ""The relationship between fractional coverage of high cloud and rainfall accumulations during GATE over the B-scale array", *Mon. Wea. Rev.*, **107**, 1382-1387.

Arkin P.A. and B.N. Meisner, 1987, "The relationship between large-scale convective rainfall and cold cloud cover over the western hemisphere during 1982-84", *Mon. Wea. Rev.*, **115**, 51-74.

Armijo L., 1969, "A theory for the determination of wind and precipitation velocities with Doppler radars", *J. Atmos. Sci.*, **26**, 570-573.

Armstrong G.M. and R.J. Donaldson Jr., 1969, "Plan shear indicator for realtime Doppler radar identification of hazardous storms", *J. Appl. Meteor.*, **8**, 376-383.

Atkinson N.C. and P.K. James, 1991, "Quality control of data from overlapping weather radars", *Preprints, 25th International Conf. Radar Meteor.*, Amer. Meteor. Soc., 206-209.

Atlas D.,1959, "Meteorological angel echoes", *J.Meteor.*, **16**, 6-11.

Atlas D.,1960, "Radar detection of the sea breeze", *J.Meteor.*, **17**, 244-258.

Atlas D.,1964, "Advances in radar meteorology", *Adv. Geophys.*, Acad. Press, **10**, 318–478.

Atlas D., 1982, "Adaptively pointing spaceborne radar for precipitation measurements", *J. Appl. Meteor.*, **21**, 429-431.

Atlas D. (Ed.), 1990, "*Radar in Meteorology*", Amer. Meteor. Soc., 806 pp.

Atlas D., 2002, "Radar calibration: some simple approaches", *Bull. Amer. Meteor. Soc.*, **83**, 1313–1316.

Atlas D. and T.L.Bell, 1992, "The relation of radar to cloud Area-Time Integrals and implications for rain measurement from space", *Mon. Wea. Rev.*, **120**, 1997-2008.

Atlas D., and A.C. Chmela, 1957, "Physical-synoptic variations of drop-size parameters", *Proc. 6th Wea. Radar Conf.*, Amer. Meteor. Soc., 21-29.

Atlas D. and F.H. Ludlam, 1961, "Multi-wavelength radar reflectivity of hailstorms", *Quart. J. Roy. Meteor. Soc.*, **87**, 523–534.

Atlas D. and R.K. Moore, 1986, "The measurement of precipitation with synthetic aperture radar", *Preprints, 23rd Conf. Radio Meteor*, Amer. Meteor. Soc., Vol **1**, 133–138.

Atlas D. and R.K. Moore, 1987, "The measurement of precipitation with synthetic aperture radar", *J. Atmos. Ocean. Technol.*, **4**, 368- 376.

Atlas D. and C.W. Ulbrich, 1974, "The physical basis for attenuation-rainfall relationships and the measurement of rainfall parameters by combined attenuation and radar methods", *J. Rech. Atmos.*, **8** (1–2), 275-298.

Atlas D. and C.W. Ulbrich, 1977, "Path- and area-integrated rainfall measurement by microwave attenuation in the 1–3 cm band", *J.Appl. Meteor*, **16**, 1322–1331.

Atlas D. and R. Wexler, 1963, "Backscatter by oblate ice spheroids", *J. Atmos. Sci.*, **20**, 48–61.

Atlas D., W.G. Harper, F.H. Ludlam and W.C. Macklin, 1960, "Radar scatter by large hail", *Quart. J. Roy. Meteor. Soc.*, **86**, 468–482.

Atlas D., M. Kerker and W. Hitschfeld, 1953, "Scattering and attenuation by non-spherical atmospheric particles", *J. Atmos. Terr. Phys.*, **3**, 108–119.

Atlas D., D. Rosenfeld and D.A. Short, 1990, "The estimation of convective rainfall by area integrals: 1. The theoretical and empirical basis", *J. Geophys. Res.*, **95**, 2153-2160.

Atlas D., D. Rosenfeld and D.B. Wolf, 1990b, "Climatologically tuned reflectivity-rain rate relations and links to area-time integrals", *J. Appl. Meteor*, **29**, 1120-1135.

Atlas D., D.A. Short, and D. Rosenfeld, 1989, "Climatologically tuned reflectivity-rain rate relations", *Preprints, 24th Conf. Radar Meteor.*, Amer. Meteor. Soc., 666-671.

Atlas D., R.C. Srivastava and R.S. Sekhon, 1973, "Doppler radar characteristics of precipitation at vertical incidence", *Rev. Geophys. Space. Phys.*, **11** (1), 1–35.

Atlas D., R. Tatehira, R.C. Srivastava, W. Marker and R.E. Carbone, 1969, "Precipitation-induced mesoscale wind perturbations in the melting layer", *Quart. J. Roy. Meteor. Soc.*, **95**, 544-560.

Atlas D., C.W. Ulbrich and R. Meneghini, 1984, The multiple-parameter remote measurement of rainfall", *Radio Sci.*, **19**, 3-22.

Atlas D, C. W. Ulbrich, F. D. Marks Jr, E. Amitai and C. R. Williams, 1999, "Systematic variation of drop size and radar-rainfall relations," *J. Geophys. Res.* **104,** 6155-6169.

Austin G.L. and A. Bellon, 1974, "The use of digital weather records for short-term precipitation forecasting", *Quart. J. Roy. Meteor. Soc.*, **100**, 658-664.

Austin G.L. and A. Bellon, 1982, "Very short range forecasting of precipitation by the objective extrapolation of radar and satellite data", pp 177-190 in "*Nowcasting*" Ed. K.A.Browning, Acad. Press., 256 pp.

Austin G.L., A. Kilambi, A. Bellon, N. Leutsarakos, A. Hausner, L. Trueman and M. Ivanich, 1986, "An operational, highspeed interactive analysis and display system for intensity radar data processing", *Preprints, 23rd Conf. Radar Meteor.*, Amer. Meteor. Soc., JP79-JP82.

Austin P.M., 1987, "Relation between measured radar reflectivity and surface rainfall", *Mon. Wea. Rev.*, **115**, 1053-1070.

Austin P.M. and A.C. Bemis, 1950, "A quantitative study of the 'bright band' in radar precipitation echoes", *J. Meteor.*, **7**, 145-151.

Aydin K., T.A. Seliga and V. Balaji, 1986, "Remote sensing of hail with dual polarization radar", *J. Climate Appl. Meteor.*, **25**, 1475-1484.

Aydin K., T.A. Seliga and V.N. Bringi, 1984, "Differential radar scattering properties of model hail and mixed-phase hydrometeors", *Radio Sci.*, **19**, 58–66.

Aydin K., Y. Zhao and T.A. Seliga, 1990, "A differential reflectivity radar hail measurement technique: Observations during the Denver hailstorm of 13 June 1984", *J. Atmos. Ocean. Technol.*, **7**, 104-113.

Bader M.J., G.S. Forbes, J.R. Grant, R.B.E. Lilley and A.J. Waters, 1995, "*Images in Weather Forecasting - A practical guide for interpreting satellite and radar imagery*", Cambridge Univ. Press, 499 pp.

Balakrishnan N. and D.S. Zrnić, 1990, "Estimation of rain and hail rates in mixed-phase precipitation", *J.Atmos.* Sci., **47**, 565–577.

Balakrishnan N., D.S.Zrnić, J. Goldhirsh and J. Rowland, 1989, "Comparison of simulated rain rates from disdrometer data employing polarimetric radar algorithms", *J. Atmos. Ocean. Technol.*, **6**, 476-486.

Balakrishnan N., D.S. Zrnić and R.J. Doviak, 1993, "Linear polarization for weather radars – trade offs", *Preprints, 26th International Conf. Radar Meteor.*, Amer. Meteor. Soc., 387–389.

Balakrishnan N., D.S. Zrnić, K. Aydin. V. Chandrasekar, V.N. Bringi and C.L. Ziegler, 1991, "Polarimetric signatures in the stratiform region of a mesoscale convective system", *Preprints, 25th International Conf. Radar Meteor.*, Amer. Meteor. Soc., 525-528.

Ball A.P., J.L. Clarke, M.J.O' Brien, A.H. Shaw, S.E.Trigg and T.A.Voller, 1979, "*The design and implementation of an operational computer-based weather radar system*", Royal Signals and Radar Establishment (RSRE), Memorandum No. 3151, RSRE, Malvern, UK., Unpublished.

Balsley B.B., 1978, "Design considerations for coherent radar systems for probing the troposphere, stratosphere and mesosphere", *Preprints, 18th Conf. Radar Meteor.*, Amer. Meteor. Soc, 387-390.

Balsley, B., 1998, "Who was Doppler?", in Brandolini A. et al., "meeting summary – Report from the International Workshop on Remote Sensing in Geophysics using Doppler Techniques", *Bull. Amer. Meteor. Soc.*, **79**, 643– 656.

Balsley B.B., D.A. Carter, A.C. Riddle, W.L. Ecklund and K.S. Gage, 1991, "On the potential of VHF wind profilers for studying convective processes in the tropics", *Bull. Amer. Meteor. Soc.*, **72**, 1355-1360.

Balsley B.B., W.L. Ecklund, D.A. Carter, A.C. Riddle, and K.S. Gage, 1988, "Average vertical motions in the tropical atmosphere observed by a radar wind profiler on Pohnpei (7°N latitude, 157°E longitude)", *J. Atmos. Sci.*, **45**, 396-405.

Banerjee A.K., K.K. Sharma and A.B. Chaudhury, 1969, "Abnormal rise of surface temperature in association with thunderstorm downdrafts", *Ind. J. Meteor. Geophys.*, **20**, 17-22.

Barclay P.A., 1972, "Radar observations of tropical cyclones Althea and Ada and estimation of maximum wind speed from radar echo motion", *Preprints, 15th Radar meteorology Conf.*, Amer. Meteor. Soc., 107-110.

Bargen D.W. and R.C. Brown, 1980, "Interactive radar velocity unfolding", *Preprints, 19th Radar Meteor. Conf.*, Amer. Meteor. Soc., 278-285.

Barnes G.M., J. F. Gamache, M.A. LeMone and G.J. Stossmeister, 1991, "A convective cell in a hurricane rainband", *Mon. Wea. Rev.*,**119**, 776-794.

Barnes G.M., E.J. Zipser, D. Jorgensen and F. Marks Jr., 1983, "Mesoscale and convective structure of a hurricane rainband", *J. Atmos. Sci.*, **40**, 2125-2137.

Barnston A.G. and J.L. Thomas, 1983, "Rainfall measurement accuracy in FACE: A comparison of gage and radar rainfalls", *J. Climate Appl. Meteor.*, **22**, 2038-2052.

Barnston A.G., W.L. Woodley, J.A. Flueck and M.H. Brown, 1983, "The Florida Area Cumulus Experiment's second phase (FACE-2). Part I: The experimental design, implementation and basic data", *J. Climate Appl. Meteor.*, **22**, 1504-1528.

Barrett E.C. and D.W. Martin, 1981, "*The use of satellite data in rainfall monitoring*", Academic Press, 340 pp.

Barratt P. and I.C. Browne, 1953, "A new method of measuring vertical air currents", *Quart.J. Roy. Meteor. Soc.*, **79**, 550.

Barth M.F., S.W. Stierle and L.A. Benjamin, 1994, "Operational performance of the wind profiler demonstration network", *Preprints, 3rd International Symposium on Tropospheric Profiling: Needs and Technologies*, August 30- September 2, 1994, Geomatikum Universität, Hamburg.

Bateman M.G., W.D. Rust and T.C. Marshall, 1994, "A balloon-borne instrument for measuring the charge and size of precipitation particles inside thunderstorms", *J. Atmos. Ocean. Technol.*, **11**, 161-169.

Battan L.J., 1953, "Observations on the formation and spread of precipitation in convective clouds", *J. Meteor.*, **10**, 311-324.

Battan L.J., 1959,"*Radar Meteorology*", Univ. Chicago Press. 161pp.

Battan L.J., 1964, "Some observations of vertical velocities and precipitation sizes in a thunderstorm", *J. Appl. Meteor.*, **3**, 415-420.

Battan L.J., 1965, "Some factors governing precipitation and lightning from convective clouds", *J. Atmos. Sci.*, **22**, 79-90.

Battan L.J., 1971, "Radar attenuation by wet ice spheres", *J.Appl. Meteor.*, **10**, 247–252.

Battan L.J., 1973, "*Radar Observation of the Atmosphere*", Univ. of Chicago Press, 324 pp.

Battan L.J., 1975, "Doppler radar observations of a hailstorm", *J. Appl. Meteor.*, **14**, 98-108.

Battan L.J., 1976, "Vertical air motions and the Z-R relation", *J. Appl. Meteor.*, **15**, 1120-1121.

Battan L.J. and J.B.Theiss, 1966, "Observations of vertical motions and particle sizes in a thunderstorm", *J. Atmos. Sci.*, **23**, 78-87.

Battan L.J. and J.B.Theiss, 1970, "Measurement of vertical velocities in convective clouds by means of pulsed-Doppler radar ", J. Atmos. Sci., **27**, 293-298.

Battan L.J., J.B. Theiss and A.R. Kassander Jr., 1964, "Some Doppler radar observations of a decaying thunderstorm", *World Conf. Radar Meteor. and 11th Wea. Radar Conf.*, Amer. Meteor. Soc., 362-365.

Bauman W.H. and S. Businger, 1996, "Nowcasting for space shuttle landings at Kennedy Space Center, Florida", *Bull. Amer. Meteor. Soc.*, **77**, 2295-2305.

Baynton H.W., 1978, "Radar design for determining the strength of tropical cyclones in the Bay of Bengal", *Preprints, 18th Conf. Radar. Meteor.*, Amer. Meteor. Soc., 350-357 + 2 pages colour pictures.

Bean B.R. and E.J. Dutton, 1968, "*Radio Meteorology*", Dover Publications, 435 pp.

Bell G.J., 1975, "Observations on the size of the typhoon eye" in "*Typhoon modification*", Proc. WMO Tech. Conf., Manila 15-18 Oct., 1974, WMO No. 408, 19-29.

Bellon A. and G.L. Austin, 1978, "The evaluation of two years of real time operation of a short-term precipitation forecasting procedure (SHARP)", *J. Appl. Meteor.*, **17**, 1778-1787.

Bellon A. and A. Kilambi, 1999, "Updates to the McGill RAPID (Radar data Analysis, Processing and Interactive Display) system", *Preprints, 29th International Radar Meteor. Conf.*, Amer. Meteor. Soc., 121-124.

Bellon A., I. Zawadzki and F. Fabry, 1997, "Measurements of melting layer attenuation at X-band frequencies", *Radio Sci.*, **32**, 943-955,

Bergen W.R. and S.C.Albers, 1988, "Two- and three-dimensional de-aliasing of Doppler radar velocities". *J. Atmos. Ocean. Technol.*, **5**, 305-319.

Berjulev G.P. and V.V. Kostarev, 1974, "Dual wavelength radar measurements of 8 mm radio wave attenuation by atmospheric precipitation and clouds", *J. Rech. Atmos.*, **8**, 359-363,

Bhandari S.M., B.M. Rao, M.S. Narayanan, S. Raghavan, D.W. Martin and B. Auvine, 1987, "*Rainfall estimation using geostationary satellite VHRR images and validation with simultaneous radar rainfall measurements*", ISRO, Bangalore, Sci. Rep., ISRO-SAC-SR-27-87, 22pp.

Bhaskara Rao N.S. and M.V. Dekate, 1967, "Effect of vertical wind shear on the growth of convective clouds", *Quart. J. Roy. Meteor. Soc.*, **93**, 363-367.

Bhaskara Rao N.S. and M.V. Dekate, 1971, "The effect of vertical wind structure on some aspects of convective activity at Bombay", *Ind. J. Meteor. Geophys.*, **22**, 59-66.

Bhattacharjee P. and A.C. De, 1976, "Cyclonic storm of 14-15 August 1974 in the Bay of Bengal – a radar study", *Ind. J. Meteor. Hydrol. Geophys.*, **27**, 9-14.

Bhattacharya A.B., 1992, "Radar echoes from inter-stroke processes in lightning discharges", *Ind. J. Radio & Space Phys.*, **21**, 294-298.

Bhattacharya A.B., 1992b, "Observations of lightning radar echoes", *J. Inst. Electronics & Telecom. Engrs.*, **38**, 216-218.

Bhattacharya A.B., 1994, "A new technique for investigating lightning radar echoes", *IETE Tech. Rev.*, **11**, 137-141.

Bhattacharya A.B., B.K. Datta and R. Bhattacharya, 1994, "Lightning echoes and atmospherics as observed by radar", *Ind. J. Radio & Space Phys.*, **23**, 323-325.

Bhattacharyya P., D.K. Rakshit and A.C. De, 1965, "Line type angel echoes observed at Dum Dum airport Calcutta", *Ind. J. Meteor. Geophys.*, **16**, 249-254.

Bieringer P. and P.S. Ray, 1996, "A comparison of tornado warning lead times with and without NEXRAD Doppler radar", *Wea. Forecasting*, **11**, 47-52.

Bieringer P.E., J.I. Ferris, and M. E. Weber, 2001, "General purpose weather surveillance with the Airport Surveillance Radar 11 (ASR-11)", *Preprints, 30th International Conf. Radar Meteor.*, Amer. Meteor. Soc., 92-94.

Biggerstaff M.I. and R.A. Houze Jr., 1991, "Kinematic and precipitation structure of the 10-11 June 1985 squall line", *Mon.Wea. Rev.*, **119**, 3034-3065.

Biggerstaff M.I. and R.A. Houze Jr., 1993, "Kinematics and microphysics of the transition zone of the 10-11 June 1985 squall line", *J. Atmos. Sci.*, **50**, 3091-3110.

Biggerstaff M.I., and S. A. Listemaa, 2000, "An Improved Scheme for Convective/Stratiform Echo Classification Using Radar Reflectivity" *J. Appl. Meteor.*, **39**, 2129–2150

Bigler S.G., R.G. McGrew and M. St. Clair, 1970, "An experiment in digitizing weather radar data from a four station network", *Preprints, 14th Conf. Radar Meteor.*, Amer. Meteor. Soc., 395-398.

Biswas B. and K. Gupta, 1989, "Heights of Cb clouds around Calcutta airport - diurnal and seasonal variations", *Mausam*, **40**, 169-174.

Biswas B. and K. Gupta, 1992, "Radar echo pattern during the onset of southwest monsoon over Gangetic West Bengal", *Mausam*, **43**, 439-441.

Biswas K.R., Bh.V. Ramanamurty and A.K. Roy, 1962, "Freezing rain at Delhi and associated melting band characteristics", *Ind.J. Meteor. Geophys.*, **13**, Spl. No., 137-142.

Biswas N.C., A.K. Sen and A.K. Hansda, 1988, "Some characteristic features of the Chandbali and Sriharikota cyclones of 1984", *Mausam*, **39**, 107-109.

Black M. and P. Dodge, 1993, "Time-lapse radar images of hurricanes Hugo(1989) and Andrew(1992)", *Preprints, 26th International Conf. Radar Meteor.*, Amer. Meteor. Soc., 97-99.

Black M.L. and R.W. Burpee, 1987, "Temporal and spatial variations of precipitation near the center of tropical cyclones", *Extended Absts., 17th Conf. Hurricanes and Tropical Meteorology*, 170-173.

Black M.L., R.W. Burpee, and F.D. Marks Jr., 1995, "Vertical motion asymmetries in the hurricane eyewall", *Preprints, 27th Conf. Radar Meteor.*, Amer. Meteor. Soc., 574-576

Black P.G., R.A. Black, J. Hallett and W.A. Lyons, 1986, "Electrical activity of the hurricane", *Preprints, 23rd Conf. Radar Meteor.*, Amer. Meteor. Soc., J277-280.

Black P.G., H.V. Senn and C.L. Courtwright, 1972, "Airborne radar observations of eye configuration changes, bright band distribution and precipitation tilt during the 1969 multiple seeding experiments in Hurricane Debbie", *Mon. Wea. Rev.*, **100**, 208-217.

Blackmer R.H. Jr. and R.O. Duda, 1972, "Application of pattern recognition techniques to digitized radar data", *Preprints, 15th Conf. Radar Meteor.*, Amer. Meteor. Soc., 138-143.

Blevis B.C. 1965, "Losses due to rain on radomes and antenna reflecting surfaces", *IEEE Trans. Antennas Propag.*, **AP-13**, 175–176.

Bluestein H.B. and D.S. Hazen, 1989, "Doppler radar analysis of a tropical cyclone over land: Hurricane Alicia (1983) in Oklahoma", *Mon. Wea. Rev.*, **117**, 2594-2611.

Bluestein H.B. and C.R. Parks, 1983, "A synoptic and photographic climatology of low-precipitation severe thunderstorms in the Southern Plains", *Mon. Wea. Rev.*, **111**, 2034-2046.

Bluestein H.B. and W.P. Unruh, 1989, "Observations of the wind field in tornadoes, funnel clouds and wall clouds with a portable Doppler radar", *Bull. Amer. Meteor. Soc.*, **70**, 1514-1525.

Bluestein H.B. and G.R. Woodall, 1990, "Doppler radar analysis of a low-precipitation severe storm", *Mon. Wea. Rev.*, **118**, 1640-1664.

Bluestein H.B., A.L. Pazmany, J.C. Galloway and R.E. McIntosh, 1995, "Studies of the substructure of severe convective storms using a mobile 3-mm-wavelength Doppler radar", *Bull. Amer. Meteor. Soc.*, **76**, 2155-2169.

Boe B.A., J.L. Stith, P.L. Smith, J.H. Hirsch, J.H. Helsdon Jr., A.G. Detwiler, H.D. Orville, B.E. Martner, R.F. Reinking, R.J. Meitin and R.A. Brown, 1992, "The North Dakota thunderstorm project: A cooperative study of high plains thunderstorms", *Bull. Amer. Meteor. Soc.*, **73**, 145-160.

Bohne A.R., 1981, "Estimation of turbulence severity in precipitation environments by radar", *Preprints, 20th Conf. Radar Meteor.*, Amer. Meteor. Soc., 446-453.

Bohren C.F. and L.J. Battan, 1980, "Radar backscattering by inhomogeneous precipitation particles", *J. Atmos. Sci.*, **37**, 1821–1827.

Bolen S. M. and V. Chandrasekar, 2000, "Quantitative Cross Validation of Space-Based and Ground-Based Radar Observations", *J. Appl. Meteor.*, **39,** 2071–2079

Bonesteele R.G. and Y.J. Lin, 1978, "A study of updraft-downdraft interaction based on perturbation pressure and single-Doppler radar data", *Mon. Wea. Rev.*, **106**, 62-68.

Booker H.E., 1948, "Some problems in Radiometeorology" (G.J. Symons memorial lecture), *Quart. J.Roy.Meteor.Soc.*, **74**, 277–315.

Boren T.A., J.R. Cruz and D. Zrnić, 1986, "An artificial intelligence approach to Doppler weather radar velocity de-aliasing", *Preprints, 23rd Conf. Radar Meteor.*, Amer. Meteor. Soc, JP107-110.

Bosart B., W.-C. Lee and R. Wakimoto, 2001, "Improved procedure to correct airborne Doppler radar data", *Preprints 30th Conf. Radar Meteorology*, 283-285.

Bosart L.F. and M.G. Landin, 1994, "An assessment of thunderstorm probability forecasting skill", *Wea. Forecasting*, **9**, 522-531.

Boucher R.J. and R. Wexler, 1961, "The motion and predictability of precipitation lines", *J. Meteor.*, **18**, 160-171.

Boyd B.F., D. E. Harms, M. S. Gremillion and M. E. Fitzpatrick, 1999, "Weather radar improvements for space launch support", *Preprints, 29th International Conf. Radar Meteor.*, Amer. Meteor. Soc.,856-859.

Brandes E.A., 1975, "Optimizing rainfall estimates with the aid of radar", *J.Appl. Meteor.*, **14**, 1339-1345.

Brandes E.A., 1984, "Relationship between radar-derived thermodynamic variables and tornadogenesis", *Mon. Wea. Rev.*, **112**, 1033-1052.

Brandes E.A. and J. Vivekanandan, 1998, "An Exploratory Study in Hail Detection with Polarimetrlc Radar", *Preprints, 14th International Conf. Interactive Information and Processing Systems (IIPS) for Meteorology, Oceanography, and Hydrology*, Amer. Meteor. Soc., 287-290.

Brandes E. A., A. V. Ryzhkov, and D. S. Zrnić, 2001, "An Evaluation of Radar Rainfall Estimates from Specific Differential Phase" *J. Atmos. Ocean. Technol.*, **18**, 363–375

Braun S.A. and R.A. Houze Jr., 1992, "The heat budget of a mid-latitude squall line determined from a thermodynamic and microphysical retrieval", *Proc. 11th International. Conf. Clouds and Precipitation*, Montreal, Canada, 750-753.

Braun S.A. and R.A. Houze Jr., 1993, "The secondary band and reflectivity trough in a mid-latitude squall line: Insights from thermodynamic and microphysical retrieval", *Preprints, 26th Conf. Radar Meteor.*, Amer. Meteor. Soc., 447-450.

Brewster K.A. and D.S. Zrnić, 1986, "Comparison of eddy dissipation rates from spatial spectra of Doppler velocities and Doppler spectral widths", *J.Atmos. Ocean. Technol.* **3**, 440-452.

Bringi V.N. and V. Chandrasekar, 2001, "*Polarimetric Doppler weather radar – Principles and applications*", Cambridge Univ. Press, 636 pp.

Bringi V.N. and A. Hendry, 1990, "Technology of polarization diversity radars in Meteorology", in "*Radar in Meteorology*", Ed. Atlas D., Amer. Meteor. Soc., 153-190.

Bringi V.N., V. Chandrasekar, N. Balakrishnan and D.S. Zrnić, 1990, "An examination of propagation effects in rainfall on radar measurements at microwave frequencies", *J. Atmos. Ocean. Technol.*, **7**, 829–840.

Bringi V.N., T.A. Seliga and E.A. Mueller, 1982, "First comparisons of rainfall rates derived from radar differential reflectivity and disdrometer measurements", *IEEE Trans. Geosci. Remote Sens.*, **GE-20**, 201-204.

Brown R.A. and V.T. Wood, 1987, "*A guide for interpreting Doppler velocity patterns*", NEXRAD Joint System Program Office, 51 pp.

Brown R.A. and V.T.Wood, 1991, "On the interpretation of single-Doppler velocity patterns within severe thunderstorms", *Wea. Forecasting*, **6**, 32-48.

Brown, R. A., J. M. Janish, and V. T. Wood, 2000, "Impact of WSR-88D scanning strategies on severe storm algorithms", *Wea. Forecasting*, **15,** 90–102.

Brown R.A., L.R. Lemon and D.W. Burgess, 1978, "Tornado detection by pulsed Doppler radar", *Mon. Wea. Rev.*, **106**, 29-38.

Brown R. A., V. T. Wood and T. W. Barker, 2002, "Improved detection using negative elevation angles for mountaintop WSR-88Ds: Simulation of KMSX near Missoula, Montana", *Wea. Forecasting*, **17**, 223–237.

Brown R. A., V. T. Wood and D. Sirmans, 2000b, "Improved WSR-88D Scanning Strategies for Convective Storms", *Wea. Forecasting*, 15, 208–220.

Brown E.N. and R.R. Braham Jr., 1959, "Precipitation-particle measurements in trade-wind cumuli", *J. Meteor.*, **16**, 609-616.

Browning K.A., 1964, "Airflow and precipitation trajectories within severe local storms which travel to the right of the winds", *J.Atmos. Sci.*, **21**, 634-639.

Browning K.A., 1965, "Some inferences about the updraft within a severe local storm", *J.Atmos. Sci.*, **22**, 669-677.

Browning K.A., 1979, "The FRONTIERS plan: a strategy for using radar and satellite imagery for very-short-range precipitation forecasting", *Meteor. Mag.*, **108**, 161-184.

Browning K.A., 1986, "Weather radar and FRONTIERS", *Weather*, **41**, 9-16.

Browning K.A. and D. Atlas, 1965, "Initiation of precipitation in vigorous convective clouds", *J.Atmos. Sci.*, **22**, 678-683.

Browning K.A. and C.G. Collier, 1982, "An integrated radar-satellite nowcasting system in the UK", pp 47-61 in "*Nowcasting*" Ed. K.A.Browning, Acad. Press., 256 pp.

Browning K.A. and R.J. Donaldson Jr., 1963, "Airflow and structure of a tornadic storm", *J.Atmos. Sci.*, **20**, 533-545.

Browning K.A. and F.H. Ludlam, 1962, "Airflow in convective storms", *Quart. J. Roy. Meteor. Soc.*, **88**, 117-135.

Browning K.A., and G. Szejwach, 1994, "Developments in observational systems for forecasting", *Meteor. Applns.*, **1**, 3-22.

Browning K.A. and R.Wexler, 1968, "A determination of kinematic properties of a wind field using Doppler radar, *J. Appl. Meteor.*, **7**, 105-113.

Browning K.A., J.C. Fankhauser, J-P Chalon, P.J. Eccles, R.C. Strauch, F.H. Merrem, D.J. Musil, E.L. May and W.R. Sand, 1976, "Structure of an evolving hailstorm. Part V: Synthesis and implications for hail growth and hail suppression", *Mon. Wea. Rev.*, **104**, 603-610.

Browning K.A., T.W. Harrold, A.J. Whyman, and J.G.D. Bymers, 1968, "Horizontal and vertical air motion and precipitation growth within a shower", *Preprints 13th Radar Meteor. Conf.*, Amer. Meteor. Soc., 122-127.

Brunkow, D., V.N . Bringi, P.C. Kennedy, S.A . Rutledge, V. Chandrasekar, E.A . Mueller, and R.K . Bowie, 2000, "A description of the CSU-CHILL National Radar Facility", *J. Atmos. Ocean. Technol.*, **17**, 1596-1608.

Burgess D.W., 1976, "Single Doppler radar vortex recognition. Part I: Mesocyclone signatures", *Preprints, 17th Conf. Radar Meteor.*, Amer. Meteor. Soc., 97-103.

Burnham J., 1970, "Atmospheric gusts - A review of the results of some recent research at the Royal Aircraft Establishment", *Mon. Wea. Rev.*, **98**, 723-734.

Byers H.R. and R.R. Braham Jr., 1949,"*The Thunderstorm*", U.S.Dept. of Commerce,Weather Bureau, Washington D.C., 287 pp.

Byers H.R. and R.D. Coons, 1947, "The "bright line" in radar cloud echoes and its probable explanation", *J.Meteor.*75-81.

Byers H.R. and H.R. Rodebush, 1948, "Causes of thunderstorms of the Florida peninsula", *J. Meteor.*, **5**, 275-280.

Calheiros R.V. and I. Zawadzki, 1987, "Reflectivity-rain rate relationships for radar hydrology in Brazil", *J. Climate Appl. Meteor.*, **26**, 118-132.

Calla O.P.N., R. Singh, P.N.R. Setty, S.S. Rana, S. Barathy and M.K. Radhakrishnan, 1986, "Rain drop size measurements at Ahmedabad", *Vayu Mandal*, **16**, 1 & 2, 6-9.

Calla O.P.N., R. Singh, S. Barathy and M.K. Radhakrishnan, 1987, "*An empirical relation between hourly and one minute rain rate*", Paper presented at IETE Symposium on Propagation Factors for Space Communication, IIT, Madras, September 1987.

Campbell S.D., 1991, "Performance results and potential operational uses for the prototype TDWR microburst prediction algorithm", *Preprints, 25th International Conf. Radar Meteor.*, Amer. Meteor. Soc., J33-J36.

Campbell S.D., M.W. Merritt and J.T. Stefano, 1989, "Microburst recognition performance of TDWR operational testbed", *3rd International Conf. Aviation Weather System*, Amer. Meteor. Soc.

Campos E.F. and I. Zawadzki, 1999, "Z-R relations from independent measurements of raindrop size distributions", *Preprints, 29th International Conf. Radar Meteor.*, Amer. Meteor. Soc., 663-665.

Caracena F., R.L. Holle, and C.A. Doswell III, 1989, "*Microbursts, A handbook for visual identification*", U.S. Dept. Commerce, National Oceanic and Atmospheric Administration (NOAA), 35 pp.

Carbone R.E. and M.J. Carpenter, 1983, "Rapid scan Doppler radar development considerations Part I: Sampling requirements in convective storms", *Preprints, 21st Conf. Radar Meteor.*, Amer. Meteor. Soc., 278-283.

Carbone R.E. and F.D. Marks Jr., 1989, "Velocity track display (VTD): a real-time application for airborne Doppler radar data in hurricanes", *Extended Absts., 18th Conf. Hurricanes and Tropical Meteorology*, Amer. Meteor. Soc., 11-12.

Carbone R.E., F.I. Harris, P.H. Hildebrand, R.A. Kropfli, L.J. Miller, W. Moninger, R.G. Strauch, R.J. Doviak, K.W. Johnson, S.P. Nelson, P.S. Ray and M. Gilet, 1980, "The multiple-Doppler radar workshop, November 1979", *Bull. Amer. Meteor. Soc.*, **61**, 1169-1203.

Carey L.D. and S. A. Rutledge, 2000, "The relationship between precipitation and lightning in tropical island convection: A C-band polarimetric radar study", *Mon. Wea. Rev.*, **128**, 2687–2710

Carlson H.C. Jr. and N. Sundararaman, 1982, "Real-time jetstream tracking; National benefit from an ST radar network for measuring atmospheric motions", *Bull. Amer. Meteor. Soc.*, **63**, 1019-1026.

Carter D.A., W.L. Ecklund, K.S. Gage, M. Spowart, H.L. Cole, E.F. Chamberlain, W.F. Dabberdt and J. Wilson, 1992, "First test of a shipboard wind profiler", *Bull. Amer. Meteor. Soc.*, **73**, 1587-1592.

Carter J.K, D. Sirmans and J. Schmidt, 1986, "Engineering description of the NSSL dual linear polarization Doppler weather radar", Preprints, 23rd Conf. Radar Meteor., Vol. 3, JP381-384.

Cavalli R., 1984, "The operational Swiss weather radar information distribution network", *Preprints, 22nd Conf. Radar Meteor.*, Amer. Meteor. Soc., 21-24.

Caya D. and I. Zawadzki, 1992, "VAD analysis of nonlinear wind fields", *J. Atmos. Ocean. Technol.*, **9**, 575-587,

Caya, A., S. Laroche, I. Zawadzki and T. Montmerle, 2002, Using Single-Doppler Data to Obtain a Mesoscale Environmental Field", *J. Atmos. Ocean. Technol.*, **19**, 21–36.

Caylor I.J. and A.J. Illingworth, 1989, "Identification of the bright band and hydrometeors using co-polar dual polarization radar", *Preprints 24th Conf. Radar Meteor.*, Amer. Meteor. Soc., 9-12.

Chadwick R.B. and K.P. Moran, 1980, "Long-term measurements of C_n^2 in the boundary layer", *Radio Sci.*, **15**, 355–361.

Chakravarty T., S.H. Damle, J.V. Chande, S. Halder, K.P. Ray and Anil Kulkarni, 1993, "Calibration of ST radar using radio source Virgo-A, *Ind. J. Radio Space Phys.*, **22**, 103-107.

Chalon J. -P., J.C. Fankhauser and P.J. Eccles, 1976, "Structure of an evolving hailstorm, Part I: General characteristics and cellular structure", *Mon. Wea. Rev.*, **104**, 564-575.

Chandrasekar V. and V.N. Bringi, 1987, "Simulation of radar reflectivity and surface measurements of rainfall", *J. Atmos. Ocean Technol.*, **4**, 464-478.

Chandrasekar V. and V.N. Bringi, 1988, "Error structure of multiparameter radar and surface measurements of rainfall. Part I: Differential reflectivity", *J. Atmos. Ocean. Technol.*, **5**, 783-795.

Chandrasekar V., C.A. Atwater and T.H. Vonder Haar, 1991, "Convective latent heating estimates from radar data", *Preprints, 25th International Conf. Radar Meteor.*, Amer. Meteor. Soc., 155–158.

Chandrasekar V., V.N. Bringi, N. Balakrishnan and D.S. Zrnić, 1990, "Error structure of multiparameter radar and surface measurements of rainfall. Part III: Specific differential phase", *J. Atmos. Ocean. Technol.*, **7**, 621-629.

Chang S. W-J., 1982, "The orographic effects induced by an island mountain range on propagating tropical cyclones", *Mon. Wea. Rev.*, **110**, 1255-1270.

Charney J.G. and A. Eliassen, 1964, "On the growth of the hurricane depression", *J. Atmos. Sci.*, **21**, 68-75.

Chatterjee G. and N.K. Sur, 1939, "Discussion of results of soundings of temperature and humidity at Jhikargacha (Bengal) in April and May 1929", *Memoirs of the Ind. Met. Dept.*, Vol **26**, Part IX, 165-187 + 20 figures.

Chatterjee K. and S. Mathur, 1982, "Elevated ducting of radio waves in association with trade wind inversion along the coasts of India", *Mausam*, **33**, 433–436.

Chatterjee R.N. and Prem Prakash, 1990, "Radar study of thunderstorms around Delhi during monsoon season", *Mausam*, **41**, 161-165.

Chatterjee R.N. and Prem Prakash, 1990b, "Some aspects of convective clouds around Delhi", *Mausam*, **41**, 475-482.

Chatterjee R.N., Kaushar Ali and Prem Prakash , 1994, "Fractal dimensions of convective clouds around Delhi", *Ind. J. Radio & Space Phys.*, **23**, 189-192.

Chatterjee R.N., Prem Prakash and Kaushar Ali, 1991, "Frequency distributions of the heights of convective clouds in different regions in India", *Atmos. Research*, **26**, 445-454.

Chatterjee R.N., Prem Prakash and Kaushar Ali, 1992, "Distribution of the areas and heights of the radar echoes from convective clouds around Delhi during monsoon season", *Mausam*, **43**, 127-130.

Chatterjee R.N., Prem Prakash and Kaushar Ali, 1993, "A radar study of vertical growth and decay of convective clouds and associated variations of surface rainfall rate", *Ind. J. Radio & Space Phys*, **22**, 230-234.

Chaudhry A.K. and D.K. Rakshit, 1970, "A radar study of pre-monsoon squall lines over Gangetic West Bengal", *Ind. J. Meteor. Geophys.*, **21**, 459-462.

Cheng L. and M. English, 1983, "A relationship between hailstone concentration and size", *J. Atmos. Sci.*, **40**, 204–213.

Cheze J-L, J.Tardieu, and M. Gilet, 1991, "The French weather radar network", *Preprints, 25th International Conf. Radar Meteor.*, Amer. Meteor. Soc., 17-20.

Chilson P.B., C.W.Ulbrich, M.F. Larsen, P.Perillat and J.E.Keener, 1993, "Observations of a tropical thunderstorm using a vertically pointing, dual-frequency, collinear beam Doppler radar", *J. Atmos. Ocean. Technol.*, **10**, 663-673.

Chong M. and C. Campos, 1996, "Extended overdetermined dual-Doppler formalism in synthesizing airborne Doppler radar data", *J. Atmos. Ocean. Technol.*, **13**, 581-597,

Chong M. and J. Testud , 1983, "Three-dimensional wind field analysis from dual-Doppler radar data. Part III: The boundary condition: An optimum determination based on a variational concept", *J.Climate Appl. Meteor.*, **22**, 1227-1241.

Chong M., F. Roux and J. Testud, 1980, "A new filtering and interpolating method for processing dual Doppler radar data: Performance in three dimensional wind restitution, ability to derive pressure and temperature fields", *Preprints, 19th Conf. Radar Meteor.*, Amer. Meteor. Soc., 286-293.

Chong M., J. Testud and F. Roux, 1983, "Three-dimensional wind field analysis from dual-Doppler radar data. Part II: Minimizing the error due to temporal evolution", *J.Climate Appl. Meteor.*, **22**, 1216-1226.

Chu Y.-H. and C.-H. Lin, 1994, "The severe depletion of turbulent echo power in precipitation observed using Chung-Li VHF Doppler radar", *Radio Sci.*, **29**, 1311-1320.

Chu Y.-H., and J. -S. Song, 1998, Observations of precipitation associated with a cold front using a VHF wind profiler and a ground-based optical rain gauge. *J. Geophys. Res.*, **103,** 11,401–11,409.

Chu Y.-H., L.-P. Chian and C.-H. Liu, 1991, "The investigation of the atmospheric precipitations by using Chung-Li VHF radar", *Radio Sci.*, **26**, 717-729.

Chu Y.-H., S-P. Shih, C-L. Su, K-L. Lee, T-H. Lin, and W-C. Liang, 1999, "A study on the relation between terminal velocity and VHF backscatter from precipitation particles using the Chung-Li VHF radar", *J. Appl. Meteor.*, **38**, 1720–1729.

Ciesielski P.E., L.M. Hartten and R.H. Johnson, 1997, "Impacts of merging profiler and rawinsonde winds on TOGA COARE Analyses", *J. Atmos. Ocean. Technol.*, **14**, 264-1279.

Cifelli R. and S.A. Rutledge, 1994, "Vertical motion structure in maritime continent mesoscale convective systems: Results from a 50-MHz profiler", *J. Atmos. Sci.*, **51**, 2631-2652.

Cifelli R., C. R. Williams, D. K. Rajopadhyaya, S. K. Avery, K. S. Gage and P. T. May, 2000, "Drop-size distribution characteristics in tropical mesoscale convective systems", *J.Appl. Meteor.*, **39**, 760–777.

Clark J.F., 1980, "The evolution of airborne weather avoidance radar toward a calibrated remote rain gauge using REACT", *Preprints, 19th Conf. Radar Meteorology*, Amer. Meteor. Soc.., 169-175.

Clarke R.H., 1988, "A statistical study of Melbourne radar rain echoes", *Aust. Meteor. Mag.*, **36**, 245-257.

Clark W.L., J.L. Green and J.M. Warnock, 1986, "The use of a vertical beam clear-air Doppler radar to measure horizontal divergence of the wind field", *Preprints, 23rd Conf. Radar Meteor.*, Amer. Meteor. Soc, 38-40.

Clothiaux E.E., M.A. Miller, B.A. Albrecht, T.P. Ackerman, J. Verlinde, D.M. Babb, R.M. Peters and W.J. Syrett, 1995, "An evaluation of a 94-GHz radar for remote sensing cloud properties", *J. Atmos. Ocean. Technol.*, **12**, 201-229.

Clothiaux E.E., R.S. Penc, D.W. Thompson, T.P. Ackerman and S.R. Williams, 1994, "A first guess feature-based algorithm for estimating wind speed in clear-air Doppler radar spectra", *J. Atmos. Ocean. Technol.*, **11**, 888-908.

Cochran W.T., J.W.Cooley, D.L. Favin, H.D. Helms, R.A. Kaenel, W.W. Lang, G.C. Maling Jr., D.E. Nelson, C.M. Rader and P.D. Welch, 1967, "What is the fast Fourier Transform?", *Proc. IEEE*, **55**, 1664-1674.

Cohn S.A., 1994, "Investigations of the wavelength dependence of radar backscatter from atmospheric turbulence", *J. Atmos. Ocean. Technol.*, **11**, 225–238.

Cohn S.A., 1995, "Radar measurements of turbulent eddy diffusion rate in the troposphere: A comparison of techniques", *J.Atmos. Ocean. Technol.*, **12**, 85-95.

Collier C.G., 1989, "*Applications of Weather Radar Systems*", John Wiley and Sons, 294 pp.

Collier C.G., 1989b, "COST-73: The development of a weather radar network in western Europe", *Preprints, COST-WMO Seminar on "Weather radar networking"*, Sept. 5–8, Brussels.

Collier C.G., 1992, "International radar networking", *Met. Mag.*, **121**, 221-239.

Collier C.G., Ed., 1992b, *International weather radar networking:* Final seminar of the COST Project 73, Kluwer Academic Publishers, 332 pp.

Collier C.G. and M. Chapuis, Eds., 1990, "*Weather radar networking*", Kluwer Academic Publishers, 580 pp.

Collier C.G. and P.K.James, 1986, "On the development of an integrated weather radar data processing system", *Preprints, 23rd Conf. Radar Meteor.*, Amer. Meteor. Soc., JP95-98.

Collier C.G., J.A. Cole and R.B. Robertson, 1980, "The North West Radar Project: the establishment of a weather radar system for hydrological forecasting", *Proc. Oxford Symposium Hydrological Forecasting*, April, 1980, IAHS Publ. No. 129, 31-40.

Collier C.G., C.A. Fair and D.H. Newsome, 1988, "International weather-radar networking in western Europe", Bull. Amer. Meteor. Soc., **69**, 16-21.

Collier C.G., D.M.Goddard and B.J.Conway, 1989, "Realtime analysis of precipitation using satellites, ground-based radars, conventional observations and numerical model output", *Meteor. Mag.*, **118**, 1-8.

Collier C.G., T.W. Harrold and C.A. Nicholass, 1975, "A comparison of areal rainfall as measured by a raingauge-calibrated radar system and raingauge networks of various densities", *Preprints, 16th Conf. Radar Meteor.*, Amer. Meteor. Soc., 467-472.

Collier C.G., P.R. Larke and B.R. May, 1983, "A weather radar correction procedure for real time estimation of surface rainfall", *Quart. J. Royal Meteor. Soc.*, **109**, 589-608.

Collinge V.K.and C. Kirby, 1987, "*Weather Radar and Flood Forecasting*", John Wiley and Sons, 296 pp.

Collis R.T.H., 1964, "Radar precipitation measurements", *Proc. 1964 World Conf. Radio Meteor. incorporating 11th Wea. Radar Conf.*, Amer. Meteor. Soc., 142-145.

Collis R.T.H., R.C. Honey and F.G. Fernald, 1963, "*Study of techniques for measuring rainfall by reference to radar attenuation*", SRI Project No. 4278, Final Report, Stanford Research Institute, Menlo Park, California, USA, 43 pp.

Colón J.A., C.R.V. Raman and V. Srinivasan, 1970, "On some aspects of the tropical cyclone of 20-29 May, 1963 over the Arabian Sea", *Ind. J. Meteor. Geophys.*, **21**, 1-22.

Coons R.D., 1947, "Guided propagation of radar in thunderstorm conditions", *Bull. Amer. Meteor. Soc.*, **28**, 324–329.

Corbet, J., C. Mueller, C. Burghart, K. Gould, and G. Granger, 1994, "Zeb: Software for integration, display, and management of diverse environmental datasets", *Bull. Amer. Meteor. Soc.*, **75**, 783–792.

Cornford S.G. and C.S. Spavins, 1973, "Some measurements of cumulonimbus tops in the pre-monsoon season in north-east India", *Meteor. Mag.*, **102**, 314-332.

Cornman L.B., and W. P. Mahoney III, 1991, "Integration of the TDWR and LLWAS wind shear detection systems", *Preprints, 25th International Conf. Radar Meteor.*, Amer. Meteor. Soc., 630-637.

Crane R.K., 1979, "Automatic cell detection and tracking", *IEEE Trans. Geosci. Electronics*, **GE-17**, 250-262.

Creutin J.D., G. Delrieu and T. Lebel, 1988, "Rain measurement by raingage-radar combination: a geostatistical approach", *J. Atmos. Ocean. Technol.*, **5**, 102-115.

Crow E.L. and K. Shimizu, 1988, "*Lognormal Distributions Theory and Applications*", Statistics:Textbooks and monographs Series/88, Marcel Dekker Inc., 270, Madison Avenue, New York, N.Y. 10016, 408 pp,

Croxton F.E., D.J. Cowden and S. Klein, 1969, "*Applied General Statistics*", Prentice-Hall of India, 754 pp.

Crozier C.L. and J.W. Scott, 1981, "A C-band weather radar system with spherics to serve multiple diverse users in the Canadian climate", *Preprints, 20th Conf. Radar Meteor.*, Amer. Meteor. Soc., 749-756.

Crozier C.L., P.I. Joe, J.W. Scott, H.N. Herscovitch and T.R. Nichols, 1989, "First experiences with an operational Doppler radar", *Preprints, 24th conf. Radar Meteor.*, Amer. Meteor. Soc., 179-185.

Crum T.D., 1995, "An update on WSR-88D Level II data availability", *Bull. Amer. Meteor. Soc.*, **76**, 2485-2487.

Crum T.D., and R.L. Alberty, 1993, "The WSR-88D and the WSR-88D Operational Support Facility", *Bull. Amer. Meteor. Soc.*, **74**, 1669-1687.

Crum T.D., R.L. Alberty and D.W. Burgess, 1993, "Recording, archiving, and using WSR-88D data", *Bull. Amer. Meteor. Soc.*, **74**, 645-653.

Crum T. D., R. E. Saffle and J. W. Wilson, 1998, "An Update on the NEXRAD Program and Future WSR-88D Support to Operations", *Wea. Forecasting*, **13**, 253–262.

Cunningham R.M., 1947, "A different explanation of the "bright line"", *J.Meteor.*, **4**, 163.

Currier P.E., S.K. Avery, B.B. Balsley, K.S. Gage and W.L. Ecklund, 1991, "Use of two wind profilers for precipitation studies", *Preprints, 25th International Conf. Radar Meteor.*, Amer. Meteor. Soc, 261-264.

Dabberdt W.F. and T.W. Schlatter, 1996, "Research opportunities from emerging atmospheric observing and modeling capabilities", *Bull. Amer. Meteor. Soc.*, **77**, 305-323.

Dance S. and R. Potts, 2002, "Microburst detection using agent networks", *J.Atmos. Ocean. Technol.* **19**, 646–653.

Danielson E.F., R. Black and D.A. Morris, 1972, "Hail growth by stochastic collection in a cumulus model", *J.Atmos. Sci.*, **29**, 135-155.

Das P.M., A.C. De and M. Gangopadhyaya, 1957, "Movement of two nor'westers of West Bengal", *Ind. J. Meteor. Geophys.*, **8**, 399-406.

Datar S.V. and D.N. Sikdar, 1964, "Angels near Bombay", *Ind.J. Meteor. Geophys.*, **15**, 459-466.

Dawson G.A., 1972, "Radar as a diagnostic tool for lightning", *J. Geophys. Res.*, **77**, 4518-4527.

Dazhang T., S.G. Geotis, R.E. Passarelli Jr., A.L. Hansen and C.L. Frush, 1984, "Evaluation of an alternating-PRF method for extending the range of unambiguous Doppler velocity", *Preprints, 22nd Conf. Radar Meteor.*, Amer. Meteor. Soc., 523-527.

De A.C., 1959, "An unusually high nor'wester radar cloud", *Ind. J. Meteor. Geophys.*, **10**, 3, 359-362.

De A.C., 1959b, "An unusual case of abnormal propagation of microwave caused by nor'wester", *Ind. J. Meteor. Geophys.*, **10**, 420-424.

De A.C., 1962, "Bright band phenomenon over Calcutta and neighbourhood", *Ind.J. Meteor. Geophys.*, **13**, Spl. No. 143-146.

De A.C., 1963, "Movement of pre-monsoon squall lines over Gangetic West Bengal as observed by radar at Dum Dum airport", *Ind. J. Meteor. Geophys.*, **14**, 37-45.

De A.C. and D.K. Rakshit, 1961, "Radar observations on the formation of cumulus clouds near Calcutta during monsoon season" *Ind. J. Meteor. Geophys.*, **12**, 289-298.

De A.C. and S.N. Sen, 1961, "A radar study of pre-monsoon thunderstorms (nor'westers) over Gangetic West Bengal", *Ind. J. Meteor. Geophys.*, **12**, 51-60.

de Elía R., and I. Zawadzki, 2000, "Sidelobe contamination in bistatic radars", *J. Atmos. Oceanic Technol.*, **17**, 1313–1329.

de Elía R. and I. Zawadzki, 2001, "Optimal layout of a bistatic radar network", *J. Atmos. Oceanic Technol.*, **18**, 1184–1194.

Decker R. and R. Ewald, 1994, "Subjective meteorological results of NWS wind profiler assessment program", *Preprints, 3rd International Symposium on Tropospheric Profiling: Needs and Technologies*, August 30- September 2, 1994, Geomatikum Universität, Hamburg, pp 14-16.

Delage D., R. Roca, F. Bertin, J. Delcourt, A. Cremieu, M. Massebeuf, R. Ney and P. Van Velthoven, 1997, "A consistency check of three radar methods for monitoring eddy diffusion and energy dissipation rates through the tropopause", *Radio Sci.*, **32**, 757-767.

Delrieu G., S. Caoudal and J. D. Creutin, 1997, "Feasibility of using mountain return for the correction of ground-based X-band weather radar", *J. Atmos. Oceanic Technol.*, **14,** 368–385.

Delrieu G., S. Serrar, E. Guardo and J.D. Creutin, 1999: Rain measurements in hilly terrain with X-band weather radar systems : Accuracy of path integrated attenuation estimates derived from mountain returns", *J. Atmos. Oceanic Technol.*J., **16**, 405-416

Desrochers P.R., 1989, "A reliable method for real-time velocity unfolding", *Proc. 24th Conf. radar Meteor.,* Amer. Meteor. Soc, 415-418.

Desrochers P.R. and R.J. Donaldson Jr., 1992, "Automatic tornado prediction with an improved mesocyclone-detection algorithm", *Wea. Forecasting*, **7**, 373-388.

Desrochers P.R., R.J. Donaldson Jr. and D.W. Burgess, 1986, "Mesocyclone rotational kinetic energy as a discriminator for tornadic and non-tornadic types", *Preprints, 23rd Conf. Radar Meteor.*, Amer. Meteor. Soc., 1-4.

Direskeneli H., K. Aydin and T.A. Seliga , 1986, "Radar estimation of rainfall rate using reflectivity factor and differential reflectivity measurements obtained during MAYPOLE '84: Comparison with ground-based raingages", *Preprints, 23rd Conf. Radar Meteor.*, Amer. Meteor. Soc., 116-120.

Dixon M. and G. Wiener, 1993, "TITAN; Thunderstorm Identification, Tracking, Analysis and Nowcasting_ A radar-based methodology", *J. Atmos. Ocean. Technol.*, **10**, 785-797.

Dobos P.H., R.J. Lind and R.L. Elsberry, 1995, "Surface wind comparisons with radar wind profiler observations near tropical cyclones", *Wea. Forecasting*, **10**, 564-575.

Doick, J. J., and A. R. Holt, 1995, "Combined three-dimensional displays of weather radar data with terrain", Preprints, *27th Conf. Radar Meteor.*, Amer. Meteor. Soc., 371–372.

Donaldson R.J. Jr., 1958, "Analysis of severe convective storms observed by radar", *J. Meteor.*, **15**, 44-50.

Donaldson R.J. Jr., 1959, "Analysis of severe convective storms observed by radar - II", *J. Meteor.*, **16**, 281-287.

Donaldson R.J. Jr., 1961, "Radar reflectivity profiles in thunderstorms", *J. Meteor.*, **18**, 292-305.

Donaldson R.J. Jr., 1965, "Methods for identifying severe thunderstorms by radar", *Bull. Amer. Meteor. Soc.*, **46**, 174-193.

Donaldson R.J. Jr., 1970, "Vortex signature detection by a Doppler radar", *J. Appl. Meteor.*, **9**, 661-670.

Donaldson R.J. Jr., 1989, "Potential vortex fit", *Preprints, 24th Conf. Radar Meteor.*, Amer. Meteor. Soc., 186-189.

Donaldson R.J. Jr., 1991, "A proposed technique of diagnosis by radar of hurricane structure", *J. Appl. Meteor.*, **30**, 1636-1645.

Donaldson R.J. Jr., and T.M. Bals, 1993, "Deformation field in hurricane Bob (1991) measured by single-Doppler radar", *Preprints, 20th Conf. Hurricanes and Trop. Meteor.*, Amer. Meteor. Soc., 451-454.

Donaldson R.J. Jr. and P.R. Desrochers, 1990, "Improvements of tornado warnings by Doppler radar measurement of mesocyclone rotational kinetic energy", *Wea. Forecasting*, **5**, 247-258.

Donaldson R.J. Jr. and P.R. Desrochers, 1991, "A statistical survey of mesocyclonic rotation provided by a new mesocyclone algorithm", *Preprints, 25th International Conf. Radar Meteor.*, Amer. Meteor. Soc., 127-130.

Donaldson R.J.Jr. and F.I. Harris, 1984, "Detection of wind field curvature and wind speed gradients by a single Doppler radar", *Preprints, 22nd Conf. Radar Meteor.*, Amer. Meteor. Soc., 514-519.

Donaldson R.J.Jr. and F.I. Harris, 1989, "Estimation by Doppler radar of curvature, diffluence and shear in cyclonic flow", *J. Atmos. Ocean. Technol.*, **6**, 26-35.

Donaldson R.J. Jr. and F.H. Ruggiero, 1986, "Wind field derivatives in hurricane Gloria estimated by Doppler radar", *Preprints, 23rd Conf. Radar Meteor.*, Amer. Meteor. Soc., Vol. **3**, JP 236-239.

Donaldson R.J. Jr. and R.Wexler, 1969, "Flight hazards in thunderstorms determined by Doppler velocity variance" *J. Appl. Meteor.*, **8**, 128-133.

Doneaud A.A., S. Ionescu-Niscov, D.L. Priegnitz and P.L. Smith, 1984, "The Area Time Integral as an indicator for convective rain volumes", *J.Climate Appl. Meteor.*, **23**, 555-561.

Doneaud A.A., J.R. Miller Jr., R. Johnson, T.H. Vonder Haar and P. Laybe, 1987, "The Area-Time-Integral to estimate convective rain volumes over areas applied to satellite data - A preliminary investigation", *J. Climate Appl. Meteor.*, **26**, 156-169.

Douglas M.W., and D.J. Stensrud, 1996, "Upgrading the North American upper-air observing network: what are the possibilities?", *Bull. Amer. Meteor. Soc.*, **77**, 907-924.

Doviak R.J. and D. Sirmans, 1973, "Doppler radar with polarization diversity", *J. Atmos. Sci.*, **30**, 737-738.

Doviak R.J. and D.S. Sirmans, 1979, "Doppler weather radar", *Proc. IEEE*, **67**, *1522-1553*.

Doviak R.J. and D.S. Zrnić, 1984, "*Doppler radar and Weather Observations*", Acad. Press, 458 pp.

Doviak R.J. and D.S.Zrnić, 1993, "*Doppler radar and Weather Observations*", Acad. Press, 2nd Edn., 562 pp.

Doviak R.J., V. Bringi, A. Ryzhkov, A. Zahrai and D. Zrnić, 2000, "Considerations for Polarimetric Upgrades to Operational WSR-88D Radars", *J. Atmos. Ocean. Technol.*, **17**, 257-278.

Doviak R.J., R.M. Rabin and A.J. Koscielny, 1983, "Doppler weather radar for profiling and mapping winds in the prestorm environment", *IEEE Trans. Geosci. Remote Sens.*, **GE-21**, 25-33.

Doviak R.J., D. Sirmans, D. Zrnić and G.B. Walker, 1976, "Resolution of pulse-Doppler radar range and velocity ambiguities in severe storms", *Preprints, 17th Conf. Radar Meteor.*, Amer. Meteor. Soc, 15-22.

Doviak R.J, D.S. Zrnić and A. Shapiro, 2001, "Phased array weather radar - benefits and challenges", *Preprints, 30th Conf. Radar Meteor.*, 202-204.

Dowell D.C., J. Wurman, and L. J. Wicker, 2001, "Centrifuging of scatterers in tornadoes", *Preprints, 30th Conf. Radar Meteor.*, 307-309.

Durden S.L. and C.L. Werner, 1996, "Application of an interferometric phase unwrapping technique to dealiasing of weather radar velocity fields", *J. Atmos. Ocean. Technol.*, **13**, 1107-1109.

Durden S.L., E. Im, F.K. Li, W. Ricketts, A. Tanner and W. Wilson, 1994, "ARMAR: An airborne rain-mapping radar", *J. Atmos. Ocean. Technol.*, **11**, 727-737.

Durden S.L, A. Kitiyakara, E. Im, A.B. Tanner, Z.S. Haddad, F.K. Li and W.J. Wilson, 1997, "ARMAR observations of the melting layer during TOGA COARE", *IEEE Trans. Geosc. Remote Sensing*, **35**, 1453-1456.

Durst C.S., 1946, "Radio Climatology", *Meteorological Factors in Radio Wave Propagation*, Phys.Soc., London, 193–215.

Duvernoy J. and J.L. Gaumet, 1996, "Precipitating hydrometeor characterization by a CW Doppler radar", *J. Atmos. Ocean. Technol.*, **13**, 620-629.

Duvernoy J., J-L. Gaumet and M. Gilet, 1991, "Measurement of hydrometeor size distributions with a vertically pointing CW radar", *Preprints, 25th International Conf. Radar Meteor.*, Amer. Meteor. Soc., 797-800.

Dvorak V.F., 1984, "*Tropical cyclone intensity analysis using satellite data*", NOAA Tech Report NESDIS 11, U.S.Dept. of Commerce, Washington D.C., 47 pp.

Dyer R., 1968, "Doppler measurements in stratiform rain", *Proc., 13th Wea. Radar Conf.* Amer. Meteor. Soc., 144-147.

East T.W.R., 1958, "Electronic constant altitude plan position indicator for a weather radar", Stormy Weather Group (Canada) Sci.Rep. MW-28.

Easterbrook C.C., 1975, "Estimating horizontal wind fields by the two-dimensional curve fitting of single Doppler radar measurements", *Preprints, 16th Conf. Radar Meteor.*, Amer. Meteor. Soc, 214-219.

Eastwood E., 1967, "*Radar ornithology*", Methuen & Co., 278 pp.

Eccles P.J. and D. Atlas, 1973, "A dual-wavelength radar hail detector", *J. Appl. Meteor.*, **12**, 847-854.

Eccles P.J. and E.A. Mueller, 1971, "X-band attenuation and liquid water content estimation by a dual-wavelength radar", *J. Appl. Meteor.*, **10**, 1252-1259.

Ecklund W.L., D.A. Carter and B.B. Balsley, 1988, "A UHF wind profiler for the boundary layer: Brief description and initial results", *J. Atmos. Ocean. Technol.*, **5**, 432-441.

Ecklund W.L., K.S. Gage and C.R. Williams, 1995, "Tropical precipitation studies using a 915-MHz wind profiler", *Radio Sci.*, **30**, 1055-1064.

Ecklund W. L., C. R. Williams, P. E. Johnston and K. S. Gage , 1999, "A 3-GHz profiler for precipitating cloud studies", *J. Atmos. Ocean. Technol.*, **16**, 309-322.

Eilts M.D, 1992, "Aviation hazard identification using Doppler radar", pp 181-193 in *Postprint Volume National weather Service Aviation Workshop*, December 1991, NOAA Tech. Memo. NWS CR-102, 411 pp.

Eilts M.D. and S.D. Smith, 1990, "Efficient dealiasing of Doppler velocities using local environment constraints", *J. Atmos. Ocean. Technol.*, **7**, 118-128.

Eilts M.D., E.D. Mitchell and K. Hondl, 1991, "The use of Doppler radar to help forecast the development of thunderstorms", *Preprints, 25th International Conf. Radar Meteor.*, Amer. Meteor. Soc., 63-66.

Eilts M.D., S.H. Olson, G.J. Stumpf. L.G. Hermes, A. Abrevaya, J. Culbert, K.W. Thomas, K. Hondl and D. Klingle-Wilson, 1991, "An improved gust front detection algorithm for the TDWR", *Preprints, 25th International Conf. Radar Meteor.*, Amer. Meteor. Soc., J37-J42.

Ekengren B. and L. Dahlberg, 1984, "A dual Doppler radar for unambiguous velocity measurements", *Preprints 22nd Conf. Radar Meteor.*, Amer. Meteor. Soc., 520-522.

Ekpenyong B.E. and R.C. Srivastava, 1970, "Radar characteristics of the melting layer – a theoretical study", *Preprints, 14th Conf. Radar Meteor.*, Amer. Meteor. Soc., 161-166.

Ellis S. M., F. Pratte and C. L. Frush, 1999, "Compensating reflectivity for clutter filter bias in the WSR-88D" *15th International Conf. Interactive Information and Processing Systems (IIPS) for meteorology, oceanography, and hydrology*, 306-309.

Ellis S.M., J. Vivekanandan, E. A. Brandes, J. Stith, and R. J. Keeler, 2000, "Aircraft icing detection using S-band polarization radar measurements, *Preprints, 9th Conference on Aviation, Range, and Aerospace Meteorology,* Amer. Meteor. Soc., Boston, MA, 35-39.

Endlich R.M. and D.E. Wolf, 1981, "Automatic cloud tracking applied to GOES and METEOSAT observations", *J. Appl. Meteor.*, **20**, 309-319.

Endlich R.M., D.E. Wolf, D.J. Hall and A.E. Brain, 1971, "Use of a pattern recognition technique for determining cloud motions from sequences of satellite photographs", *J. Appl. Meteor.*, **10**, 105-117.

English E.J., 1973, "An objective method of calculating areal rainfall", *Meteor. Mag.*, **102**, 292-298.

Evans J.E., 1991, "Status of the Terminal Doppler Weather Radar one year before deployment", *Preprints, 25th International Conf. Radar Meteor.*, Amer. Meteor. Soc., J1-J6.

Fabry F. and I. Zawadzki, 1995, "Long-term radar observations of the melting layer of precipitation and their interpretation, J. Atmos. Sci., **52**, 838-851.

Fankhauser J.C., 1976, "Structure of an evolving hailstorm, Part II: Thermodynamic structure and airflow in the near environment", *Mon. Wea. Rev.*, **104**, 576-587.

Federer B., 1977, "Methods and results of hail suppression in Europe and in the USSR", pp 215-223 in "*Hail: A Review of Hail Science and Hail Suppression*", G.B. Foote and C.A. Knight Eds., Amer. Meteor. Soc. Meteor. Monograph No. 38.

Feingold G. and Z. Levin, 1987, "Application of the log-normal raindrop distribution to differential reflectivity radar measurement (ZDR)", *J. Atmos. Ocean. Technol.,* **4**, 377–382.

Ferraro R.R. and G.F. Marks, 1995, "The development of SSM/I rain-rate retrieval algorithms using ground-based radar measurements", *J. Atmos. Ocean. Technol.*, **12**, 755-770.

Fiore J.V. Jr., R.K. Farnsworth and G. Huffman, 1986, "Quality control of radar-rainfall data with VISSR satellite data", *Preprints, 23rd Conf. Radar Meteor.*, Amer. Meteor. Soc, JP15-JP18.

Firda J. M., S. M. Sekelsky and R. E. Mcintosh, 1999, "Application of dual-frequency millimeter-wave Doppler spectra for the retrieval of drop size distributions and vertical air motion in rain", *J. Atmos. Ocean. Technol.*, **16**, 216-236.

Foote G.B. and P.S Du Toit, 1969, "Terminal velocity of raindrops aloft" *J. Appl. Meteor.* **8**, 249–253.

Foote G.B. and C.A. Knight, 1979, "Results of a randomized hail suppression experiment in northeast Colorado. Part I: Design and conduct of the experiment", *J. Appl. Meteor.*, **18**, 1526-1537.

Forbes G.S., 1981, "On the reliability of hook echoes as tornadic indicators", *Mon. Wea. Rev.*, **109**, 1457-1466.

Forsyth D.E., D.W. Burgess, M.H. Jain and L.E. Mooney, 1989, "DOPLIGHT '87: Application of Doppler radar technology in a National Weather Service Office", *Preprints, 24th Conf. Radar Meteor.*, Amer. Meteor. Soc., 198-202.

Frank N.L., P.L. Moore and G.E. Fisher, 1967, "Summer shower distribution over the Florida peninsula as deduced from digitized radar data", *J. Appl. Meteor.*, **6**, 309-316.

Frankel M.S., N.J.F. Chang and M.J. Sanders Jr., 1977, "A high-frequency radio-acoustic sounder for remote measurement of atmospheric winds and temperature", *Bull. Amer. Meteor. Soc.*, **58**, 928-934.

Frisch A.S. and S.F. Clifford, 1974, "A study of convection capped by a stable layer using Doppler radar and acoustic echo sounders", *J. Atmos. Sci.*, **31**, 1622-1628.

Frisch A.S. and R.G. Strauch, 1976, "Doppler radar measurements of turbulent kinetic energy dissipation rates in a northeastern Colorado convective storm", *J. Appl. Meteor.*, **15**, 1012-1017.

Frisch A.S., and B.L.Weber, 1992, "The distribution of C_n^2 as measured by 50-, 405- and 915-MHz wind profilers", *J. Atmos. Ocean. Technol.*, **9**, 318-322.

Frisch A.S., C.W. Fairall and J.B. Snider, 1995, "Measurements of stratus cloud and drizzle parameters in ASTEX with a Ka band Doppler radar and a microwave radiometer", *J. Atmos. Sci.*, **52**, 2788-2799.

Frisch A.S., D.H. Lanschow, C.W. Fairall, W.H. Schubert and J.S. Gibson, 1995b, "Doppler radar measurements of turbulence in marine stratiform cloud during ASTEX, *J. Atmos. Sci.*, **52**, 2803-2808.

Frisch A.S., B.L. Weber, R.G. Strauch, D.A. Merritt and K.P. Moran, 1986, "The altitude coverage of the Colorado wind profilers at 50, 405 and 915 MHz", *J. Atmos. Ocean. Technol.*, **3**, 680-692.

Frisch A.S., B.L. Weber, D.B. Wuertz, R.G. Strauch and D.A. Merritt, 1990, "The variations of C_n^2 between 4 and 18 km above sea leval as measured over 5 years", *J. Appl. Meteor.*, **29**, 645-651.

Frost I.R., J.W.F. Goddard and A.J. Illingworth, 1991, "Hydrometeor identification using cross polar radar measurements and aircraft verification", *Preprints, 25th International Conf. Radar Meteor.*, Amer. Meteor. Soc., 658-661.

Frost W., H-P. Chang, J. McCarthy and K. Elmore, 1983, "Aircraft performance in a JAWS microburst", *Preprints, 21st Conf. Radar Meteorology*, Amer. Meteor. Soc., 630-637.

Frush C.L., 1984, "Using the sun as a calibration aid in mutiple parameter meteorological radars", *Preprints, 22nd Conf. Radar Meteor.*, Amer. Meteor. Soc., 306-311.

Frush C. L. and J. Daughenbaugh, 1999, "Range/velocity ambiguity reduction: evaluation of the SZ(8/64) phase code on a WSR-88D radar", *Preprints, 29th International Conf. Radar Meteor.*, Amer. Meteor. Soc., 513-516.

Frush C., P. Hildebrand and C. Walther, 1986, "The NCAR airborne Doppler radar, Part II: System design considerations", *Preprints, 23rd Conf. Radar Meteor.*, Amer. Meteor. Soc., 151-154.

Fujita T.T., 1959, "Precipitation and cold air production in mesoscale thunderstorm systems", *J. Meteor.*, **16**, 454-466.

Fujita T.T., 1985, "*The downburst, microburst and macroburst*", SMRP research paper no. 210, Univ. of Chicago, 122 pp.

Fujita T.T., 1981, "Tornadoes and downbursts in the context of generalized planetary scales", *J.Atmos.Sci.*, **38**, 1511-1534.

Fujita T.T and J. McCarthy, 1990, "The application of weather radar to aviation meteorology", pp 657-681 in "*Radar in Meteorology*", (Ed., D. Atlas), Amer. Meteor. Soc., 806 pp.

Fujiwara M., 1965, "Raindrop size distribution in individual storms", *J. Atmos. Sci.*, **22**, 585-591.

Fukao S., M. Inaba, I. Kimura, P.T. May, T. Sato, T. Tsuda, and S. Kato, 1988b, "A systematic error in MST/ST radar wind measurement induced by a finite range volume effect. 2. Numerical considerations", *Radio Sci.*, **23**, 74-82.

Fukao S., T.Sato, P.T. May, T. Tsuda, S. Kato, M. Inaba and I. Kimura, 1988, "A systematic error in MST/ST radar wind measurement induced by a finite range volume effect. 1. Observational results", *Radio Sci.*, **23**, 59-73.

Fukao S., T.Sato, T.Tsuda, S. Kato, M. Inaba and I. Kimura, 1988c, "VHF Doppler radar determination of the momentum flux in the upper troposphere and lower stratosphere: Comparison between the three- and four-beam methods", *J. Atmos. Ocean. Technol.*, **5**, 57-69.

Fukao S., T. Sato, N. Yamasaki, R.N. Harper and S. Kato, 1982, "Winds measured by a UHF Doppler radar and rawinsondes: comparisons made on 26 days (August-September 1977) at Arecibo, Puerto Rico", *J. Appl. Meteor.*, **21**, 1357-1363.

Fukao S., K. Wakasugi, T.Sato, S. Morimoto, T.Tsuda, I. Hirota, I. Kimura and S.Kato, 1985, "Direct measurement of air and precipitation particle motion by very high frequency Doppler radar", *Nature*, **316**, 712-714.

Fulton R. A., J. P. Breidenbach, D-J. Seo, D. A. Miller and T. O'Bannon, 1998, "The WSR-88D Rainfall Algorithm", *Wea. Forecasting*, **13**, 377–395.

Fulton R., D.S. Zrnić and R.J. Doviak, 1986, "Doppler radar and in-situ measurements of a nocturnal thunderstorm gust front", *Preprints, 23rd Conf. Radar Meteor.*, Amer. Meteor. Soc., J85-J88.

Furumoto J., K. Kurimoto and T. Tsuda, 2003, "Continuous Observations of Humidity Profiles with the MU Radar– RASS Combined with GPS and Radiosonde Measurements", *J. Atmos. Ocean.Technol.*, **20**, 23–41.

Gage K.S., 1990, "Radar observations of the free atmosphere: Structure and dynamics", pp534-565 in *"Radar in Meteorology"*, Ed. D. Atlas, Amer. Meteor. Soc, 806 pp.

Gage K.S. and J.L. Green, 1979, "Tropopause detection by partial specular reflection using VHF radar", *Science*, **203**, 1238-1240.

Gage K.S. and J.L. Green, 1982, "A technique for determining the temperature profile from VHF radar observations", *J. Appl. Meteor.*, **21**, 1146-1149.

Gage K.S. and J.L. Green, 1982b, "An objective method for the determination of tropopause height from VHF radar observations", *J. Appl. Meteor.*, **21**, 1150-1154.

Gage K.S., B.B. Balsley and J.L. Green, 1981, "Fresnel scattering model for the specular echoes observed by VHF radar", *Radio Sci.*, **16**, 1447-1453.

Gage K.S., W.L. Ecklund and B.B. Balsley, 1985, "A modified Fresnel scattering model for the parameterization of Fresnel returns", *Radio Sci.*,**20**, 1493–1501.

Gage K.S., W.L. Ecklund, A.C. Riddle and B.B. Balsley, 1986, "Objective tropopause height determination using low-resolution VHF radar observations", *J. Atmos. Ocean. Technol.*, **3**, 248-254.

Gage K.S., J.R. McAfee, W.L. Ecklund, D.A. Carter, C.R. Williams, P.E. Johnston and A.C. Riddle, 1994, "The Christmas Island wind profiler: A prototype VHF wind-profiling radar for the tropics", *J. Atmos. Ocean. Technol.*, **11**, 22-31.

Gage K.S., C.R. Williams and W.L.Ecklund, 1994b, "UHF wind profilers: A new tool for diagnosing tropical convective cloud systems", *Bull. Amer. Meteor. Soc.*, **75**, 2289-2294.

Gage, K. S., C. R. Williams, W. L. Ecklund and P. E. Johnston, 1999, "Use of two profilers during MCTEX for unambiguous identification of Bragg scattering and Rayleigh scattering", *J. Atmos. Sci.*, **56**, 3679–3691.

Gage K. S., C. R. Williams, P. E. Johnston, W. L. Ecklund, R. Cifelli, A. Tokay and D. A. Carter. 2000, "Doppler Radar Profilers as Calibration Tools for Scanning Radars", *J. Appl. Meteor.*, **39**, 2209–2222.

Gagin A., D. Rosenfeld and R.E. López, 1985, "The relationship between height and precipitation characteristics of summertime convective cells in Florida", *J.Atmos. Sci.*, **42**, 84-94.

Gal-Chen T., 1978, "A method for initialization of the anelastic equations: Implications for matching models with observations", *Mon. Wea. Rev.*, **106**, 587-606.

Gal-Chen, T., 1982, "Errors in fixed and moving frame of references: Applications for conventional and Doppler radar analysis", *J. Atmos. Sci.*, **39,** 2279–2300.

Gal-Chen, T., and J. Zhang, 1993, "On the optimal use of reflectivities and single-Doppler radar velocities to deduce 3D motions", *Preprints, 26th International Conf. Radar Meteor.*, Amer. Meteor. Soc., 414–416.

Gall R., J. Tuttle and P. Hildebrand, 1998, "Small-Scale Spiral Bands Observed in Hurricanes Andrew, Hugo, and Erin", *Mon. Wea. Rev.*, . **126**, 1749–1766

Gallagher III F.W., 1989, "The next generation of sensors of ASOS", *WMO Instruments and Observing Methods Report No. 35*, WMO/TD - No. 303, WMO, Geneva, pp 105-109.

Galli G., 1984, "Generation of the Swiss radar composite", *Preprints, 22nd Conf. Radar Meteor.*, Amer. Meteor. Soc., 208-213.

Galloway J., A. Pazmany, I. Mead, R.E. McIntosh, D. Leon, J. French, R. Kelly and G. Vali, 1997, "Detection of ice hydrometeor alignment using an airborne W- band polarimetric radar", *J. Atmos. Ocean. Technol.*, **14**, 3-12.

Gamache J.F., F.D. Marks Jr. and F. Roux, 1995, "Comparison of three airborne Doppler sampling techniques with airborne in situ wind observations in Hurricane Gustav (1990)", *J. Atmos. Ocean. Technol.*, **12**, 171-181.

Gao Ding Xi, 1985, "Radar study of typhoon Vera", in "*Report of the seminar on the application of radar data to tropical cyclone forecasting, Bangkok, 1983*", World Meteorological Organization(WMO), Rep. No. TCP-19, WMO/TD No. 26.

Gao J., M. Xue, A. Shapiro, Q. Xu and K. K. Droegemeier, 2001, "Three-dimensional simple adjoint velocity retrievals from single-Doppler radar", *J. Atmos. Ocean. Technol.*, **18**, 26–38.

Garner T. and T. D. Oram., 2000, "Natural and triggered lightning forecasts for space shuttle landings", *Preprints, Ninth Conf. Aviation, Range, and Aerospace Meteor.*, Amer. Meteor. Soc., 409-414.

Gentry R.C., 1969, "Project Stormfury", *WMO Bulletin*, **18**, 146-154.

Gentry R.C., 1983, "Genesis of tornadoes associated with hurricanes", *Mon. Wea. Rev.*, **111**, 1793-1805.

Georges T.M., and G.D. Thome, 1990, "An opportunity for long-distance oceanographic and meteorological monitoring using over-the-horizon defense radars", *Bull. Amer. Meteor. Soc.*, **71**, 1739-1745.

Georges T.M., J.A. Harlan, L.R. Meyer, and R.G. Peer, 1993," Tracking hurricane Claudette with the U.S.air force over-the-horizon radar", *J. Atmos. Ocean. Technol.*, **10**, 441-451.

Geotis S.G., 1963, "Some radar measurements of hailstorms", *J. Appl. Meteor.*, **2**, 270-275.

Geotis S.G., 1991, "Relationship between echo area and rain flux in tropical rains", *Preprints, 25th Imternational Conf. Radar Meteor.*, Amer. Meteor. Soc., 366-368.

Germann U. and J.Joss, 1999, "Meso-gamma reflectivity profiles correlate with horizontal features, a help to improve precipitation estimates?", *Preprints, 29TH Conf. Radar Meteor.*, Amer. Meteor. Soc., 848-851

Ghosh A.K., V. Siva Kumar, K. Kishore Kumar and A.R. Jain, 2001, "VHF radar observation of atmospheric winds, associated shears and C_n^2 at a tropical location: Interdependence and seasonal pattern", *Ann. Geophysicae*, **19**, 965-973.

Ghosh S.K., 1977, "Prediction of storm surges on the east coast of India", *Ind. J. Meteor. Hydrol. Geophys.*, **28**, 2, 157-168.

Gibble D., 1964, "*Effects of rain on transmission performance of a satellite communications system*", Paper presented at IEEE International convention, New York, N.Y., March 23–26, 1964.

Gilet. M.,1994, "Standard specifications of wind profilers", in COST-74 final report "*Utilization of UHF/VHF radar wind profiler networks for improving weather forecasting in Europe*", Ed., C. Lafaysse, Directorate-General, Science Research & Development, European Commission, pp 175-183.

Giuli D., M.Gherardelli, A.Freni, T.A. Seliga and K. Aydin, 1991, "Rainfall and clutter discrimination by means of dual-linear polarization radar measurements". *J. Atmos. Ocean. Technol.*, **8**, 777-789.

Given T.J., 1989, "*An evaluation of dual-Doppler interpolation algorithms*", M.S.Thesis, Florida State University, Tallahassee, FL.,USA., 139 pp.

Glickman T.S., (Ed.), 2000, "*Glossary of Meteorology*", 2nd Edition, Amer. Meteor. Soc, Boston, 855 pp.

Glover K.H., and D.E. Forsyth, 1986, "Doppler wind analysis of hurricane Gloria" *Preprints, 23rd Conf. Radar Meteor.*, Amer. Meteor. Soc., Vol. 3, JP 233-235.

Goddard J.W.F., S.M. Cherry and V.N. Bringi, 1982, "Comparison of dual-polarization radar measurements of rain with ground-based disdrometer measurements", *J. Appl. Meteor.*, **21**, 252-256.

Golden J.H., 1990, "Meteorological data from hurricane Hugo", *Proc. 22nd joint UJNR Panel meetings - wind and seismic effects*, May 14-18.

Goldhirsh J., 1975, "Improved error analysis in estimation of raindrop spectra, rain rate liquid water content using multiple wavelength radars", *IEEE Trans. Antennas Propag.*, **AP-23**, 718-720.

Goldhirsh J., and I. Katz, 1974, "Estimation of raindrop size distribution using multiple wavelength radar systems", *Radio Sci.*, **9**, 439-446.

Golding B. W., 1998, "Nimrod : A system for generating automated very short range forecasts" *Meteor. Applns.* 5, 1-16.

Golestani Y., V. Chandrasekar and V.N. Bringi, 1989, "Intercomparison of multiparameter radar measurements", *Preprints, 24th Conf. Radar Meteor.*, Amer. Meteor. Soc., 309–313.

Golestani Y., V. Chandrasekar and R.J. Keeler, 1995, "Dual polarized staggered PRT scheme for weather radars: Analysis and applications", *IEEE Trans. Geosci.Remote Sens.*, **33**, 239-246.

Goodrich R.K., F.W. Wilson Jr. and D. Rogers, 1991 "A thin line detection algorithm", *Preprints, 25th International Conf. Radar Meteor.*, Amer. Meteor. Soc., 111-114.

Goodman S.J., D.E. Buechler, P.D. Wright, W.D. Rust and K.E. Nielsen, 1989, "Polarization radar and electrical observations of microburst producing storms during COHMEX", *Preprints 24th Conf. Radar Meteor.*, Amer. Meteor. Soc., 109-112.

Gorelik A.G. and Yu.V. Mel'nichuk, 1968, "A new method for measuring dissipation rate of turbulence in clouds and precipitation using conventional radar", pp. 150-156 in "*Radar meteorology, Proc. 3rd all-Union conference*", Moskva, 1968, Ed. Kostarev V.V., A.A.Chernikov and A.B. Shupyatskii, Israel Programme for Translations, Jerusalem, 1971.

Gorgucci E, V. Chandrasekar, and G. Scarchilli, 1991, "Radar receiver transfer function estimate from correlated meteorological echoes", *J. Atmos. Ocean. Technol.*, **8**, 191-197.

Gorgucci E., G. Scarchilli and V. Chandrasekar, 1994, "A robust estimator of rainfall rate using differential reflectivity", *J. Atmos. Ocean. Technol.*, **11**, 586-592.

Gorgucci E., G Scarchilli, V. Chandrasekar and V. N. Bringi, 2001, "Rainfall Estimation from Polarimetric Radar Measurements: Composite Algorithms Immune to Variability in Raindrop Shape–Size Relation", *J. Atmos. Ocean. Technol.*, . **18**,1773–1786.

Gorgucci, E., G. Scarchilli, V. Chandrasekar, and V. N. Bringi, 2000, Measurement of mean raindrop shape from polarimetric radar observations. *J. Atmos. Sci.*, **57**, 3406–3413.

Gorgucci E., G. Scarchilli and R. M. Leonardi, 1984, "Radar calibration by means of bias measurement", *Preprints, 22nd Conf. Radar Meteor.*, Amer. Meteor. Soc., 312-317.

Gorgucci E., G. Scarchilli and R. Leonardi, 1989, "Radar receiver calibration by means of multiple processing of weather echoes", *J. Atmos. Ocean. Technol.*, **6**, 487-499.

Gossard E.E., 1988, "Measuring drop-size distributions in clouds with a clear-air-sensing Doppler radar", *J. Atmos. Ocean. Technol.*, **5**, 640-649,

Gossard E.E., 1990, "Radar research on the atmospheric boundary layer" pp 477-527 in "*Radar in Meteorology*", Ed. D. Atlas, Amer. Meteor. Soc, 806 pp.

Gossard E.E., 1994, "Measurement of cloud droplet size size spectra by Doppler radar", *J. Atmos. Ocean. Technol.*, **11**, 712-726.

Gossard E.E. and R.G. Strauch, 1983, "*Radar observation of clear air and clouds*", Developments in Atmospheric Science, **14**, Elsevier, 280 pp.

Gossard E.E., S. Gutman, B. B. Stankov and D. E. Wolfe, 1999, "Profiles of radio refractive index and humidity derived from radar wind profilers and the Global Positioning System", *Radio Sci.*, 34, 371-383.

Gossard E.E., J.B. Snider, E.E. Clothiaux, B. Martner, J.S. Gibson, R.A. Kropfli and A.S. Frisch, 1997, "The potential of 8-mm radars for remotely sensing cloud drop size distributions", *J. Atmos. Ocean. Technol.*, **14**, 76-87.

Gossard E.E., R.G.Strauch, D.C.Welsh and S.Y. Matrosov, 1992, "Cloud layers, particle identification, and rain-rate profiles from ZRVf measurements by clear-air Doppler radars", *J. Atmos. Ocean. Technol.*, **9**, 108-119.

Gossard E.E., D.E. Wolfe, K.P. Moran, R.A. Paulus, K.D. Anderson and L.T. Rogers, 1998, "Measurement of clear-air gradients and turbulence properties with radar wind profilers", *J. Atmos. Ocean. Technol.*, **15**, 321-342.

Gossard E.E., D.E. Wolfe and B. B. Stankov, 1999b, "Measurement of Humidity Profiles in the Atmosphere by the Global Positioning System and Radar Wind Profilers", *J. Atmos. Ocean. Technol.*, **16**, 156–164.

Gosset M., and H. Sauvageot, 1992, "A dual-wavelength radar method for ice-water characteristics in mixed-phase clouds", *J. Atmos. Ocean. Technol.*, **9**, 538-547.

Gourley J. J., R. A. Maddox, K. W. Howard and D. W. Burgess, 2002, "An exploratory multisensor technique for quantitative estimation of stratiform rainfall", *J. Hydrometeor.*, **3**, 166–180.

Grecu, M., and W. F. Krajewski, 1999, "Anomalous pattern detection in radar echoes by using neural networks", *IEEE Trans. Geosci.Remote Sens.*, **37** (1), 287–296.

Grecu, M., and W. F. Krajewski, 2000, "An Efficient Methodology for Detection of Anomalous Propagation Echoes in Radar Reflectivity Data Using Neural Networks", *J. Atmos. Ocean. Technol.*, **17**, 121-129.

Greene D.R., 1972, "A comparison of echo predictability: constant elevation vs. VIL radar-data patterns", *Preprints, 15th Conf. Radar Meteor.*, Amer. Meteor. Soc., 111–116.

Greene D.R. and R.A. Clark, 1972, "Vertically integrated liquid water — a new analysis tool", *Mon. Wea. Rev.*, **100**, 548-552.

Greene D.R., J.D. Nilsen, R.E. Saffle, D.W. Holmes, M.D. Hudlow and P.R. Ahnert, 1983, "RADAP-II, an interim radar data processor", *Preprints, 21st Conf. Radar Meteor.*, Amer. Meteor. Soc., 404-408.

Green J.L., W.L. Clark, J.M. Warnock, and K.J. Ruth, 1983, "Absolute calibration of MST/ST radars", *Preprints, 21st Conf. Radar Meteor.*, Amer. Meteor. Soc, 144-147.

Green J.L., K.S. Gage and T.E. VanZandt, 1979, "Atmospheric measurements by VHF pulse Doppler radar", *IEEE Trans. Geosci. Electronics*, GE-17, 262-280.

Green J.L., R.H. Winkler, J.M. Warnock, W.L. Clark, K.S. Gage and T.E. VanZandt, 1978, "Observations of enhanced clear air refractivity associated with convective clouds", *Preprints, 18th Conf. Radar Meteor.*, Amer. Meteor. Soc, 88-93.

Gremillion M.S. and R. E. Orville, 1999, "Thunderstorm characteristics of cloud-to-ground lightning at the Kennedy Space Center, Florida: A study of lightning initiation signatures as indicated by the WSR-88D", *Wea. Forecasting*, **14**, 640–649.

Griffin J.S., R.W. Burpee, F.D. Marks Jr. and J.L. Franklin, 1992, "Real-time airborne analysis of aircraft data supporting operational hurricane forecasting", Wea. Forecasting, **7**, 480-490.

Griffith C.G., W.L. Woodley, P.G. Grube, D.W. Martin, J. Stout and D.N. Sikdar, 1978, "Rain estimation from geosynchronous satellite imagery - Visible and infrared studies", *Mon. Wea. Rev.*, **106**, 1153-1171.

Groginsky H.L. and K.M. Glover, 1980, "Weather radar canceller design", *Preprints, 19th Conf. Radar Meteor.*, Amer. Meteor. Soc, 192-198.

Gunn K.L.S. and T.W.R. East, 1954, "The microwave properties of precipitation particles", *Quart. J. Roy. Meteor. Soc.*, **80**, 522-545. (See also Corrigendum *Quart. J. Roy. Meteor. Soc.*, **81**, 653).

Gunn K.L.S. and J.S. Marshall, 1958, "The distribution with size of aggregate snowflakes", *J. Meteor.*, **15**, 452-461.

Gunn R. and G.D. Kinzer, 1949, "The terminal velocity of fall for water droplets in stagnant air", *J. Meteor.*, **6**, 243-248.

Gupta A., D.P. Mishra and V.K. Rajeev, 1990, "Incidence of a mesoscale vortex near Madras", *Vayu Mandal*, **20**, 53-57.

Gupta B.K., A. Mani and S.P. Venkiteshwaran, 1955, "Radar observations of rain at Poona", *Ind.J. Meteor. Geophys.*, **6**, 31-40.

Gupta B.K., A. Mani and S.P. Venkiteshwaran, 1961, "Some observations of melting band in radar precipitation echoes at Poona", *Ind. J. Meteor. Geophys.*, **12**, 317-322.

Gupta H.N. and S.K. Ghosh, 1980, "North Delhi tornado of 17 March 1978", *Mausam*, **31**, 93-100.

Haddad Z.S. and E. Im, 1993, "Optimal estimation of rain rate profiles from radar returns", *Preprints, 26th Imternational Conf. Radar Meteor.*, Amer. Meteor. Soc., 699-701.

Hagelberg C. and J. Helland, 1995, "Thin-line detection in meteorological radar images using wavelet transforms", *J. Atmos. Ocean. Technol.*, **12**, 633-642.

Hagen M., J. Hubbert, C. Richter, V.N. Bringi and P. Meischner, 1993, "Bright band observations with radar and aircraft", *Preprints, 26th International Conf. Radar Meteor.*, Amer. Meteor. Soc., 304-305.

Hall M.P.M., 1984, "A review of the application of multiple-parameter radar measurement of precipitation", *Radio Sci.*, **19**, 37-43.

Hamann D.J., 1991, "Extraction of fronts from Doppler radar images", *Preprints, 25th International Conf. Radar Meteor.*, Amer. Meteor. Soc., 119-122.

Hamilton R.E., 1970, "Use of detailed intensity radar data in meso-scale surface analysis of the July 4, 1969 storm in Ohio", *Preprints, 14th Conf. Radar Meteor.*, Amer. Meteor. Soc., 339-342.

Hane C.E. and P.S. Ray, 1985, "Pressure and Buoyancy fields derived from Doppler radar data in a tornadic thunderstorm", *J. Atmos. Sci.*, **42**, 18-35.

Hane C.E. and B.C. Scott, 1978, "Temperature and Pressure perturbations within convective clouds derived from detailed air motion information: Preliminary testing", *Mon. Wea. Rev.*, **106**, 654-661.

Harasti, P. R., and R. List, 1995, "A single-Doppler radar analysis method for intense axisymmetric cyclones", *Preprints, 27th Conf. Radar Meteor.*, Amer. Meteor. Soc., 221–223.

Hardaker P.J., A.R. Holt and C.G. Collier, 1991, "A theoretical study of scattering effects caused by the melting layer", *Preprints, 25th International Conf. Radar Meteor.*, Amer. Meteor. Soc., 725-728.

Hardaker P.J., A.R. Holt and C.G. Collier, 1993, "Correcting for the radar bright band in single polarisation radar echoes", *Preprints, 26th International Conf. Radar Meteor.*, Amer. Meteor. Soc., 763-764.

Hardaker, P. J., A. R. Holt, and C. G. Collier, 1995, A melting layer model and its use in correcting for the bright-band in single polarisation radar echoes. *Quart. J. Roy. Meteor. Soc.*, **121**, 495–525

Hardaker P.J., A.R. Holt and J.W.F. Goddard, 1997, "Comparing model and measured rainfall rates obtained from a combination of remotely sensed and in situ observations", *Radio Sci.*, **32**, 1785-1796.

Hardy K.R., 1963, "The development of raindrop-size distribution and implications related to the physics of precipitation", *J. Atmos. Sci.*, **20**, 299-312.

Hardy K.R. and K.S. Gage, 1990, "The history of radar studies of the clear atmosphere", pp130-142 in "*Radar in Meteorology*", Ed. D. Atlas, Amer. Meteor. Soc, 806 pp.

Harris F.I., T. Bals, R.J. Donaldson Jr., D. Hamann, and D.J. Smalley, 1993, "*Doppler weather radar studies of precipitation systems*", Philips Lab., Hanscom Air Force Base, USA, Report No. PL-TR-93-2168, 64 pp.

Harris F.I., R.J. Donaldson Jr., D.J. Smalley and Shu-Lin Tung, 1994, "*Radar studies of aviation hazards, Scientific Report No. 1*", Phillips Lab., Hanscom Air Force Base, USA, Report No. PL-TR-94-2146, 84 pp.

Harris I., D.J. Hamann and R.J. Donaldson Jr., 1989, "Hurricane monitoring with Doppler radar: a simulation", *Preprints, 24th Conf. Radar Meteor.*, Amer. Meteor. Soc., 203-206.

Harrison Mathews L., 1982, "Biologist at war", *New Scientist*, 28 Jan. p253, 4 Feb. pp 326-327, 11 Feb. pp 392-393, 18 Feb. pp 458-459, 25 Feb. 530-531, 4 Mar. pp 598-599, 18 Mar. pp738-739, 25 Mar. pp801-803.

Harrold T.W., 1967, "Attenuation of 8.6. mm-wavelength radiation in rain", *Proc. Inst. Elec. Engrs.*, **114**, 201-203.

Harrold T.W. and P.G. Kitchingman, 1975, "Measurement of surface rainfall using radar when the beam intersects the melting layer", *Preprints, 16th Conf. Radar Meteor.*, Amer. Meteor. Soc., 473-478.

Harrold T.W., E.J. English and C.A. Nicholass, 1972, "Some results of radar measurements of precipitation in hilly terrain", *Preprints, 15th Radar Meteor. Conf.*, Amer. Meteor. Soc., 161-166.

Harrold T.W., E.J. English and C.A. Nicholass, 1974, "The accuracy of radar-derived rainfall measurements in hilly terrain", *Quart. J. Roy. Meteor. Soc.*, **100**, 331-350.

Hartmann D.L., H.H. Hendon and R.A. Houze Jr., 1984, "Some implications of the mesoscale circulation in tropical cloud clusters for large-scale dynamics and climate", *J. Atmos. Sci.*, **41**, 113-121.

Hasler A.F. and K.R. Morris, 1986, "Hurricane structure and wind fields from stereoscopic and infrared satellite observations and radar data", *J. Climate Appl. Meteor.*, **25**, 709-727.

Hatcher R.W. and J.S. Sawyer, 1947, "Sea breeze structure with particular reference to temperature and water vapour gradients and associated radio-ducts", *Quart. J. Roy. Meteor. Soc.*, **73**, 391-406.

Hauser D. and P. Amayenc, 1981, "A new method for deducing hydrometeor size distribution and vertical air motions from Doppler radar measurements at vertical incidence", *J. Appl. Meteor.*, **20**, 547-555.

Hauser D. and P. Amayenc, 1983, "Exponential size distributions of raindrops and vertical air motions deduced from vertically pointing Doppler radar data using a new method", *J. Climate Appl. Meteor.* **22**, 407-418.

Hauser D. and P. Amayenc, 1984, "Raindrop size distributions and vertical air motions as inferred from zenith-pointing Doppler radar with the RONSARD system", *Radio Sci.*, **19**, 185-192.

Hauser D. and P. Amayenc, 1986, "Retrieval of cloud water and water vapor contents from Doppler radar data in a tropical squall line", *J. Atmos. Sci.*, **43**, 823-838.

Hauser, D., P. Amayenc, B. Nutten, and P. Waldteufel, 1984, "A New optical instrument for simultaneous measurement of raindrop diameter and fall speed distribution." *J. Atmos. Ocean. Technol.*, **1**, 256-269.

Hauser D., F. Roux and P. Amayenc, 1988, "Comparison of two methods for the retrieval of thermodynamic and microphysical variables from Doppler-radar measurements: Application to the case of a tropical squall line", *J. Atmos. Sci.*, **45**, 1285-1303.

Heiss W.H., D.L. McGrew and D. Sirmans, 1990, "Nexrad: next generation weather radar (WSR-88D)", *Microwave J.*, **33**, 80-98.

Helfrick A., 1984, "*Modern aviation electronics*", Prentice-Hall, New Jersey, USA., 312 pp.

Hennington L., 1981, "Reducing the effects of Doppler radar ambiguities", *J. Appl. Meteor.*, **20**, 1543-1546.

Hennington L.D. and D.W. Burgess, 1981, "Automatic recognition of mesocyclones from single Doppler radar data", *Preprints, 20th Conf. Radar Meteor.*, Amer. Meteor. Soc., 704-706.

Herman B.M. and L.J. Battan, 1961, "Calculations of Mie backscattering from melting ice spheres", *J. Meteor.*, **18**, 468-478.

Hermes L.G., A.Witt, S.D. Smith, D. Klingle-Wilson, D. Morris, G.J. Stumpf and M.D. Eilts, 1993, "The gust-front detection and wind-shift algorithms for the Terminal Doppler Weather Radar system", *J. Atmos. Ocean. Technol.*,**10**, 693-709.

Herzegh P.H. and A.R. Jameson, 1992, "Observing precipitation through dual-polarization radar measurements, *Bull. Amer. Meteor. Soc.*, **73**, 1365-1374.

Heymsfield A.J. and A.R. Jameson, 1981, "Hail growth investigations through use of dual-wavelength radar observations", pp 58-71, *"Recent progress in radar meteorology", Atmospheric Technology*, No. **13**, Ed. R. Carbone, National Center for Atmospheric Research (NCAR), Boulder, Colorado, USA.

Heymsfield G.M., S.W. Bidwell, I.J. Caylor, S. Ameen, S. Nicholson, W. Boncyk, L. Miller, D. Vandmark, P.E. Racette and L.R. Dod, 1996, "The EDOP radar system on the high-altitude NASA ER-2 aircraft", *J. Atmos. Ocean. Technol.*, **13**, 795-809.

Heymsfield G.M., K.K. Ghosh and L.C. Chen, 1983, "An interactive system for compositing digital radar and satellite data", *J. Climate Appl. Meteor.*, **22**, 705-713.

Heymsfield G.M., J. B. Halverson. J. Simpson, L. Tian and T.P. Bui, 2001, "ER-2 Doppler Radar Investigations of the Eyewall of Hurricane Bonnie during the Convection and Moisture Experiment-3", *J. Appl. Meteor.*, **40**, 1310–1330.

Heymsfield G.M., C. Parsons, L.R. Dod and L. Miller, 1989, "Planned ER-2 Doppler radar (EDOP) for studying convective storms and mesoscale phenomena", *Preprints, 24th conf. Radar Meteor.*, Amer. Meteor. Soc, 581-584.

Hildebrand P.H., 1978, "Iterative correction for attenuation of 5 cm radar in rain", *J. Appl. Meteor.*,**17**, 508-514.

Hildebrand P.H. and R.K. Moore, 1990, "Meteorological radar observations from mobile platforms", in *"Radar in Meteorology"*, Ed. D.Atlas, Amer. Meteor. Soc, 287-314.

Hildebrand P.H., Lee W-C., C. A. Walther, C. Frush, M. Randall, E. Loew, R. Neitzel, R. Parsons, J. Testud, F. Baudin and A. LeCornec, 1996, "The ELDORA/ASTRAIA Airborne Doppler weather radar: High resolution observations from TOGA COARE", *Bull. Amer. Meteor. Soc.*, **76**, 213-232.

Hildebrand P.H., N. Towery and M.R. Snell, 1979, "Measurement of convective mean rainfall over small areas using high-density rain gages and radar", *J. Appl. Meteor.*, **18**, 1316-1326.

Hildebrand P., C. Walther and C. Frush , 1986, "The NCAR ELECTRA Doppler radar, Evaluation of scientific needs", *Preprints, 23rd Conf. Radar Meteor.*, Amer. Meteor. Soc., 147-150.

Hildebrand P.H., C.A. Walther, C.L. Frush and C.K. Mueller, 1983, "Airborne Doppler weather radar: evaluation and discussion of applications", *Preprints, 21st Conf. Radar Meteor.*, Amer. Meteor. Soc., 270-277.

Hitschfeld W.F., 1986, "The invention of radar meteorology", *Bull. Amer. Meteor. Soc.*, **67**, 33-37.

Hitschfeld W.F. and J. Bordan, 1954, "Errors inherent in the radar measurement of rainfall at attenuating wavelengths", *J. Meteor.*, **11**, 58-67.

Hjelmfelt M.R., 1988, "Structure and life cycle of microburst outflows observed in Colorado", 1988, *J. Appl. Meteor.*, **27**, 900-927.

Hobbs S.E. and W.W. Wolf, 1996, "Developments in airborne entomological radar", *J. Atmos. Ocean. Technol.*, **13**, 58-61.

Hocking W.K., 1985, "Measurement of turbulent eddy dissipation rates in the middle atmosphere by radar techniques: a review", *Radio Sci.*, **20**, 1403-1422.

Hocking W.K., 1997, "System design, signal processing procedures and preliminary results for the Canadian (London, Ontario) VHF atmospheric radar", *Radio Sci.*, **32**, 687-706.
Hocking W.K., 1997b, "Recent advances in radar instrumentation and techniques for studies of the mesosphere, stratosphere and troposphere, *Radio Sci.*, **32**, 2241-2270.
Hoddinott M.H.O., 1986, "Thunderstorm observations in West Bengal, 1945-46", *Weather*, **41**, 2-5.
Holland G.J., 1987, "Mature structure and structure change", pp13-52 in "*A global view of tropical cyclones*", Ed. R.L. Elsberry, Office of Naval Research, USA., 185 pp.
Holland G.J., 1993, in World Meteorological Organization (WMO), 1993,"*Global guide to Tropical Cyclone Forecasting*", WMO TD 560, Report No. TCP-31, Section 3.2.1, Page 3.6.
Holland G.J., 1993b, in World Meteorological Organization (WMO), 1993,"*Global guide to Tropical Cyclone Forecasting*", WMO TD 560, Report No. TCP-31, Section 9.8, Page 9.29.
Holle R.L., R.E. López, W.L. Hiscox and D. Rosenfeld, 1984, "Cloud-ground lightning associated with radar returns from south Florida", Preprints, *15th Conf. Hurricanes and Tropical Meteor.*, Amer. Meteor. Soc., 479-484.
Holloway C.L. and R.J. Keeler, 1993, "Rapid scan Doppler radar: The antenna issues", *Preprints, 26th International Conf. Radar Meteor.*, Amer. Meteor. Soc., 393-395.
Horn J.D., 1951, "On irregular movements of tropical cyclones in the Pacific", *Bull. Amer. Meteor. Soc.*, **32**, 344-345.
Houze R.A. Jr., 1982, "Cloud clusters and large scale vertical motions in the tropics", *J. Meteor. Soc. Japan*, **60**, 396-410.
Houze R.A. Jr., 1989, "Observed structure of mesoscale convective systems and implications for large-scale heating", *Quart. J. Roy. Meteor. Soc.*, **115**, 425-461.
Houze R.A. Jr., 1997, "Stratiform precipitation in regions of convection: A meteorological paradox?", *Bull. Amer. Meteor. Soc.*, **78**, 2179-2196.
Houze R.A.Jr., and A.K. Betts, 1981, "Convection in GATE", *Rev. Geophys. Space Phys.*, **19**, 541-576.
Houze R.A. Jr., and P.V.Hobbs, 1982, "Organization and structure of precipitating cloud systems", *Adv. Geophys.*, Acad.Press, **24**, 225-315.
Houze R.A.Jr., and E.N. Rappaport, 1984, "Air motions and precipitation structure of an early summer squall line over the eastern tropical Atlantic", J. *Atmos. Sci.*, **41**, 553-574.
Houze R.A. Jr., and D.D. Churchill, 1987, "Mesoscale organization and cloud microphysics in a Bay of Bengal depression", *J. Atmos. Sci.*, **44**, 1845-1867.
Houze R.A. Jr., S.G. Geotis, F.D. Marks and A.K. West, 1981, "Winter monsoon convection in the vicinity of North Borneo. Part I: Structure and time variation of the clouds and precipitation", *Mon. Wea. Rev.*, **109**, 1595-1614.
Houze R.A. Jr., Hobbs P.V., P.H. Herzegh and D.B. Parsons, 1979, "Size distribution of precipitation particles in frontal clouds", *J. Atmos. Sci.*, **36**, 156-162.
Houze R.A. Jr., S.A. Rutledge, M.I. Biggerstaff and B.F. Smull, 1989, "Interpretation of Doppler weather radar displays of midlatitude mesoscale convective systems", *Bull. Amer. Meteor. Soc.*, **70**, 608-619.
Houze R.A.Jr., W. Schmid, R.G. Fovell and H.-H. Schiesser, 1993, "Hailstorms in Switzerland: Left movers, right movers, and false hooks", *Mon. Wea. Rev.*, **121**, 3345-3370.
Huaman M.M and B.B. Balsley, 1996, "Long-term average vertical motions observed by VHF wind profilers: The effect of slight antenna-pointing inaccuracies", *J. Atmos. Ocean. Technol.*, **13**, 560-569.
Hubbert J., V. Chandrasekar, V.N. Bringi and P. Meischner, 1993, "Processing and interpretation of coherent dual-polarized radar measurements", *J. Atmos. Ocean. Technol.*, **10**, 155-164.
Hudlow M.D., 1979, "Mean rainfall patterns for the three phases of GATE", *J. Appl. Meteor.*, **18**, 1656-1669.

Hudlow M.D., 1985, in "*Report of the seminar on the application of radar data to tropical cyclone forecasting, Bangkok, 1983*", World Meteorological Organization(WMO), Rep. No. TCP-19, WMO/TD No. 26.

Hudlow M.D. and R.E. Arkell, 1978, "Effect of temporal and spatial sampling errors and Z-R variability on accuracy of GATE radar rainfall estimates", *Preprints, 18th Conf. Radar Meteor.*, Amer. Meteor. Soc.., 342-350.

Hudlow M.D., R. Arkell, V. Patterson, P. Pytlowany, F. Richards and S. Geotis, 1979, "*Calibration and intercomparison of the GATE C-band radars*", NOAA Tech. Report EDIS 31, U.S. Dept. of Commerce, NOAA, Environmental Data and Information Service, 98 pp.

Huff F.A., 1970, "Sampling errors in measurement of mean precipitation", *J. Appl. Meteor.*, **9**, 35-44.

Humphries R.G. and B.L. Barge, 1979, "Polarization and dual wavelength radar observations of the bright band", *IEEE Trans. Geosci. Electronics*, **GE-17**, 190-195.

Iguchi T. and R. Meneghini, 1994, "Intercomparison of single-frequency methods for retrieving a vertical rain profile from airborne or spaceborne radar data", *J. Atmos. Ocean. Technol.*, **11**, 1507-1516.

Iguchi, T., Kozu, R. Meneghini, J. Awaka, and K. Okamoto, 2000, "Rain profiling algorithm for the TRMM precipitation radar," *J. Appl. Meteor.*, **39,** 2038–2052.

Illingworth A.J. and I.J. Caylor , 1989, "Cross-polar observations of the bright band", *Preprints 24th Conf. Radar Meteor.*, Amer. Meteor. Soc., 323-327.

Illingworth A.J. and D.S. Zrnić, 1995, "Workshop on weather radar polarimetry for research and operational applications", *Bull. Amer. Meteor. Soc.*, **76**, 555-558.

Im E. and Z.S. Haddad, 1993, "Simultaneous retrieval of surface and rain backscattering parameters from spaceborne radar measurements", *Preprints, 26th Imternational Conf. Radar Meteor., Amer. Meteor. Soc.*, 702-704.

IMD (India Meteorological Department), 1970, "Rainfall and floods during 1969 southwest monsoon period", *Ind. J. Meteor. Geophys.*, **21**, 347-354.

IMD (India Meteorological department), 1944, "*Nor'westers of Bengal*", Tech. Note No. 10, 11 pp.

International Civil Aviation Organization (ICAO), 1987, "*Wind Shear*", ICAO Circular 186-AN/122, 183 pp.

Ishihara M., H. Sakakibara, A. Tabata, K. Akaeda and H. Okamura, 1991, "Doppler radar analysis of a typhoon eyewall", *Preprints, 25th International Conf. Radar Meteor.*, Amer. Meteor. Soc., 702-705.

Ishihara M., Z. Yanagisawa, H. Sakakibara, K. Matsuura and J. Aoyagi, 1986, "Structure of a typhoon rainband observed by two Doppler radars", *J.Meteor. Soc. Japan*, Ser II, **64**, 923-938.

Istok M., 1981, "Analysis of Doppler spectrum broadening mechanisms in thunderstorms", *Preprints, 20th Conf. Radar Meteor.*, Amer. Meteor. Soc., 454-458.

Istok M.J. and R.J. Doviak, 1986, "Analysis of the relation between Doppler spectral width and thunderstorm turbulence", *J.Atmos. Sci.*, **43**, 2199-2214.

Ivanov A.A., U.V. Mel'nichuk and A.A. Chernikov, 1989, "The automated radar subsystem for very short-range weather forecast", *WMO Instruments and Observing Methods Report No. 35*, WMO/TD - No. 303, WMO, Geneva, 287-288.

Jain A.R., Y. Jaya Rao, A.K. Patra, P.B. Rao, G. Viswanathan and S.K. Subramanian, 2000b, "Observations of tropical convection events using Indian MST radar: first results", *Quart. J. Roy. Meteor. Soc.*, **126**, 3097-3115.

Jain A.R., Y. Jaya Rao and P.B. Rao, 1997, "Aspect sensitivity of the received radar backscatter at VHF: Preliminary observations using Indian MST radar", *Radio Sci.*, **32**, 1249-1260.

Jain A.R., Y. Jaya Rao, P.B. Rao, V.K. Anandan, S.H. Damle, P. Balamuralidhar, A. Kulkarni and G. Viswanathan 1995, "Indian MST radar 2. First results in the ST mode", *Radio Sci.*, **30**, 1139-1158.

Jain A.R., D. Narayana Rao and P.B. Rao, 2000, "Indian MST radar – An overview of the scientific programmes and results", *Ind. J. Radio Space Phys.*, **29**, 149-171.

Jain P.S., 1989, "Fractal dimensions of clouds around Madras", *Mausam*, **40**, 311-316.

James C.N.and R. A. Houze Jr., 2001, "A Real-Time Four-Dimensional Doppler Dealiasing Scheme", *J. Atmos. Ocean. Technol.*, **18**, 1674–1683.

James C.N., S R. Brodzik, H. Edmon, R. A. Houze Jr., and S. E. Yuter, 2000, "Radar Data Processing and Visualization over Complex Terrain", *Wea. Forecasting*, **15**, 327-338.

Jameson A.R., 1985, "Microphysical interpretation of multiparameter radar measurements in rain. Part III: Interpretation and measurement of propagation differential phase shift between orthogonal linear polarizations", *J. Atmos. Sci.*, **42**, 607-614.

Jameson A.R., 1991, "A comparison of microwave techniques for measuring rainfall", *J. Appl. Meteor.*, **30**, 32-54.

Jameson A.R., 1994, "An alternative approach to estimating rainfall rate by radar using propagation differential phase shift", *J. Atmos. Ocean. Technol.*, **11**, 122-131.

Jameson A.R. and A.B. Kostinski, 2001, "What is a Raindrop Size Distribution?", *Bull. Amer. Meteor. Soc.*, **82**, 1169–1178.

Jameson A.R. and A.B. Kostinski, 2001b, "Reconsideration of the physical and empirical origins of Z-R relations in Radar Meteorology", *Quart. J. Roy. Meteor. Soc.*, **127**, 517-538.

Jameson A.R. and R.C. Srivastava, 1978, "Dual-wavelength Doppler radar observations of hail at vertical incidence", *J. Appl. Meteor.*, **17**, 1694-1703.

Jaya Rao Y., A.R. Jain, V.K. Anandan, P. B. Rao, G. Viswanathan and R. Arvindan, 1994, "Some observations of tropical tropopause using ST mode of the Indian MST radar: Multiple stable layer structure", *Ind. J. Radio Space Phys.*, **23**, 75-85.

Jayanthi N., and R. Asokan, 1994, "An objective method of forecasting landfall point of cyclonic storms using radar fixes.", *SAARC Seminar on Radar Meteorology*, IMD, New Delhi, December 19-22, 1994, 56-68.

Jayanthi N., Akhilesh Gupta and A. Mary Selvam, 1990, "The fractal geometry of winter monsoon clouds over the Indian region", *Mausam*, **41**, 579-582.

Jayanthi N., J.V. Narayana and V. Balasubramanian, 1980, "A climatological study of the heights of radar cloud tops", *Mausam*, **31**, 613-615.

Jing Z. and G. Wiener, 1993, "Two-dimensional dealiasing of Doppler velocities", *J. Atmos. Ocean. Technol.*, **10**, 798-808.

Joe P., 1991, "Classification of radar imagery using texture analysis", *Preprints, 25th International Conf. Radar Meteor.*, Amer. Meteor. Soc., 107-110.

Joe P., J. Abraham and K. MacDonald, 1991, "A Doppler radar study of post-tropical HUGO interacting with a cold front", *Preprints, 25th International Conf. Radar Meteor.*, Amer. Meteor. Soc., 678-682.

Johns R.H., 1993, "Meteorological conditions associated with bow echo development in convective storms", *Wea. Forecasting*, **8**, 294-299.

Johns R.H. and W.D. Hirt 1987, "Derechos: Widespread convectively induced windstorms", *Wea. Forecasting*, **2**, 32-49.

Johnson D.B., W.C. Lee, C.S. Cheng, and J.D. Tuttle, 1993, "Chaff studies during the Hawaiian rainband project", *Preprints, 26th International Conf. Radar Meteor.*, Amer. Meteor. Soc., 267-269.

Johnson G.H., P.L.Smith Jr., F.E.Nathanson and L.W.Brooks, 1975, "An analysis of techniques for dealing with anomalous propagation", *Preprints, 16th Radar Meteor. Conf.*, Amer. Meteor. Soc, 374-377.

Johnson R.H. and R.A. Houze, Jr., 1987, "Precipitating cloud systems of the Asian monsoon", pp 298-353, Chapter 10 of "*Monsoon Meteorology*", C.-P. Chang and T.N.Krishnamurti, Eds., Oxford Univ. Press, 544 pp.

Johnson R.H. and G.S. Young, 1983, "Heat and moisture budgets of tropical mesoscale anvil clouds", *J. Atmos. Sci.*, **40**, 2138-2147.

Johnston B.W., J.D.Marwitz, and R.E. Carbone, 1990, "Single-Doppler radar analysis of banded precipitation structures", *J. Atmos. Ocean. Technol.*, **7**, 6, 866-875.

Jones D.M.A., 1959, "The shape of raindrops", *J. Meteor.*, **16**, 504-510.

Jones R.F., 1958, "Radar echoes from atmospheric inhomogeneities", *Quart. J. Roy. Meteor. Soc.*, **84**, 437-442.

Jordan E.C., 1967, "*Electromagnetic waves and radiating systems*", Prentice-Hall of India, 710 pp.

Jorgensen D.P., 1984, "Mesoscale and convective-scale characteristics of mature hurricanes. Part I: General observations by research aircraft", *J. Atmos. Sci.*, **41**, 1268-1285.

Jorgensen D.P., 1984b, "Mesoscale and convective-scale characteristics of mature hurricanes. Part II: Inner core structure of hurricane Allen (1980)", *J. Atmos. Sci.*, **41**, 1287-1311.

Jorgensen D.P. and J.D. DuGranrut, 1991, "A dual-beam technique for deriving wind fields from airborne Doppler radar", *Preprints, 25th International Conf. Radar Meteor.*, Amer. Meteor. Soc., 458-461.

Jorgensen D.P. and F.D. Marks Jr., 1984, "Airborne Doppler radar study of the structure and three-dimensional airflow within a hurricane rainband", *Preprints, 22nd Conf. Radar Meteor.*, Amer. Meteor. Soc., 572-577.

Jorgensen D.P. and B.F. Smull, 1993, "Mesovortex circulations seen by airborne Doppler radar within a bow-echo meso-scale convective system", *Bull. Amer. Meteor. Soc.*, **74**, 2146-2157.

Jorgensen D.P. and P.T. Willis, 1982, "A Z-R relationship for hurricanes", *J. Appl. Meteor.*, **21**, 356-366.

Jorgensen D.P., P.H. Hildebrand and C.L. Frush, 1983, "Feasibility test of an airborne pulse-Doppler meteorological radar", *J.Climate Appl. Meteor.*, **22**, 744-757.

Jorgensen D.P., T. R. Shepherd, and A. S. Goldstein, 2000, "A dual-pulse repetition frequency scheme for mitigating velocity ambiguities of the NOAA P-3 airborne Doppler radar", *J. Atmos. Ocean. Technol.* **17**, 585–594.

Jorgensen D.P., E.J. Zipser and M.A. LeMone, 1985, "Vertical motions in intense hurricanes", *J. Atmos. Sci.*, **42**, 839-856.

Josefsson L., 1991, "Phased array antenna technology for weather radar applications", *Preprints, 25th International Conf. Radar Meteor.*, Amer. Meteor. Soc., 752-755.

Joseph P.V., 1982, "A tentative model of Andhi", *Mausam*, **33**, 417-422.

Joseph P.V., D.K. Raipal and S.N. Deka, 1980, "'Andhi' the convective duststorm in northwest India", *Mausam*, **31**, 431-442.

Joss J. and C.G. Collier, 1991, "An electronically scanned antenna for weather radar", *Preprints, 25th International Conf. Radar Meteor.*, Amer. Meteor. Soc., 748-751.

Joss J. and R. Lee, 1995, "The application of radar-gauge comparisons to operational precipitation profile corrections", *J. Appl. Meteor.*, 34, 2612-2630.

Joss J. and A. Waldvogel, 1967, "Ein Spektrograph für Niederschlagstropfen mit automatischer Auswertung", *Pure Appl. Geophys.*, **68**, 240-246.

Joss J. and A. Waldfogel, 1970, "Raindrop size distributions and Doppler velocities", *Preprints, 14th Conf. Radar Meteor.* Amer. Meteor. Soc., 153-156.

Joss J. and A. Waldvogel, 1970b, "A method to improve the accuracy of radar measured amounts of precipitation", *Preprints, 14th Radar Conf., Meteor.* Amer. Meteor. Soc. , 237-238.

Joss, J. and A. Waldvogel, 1990, "Precipitation measurement and hydrology" pp 577–606, in "*Radar in Meteorology*", D. Atlas (Ed.), Amer. Meteor. Soc., 806 pp.

Kabêche A. and J. Testud, 1995, "Stereoradar meteorology: A new unified approach to process data from airborne or ground-based meteorological radars", *J. Atmos. Ocean. Technol.*, **12**, 783-799.

Kalsi S.R., 2002, "*Use of satellite imagery in tropical cyclone intensity analysis and forecasting; A guide book for forecasters*", Meteorological Monograph, Cyclone Warning Division No. 1/2002, India Meteorological Department, New Delhi, 44pp + figures.

Keeler R.J. and R.E. Carbone, 1986, "A modern pulsing/processing technique for meteorological Doppler radars" *Preprints, 23rd Conf. Radar Meteor.*, Amer. Meteor. Soc, JP357-360.

Keeler R.J. and C.L. Frush, 1983, "Rapid scan Doppler radar development considerations Part II: Technology assessment", *Preprints, 21st Conf. Radar Meteor.*, Amer. Meteor. Soc., 284-290.

Keeler R.J. and R.E. Passarelli, 1990, "Signal processing for atmospheric radars", pp 199-229 in "*Radar in Meteorology*", (Ed., D. Atlas) Amer. Meteor. Soc, 806 pp.

Keeler R.J., C. Hwang, V. Chandrasekar and R. Xiao, 1993, "Pulse compression polarization waveforms for rapid scan Doppler radar", *Preprints, 26th International Conf. Radar Meteor.*, Amer. Meteor. Soc., 255-257.

Keeler, R.J., C. Kessinger, S. Ellis and J. VanAndel, 1999b, "Implementation of NEXRAD AP Clutter Mitigation Scheme", Preprints, 29th *International Conf. Radar Meteor.*, Amer. Meteor. Soc, 580-581.

Keeler R. J., D. S. Zmić, and C. L. Frush, 1999, "Review of Range Velocity Ambiguity Mitigation Techniques", *Preprints, 29th International Conf. Radar Meteor.*, Amer. Meteor. Soc, 158-163.

Keenan T.D., and S.J. Anderson, 1987, "Some examples of surface wind analysis based on Jindalee skywave radar data", *Aust. Meteor. Mag.*, **35**,153-161.

Keenan T.D., K. Glasson, F. Cummings, T. S. Bird, J. Keeler, and J. Lutz, 1998 "The BMRC/ NCAR C-band polarimetric (C-POL) radar system", *J. Atmos. Oceanic Technol.*, 15, 871–886.

Keenan T., R. Potts and T. Stevenson, 1992, "An evaluation of the Darwin Area forecast experiment storm occurrence forecasts", *Wea. Forecasting*, **7**, 515-523.

Kerr D.E., (Ed.), 1951, "*The Propagation of Short Radio Waves*", **13**, MIT Radiation Laboratory Series, McGraw-Hill, 728 pp.

Keshav Kumar, S. Singh, K.K. Mahajan, and S.K. Srivastava, 1994, "Short-period waves in the vertical wind using the Indian MST radar", *Ind. J. Radio Space Phys.*, **23**, 21-29.

Keshav Kumar, S. Singh, K.K. Mahajan, and N.K. Sethi, 1994b, "A study of tropopause and stable layers in lower atmosphere using the Indian MST radar", *Ind. J. Radio Space Phys.*, **23**, 71-74.

Kessinger, C, S. Ellis and J.H. VanAndel, 1999, A Fuzzy Logic AP Clutter Detection Algorithm for the WSR-88D, *Preprints, 29th International Conf. Radar Meteor.*, Amer. Meteor. Soc, 576-579.

Kessinger C.J., D.B. Parsons and J.W. Wilson, 1988, "Observations of a storm containing misocyclones, downbursts and horizontal vortex circulations", *Mon. Wea. Rev.*, **116**, 1959-1982.

King P., T-C. Yip and J.D. Steenbergen, 1989, "RAINSAT: A one year evaluation of a bispectral method for the analysis and short-range forecasting of precipitation areas", *Wea. Forecasting*, **4**, 210-221.

Kinzer G.D., 1974, "Cloud-to-ground lightning versus radar reflectivity in Oklahoma thunderstorms", *J. Atmos. Sci.*, **31**, 787-799.

Kishore P., K. Krishna Reddy, D. Narayana Rao, P.B. Rao, A.R. Jain, G.V. Rama and S. Sankar, 2000, "A statistical comparison of Indian MST radar and rawinsonde wind measurements", *Ind. J. Radio Space Phys.*, **29**, 102-114.

Kishore P., P.B. Rao, A.R. Jain, P. Balamuralidhar, S.H. Damle, G. Viswanathan and G.V. Rama, 1994, "High resolution measurements of horizontal winds using ST mode of the Indian MST radar: A comparison with balloon measurements", *Ind. J. Radio Space Phys.*, **23**, 41-51.

Kitchen M., R. Brown and A.G. Davies, 1994, "Real-time correction of weather radar data for the effects of bright band, range and orographic growth in widespread precipitation", *Quart. J. Roy. Meteor. Soc.*, **120**, 1231-1254.

Klaassen W., 1983, "Accurate determination of vertical air velocities in rain by Doppler-radar", *J. Climate Appl. Meteor.*, **22**, 1788-1793.

Klaassen W., 1988, "Radar observations and simulation of the melting layer of precipitation", *J. Atmos. Sci.*, **45**, 3741-3752.

Klaassen W., 1989, "Determination of rain intensity from Doppler spectra of vertically scanning radar", *J. Atmos. Ocean. Technol.*, **6**, 552-562.

Klaus V., M. Mauprivez and I. Seloyan, 2000, "The Météo-France wind profiler of la Ferté-Vidame: Towards a validated instrument for the observation network", *Final Workshop of COST-76 Action, combined with the 9th International Workshop on Technical and Scientific Aspects of MST Radar*, 4pp.

Klazura G.E. and D.A. Imy, 1993, "A description of the initial set of analysis products available from the NEXRAD WSR-88D system", *Bull. Amer. Meteor. Soc.*, **74**, 1293-1311.

Klimowski B.A., and J.D. Marwitz, 1992, "The synthetic dual-Doppler analysis technique", *J. Atmos. Ocean. Technol.*, **9**, 728-745,

Klingle D.L., D.R. Smith and M.M. Wolfson, 1987, "Gust front characteristics as determined by Doppler radar", *Mon. Wea. Rev.* **115**, 905-918.

Klingle-Wilson D., S.H. Olson, W. Wilson, W.P. Mahoney III, S.D. Smith, A. Witt, and M.D. Eilts, 1989, "Gust front detection algorithm for the Terminal Doppler Weather Radar: Part 2, performance assessment", *3rd International Conf. Aviation Weather System*, Amer. Meteor. Soc.

Klugmann D., 2001, "Retrieval of atmospheric parameters from hydrometeor backscatter measured with Doppler radar profiler", *J. Atmos. Ocean. Technol.*, **18**, 56-76.

Knight C.A. and L.J. Miller, 1993, "First radar echoes from cumulus clouds", *Bull. Amer. Meteor. Soc.*, **74**, 179-188.

Kobayashi F., K. Kikuchi and H. Uyeda, 1996, "Life cycle of the Chitose tornado of September 22, 1988", *J. Meteor. Soc. Japan*, **74**, 125-140.

Kodaira N., 1964, "Radar meteorology", *Inter-regional seminar on advanced tropical meteorology*, Manila, World Meteorological Organization.

Kodaira N., 1963, "The radar wave attenuation caused by precipitation particles", *Papers. Meteor.Geophys.*, **XIV**, 181-189.

Koistinen J., 1991, "Operational correction of radar rainfall errors due to the vertical reflectivity profile", *Preprints, 25th Imternational Conf. Radar Meteor.*, Amer. Meteor. Soc., 91-94.

Kollias P., B.A. Albrecht and F.D. Marks Jr., 2001, "Raindrop sorting induced by vertical drafts in convective clouds", *Geophys. Res. Letters*, **28**, 2787-2790.

Kollias P., B.A. Albrecht and F.D. Marks Jr., 2002, "Why Mie? Accurate observations of vertical air velocities and raindrops using a cloud radar", *Bull. Amer. Meteor. Soc.*, **83**, 1471-1483.

Kollias P., R. Lhermitte and B.A. Albrecht, 1999, "Vertical air motion and raindrop size distributions in convective systems using a 94 GHz radar", *Geophys. Res. Lett.*, **26**, 3109-3112.

Koscielny A.J., R.J.Doviak and R. Rabin, 1982, "Statistical considerations in the estimation of divergence from single-Doppler radar and applications to prestorm boundary-layer observations", *J.Appl. Meteor.*, **21**, 197- 210.

Kostarev V.V., A.A. Chernikov and A.B. Shupyatskii (Ed.), 1968, "*Radar meteorology, Proc. 3rd all-Union conference*", Moskva, 1968, Israel Programme for Translations, Jerusalem. 1971.

Kostinski A.B., J.M. Kwiatkowski and A.R. Jameson, 1993, "Spaceborne radar sensing of precipitation above an ocean surface: Polarization contrast study", Preprints, 26th International Conf. Radar Meteor., 705-706.

Koteswaram P. and A.C. De, 1959, "Study of pre-monsoon thunderstorms over Gangetic West Bengal by radar", *Ind. J. Meteor. Geophys.*, **10**, 275-282.

Koteswaram P. and S. Gaspar, 1956, "The surface structure of tropical cyclones in the Indian area", *Ind. J. Meteor. Geophys.*, **7**, 339-352.

Kozu T. and K. Nakamura, 1991, "Rainfall parameter estimation from dual radar measurements combining reflectivity profile and path-integrated attenuation", *J. Atmos. Ocean. Technol.*, **8**, 259-270.

Kozu T., K. Nakamura and R. Meneghini, 1991, "Estimation of raindrop size distribution parameters from a dual-parameter spaceborne radar measurement", *Preprints, 25th Imternational Conf. Radar Meteor.*, Amer. Meteor. Soc., 384-387.

Kozu T., K. Nakamura, R. Meneghini and W.C. Boncyk, 1991b, "Dual-parameter radar rainfall measurement from space: a test result from an aircraft experiment", *IEEE Trans. Geosci. Remote Sens* **29**, 690-703.

Krajewski W. F. and B. Vignal, 2001, "Evaluation of anomalous propagation echo detection in WSR-88D Data: A large sample case study", *J. Atmos. Ocean. Technol.*, 18, 807-814

Kraus J.D., 1950, "*Antennas*", McGraw-Hill, 553 pp.

Krehbiel P.R., W. Rison, S. McCrary, T. Blackman and M. Brook, 1991, "Dual-polarization radar observations of lightning echoes and precipitation alignment at 3 cm wavelength", *Preprints 25th International Conf. Radar Meteor.*, Amer. Meteor. Soc., 901-904.

Krishna Reddy K., T. Kozu, Y. Ohno, K. Nakamura, A. Higuchi, K. Madhu Chandra Reddy, V.K. Anandan, P. Srinivasulu, A.R. Jain, P.B. Rao,R. Ranga Rao, G. Viswanathan and D. Narayana Rao, 2002, "Planetary boundary layer and precipitation studies using lower atmospheric wind profiler over tropical India", *Radio Sci.*, **37**, 14-1 to 14-17.

Kropfli R.A., 1986, "Single Doppler radar measurements of turbulence profiles in the convective boundary layer", *J. Atmos. Ocean. Technol.*, **3**, 305-314.

Kropfli R.A., I. Katz, T.G. Konrad and E.B.Dobson, 1968, "Simultaneous radar reflectivity measurements and refractive index spectra in the clear atmosphere", *Radio Sci.*, **3**, 991-994.

Kruger A. and W. Krajewski, 2002, "Two-dimensional video disdrometer: A description", *J. Atmos. Ocean. Technol.*, **19**, 602–617.

Krumm W.R., 1954, "On the cause of downdrafts from dry thunderstorms over the plateau of the United States", *Bull. Amer. Meteor. Soc.*, **35**, 122-126.

Kulshrestha S.M., 1988, "Radar and satellite signatures of monsoon depressions", *Mausam*, 39, 229-230.

Kulshrestha S.M., 1962, "Heights of cumulonimbus cloud tops over North India - a radar study", *Ind. J. Meteor. Geophys.*, **13**, 167-172.

Kulshrestha S.M., 1962b, "Hot weather angels associated with high level temperature inversions", *Ind. J. Meteor. Geophys.*, **13**, 218-226.

Kulshrestha S.M., 1966, "Radar-hydrology and its application in India", *Ind. J. Meteor. Geophys.*, **17**, Spl. No., 245-256.

Kulshrestha S.M., 1970, "Spiral bands in monsoon depressions over North India", *Preprints, 14th Conf. Radar Meteor.*, Amer. Meteor. Soc., 367-368.

Kulshrestha S.M. and M.G. Gupta, 1964, "Satellite study of an inland monsoon depression", *Ind. J. Meteor.* Geophys., **15**, 175-182.

Kulshrestha S.M. and P.S. Jain, 1967, "Radar climatology of Delhi and neighbourhood - occurrence of severe weather", *Ind. J. Meteor. Geophys.*, **18**, 105-110.

Kulshrestha S.M. and B.L. Sharma, 1961, "Very fast moving radar echoes as observed on CPS-9 radar", *Ind. J. Meteor. Geophys.*, **12**, 629-636.

Kumagai H., R. Meneghini and T. Kozu, 1993, "Preliminary results from multiparameter airborne rain radar measurement in the western Pacific", *J. Appl. Meteor.*, **32**, 431-440.

Kumagai H., R. Meneghini and J. R. Wang, 1993b, "Combined analysis of airborne single-frequency radar and multi-frequency radiometer observations in the TRMM-1 experiment", *Preprints, 26th Imternational Conf. Radar Meteor.*, Amer. Meteor. Soc., 696-698.

Kumar A., 1992, "A climatological study of thunderstorms at Lucknow airport", *Mausam*, **43**, 441-444.

Kummerow C., W. Barnes, T. Kozu, J. Shiue and J. Simpson, 1998, "The Tropical Rainfall Measuring Mission (TRMM) Sensor Package", *J. Atmos. Ocean. Technol.*, **15**, 809–817.

Kummerow C., R.A. Mack and I.M. Hakkarinen, 1989, "A self-consistency approach to improve microwave rainfall estimation from space", *J. Appl. Meteor.*, **28**, 869-884.

Kummerow C., J. Simpson, O. Thiele, W. Barnes, A. T. C. Chang, E. Stocker, R. F. Adler, A. Hou, R. Kakar, F. Wentz, P. Ashcroft, T. Kozu, Y. Hong, K. Okamoto, T. Iguchi, H. Kuroiwa, E. Im, Z. Haddad, G. Huffman, B. Ferrier, W. S. Olson, E. Zipser, E. A. Smith, T. T. Wilheit, G. North, T. Krishnamurti and K. Nakamura, 2000, "The status of the Tropical Rainfall Measuring Mission (TRMM) after two years in orbit", *J. Appl. Meteor.*, **39**, 1965–1982.

Kundu M.M. and A.C. De, 1971, "On the formation of line type angel echo", *Ind. J. Meteor. Geophys.*, **22**, 227-230.

Kuo H.L., 1969, "Motions of vortices and circulating cylinder in shear flow with friction", *J.Atmos. Sci.*, **26**, 390-398.

Kuo Y.-H. and Y.R. Guo, 1989, "Dynamic initialization using observations from a netork of profilers", *Mon. Wea. Rev.*, **117**, 1975-1989.

Kuo Y.-H., E.G. Donall and M.A. Shapiro, 1987, "Feasibility of short-range numerical weather prediction using observations from a network of profilers", *Mon. Wea. Rev.*, **115**, 2402-2427.

Kuo Y. -H., D.O. Gill and L. Cheng, 1987b, "Retrieving temperature and geopotential fields from a network of wind profiler observations", *Mon. Wea. Rev.*, **115**, 3146-3165.

Lafaysse C. Ed., , 1994, COST-74 final report "*Utilization of UHF/VHF radar wind profiler networks for improving weather forecasting in Europe*", Directorate-General, Science Research & Development, European Commission.

Lam C.Y., 1996, "A microburst in Hong Kong", *Weather*, **51**, 101-104.

Langleben M.P. and W.D. Gaherty, 1957, "*An optical system for automatic synthesis of constant altitude radar maps*", Stormy Weather Group (Canada), Sci.Rep. MWT-3.

Lapczak S., E. Aldcroft, M. Stanley-Jones, J. Scott, P. Joe, P. Van Rijn, M. Falla, A. Gagne, P. Ford, K. Reynolds, and D.R. Hudak , 1999, "The Canadian national radar project" *29th International Conf. Radar Meteor.*, Amer. Meteor. Soc., Montreal, Canada. 327-330.

Laroche S. and I. Zawadzki, 1994, "A variational analysis method for retrieval of three-dimensional wind field from single-Doppler radar data", *J. Atmos. Sci.*, **51**, 2664-2682.

Laroche S. and I. Zawadzki, 1995, "Retrieval of horizontal winds from single-Doppler clear-air data by methods of cross-correlation and variational analysis", *J. Atmos. Ocean Technol.*, **12**, 721-738.

Larsen M.F., 1983, "Can a VHF Doppler radar provide synoptic wind data? A comparison of 30 days of radar and radiosonde data", *Mon. Wea. Rev.*, **111**, 2047-2057.

Larsen M.F. and J. Röttger, 1987, "Observations of thunderstorm reflectivities and Doppler velocities measured at VHF and UHF", *J. Atmos. Ocean. Technol.*, **4**, 151-159.

Larsen M.F. and J. Röttger, 1989, "The spaced antenna technique for radar wind profiling", *J. Atmos. Ocean. Technol.*, **6**, 920-938.

Larsen M.F. and J.Röttger, 1991, "VHF radar measurements of in-beam incidence angles and associated vertical-beam radial velocity corrections", *J. Atmos. Ocean. Technol.*, **8**, 477-490.

Larsen M.F., S. Fukao, O. Aruga, M.D. Yamanaka, T. Tsuda and S. Kato, 1991, "A comparison of VHF radar vertical velocity measurements by a direct vertical-beam method and by a VAD technique", *J. Atmos. Ocean. Technol.*, **8**, 766-776.

Larsen M.F., R.D. Palmer, S. Fukao, R.F. Woodman, M. Yamamoto, T. Tsuda and S. Kato, 1992, "An analysis technique for deriving vector winds and in-beam incidence angles from radar interferometer measurements", *J. Atmos. Ocean. Technol.*, **9**, 3-14.

Larsen M.F., J. Röttger and T.S. Dennis, 1988, "A comparison of operational analysis and VHF wind profiler vertical velocities", *Mon. Wea. Rev.*, **116**, 48-59.

Laws J.O., and D.A.Parsons., 1943, "The relation of raindrop size to intensity", *Trans. Amer. Geophys. Union*, **24**, Part II, 452-459.

Lazarus, S., A. Shapiro, and K. Droegemeier, 1999, "Analysis of the Gal-Chen–Zhang single-Doppler velocity retrieval", *J. Atmos.Ocean. Technol.*, **16,** 5–18.

Leary C.A., and R.A. Houze Jr., 1979, "Melting and evaporation of hydrometeors in precipitation from the anvil clouds of deep tropical convection", *J. Atmos. Sci.*, **36**, 669-679.

Leary C.A., and R.A. Houze Jr., 1980, "The contribution of mesoscale motions to the mass and heat fluxes of an intense tropical convective system", *J. Atmos. Sci.*, **37**, 784-796.

Lee C-W. and Y-C. Tsai, 1991, "An analysis of `YANCY' typhoon's (1990) structure by using single Doppler radar observation", *Preprints, 25th International Conf. Radar Meteor.*, Amer. Meteor. Soc., 698-701.

Lee W-C.and F.D. Marks Jr., 2000, "Tropical Cyclone Kinematic Structure Retrieved from Single-Doppler Radar Observations. Part II: The GBVTD-Simplex Center Finding Algorithm", *Mon. Wea. Rev.*, **128**, 1925-1936.

Lee W-C., P. Dodge, F.D. Marks Jr. and P.H. Hildebrand, 1994, "Mapping of airborne Doppler radar data", *J. Atmos. Ocean. Technol.*, **11**, 572-578.

Lee W-C., B.J.-D. Jou and B.-L.Chang, 1993, "Nowcasting typhoon circulation by a ground-based Doppler radar - the ground-based VTD (GBVTD) technique", *Preprints, 26th International Conf. Radar Meteor.*, Amer. Meteor. Soc., 83-87.

Lee W-C., B.J.-D.Jou and P-L.Chang, S-M. Deng, 1999, "Tropical cyclone kinematic structure retrieved from single-Doppler radar observations. Part I: Interpretation of Doppler Velocity Patterns and the GBVTD Technique", *Mon. Wea. Rev.*, **127**, 2419–2439.

Lee W-C., B.J.-D.Jou and P-L.Chang and F.D. Marks Jr., 2000, "Tropical cyclone kinematic structure retrieved from single-Doppler radar observations. Part III: Evolution and Structures of Typhoon Alex (1987), *Mon. Wea. Rev.*, **128**, 3982–4001

Lee W-C., F.D. Marks Jr., and R.E. Carbone, 1994b, "Velocity Track Display - A technique to extract real-time tropical cyclone circulations using a single airborne Doppler Radar", J. Atmos. Ocean. Technol., **11**, 337-356.

Lee W.-C, R.M.Wakimoto and R.E. Carbone, 1992, "The evolution and structure of a `bow echo-microburst' event: Part II The Bow echo", *Mon. Wea. Rev.*, **120**, 2211-2225.

Leese J.A., C.S. Novak and B.B. Clark, 1971, "An automated technique for obtaining cloud motion from geosynchronous satellite data using cross correlation", *J. Appl. Meteor.*, **10**, 118-132.

Lemaître Y., P. Neveux and G. Scialom, 1994, "A new analysis to diagnose ageostrophic winds from wind and temperature measurements made by an observational network", *J. Atmos. Ocean. Technol.*, **11**, 637-660.

Lemon L.R., 1978, "On the use of storm structure for hail identification", *Preprints, 18th Conf. Radar Meteor.*, Amer. Meteor. Soc., 203-206.

Leone D.A., R.M. Endlich, J. Petriceks, R.T.H. Collis and J.R. Porter, 1989, "Meteorological considerations used in planning the NEXRAD network", *Bull. Amer. Meteor. Soc.*, **70**, 4-13.

Lewis B.M. and H.F. Hawkins, 1982, "Polygonal eye walls and rainbands in hurricanes", *Bull. Amer. Meteor. Soc.*, **63**, 1294-1300.

Lewis W., 1970, "Development of procedures for vectoring aircraft around thunderstorms", *Preprints, 14th Conf. Radar Meteorology*, Amer. Meteor. Soc., 301-304.

Lewis W., 1978, "Some measurements of digitized severe thunderstorm echoes with application to improved use of air traffic radar weather information", *Preprints, 18th Conf. Radar Meteorology*, Amer. Meteor. Soc., 417-423.

Lewis W., 1981, "Doppler radar and aircraft measurements of thunderstorm turbulence", *Preprints, 20th Conf. Radar Meteor.*, Amer. Meteor. Soc., 440-445.

Lhermitte R.M., 1960, "New developments of the echo fluctuation theory and measurements", *Proc. 8th Wea. Radar Conf.*, Amer. Meteor. Soc., 263-268.

Lhermitte R.M., 1966, "Probing air motion by Doppler analysis of radar clear air returns", *J. Atmos. Sci.*, **23**, 575-591.

Lhermitte R.M., 1968, "Turbulent air motion as observed by Doppler radar", *Proc. 13th Conf. Radar Meteor.*, Amer. Meteor. Soc.., 498-503.

Lhermitte R.M., 1970, "Dual-Doppler radar observation of convective storm circulation", *Preprints, 14th Radar Meteor. Conf.*, Amer. Meteor. Soc., 139-144

Lhermitte R.M., 1975, "Real time velocity observations by multi-Doppler radar", *Preprints, 16th Radar Meteor.* Conf., Amer. Meteor. Soc., 107-114.

Lhermitte R. M., 1987, "A 94-GHz. Doppler radar for cloud observations". *J. Atmos. Ocean. Technol.*, **4**, 36-48.

Lhermitte, R. M., 1988, "Observations of rain at vertical incidence with a 94-GHz Doppler radar: An insight on Mie scattering," *Geophys. Res. Lett.*, **15,** 1125–1128.

Lhermitte R.M., 1990, "Attenuation and scattering of millimeter wavelength radiation by clouds and precipitation", *J. Atmos. Ocean. Technol.*, **7**, 464-479.

Lhermitte R.M. and D. Atlas, 1961, "Precipitation motion by pulsed Doppler radar", *Proc. 9th Wea. Radar Conf.*, Amer. Meteor. Soc., 218-223.

Lhermitte R.M. and D.Atlas, 1963, "Doppler fall speed and particle growth in stratiform precipitation", *Proc. 10th Wea. Radar Conf.*, Amer. Meteor. Soc.., 297-302.

Lhermitte R.M. and J. Dooley, 1966, "Doppler study of the motion of clear air targets", *Proc. 12th Wea. Radar Conf.*, Amer. Meteor. Soc., 293-299.

Lhermitte R.M. and E. Kessler, 1966, "Estimation of the average intensity of precipitation targets", *Proc. 12th Wea. Radar Conf.*, Amer. Meteor. Soc., 23-27.

Lhermitte R.M. and L.J. Miller, 1970, "Doppler radar methodology for the observation of convective storms", *Preprints 14th Radar Meteor. Conf.*, Amer. Meteor. Soc., 133-138.

Li L., S. M. Sekelsky, S. C. Reising, C. T. Swift, S. L. Durden, G. A. Sadowy, S. J. Dinardo, F. K. Li, A. Huffman, G. Stephens, D. M. Babb and H. W. Rosenberger, 2001, "Retrieval of atmospheric attenuation using combined ground-based and airborne 95-GHz cloud radar measurements", *J. Atmos. Ocean. Technol.*, **18**, 1345–1353.

Ligda M.G.H., 1950, "Lightning detection by radar", *Bull. Amer. Meteor. Soc.*, **31**, 279-283.

Ligda M.G.H., 1951, "Radar storm observation", *Compendium of Meteorology*, Amer. Meteor. Soc., 1265-1282.

Ligda M.G.H., 1953, "*The horizontal motion of small precipitation areas as observed by radar*", Dept. of Meteorology, M.I.T., Tech. Rept. 21, 60 pp.

Ligda M.G.H., 1955, "*Analysis of motion of small precipitation areas in the hurricane, August 23-28, 1949*", Dept. of Meteorology, M.I.T.,Tech. Note 3, 41 pp.

Ligda M.G.H., 1956, "Radar observation of lightning", J. Atmos. Terr. Phys., **9**, 329-346.

Lilly D.K. and C.-H. Moeng, 1984, "A technique for recovering the total velocity field from single Doppler measurements", *Preprints, 22nd Conf. Radar Meteor.*, Amer. Meteor. Soc., 506-508.

Liou, Y.-C., 1999, "Single radar recovery of cross-beam wind components using a modified moving frame of reference technique", *J. Atmos. Ocean. Technol.*, **16,** 1003–1016.

Liou Y-C., 2001, "The Derivation of Absolute Potential Temperature Perturbations and Pressure Gradients from Wind Measurements in Three-Dimensional Space", *J. Atmos. Ocean. Technol.*, **18**, 577–590.

Liou Y-C. and I-S. Luo, 2001, "An investigation of the moving-frame single-Doppler wind retrieval technique using Taiwan Area Mesoscale Experiment low-level data", *J. Appl. Meteor.*, **40**, 1900–1917.

Liou, Y.-C., T. Gal-Chen, and D. K. Lilly, 1991, "Retrievals of wind temperature and pressure from single Doppler radar and a numerical model", *Preprints, 25th Int. Conf. on Radar Meteorology,* Amer. Meteor. Soc., 151–154.

Liu C.H., J. Röttger, G. Dester, S.J. Franke and C.-J. Pan, 1991b, "The oblique spaced antenna method for measuring the atmospheric wind field", *J. Atmos. Ocean. Technol.*, **8**, 247-258.

Liu H. and V. Chandrasekar , 2000, "Classification of hydrometeors based on polarimetric radar measurements: Development of fuzzy logic and neuro-fuzzy Systems, and in situ verification" *J. Atmos. Ocean. Technol.*, **17**, 140–164.

Liu H,, V. Chandrasekar and G. Xu, 2001, "An Adaptive Neural Network Scheme for Radar Rainfall Estimation from WSR-88D Observations" *J. Appl. Meteor:* **40**, 2038–2050.

Liu K.-Y., T.-Y. Shyu and T.-W. Lin, 1991, "The interaction between typhoon and Taiwan island", *Preprints, 25th International Conf. Radar Meteor.*, Amer. Meteor. Soc., 690-693.

Lohmeier S.P., S.M. Sekelsky, J.M. Firder, and R. E, McIntosh, 1997, "Classification of particles in stratiform clouds using the 33 and 95 GHz, polarimetric cloud profiling radar system (CPRS)", *IEEE Trans. Geophys. Remote Sensing*, **35**, 256-270.

Longley R.W., 1949, "On the energy in the hurricane", *Bull. Amer. Meteor. Soc.*, **30**, 194-195.

Longtin D.R., C.F. Bohren and L.J. Battan, 1987, "Radar backscattering by large, spongy ice oblate spheroids", *J. Atmos. Ocean. Technol.*, **4**, 355-358.

López R.E., 1976, "Radar characteristics of the cloud populations of tropical disturbances in the North Atlantic", *Mon. Wea. Rev.*, **104**, 268-283.

López R.E., 1977, "The log-normal distribution and cumulus cloud populations", *Mon. Wea. Rev.*, **105**, 865-872.

López R.E., 1978, "Internal structure and development processes of C scale aggregates of cumulus clouds", *Mon. Wea. Rev.*, **106**, 1488-1494.

López R.E., D. Atlas, D. Rosenfeld, J.L. Thomas, D.O. Blanchard and R.L. Holle, 1989, "Estimation of areal rainfall using the radar echo Area Time Integral", *J. Appl. Meteor.*,**28**, 1162-1175.

López R.E., J.R. Daugherty and R.L. Holle, 1986, "Some relationships between radar echoes and lightning from mapping systems", *Preprints, 23rd Conf. Radar Meteor.*, Amer. Meteor. Soc., J290-293.

López R.E., R. Ortiz, W.D. Otto and R.L. Holle, 1991, "Lightning activity and precipitation yield of convective cloud systems in central Florida", *Preprints, 25th International Conf. Radar Meteor.*, Amer. Meteor. Soc., 907-910.

López R.E., J. Thomas, D.O. Blanchard and R.L. Holle, 1983, "Estimation of rainfall over an extended region using only measurements of the area covered by radar echoes", *Preprints, 21st Conf. Radar Meteor.*, Amer. Meteor. Soc., 681-686.

Lovejoy S., 1981, "The statistical characterization of rain areas in terms of fractals", *Preprints, 20th Conf. Radar Meteor.*, Amer. Meteor. Soc., 476-483.

Lovejoy S., 1982, "The area-perimeter relation for rain and cloud areas", *Science*, **216**, 185-187.

Lovejoy S. and G.L. Austin, 1979, "The sources of error in rain amount estimating schemes from GOES visible and IR satellite data", *Mon. Wea. Rev.*, **107**, 1048-1054.

Luce H., S. Fukao, M. Yamamoto, C. Sidi and F. Dalaudier, 2001, "Validation of Winds Measured by MU Radar with GPS Radiosondes during the MUTSI Campaign", *J. Atmos. Ocean. Technol.*, **18**, 817-829.

Lutz J., B. Rilling, J. Wilson, T. Weckworth and J. Vivekanandan, 1997, "S-Pol after three operational deployments, technical performance, siting experiences and some data examples", *Preprints, 28th Conf. Radar Meteor.,* Amer. Meteor. Soc, 286-287.

Lutz J.S., S.E. Yuter and J.W. Wilson, 1989, "The nowcaster workstation for analyzing four-dimensional radar data", *Preprints, 24th Conf. Radar Meteor.*, Amer. Meteor. Soc., 216-219.

Maddox R.A., 1980, "Mesoscale convective complexes", *Bull. Amer. Meteor. Soc.*, **61**, 1374-1387.

Maddox R.A., J. Zhang, J. J. Gourley and K. W. Howard, 2002, "Weather radar coverage over the contiguous United States", *Weather and Forecasting*, **17**, 927–934.

Maguire II W.B. and S.K. Avery, 1994, "Retrieval of raindrop size distributions using two Doppler wind profilers: Model sensitivity testing", *J. Appl. Meteor.*, **33**, 1623-1635.

Mahapatra P.R., 1999, *"Aviation weather surveillance systems: Advanced radar and surface sensors for flight safety and air traffic management"*, Institution of Electrical Engineers and American Institute of Aeronautics and Astronautics, 453 pp.

Mahapatra P.R. and D.S. Zrnić, 1991, "Sensors and systems to enhance aviation safety against weather hazards", *Proc. IEEE*, **79**, 1234-1267.

Mahoney III W.P., 1988, "Gust front characteristics and the kinematics associated with interacting thunderstorm outflows", *Mon. Wea. Rev.*, **116**, 1474-1491.

Mahoney W., W. Wilson, K. Brislawn and M.K. Politovich, 1989, "Microburst detection by TDWR: performance assessment", *3rd International Conf. Aviation Weather System*, Amer. Meteor. Soc.

Maki M., T. D. Keenan, Y. Sasaki and K. Nakamura, 2001, "Characteristics of the raindrop size distribution in tropical continental squall Lines observed in Darwin, Australia", *J. Appl. Meteor.*, **40**, 1393–1412.

Mandelbrot B., 1977, "*Fractals: form, chance and dimension*", W.H. Freeman & Co., San Francisco, USA, 365 pp.

Mani A. and S.P. Venkiteshwaran, 1961, "Radar observations of thunderstorms at Poona", *Ind.J. Meteor. Geophys.*, **12**, 61-70.

Mani A., B.B. Huddar and G.P. Srivastava, 1962, "Studies of "melting band" in radar precipitation echoes at Poona and Bombay during the monsoon", *Ind.J. Meteor. Geophys.*, **13**, Spl. No., 127-136.

Mapes B. and R.A. Houze Jr.,1993, "An integrated view of the 1987 Australian monsoon and its mesoscale convective systems", *Quart. J. Roy. Meteor. Soc.*, **119**, 733-754.

Marécal V. and D. Hauser, 1991, "Retrieval of thermodynamic and microphysical variables from dual-Doppler radar measurements in frontal rainbands", *Preprints, 25th International Conf. Radar Meteor.*, Amer. Meteor. Soc., 143-146.

Maresca J.W. and J.R. Barnum, 1979, "Remote measurement of the position and surface circulation of hurricane Eloise by HF skywave radar", *Mon. Wea. Rev.*, **107**, 1648-1652.

Marks F.D. Jr., 1985, "Evolution of the structure of precipitation in hurricane Allen (1980)", *Mon. Wea. Rev.*, **113**, 909-930.

Marks F.D.Jr., 1990, "Radar observations of tropical weather systems", chapter 25a in "*Radar in Meteorology*", Ed. D. Atlas, Amer. Meteor. Soc., 401-425.

Marks F.D. Jr. and R.A. Houze Jr., 1984, "Airborne Doppler radar observations in hurricane Debby." *Bull. Amer. Meteor. Soc.*, **65**, 569-582.

Marks F.D. Jr. and R.A. Houze Jr., 1987, "Inner core structure of hurricane Alicia from Doppler Radar observations." *J. Atmos.Sci.*, **44**, 1296-1317.

Marks F.D. Jr. and M.D. Hudlow, 1976, "A study of the precipitation cloud population during the GATE", *Preprints, 17th Conf. Radar Meteor.*, Amer. Meteor. Soc., 498-504.

Marks F.D. Jr., R.A.Houze Jr. and J.F. Gamache, 1992, "Dual-aircraft investigation of the inner core of hurricane Norbert. Part I: Kinematic structure", J.Atmos.Sci.,**49**, 919-942.

Marshall J.S., and W.McK. Palmer., 1948, "The distribution of raindrops with size", *J.Meteor.*, **5**, 165-166.

Marshall J.S. and S. Radhakant, 1978, "Radar precipitation maps as lightning indicators", *J. Appl. Meteor.*, **17**, 206-212.

Marshall J.S., W. Hitschfeld and K.L.S. Gunn, 1955, "Advances in radar weather", *Adv. Geophys.*, Acad. Press, 1-56.

Marshall J.S., R.C. Langille and W.McK. Palmer., 1947, "Measurement of rainfall by radar", *J.Meteor.*, **4**, 186-192.

Martin D.W. and M.R. Howland, 1986, "Grid-history: A geostationary satellite technique for estimating daily rainfall in the tropics, *J. Climate Appl. Meteor*, **25**, 184-195.

Martner B.E., J.D. Marwitz and R.A. Kropfli, 1992, "Radar observations of transport and diffusion in clouds and precipitation using TRACIR", *J. Atmos. Ocean. Technol.*, **9**, 226-241.

Martner B.E., D.B. Wuertz, B.B. Stankov, R.G. Strauch, E.R. Westwater, K.S. Gage, W.L. Ecklund, C.L. Martin and W.F. Dabberdt, 1993, "An evaluation of wind profiler, RASS, and microwave radiometer performance", *Bull. Amer. Meteor. Soc.*, **74**, 599-613.

Marwitz J.D., 1973, "Trajectories within the weak echo regions of hailstorms', *J. Appl. Meteor.*, **12**, 1174-1182.

Marwitz J.D. and E.X. Berry, 1971, "The airflow within the weak echo region of an Alberta hailstorm", *J. Appl. Meteor.*, **10**, 487-492.

Marzoug M. and P. Amayenc, 1991, "Improved range-profiling algorithm of rainfall rate from a spaceborne radar with path-integrated attenuation constraint", *IEEE Trans. Geosci. Remote Sens.*, **29**, 584-592.

Marzoug M. and P. Amayenc, 1991b, "Experimental tests of an improved single-frequency algorithm for rain rate profiling using airborne radar data", *Preprints, 25th Imternational Conf. Radar Meteor.*, Amer. Meteor. Soc., 396-399.

Marzoug M. and P. Amayenc, 1993, "Algorithms for range-profiling of rain rate from a spaceborne radar: overall validation using airborne radar data", *Preprints, 26th Imternational Conf. Radar Meteor.*, Amer. Meteor. Soc., 690-692.

Marzoug M. and P. Amayenc, 1994, "A class of single- and dual-frequency algorithms for rain-rate profiling from a spaceborne radar. Part I: Principle and tests from numerical simulations", *J. Atmos. Ocean. Technol.*, **11**, 1480-1506.

Mason B.J., 1955, "Radar evidence for aggregation and orientation of melting snowflakes", *Quart. J. Roy. Meteor. Soc.*, **81**, 262-264.

Mason B.J. and J.B Andrews, 1960, "Dropsize distribution for various types of rain", *Quart. J. Roy. Meteor. Soc.*, **86**, 346-353.

Mason B.J. and R. Ramanadham, 1954, "Modification of the size distribution of falling raindrops by coalesence", *Quart. J. Roy. Meteor. Soc.*, **80**, 388-394.

Masuda Y., 1988, "Influence of wind and temperature on the height limit of a radio acoustic sounding system", *Radio Sci.*, **23**, 647-654.

Mather G.K., D. Treddnick and R. Parsons, 1976, "An observed relationship between the height of the 45 dBZ contours in storm profiles and surface hail reports", *J. Appl. Meteor.*, **15**, 1336-1340.

Mathur L.S.,1958 "Radar study of thunderstorms", *Proc. Symposium "Meteorology in relation to high level aviation in India and surrounding areas"*, India Meteor. Dept., New Delhi, 63-72.

Mathur L.S. and S.M. Kulshrestha, 1961, "On superrefraction after the passage of a thunderstorm", *Ind. J. Meteor. Geophys.*, **12**, 71-77.

Mathur L.S.and S.M. Kulshrestha, 1966, "Classification and interpretation of radar weather echoes in India", *Ind. J. Meteor. Geophys.*, **17**, 1-16.

Matrosov S.Y., and R.A. Kropfli, 1993, " Cirrus cloud studies with elliptically polarized Ka-band radar signals: A suggested approach", *J. Atmos. Ocean. Technol.*, **10**, 684-692.

Matrosov S.Y., R.F. Reinking, R.A. Kropfli, and B.W. Bartram, 1996, "Estimation of ice hydrometeor types and shapes from radar polarization measurements, *J. Atmos. Ocean. Technol.,* **13**, 85-96.

Matuura N., Y. Masuda, H. Inuki, S.Kato, S. Fukao, T. Sato and T. Tsuda, 1986, "Radio acoustic measurement of temperature profile in the troposphere and stratosphere", *Nature*, **323**, 426-428.

May P.T., 1993, "Comparison of wind-profiler and radiosonde measurements in the tropics", *J. Atmos. Ocean. Technol.*, **10**, 122-127.

May P.T., 2001, "Mesocyclone and Microburst Signature Distortion with Dual PRT Radars", *J. Atmos. Ocean. Technol.*, **18**, 1229–1233.

May P.T and R.G. Strauch, 1989, "An examination of wind profiler signal processing algorithms", *J. Atmos. Ocean. Technol.*, **6**, 731-735.

May P.T., S. Fukao, T. Tsuda, T. Sato, and S. Kato, 1988, "The effect of thin scattering layers on the determination of wind by Doppler radars", *Radio Sci.*, **23**, 83-94.

May P.T., G.J. Holland and W.L. Ecklund, 1994, "Wind profiler observations of tropical storm FLO at Saipan", *Wea. Forecasting,* **9**, 410-426.

May P.T., A. R. Jameson, T. D. Keenan and P. E. Johnston, 2001, "A Comparison between Polarimetric Radar and Wind Profiler Observations of Precipitation in Tropical Showers", *J. Appl. Meteor.*, **40**, 1702–1717.

May P.T., K.P. Moran and R.G. Strauch, 1989b, "The accuracy of RASS temperature measurements", *J. Appl. Meteor.*, **28**, 1329-1335.

May P.T, R.G. Strauch, K.P. Moran and W.L. Ecklund, 1990, "Temperature sounding by RASS with wind profiler radars: A preliminary study", *IEEE Trans. Geosci. Remote sensing*, **28**, 19-28.

May P.T., T. Sato, M.Yamamoto, S. Kato, T. Tsuda and S. Fukao, 1989, "Errors in the determination of wind speed by Doppler radar", *J.Atmos. Ocean. Technol.*, **6**, 235-242.

Maynard R.H. 1945, "Radar and Weather", *J. Meteor.*, **2**, 214-226.

Mazur V., D. Zrnić and W.D. Rust, 1984, "Lightning and precipitation velocity spectra determined with a vertically looking radar", *Preprints, 22nd Conf. Radar Meteor.*, Amer. Meteor. Soc., 91-96.

Mazur V., D. Zrnić and W.D.Rust, 1986, "Transient changes in Doppler spectra of precipitation associated with lightning", *Preprints, 23rd Conf. Radar Meteor.* Amer. Meteor. Soc., J268-272.

McCarthy J. and R.J. Serafin, 1990, "An advanced aviation weather system based on new weather sensing technologies", *Proc. 43rd International Air Safety Seminar*, 19-22 November 1990, Rome, Italy, 15 pp.

McCarthy J., J. Wilson, and T.T. Fujita, 1982, "The Joint Airport Weather Studies Project", *Bull. Amer. Meteor. Soc.*, **63**, 15-22.

McGrew R.G, 1972, "Project D/RADEX (Digitized Radar Experiments)", *Preprints, 15th Conf. Radar Meteor.*, Amer. Meteor. Soc., 101-106.

Mead J.B., R.E. McIntosh, D. Vandemark and C.T. Swift, 1989, "Remote sensing of clouds and fog with a 1.4 mm radar", *J. Atmos. Ocean. Technol.*, **6**, 1090-1097.

Meighen P.J., 1976, "On the determination of tropical cyclone tracks from land-based radar observations", *Preprints, 17th Conf. Radar Meteor.*, Seattle, Washington, USA., Amer. Meteor. Soc., 543-546.

Meighen P.J., 1985, in "*Report of the seminar on the application of radar data to tropical cyclone forecasting, Bangkok, 1983*", World Meteorological Organization(WMO), Rep. No. TCP-19, WMO/TD No. 26.

Meischner P., R. Baumann, H. Höller, and T. Jank, 2001, "Eddy Dissipation Rates in Thunderstorms Estimated by Doppler Radar in Relation to Aircraft In Situ Measurements", *J. Atmos. Ocean. Technol.*, **18**, 1609–1627.

Meischner, P., V. N. Bringi, M. Hagen, and H. Holler, 1991: "Multi-parameter radar characterization of a melting layer compared with in situ measurements", *Preprints, 25th International Conf. on Radar Meteorology,* Amer. Meteor. Soc., 721–724.

Meischner P., C. Collier, A. Illingworth, J. Joss and W. Randeu, 1997, "Advanced weather radar systems in Europe: The COST 75 action", *Bull. Amer. Meteor. Soc.*, **78**, 1411-1430.

Meneghini R., 1993, "Description of a radar/radiometer algorithm for stratiform rain over the ocean", *Preprints, 26th Imternational Conf. Radar Meteor.*, Amer. Meteor. Soc., 693-695.

Meneghini R. and J.A. Jones, 1993, "An approach to estimate the areal rain-rate distribution from spaceborne radar by the use of multiple thresholds", *J. Appl. Meteor.*, **32**, 386-398.

Meneghini R. and T. Kozu, 1990, "*Spaceborne Weather Radar*", Artech House Inc., Norwood, MA.,USA., 1990, 199 pp.

Meneghini R., and T. Kozu, 1991, "A potential method for the estimation of drop size distribution from a dual-wavelength airborne radar", *Preprints, 25th Imternational Conf. Radar Meteor.*, Amer. Meteor. Soc., 380-383.

Meneghini R. and H. Kumagai, 1994, "Characteristics of the vertical profiles of dual-frequency, dual-polarization radar data in stratiform rain, *J. Atmos. Ocean. Technol.*, **11**, 701-711.

Meneghini R., J. Eckerman and D. Atlas, 1983, "Determination of rain rate from a spaceborne radar using measurements of total attenuation", *IEEE Trans. Geosci. Remote Sens.*, **GE-21**, 34-43.

Meneghini, R., T. Iguchi, T. Kozu, L. Liao, K. Okamoto, J. A. Jones, and J. Kwiatkowski, 2000: Use of the surface reference technique for path attenuation estimates from TRMM precipitation radar. *J. Appl. Meteor.,* **39,** 2053–2070.

Meneghini R., J.A. Jones and L.H. Gesell, 1987, Analysis of a dual-wavelength surface reference radar technique", *IEEE Trans. Geosci Remote Sensing*, **GE-25**, 456-471.

Meneghini R., T. Kozu, H. Kumagai and W.C. Boncyk, 1992, "A study of rain estimation methods from space using dual-wavelength radar measurements at near-nadir incidence over ocean", *J. Atmos. Ocean. Technol.*, **9**, 364-382.

Meneghini R., H. Kumagai and J.R. Wang, 1997, "Microphysical retrievals over stratiform rain using measurements from an airborne dual-wavelength radar-radiometer", *IEEE Trans. Geosci Remote Sens.*, **GE-35**, 487-506.

Meneghini R., K. Nakamura, C.W. Ulbrich, D. Atlas, T.M. Klem and S.R. Kostic, 1986, "Comparisons of methods using data from an air-borne dual-wavelength radar", *Preprints, 23rd Conf. Radar Meteor., Amer. Meteor.* Soc., **1**, 100-103.

Merritt D.A., 1995, "A statistical averaging method for wind profiler Doppler spectra", *J. Atmos. Ocean. Technol.*, **12**, 985-995.

Merritt L.P., 1969, "Comparison of airborne and ground-based weather radars", *J.Appl. Meteor.*, **8**, 963-974.

Merritt M.W., 1984, "Automatic velocity de-aliasing for all real-time applications", *Preprints, 22nd Conf. Radar Meteor.*, Amer. Meteor. Soc., 528-533.

Metcalf J.I., 1980, "Airborne weather radar and severe weather penetration", *Preprints, 19th Conf. Radar Meteorology*, Amer. Meteor. Soc., 125-129.

Metcalf J.I., A.W. Bishop, R.C. Chanley, T.C. Hiett, and P.J. Petrocchi, 1993, "An 11-cm coherent polarimetric radar for meteorological research", *J. Atmos. Ocean. Technol.*, **10**, 249-261.

Mewes J. J. and A. Shapiro, 2002, "Use of the Vorticity Equation in Dual-Doppler Analysis of the Vertical Velocity Field", *J. Atmos. Ocean. Technol.* **19**, 543–567.

Michelson M., W.W. Schrader and J.G. Wieler, 1990, "Terminal Doppler Weather Radar (TDWR)", *Microwave J.*, **33**, 139-148.

Mie G., 1908, "Beiträge zur Optik trüber Medien, speziell kolloidaler Metallösungen", *Ann. Phys.*, **25**, 377-445.

Miller J.R., 1972, "A climatological Z-R relationship for convective storms in the northern great plains", *Preprints, 15th Conf. Radar Meteor.*, Amer. Meteor. Soc., 153-154.

Miller M.A., J. Verlinde, C. V. Gilbert, G. J. Lehenbauer, J. S. Tongue, E. E. Clothiaux, 1998, "Detection of nonprecipitating clouds with the WSR-88D: a theoretical and experimental survey of capabilities and limitations", *Wea. Forecasting*, **13,** 1046–1062.

Miller, P.A., M.F. Barth, J.R. Smart, L.A. Benjamin, 1997, "The extent of bird contamination in the hourly winds measured by the NOAA profiler network: Results before and after implementation of the new bird contamination quality control check", *Preprints, 1st Symposium on Integrated Observing Systems,* Amer. Meteor. Soc., 138-144.

Miller P.A., T.W. Schlatter, D.W. Van de Kamp, M.F. Barth and B.L. Weber, 1994, "An unfolding algorithm for profiler winds", *J. Atmos. Ocean. Technol.*, **11**, 32-41.

Misra P.K. and S.K. Prasad, "Forecasting hailstorms over India", *Mausam*, **31**, 385-396.

Mohan K., D. Narayana Rao, T. Narayana Rao, and S. Raghavan, 2001, "Estimation of temperature and humidity from MST radar observations", *Ann. Geophysicae*, **19**, 855-861.

Mohr C.G. and L.J. Miller, 1983, "CEDRIC __ A software package for Cartesian space editing, synthesis and display of radar fields under interactive control", *Preprints, 21st Conf. Radar Meteor.*, Boston, Amer. Meteor. Soc., 569-574.

Mohr C.G., L.J. Miller and R.L. Vaughan, 1981, "An interactive software package for the rectification of radar data to three-dimensional Cartesian coordinates", *Preprints, 20th Conf. Radar Meteor.*, Boston, Amer. Meteor. Soc., 690-695.

Mohr C.G., L.J. Miller, R.L. Vaughan and H.W. Frank, 1986, "The merger of mesoscale datasets into a common Cartesian format for efficient and systematic analyses", *J. Atmos. Ocean. Technol.*, **3**, 143-174.

Moninger W.R., 1986, "A computer expert system for interpreting Doppler radar displays", *Preprints, 23rd Conf. Radar Meteor.*, Amer. Meteor. Soc, JP103-106.

Moninger W.R. and R.A. Kropfli, 1987, "A technique to measure entrainment in cloud by dual-polarization radar and chaff", *J. Atmos. Ocean. Technol.*, **4**, 75-83,

Moninger W.R., V.N. Bringi, T.R. Detman, J.R. Jordan, T.A. Seliga and K. Aydin, 1984, "Melting layer observations during MAYPOLE", *Preprints, 22nd Conf. Radar Meteor.*, Amer. Meteor. Soc., 364-369.

Mooley D.A. and C.M. Mohile, 1986, "Some aspects of rainfall associated with cyclonic storms of the Bay of Bengal", *J. Climatol.*, **6**, 149-160.

Moore J.A., L. Cornman and C. Biter, 1991, "Terminal Doppler weather radar program at Denver's Stapleton International Airport during 1989 and 1990", *Preprints, 25th International Conf. Radar Meteor.*, Amer. Meteor. Soc., J21-J26.

Moores W. and T.W. Harrold, 1975, "Estimating the distribution and intensity of ground clutter at possible radar sites in hilly terrain", *Preprints, 16th Radar Meteor. Conf.*, Amer. Meteor. Soc, 370-373.

Moran K.P. and R.G. Strauch, 1994, "The accuracy of RASS temperature measurements corrected for vertical air motion", *J. Atmos. Ocean. Technol.*, **11**, 995-1001.

Moran, K.P., B. E. Martner, M. J. Post, R. A. Kropfli, D. C. Welsh and K. B. Widener, 1998, "An unattended cloud-profiling radar for use in climate research", *Bull. Amer. Meteor. Soc.*, **79**, 443-455.

Moszkowicz S., G.J. Ciach and W.F. Krajewski, 1994, "Statistical detection of anomalous propagation in radar reflectivity patterns", *J. Atmos. Ocean. Technol.*, **11**, 1026-1034.

Mueller E.A., 1984, "Calculation procedures for differential phase shift", *Preprints, 22nd Conf. Radar Meteor.*, Amer. Meteor. Soc, 397-399.

Mueller E.A., 1989, "Description of the CHILL radar system", *Preprints, 24th Conf. Radar Meteor.*, Amer. Meteor. Soc, 609-614.

Mueller E.A. and E.J. Silha, 1978, "Unique features of the CHILL radar system", 1978, *Preprints 18th conf. Radar Meteor.*, Amer. Meteor. Soc, 381-382.

Mukherjee A.K. and P.B. Bhattacharya, 1972, "An early morning tornado at Diamond Harbour on 21 March 1969", *Ind. J. Meteor. Geophys.*, **23**, 227-230.

Mukherjee A.K., and K.P. Padmanabham, 1980, "Rainfall over Bombay associated with cyclonic storms striking Saurashtra coast", *Mausam*, **31**, 45-50.

Mukherjee A.K., S. Kumar and G. Krishnamurthy, 1977, "A radar study of growth and decay of thunderstorms around Bombay during the pre-monsoon season", *Ind. J. Meteor. Hydrol. Geophys.*, **28**, 475-478.

Mukherjee A.K., D.K. Rakshit and A.K. Chaudhury, 1972, "On the very high tops of Cb clouds around Calcutta", *Ind. J. Meteor. Geophys.*, **23**, 217-218.

Mukherjee A.K., A.R. Ramakrishnan and R. Jambunathan., 1981, "Cyclones and depressions over the Indian seas in 1979", *Mausam*, **32**, 115-126.

Mukherjee A.K., D.V. Subramanian and C.E.J. Daniel, 1977, "Possibility of diurnal variation and formation of secondary vortices in the cyclone field as revealed by the radar study of the Arabian sea cyclonic storm of June 1976", *Ind. J. Meteor. Hydrol. Geophys.*, **28**, 311-320.

Mull S., and S.M. Kulshrestha, 1962, "The severe hailstorm of 27 May 1959 near Sikar (Rajasthan): A synoptic and radar study", *Ind. J. Meteor. Geophys.*, **13**, Spl. No., 81-94.

Mull S. and Y.P. Rao, 1950, "Dynamics of thunderstorms, Part II, Origin of downdrafts in a thunderstorm and the circulation in a full-fledged thunderstorm cell", *Ind. J. Meteor. Geophys.*, **1**, 291-297.

Mull S., H. Mitra and S.M. Kulshrestha, 1963, "Tropical thunderstorms and radar echoes", *Ind. J. Meteor. Geophys.*, **14**, 23-36.

Muramatsu T., 1986, "Trochoidal motion of the eye of typhoon 8019", *J. Meteor. Soc. Japan*, Ser. II, **64**, 259-272.

Muramatsu T., 1986b, "The structure of polygonal eye of a typhoon", J. Meteor. Soc. Japan, **64**, 913-920.

Nagpal O.P., D. Praveen Kumar and S.K. Dhaka, 1994, "Characteristics of tropospheric gravity waves by the Indian MST radar: ST mode operation", *Ind. J. Radio Space Phys.*, **23**, 6-11.

Naidu J.V.M., C.N.Murthy and M. Shankar, 1992, "The radar fixes of tropical cyclonic storms", *Mausam*, **43**, 212-214.

Nakamura K. and H. Inomata, 1991, "Non-Rayleigh scattering effect in rain observations by an X and Ka band dual-wavelength radar", *J. Atmos. Ocean. Technol.*, **8**, 352-362.

Nakamura K., H. Inomata, T. Kozu, J. Awaka, and K. Okamoto, 1990, "Rain observation by an X- and Ka-band dual-wavelength radar", *J. Meteor. Soc. Japan*, **68**, 509-521.

Nania A., 2002, "Remarks on the Italian rain enhancement project", *WMO Bull.*, **51**, 58-61.

Narayana Rao D., P. Kishore and K. Krishna Reddy, 1996, "Refractive turbulence structure constant (C_n^2) measurements using Indian MST radar", *Ind. J. Radio & Space Phys.*, **25**, 237-244.

Narayana Rao D., P. Kishore, K. Krishna Reddy, M. Hareesh, T. Narayana Rao, and S. V. Bhaskara Rao, 1998, "Application of quality control algorithms to the Indian MST radar wind data and comparison with rawinsonde wind data", *Ind. J. Radio & Space Phys.*, **27**, 264-273.

Narayana Rao D., P. Kishore, T. Narayana Rao, S. Vijayabhaskara Rao, K. Krishna Reddy, M. Yarriah and M. Hareesh, 1997, "Studies on refractivity structure constant, eddy dissipation rate, and momentum flux at a tropical latitude", *Radio Sci.*, **32**, 1375-1389.

Narayana Rao D., T. Narayana Rao, M.V. Ratnam, S. Thulasiraman, S.V.B. Rao, P.B. Rao and S. Srinivasalu, 2001c, "Diurnal and seasonal variability of turbulence parameters observed with Indian MST radar", *Radio Sci.*, **36**, 1439-1457.

Narayana Rao D., R. Randhir Singh and S. Vijaya Bhaskara Rao, 2001b, "Vertical velocity, horizontal divergence and turbulence associated with tropical mesoscale convective system", *Ind. J. Radio & Space Phys.*, **30**, 91-97.

Narayana Rao T., D. Narayana Rao and S. Raghavan, 1999, "Tropical precipitating systems observed with Indian MST radar", *Radio Sci.*, **34**, 1125-1139.

Narayana Rao T., D. Narayana Rao, K. Mohan and S. Raghavan, 2001, "Classification of tropical precipitating systems and associated Z-R relationships", *J. Geophys. Res.*, **106**, D16, 17,699-17,711.

Nastrom G.D. and K.S. Gage, 1983, "A brief climatology of vertical air motions from MST radar data at Poker Flat, Alaska", *Preprints, 21st Conf. Radar Meteor.*, Amer. Meteor. Soc, 135-140.

Nastrom G.D., B.R. Balsley and K.S. Gage, 1981, "Changes with season of C_n^2 at Poker Flat, Alaska, from MST Doppler radar observations", *Preprints, 20th Conf. Radar Meteor.*, Amer. Meteor. Soc., 136-141.

Nastrom G.D., W.L. Ecklund and K.S. Gage, 1985, "Direct measurement of large-scale vertical velocities using clear-air Doppler radars", *Mon. Wea. Rev.*, **113**, 708-718.

Nathanson F.E., 1969, "*Radar design principles, signal processing and the environment*", McGraw-Hill, 626 pp.

Nehrkorn T., 2000, "Analysis and quality control of profiler data using Optimum Interpolation", *J. Atmos. Ocean. Technol.*, **17**, 651–655.

Neiman P.J., P.T.May and M.A.Shapiro, 1992, "Radio-acoustic sounding system (RASS) and wind profiler observations of lower- and mid-tropospheric weather systems", *Mon, Wea. Rev.*, **120**, 2298-2313.

Nielsen-Gammon J.W. and W.L. Read, 1995, "Detection and interpretation of left-moving severe thunderstorms using the WSR-88D: A case study", *Wea. Forecasting*, **10**, 127-140.

Niranjan S., S.K. Banerjee, P. Rajesh Rao and S.K. Kundu, 1990, "On relation between radar super-refraction echoes and height of ducting layers along west coast of India", *Mausam*, **41**, 589-592.

Nissen R., D. Hudak, S. Laroche, R. de Elía, I. Zawadzki and Y. Asuma, 2001, "3D wind field retrieval applied to snow events using Doppler radar", *J. Atmos. Ocean. Technol.*, **18**, 348-362.

NOAA (National Oceanic and Atmospheric Administration, USA), 1988, "*RADAP II, Processing and enhancement of weather radar data*", Circular dated June 1988, 12 pp.

NOAA (National Oceanic and Atmospheric Administration), 1992, "*Postprint Volume, National Weather Service Aviation Workshop*, Kansas City, Missouri, 10-13 December 1991", NOAA Tech. Memorandum NWS CR-102, 411 pp.

Nolen R.H., 1959, "A radar pattern associated with tornadoes", *Bull. Amer. Meteor. Soc.*, **40**, 277-279.

North E.M., A.M. Peterson and H. Dean Parry, 1973, "RASS, a remote sensing system for measuring low-level temperature profiles", *Bull. Amer. Meteor. Soc.*, **54**, 912-919.

Novlan D.J. and W.M. Gray, 1974, "Hurricane-spawned tornadoes", *Mon. Wea. Rev.*, **102**, 476-488.

NTIS (National Technical Information Service), 1988, "*Meteorological support for space operations: review and recommendations*", US. Dept. of Commerce, PB88-246376, 77 pp.

Nutten B., P. Amayenc, M. Chang, D. Hauser, F. Roux and J. Testud, 1979, "The RONSARD radars: A versatile C-band dual Doppler facility", *IEEE Trans. Geosci. Electronics*, **GE-17**, 281-287.

Nutten B., D. Hauser, F. Roux and G. Scialom, 1979b, "The RONSARD radars: Internal calibration techniques using coherent and noise sources", *IEEE Trans. Geosci. Electronics*, **GE-17**, 288-295.

Oguchi T., 1981, "Scattering from hydrometeors: a survey", *Radio Sci.*, **16**, 691-730.

Orlanski I., 1975, "A rational subdivision of scales for atmospheric processes", *Bull. Amer. Meteor. Soc.*, **56**, 527-530.

Orr B.W. and B.E. Martner, 1996, "Detection of weakly precipitating winter clouds by a NOAA 404-MHz wind profiler", *J. Atmos. Ocean. Technol.*, **13**, 570-580.

Ottersten H., 1969, ""Atmospheric structure and radar backscattering in clear air", *Radio Sci.*, **4**, 1179-1193.

Oye R., and R. Carbone, 1981, "Interactive Doppler editing software", *Preprints, 20th Conf. Radar Meteor.*, Boston, Amer. Meteor. Soc., 683-689.

Palmer R.D, M.F. Larsen, R.F. Woodman, S. Fukao, M. Yamamoto, T. Tsuda and S. Kato, 1991, "VHF radar interferometry measurements of vertical velocity and the effect of tilted refractivity surfaces on standard Doppler measurements", *Radio Sci.*, **26**, 417-427.

Pamment J.A.and B J. Conway, 1998, "Objective identification of echoes due to anomalous propagation in weather radar data", *J. Atmos. Ocean. Technol.*, **15**, 98–113

Panofsky H.A. and G.W. Brier, 1965, "*Some applications of statistics to meteorology*", Penn. Stat. Univ. Press, 224 pp.

Pantley K.C. and P.F. Lester, 1990, "Observations of severe turbulence near thunderstorm tops", *J. Appl. Meteor.*, **29**, 1171-1179.

Parrish J.R., 1989, "New NOAA OAO WP-3D Doppler radar system", *Preprints, 24th conf. Radar Meteor.*, Amer. Meteor. Soc, 615-618.

Parrish J.R., R.W. Burpee and F.D. Marks Jr., 1982, "Rainfall patterns observed by digitized radar during the landfall of hurricane Frederic (1979)", *Mon. Wea. Rev.*, **110**, 1933-1944.

Parsons D.B. and R.A. Kropfli, 1990, "Dynamics and fine structure of a microburst", *J. Atmos. Sci.*, **47**, 1674-1692.

Pasqualucci F., 1984, "Drop size distribution measurements in convective storms with a vertically pointing 35-GHz Doppler radar", *Radio Sci.*, **19**, 177-183.

Pasqualucci F., B.W. Bartram, R.A. Kropfli and W.R. Moninger, 1983, "A millimeter-wavelength dual-polarization radar for cloud and precipitation studies", *J. Climate Appl. Meteor.*, **22**, 758-765.

Passarelli R.E. Jr., 1983, "Parametric estimation of Doppler spectra: An alternative ground-clutter rejection technique", *J. Climate Appl. Meteor.*, **22**, 850-857.

Passarelli R.E. Jr. and R.C. Srivastava, 1980, "A new aspect of the vertical incidence Doppler radar spectrum of ice particle fallspeeds", *J. Appl. Meteor.*, **19**, 1318-1322.

Patterson V., M.D. Hudlow, P.J. Pytlowany, F.P. Richards and J.D. Hoff, 1979, "*GATE radar rainfall processing system*", NOAA Tech. Memorandum EDS 26.

Pazmany A.L., J. Galloway, I. Popstefnija, R.E. McIntosh, R. Kelly and G. Vali, 1993, "A three millimeter airborne radar for high resolution polarimetric cloud measurements", *Preprints, 26th International Conf. Radar Meteor.*, Amer. Meteor. Soc., 376-377.

Pazmany, A. L., J. C. Galloway, J. B. Mead, I. Popstefanija, R. E. McIntosh, and H. B. Bluestein, 1999b, "Polarization diversity pulse-pair technique for millimeter-wave Doppler radar measurements of severe storm features", *J. Atmos. Ocean. Tech.*, **16**, 1900–1911.

Pazmany A., J. Mead, R.E. McIntosh, M. Hervig, R. Kelly and G. Vali, 1994, "95-GHz polarimetric radar measurements of orographic cap clouds", *J. Atmos. Ocean. Technol.*, **11**, 140-153.

Pazmany A.L., C. T. Swift and H. B. Bluestein, 1999, "A 3 mm wavelength mobile Doppler radar for the observations of Tornadoes", *Preprints, 29th International Conf. Radar Meteor.*, Amer. Meteor. Soc., 827-829.

Peace R.L.Jr., R.A.Brown and H.G.Camnitz, 1969, "Horizontal motion field observations with a single pulse Doppler radar", *J. Atmos. Sci.*, **26**, 1096-1103.

Pestaina-Haynes M. and G.L. Austin, 1976, "Comparison between maritime tropical (GATE and Barbados) and continental mid-latitude precipitation lines", *J. Appl. Meteor.*, **15**, 1077-1082.

Peters G. and H.J. Kirtzel, 1994, "Measurements of momentum flux in the boundary layer by RASS", *J. Atmos. Ocean. Technol.*, **11**, 63-75.

Peterson R.E. and K.C. Mehta, 1981, "Climatology of tornadoes of India and Bangladesh", *Arch. Met. Geoph. Biokl.*, Ser. B, **29**, 345-356.

Petersen W. A., S. W. Nesbitt, R. J. Blakeslee, R. Cifelli, P. Hein and S. A. Rutledge, 2002, "TRMM observations of intraseasonal variability in convective regimes over the Amazon", *J. Climate*, **15**, 1278–1294.

Piddington H., 1848, "*The sailor's horn-book for the law of storms*", Smith, Elder & . Co, London.

Pilon J., 1989, "Vers l'automatisation complete de l'observation météorologique de surface: le Projet SOLFEGE", *WMO Instruments and Observing Methods Report No. 35*, WMO/TD - No. 303, WMO, Geneva, pp101-104.

Pilon J., 1992, "Prise au compte opérationelle des mesures d'un réseau de profileurs de vent; problèmes soulèves, solutions attendues", in *Instruments and Observing Methods Report No. 49*, 366-369, WMO/TD. No. 462.

Pilon J., Ph. Currier and V. Klaus, 1995, "A new wind profiler for operational meteorology needs", *Seventh Workshop on Technical and Scientific Aspects of MST/ST radar*, November 1995, USA.

Polger P.D., B.S. Goldsmith, R.C. Przywarty and J.R. Bocchieri, 1994, "National Weather Service warning performance based on the WSR-88D", *Bull. Amer. Meteor. Soc.*, **75**, 203-214.

Potts R., 1989, "Microburst precursors observed with Doppler radar", *Preprints, 24th Conf. Radar Meteor.*, Amer. Meteor. Soc., 158-162.

Powell M.D., 1990, "Boundary layer structure and dynamics in outer hurricane rainbands. Part I: Mesoscale rainfall and kinematic structure", *Mon. Wea. Rev.*, **118**, 891-917.

Powell M.D., P.P. Dodge, and M.L. Black, 1991, "The landfall of hurricane Hugo in the Carolinas: Surface wind distribution", *Wea. Forecasting*, **3**, 379-399.

Pratte J.F. and D.G. Ferraro, 1989, "Automated solar gain calibration", *Preprints, 24th Conf. Radar Meteor.*, Amer. Meteor. Soc., 619-622.

Pratte J.F. and R.J. Keeler, 1986, "Sensitivities of operational weather radars", *Preprints, 23rd Conf. Radar Meteor.*, Amer. Meteor. Soc., JP333-336.

Pratte F., D. Ecoff, J. VanAndel and R.J. Keeler, 1997, "AP clutter mitigation in the WSR-88D", *Preprints, 28th conf. Radar Meteor.*, Amer. Meteor. Soc, 504-505.

Probert-Jones J.R., 1962, "The radar equation in meteorology", *Quart. J. Roy. Meteor. Soc.*, **88**, 485-495.

Probert-Jones J.R., 1984, "Resonance component of backscattering by large dielectric spheres", *J. Opt. Soc., Am.*, **A1**, 822-830.

Probert-Jones P.R. and W.G. Harper, 1961, "Vertical air motion in showers as revealed by Doppler radar", *Proc. 9th Wea. Radar Conf.* Amer. Meteor. Soc., 225-232.

Pruppacher H.R. and K.V. Beard, 1970, "A wind-tunnel investigation of the internal circulation and shape of water drops falling at terminal velocity in air ", *Quart. J. Roy. Meteor. Soc.*, **96**, 247-256.

Qiu C-J. and Q. Xu, 1992, "A simple adjoint method of wind analysis for single-Doppler data", *J. Atmos. Ocean. Technol.*, **9**, 588-598.

Rabiner L.R. and B.Gold, 1978, "*Theory and application of digital signal processing*", Prentice-Hall of India, 762 pp.

Racette P., G.M. Heymsfield, R. Meneghini and L. Miller, 1993, "Design of a 94 GHz. airborne cloud radar system", *Preprints, 26th International Conf. Radar Meteor.*, Amer. Meteor. Soc., 381-383.

Raghava Reddi C. and S. Muraleedharan Nair, 1994, "Processing early observations of Indian MST radar data in its ST mode of operation", *Ind. J. Radio & Space Phys.*, **23**, 361-374.

Raghavan S., 1962, "Some characteristics of convective cells passing over Madras - a radar study", *Ind. J. Meteor. Geophys.*, **13**, Spl. No., 119-126.

Raghavan S., 1964, "3 cm propagation in the rear of a thunderstorm", *World Conf. Radio Meteorology and 11th Weather Radar Conf.*, Amer. Meteor. Soc., 472-475.

Raghavan S., 1965, "An explanation of anomalous radar propagation following thunderstorm activity", *Ind. J. Meteor. Geophys.*, **16**, 91-96.

Raghavan S., 1977, "Structure of an immature cyclonic storm in the Bay of Bengal as revealed by radar", *Ind. J. Meteor. Hydrol. Geophys.*, **28**, 223-232.

Raghavan S., 1984, "*Use of solar observations for checking antenna settings of cyclone detection radars*", IMD Pre-published Scientific Report No. 84/2, 6 pp., India Meteorological Department.

Raghavan S., 1985, in "*Report of the seminar on the application of radar data to tropical cyclone forecasting, Bangkok, 1983*", World Meteorological Organization(WMO), Rep. No. TCP-19, WMO/TD No. 26.

Raghavan S., 1987, "*Derivation of input parameters for storm surge prediction from radar observations*", WMO Second Regional Workshop on Cyclone Storm Surge, Calcutta, 14-19 December, 7 pp + 14 figs.

Raghavan S., 1990, "Structure of tropical cyclones in the Bay of Bengal",*Mausam*, **41**, 325-328.

Raghavan S., 1991, "The May 1990 cyclone and its predecessors", *J. Appl. Hydrol.*, Visakhapatnam, **IV**, 1-8.

Raghavan S., 1993, "Some recent results on tropical cyclone core structure from modern observational techniques", in '*Advances in Tropical Meteorology*', Ed. R.N.Keshavamurty and P.C.Joshi, Tata McGraw-Hill, New Delhi, pp 277-288.

Raghavan S., 1997, "Radar observations of tropical cyclones over the Indian seas", *Mausam*, **48**, 169-188.

Raghavan S., 1999, "The tropical cyclone with the smallest eye", *Mausam*, **50**, 105.

Raghavan S., 1999b, "Observational Techniques and Strategies beyond 2000", pp 143-152 in "*Meteorology beyond 2000, Proc. National symposium, TROPMET-99,* Chennai, India, Ind. Meteor. Soc., Chennai Chapter, 658 pp.

Raghavan S. and N.S. Rajagopalan, 1980, "The Andhra cyclone of 12 May 1979: the radar meteorologist's point of view", *Mausam*, **31**, 573-580.

Raghavan S. and B. Ramakrishnan, 1981, "Streamers giving heavy rainfall in the rear of cyclones", *Mausam*, **36**, 111-112.

Raghavan S. and T.R. Sivaramakrishnan, 1982, "Radar estimation of precipitation around Madras", *Mausam*, **33**, 21-28.

Raghavan S. and V.M.Varadarajan, 1981, "Radar estimate of rainfall and latent heat release in tropical cyclones of the Bay of Bengal", *Mausam*, **32**, 247-252.

Raghavan S. and K. Veeraraghavan, 1979, "Radar-synoptic study of the Nagappattinam cyclone of 12 November 1977", *Mausam*, **30**, 21-30.

Raghavan S. and G. Viswanathan, 1995, "Feasibility of the deployment of an operational wind profiler network in India", pp 253-260 in '*Advances in Tropical Meteorology*', Ed. R.K. Datta, Concept Publishing Company, New Delhi, 595 pp.

Raghavan S., S. Rengarajan, V. Natarajan and V. Ramaswami, 1985, "Accuracy of radar fixes of a tropical cyclone – a case study", *Mausam*, **36**, 111-112.

Raghavan S., S. Rengarajan, V. Ramaswami and S.W. Premkumar, 1989, "Some structural features of a Bay of Bengal tropical cyclone",*Mausam*, **40**, 65-72.

Raghavan S., S. Rengarajan and V.M. Varadarajan, 1980, "Some experiences of radar observation of cyclonic storms in the southern Bay of Bengal", *Mausam*, **31**, 81-92.

Raghavan S., S. Rengarajan and V.M.Varadarajan, 1980b, "Radar study of the Bay of Bengal cyclone of 19 November 1977", *Mausam*, **31**, 229-240.

Raghavan S., T.R. Sivaramakrishnan and B. Ramakrishnan, 1983, "Size distribution of radar echoes as an indicator of growth mechanisms in monsoon clouds around Madras", *J. Atmos. Sci.*, **40**, 428-434.

Raghavan S., T.R. Sivaramakrishnan and V. Ramaswamy, 1981, "Radar study of convective clouds around Madras during MONEX", *National Symposium on early results of MONEX-79 experiment*, New Delhi, March 1981.

Raghavan S., T.R. Sivaramakrishnan, S. Rengarajan and S.W. Premkumar, 1987, "A radar reflectivity-rainfall rate relationship for the southwest monsoon season for the Madras area", *Mausam* **38**, 335-340.

Raghavan S., R.C. Saxena, E.V. Subba Reddy, S. Rengarajan, V. Ramaswami, S.W. Premkumar and B. Ramakrishnan, 1984, "Digital radar study of the core structure of a Bay of Bengal cyclone - results and operational applications", *Vayu Mandal*, **14**, 39-45.

Raghavan S., V.M. Varadarajan and P.P. Narayanan, 1976, "Spurious radar echoes produced by reflections from buildings", *Vayu Mandal*, **6**, 69-71.

Rai D.B., 1959, ""A preliminary study of angel activity near Bombay" *Ind. J. Meteor. Geophys.*, **10**, 313-320.

Rai D.B., 1960, "Angel observation in association with sea breeze" *Ind. J. Meteor. Geophys.*, **11**, 412-415.

Rai D.B., 1961, "A study of radar angels near west coast of India", *Ind. J. Meteor. Geophys.*, **12**, 307-316.

Rai D.B., 1961b, "A turbulent convection theory of radar angels", *Ind. J. Meteor. Geophys.*, **12**, 439-446.

Rajagopalan N.S and S. Raghavan, 1980, "Tropospheric radiowave propagation over the Bay of Bengal", *Mausam*, **31**, 415-420.

Rajesh Rao P., R.C. Saxena and S.K. Banerjee, 1992, "Some aspects of accuracy of radar/satellite fixes of tropical cyclone over Bay of Bengal", *Mausam*, **43**, 379-384.

Rajopadhyaya D.K., S. K. Avery, P. T. May and R. C. Cifelli, 1999, "Comparison of precipitation estimation using single- and dual-frequency wind profilers: simulations and experimental results", *J. Atmos. Ocean. Technol.*, **16**, 165–173.

Rajopadhyaya D.K., P.T.May, and R.A. Vincent, 1993, "A general approach to the retrieval of raindrop size distributions from wind profiler Doppler spectra: Modelling results", *J. Atmos. Ocean. Technol.*, **10**, 710-717.

Rajopadhyaya D.K., P.T. May and R.A. Vincent, 1994, "The retrieval of ice particle size information from VHF wind profiler Doppler spectra", *J. Atmos. Ocean. Technol.*, **11**, 1559-1568.

Rakshit H. and I.N. Bhar, 1936, "Some observations in the C layers of the ionosphere", *Nature*, **138**, 283-284.

Ralph F.M., 1995, "Using radar-measured radial vertical velocities to distinguish precipitation scattering from clear-air scattering", *J. Atmos. Ocean. Technol.*, **12**, 257-267.

Ralph F.M., P.J. Neiman, D. Ruffieux, 1996, "Precipitation identification from radar wind profiler spectral moment data: Vertical velocity histograms, velocity variance and signal power-vertical velocity correlations", *J. Atmos. Ocean. Technol.*, **13**, 545-559.

Ralph F.M., P.J. Neiman, D.W. van de Kamp and D.C. Law, 1995, "Using spectral moment data from NOAA's 404-MHz radar wind profilers to observe precipitation", *Bull. Amer. Meteor. Soc.*, **76**, 1717-1739.

Rama Verma Raja, M.K., G.C. Asnani, P.S. Salvekar, A.R. Jain, D. Narayana Rao, S. Venkoba Rao, P. Kishore and M. Hareesh, 1999, *Proc. Indian Acad. Sci. (Earth and Planet. Sci.)*, **108**, 287-295.

Ramakrishnan K.P., 1937, "*The rainfall in the Indian peninsula associated with cyclonic storms from the Bay of Bengal during the post-monsoon and early winter seasons*", IMD Sci. Notes, Vol. VII, No. 74.

Ramana Murthy Bh.V. and K.R. Biswas, 1961, "Vertical growth and decay of convective cloud cells and associated precipitation rates at different levels", *Ind. J. Meteor. Geophys.*, **12**, 87-92.

Ramanamurty Bh.V., A.K. Roy and K.R. Biswas, 1965, "Radar echo intensity below bright band", *J.Atmos.Sci.*, **22**, 91-94.

Ramana Murthy Bh.V., K.R. Biswas and B.K. Ghosh Dastidar, 1960, Incidence of warm and cold rain in and around Delhi and their contribution to season's rainfall", *Ind. J. Meteor. Geophys.*, **11**, 4331-346.

Randall, M., J. Lutz, and J. Fox, 1997: S-Pol's high isolation mechanical polarization switch. *Preprints, 28th Conf. Radar Meteor.*, Amer. Meteor. Soc., 252–253.

Rankine W.J.M., 1901, "*A manual of applied mechanics*", Charles Griff and Company, 574-578.

Rao A.V.R.K., R.R. Kelkar and P.A. Arkin, 1989, "Estimation of precipitation and outgoing longwave radiation from INSAT-1B radiance data", *Mausam* **40**, 123-130.

Rao D.Venkateswara, 1967, "On the ellipticity and gyration of the radar eye of a Bay storm", *Ind. J. Meteor. Geophys.*, 18, 491-496.

Rao D. Venkateswara, S. Raghavan and K. Soundararajan, 1961, "A radar study of the development and movement of thunderstorms in the vicinity of Madras" *Ind. J. Meteor. Geophys.*, **12**, 79-86.

Rao P.B., 1990, "Radars for atmospheric sciences: MF/HF and MST radars", *Ind. J. Radio Space Phys.*, **19**, 326-342.

Rao P.B., A.R. Jain, P. Kishore, P. Balamuralidhar, S.H. Damle, and G. Viswanathan, 1995, "Indian MST radar 1. System description and sample vector wind measurements in ST mode", *Radio Sci.*, **30**, 1125-1138.

Rao P.B., A.R. Jain, G. Viswanathan and S.H. Damle, 1994, "Indian MST radar in ST mode - an overview", *Ind. J. Radio Space Phys.*, **23**, 1-5.

Ray P.S., Ed., 1986, "*Mesoscale Meteorology and Forecasting*", Published by Amer. Meteor. Soc., 793 pp.

Ray P.S., 1990, "Convective dynamics", Chapter 24a, pp 348-390, in Atlas D. (Ed.), 1990, "*Radar in Meteorology*", Amer. Meteor. Soc., 806 pp.

Ray P.S. and D.P. Jorgensen, 1988, "Uncertainties associated with combining airborne and ground-based Doppler radar data", *J. Atmos. Ocean. Technol.*, **5**, 177-196.

Ray P.S. and K.L. Sangren, 1983, "Multiple-Doppler radar network design", *J. Climate Appl. Meteor.*, **22**, 1444-1454.

Ray P.S., B.C. Johnson, K.W. Johnson, J.S. Bradberry, J.J. Stephens, K.K. Wagner, R.B. Wilhelmson and J.B. Klemp, 1981, "The morphology of several tornadic storms on 20 May 1977", *J. Atmos. Sci.*, **38**, 1643-1663.

Ray P.S., D.P. Jorgensen and S-L Wang, 1985, "Airborne Doppler radar observations of a convective storm", *J. Climate Appl. Meteor.*, **24**, 687-698.

Ray P.S., L.R. Lemon, D.A. Miller and B.D. Whissel, 1991, "Operational Doppler radar in the tropics", *Preprints, 25th International Conf. Radar Meteor.*, Amer. Meteor. Soc., 35-38.

Ray T.K. and A.C. De, 1971, "Radar study of pre-monsoon squall lines as observed at Gauhati airport", *Ind. J. Meteor. Geophys.*, **22**, 223-226.
Rayleigh (Lord), 1871, "On the scattering of light by small particles", *Phil. Mag.*, **41**, 447-452.
Reap R.M., 1986, "Evaluation of cloud-to-ground lightning data from the western United States for the 1983-84 summer seasons", *J. Climate Appl. Meteor.*, **25**, 785-799.
Revathy K., Mini Praseedom and S.R. Prabhakaran Nayar, 1994, "Short-period fluctuations at low latitude tropospheric heights using MST radar", *Ind. J. Radio Space Phys.*, **23**, 12-20.
Revathy K., S.R. Prabhakaran Nair and B.V. Krishna Murthy, 1996, "Deduction of temperature profile from MST radar observations of vertical wind", *Geophys. Res. Lett.*, **23**, 285-288.
Richard P., 1991, "Localization of atmospheric discharges, a new way for severe weather forecasting", *Preprints, 25th International Conf. Radar Meteor.*, Amer. Meteor. Soc., 911-915.
Richter C., 1993, "A combined Doppler radar/RASS system as rain gauge", *Preprints, 26th Imternational Conf. Radar Meteor.*, Amer. Meteor. Soc., 646-647.
Riehl H., 1954, pp 281-357 in "*Tropical Meteorology*", McGraw Hill, pp 392.
Rinehart R.E., 1989, "The Rinehart projection: a new map projection for range-aliased data", *Preprints, 24th Conf. Radar Meteor.*, Amer. Meteor. Soc, 697-700.
Rinehart R.E., 1991, "Spurious velocities in Doppler radar data caused by a moving antenna feedhorn", *J.Atmos. Ocean. Technol.*, **8**, 733-745.
Rinehart R.E., 1999, "*Radar for meteorologists*", 3rd Edition, Rinehart Publications, P.O.Box 6124, Grand Forks, ND 58206-6124, USA, 428 pp.
Rinehart R.E., and P.J. Eccles, 1976, "Use of a nodding dihedral target for antenna gain measurements", *Preprints, 17th Conf. Radar Meteor.*, Amer. Meteor. Soc., 66-71.
Rinehart R.E. and E.T. Garvey, 1978, "Three-dimensional storm motion detection by conventional weather radar", *Nature*, **273**, 287-289.
Rinehart R.E., and J.D. Tuttle, 1981, "A technique for determining antenna beam patterns using a ground target", *Preprints, 20th Conf. Radar Meteor.*, Amer. Meteor. Soc., 672-675.
Rinehart R.E., and J.D. Tuttle, 1982, "Antenna beam patterns and dual-wavelength processing", *J. Appl. Meteor.*, **21**, 1865-1880.
Rinehart R.E., and J.D. Tuttle, 1984, "Dual-wavelength processing: Some effects of mismatched antenna beam patterns", *Radio. Sci.*, **19**, 123-131.
Rinehart R.E., A. Borho and C. Curtiss, 1993, "Observations of microburst rotation", *Preprints, 26th Conf. Radar Meteor.*, Amer. Meteor. Soc., 676-678.
Rison W., P.R. Krehbiel, T. Chen and P. Gondlia (1991), "Design of a PC-based real-time radar display", *Preprints 25th International Conf. Radar Meteor.*, Amer. Meteor. Soc., 227-228.
Roberts R. and J. Wilson, 1989, "A proposed microburst nowcasting procedure using single Doppler radar", *J.Appl. Meteor.*, **28**, 285-303.
Robinson I.S., 1985, "*Satellite Oceanography*", John Wiley/Ellis Harwood, 455 pp.
Rocken C., R. Anthes, M. Exner, D. Hunt, S. Sokolovskiy, R. Ware, M. Gorbunov, W. Schreiner, D. Feng, B. Herman, Y.-H. Kuo and X. Zou, 1997, "Analysis and validation of GPS/MET data in the neutral atmosphere", *J. Geophys. Res.*, **102**, 29 849-29 866.
Rockney V.D., 1956, "Hurricane detection by radar and other means", *Proc. Tropical Cyclone Symp.*, Brisbane, Bureau of Meteorology, Australia, 179-197.
Roeder W.P., and C.S. Spinder, 1998, "Lightning forecasting empirical techniques for central Florida in support of America's space program", *Preprints, 16th Conf. Weather analysis and forecasting*, Amer. Meteor. Soc., 475-477.
Roeder W.P., J. E. Sardonia, S. C. Jacobs, M. S. Hinson, A. A. Guiffrida, J. T. Madura , 1999, "Avoiding triggered lightning threat to space launch from the Eastern Range/Kennedy Space Center", *Preprints, Eighth Conf. Aviation, Range, and Aerospace Meteor.*, 120-124.
Rogers D., F.W. Wilson Jr. and R.K. Goodrich , 1991 "A feature identification algorithm", *Preprints, 25th International Conf. Radar Meteor.*, Amer. Meteor. Soc., 115-118.

Rogers R.R., 1964, "An extension of the Z-R relationship for Doppler radar", *World Conf. Radar Meteor. and 11th Wea. Radar Conf.* Amer. Meteor. Soc., 158-160.
Rogers R.R., 1967, "Doppler radar investigation of Hawaiian rain", *Tellus*, **19**, 432-435.
Rogers R.R., 1971, "The effect of variable target reflectivity on weather radar measurements", *Quart. J. Roy. Meteor. Soc.*, **97**, 154-167.
Rogers R.R., 1984, "A review of multiple-parameter radar observations of precipitation", *Radio Sci.*, **19**, 23-36.
Rosenfeld D., 1987, "Objective method for analysis and tracking of convective cells as seen by radar", *J. Atmos. Ocean. Technol.*, **4**, 422-434.
Rosenfeld D., and A. Gagin, 1989, "Factors governing the total rainfall yield from continental convective clouds ", *J. Appl. Meteor.*, **28**, 1015-1030.
Rosenfeld D. and W. L. Woodley, 2000, "Deep convective clouds with sustained supercooled liquid water down to –37.5 °C", *Nature*, **405**, 440-442.
Rosenfeld D., D. Atlas and D.A. Short, 1990, "The estimation of convective rainfall by area integrals: 2. The Height Area Rainfall Threshold (HART) method", *J. Geophys. Research*, **95**, 2161-2176.
Rosenfeld D., E. Amitai and D.B. Wolff, 1995, "Classification of rain regimes by the three-dimensional properties of reflectivity fields", *J. Appl. Meteor.*, **34**, 198-211.
Rosenfeld D., E. Amitai and D.B. Wolff, 1995b, "Improved accuracy of radar WPMM estimated rainfall upon application of objective classification criteria", *J. Appl. Meteor.*, **34**, 212-223.
Röttger J.. 1986, "Determination of the Brunt-Väisälä Frequency from vertical velocity spectra", *Middle Atmosphere Program, Handbook for MAP*, Vol. 20, 168-172 Eds. S.A. Bowhill and Belva Edwards, 505 pp. (Copies available from SCOSTEP Secretariat, University of Illinois, 1406 W. Green St., Urbana, Illinois-61801, USA).
Röttger J. and M.F. Larsen, 1990, "UHF/VHF radar techniques for atmospheric research and wind profiler applications", pp 235-281 in *"Radar in Meteorology"*, Ed. D. Atlas, Amer. Meteor. Soc., 806 pp.
Roux F., 1985, "Retrieval of thermodynamic fields from multiple-Doppler radar data using the equations of motion and the thermodynamic equation", *Mon. Wea. Rev.*, **113**, 2142-2157.
Roux F., 1988, "The West African squall line observed on 23 June 1981 during COPT 81: Kinematics and thermodynamics of the convective region", *J. Atmos. Sci.*, **45**, 406–426.
Roux F. and F.D. Marks Jr., 1996, "Extended Velocity Track Display (EVTD): An improved processing method for Doppler radar observations of tropical cyclone", *J. Atmos. Ocean. Technol.*, **13**, 875-899..
Roux F. and J. Sun, 1990, "Single-Doppler observations of a West African squall line on 27-28 May 1981 during COPT-81: Kinematics, thermodynamics and water budget", *Mon. Wea. Rev.*, **118**, 1826-1854.
Rubin W.L., 2000, "Radar–Acoustic Detection of Aircraft Wake Vortices" *J. Atmos. Ocean. Technol.*, **17**, 1058–1065.
Ruf C.S., K. Aydin, S. Mathur and J.P. Bobak, 1996, "35-GHz dual-polarization propagation link for rain-rate estimation", *J. Atmos. Ocean. Technol.*, **13**, 419-425.
Ruggiero, F.H., and Donaldson R.J. Jr., 1987, "Wind field derivatives: a new diagnosis tool for analysis of hurricanes by a single Doppler radar", *Extended Abst., 17th Conf. Hurricanes and Trop. Meteor.*, Amer. Meteor. Soc., 178-181.
Russell R.W. and J.W. Wilson, 1997, "Radar-observed fine lines in the optically clear boundary layer. Reflectivity contributions from aerial plankton and its predators", *Bound.-Layer Meteor.*, **82**, 235-262.
Rust W.D. and R.J. Doviak, 1982, "Radar research on thunderstorms and lightning", *Nature*, **297**, 461-468.

Rüster R. and P. Czechowsky, 1980, "VHF radar measurements during a jet stream passage", *Radio Sci.*, **15**, 363-369.
Rutledge S.A., 1986, "A diagnostic modeling study of the stratiform region associated with a tropical squall line", *J.Atmos. Sci.*, **43**, 1356-1377.
Rutledge S.A. and P.V. Hobbs, 1983, "The mesoscale and microscale structure and organization of clouds and precipitation in mid-latitude cyclones. VII: A model for the 'seeder-feeder' process in warm frontal rainbands", *J.Atmos. Sci.*, **40**, 1185-1206.
Rutledge S.A. and P.V. Hobbs, 1984, "The mesoscale and microscale structure and organization of clouds and precipitation in mid-latitude cyclones. XII: A diagnostic modeling study of precipitation development in narrow cold frontal rainbands", *J.Atmos. Sci.*, **41**, 2949-2972.
Ryan B.F., G.M. Barnes, and E.J. Zipser, 1992, "A wide rainband in a developing tropical cyclone", *Mon.Wea. Rev.*, **120**, 431-447.
Ryde J.W., 1946, "The attenuation and radar echoes produced at centimetre wavelengths by various meteorological phenomena", *Meteorological Factors in Radio Wave Propagation*, Phys. Soc., London, 169-189.
Rys F.S. and A. Waldvogel, 1986, "Fractal shape of hail clouds", *Phys. Rev. Letters*, **56**, 784-787.
Ryzhkov A.V. and D.S. Zrnić, 1995, "Comparison of dual-polarization radar estimators of rain", *J. Atmos. Ocean. Technol.*, **12**, 249-256.
Ryzhkov A.V. and D. S. Zrnić, 1998, "Discrimination between Rain and Snow with a Polarimetric Radar" *J. Appl. Meteor.*, **37**, 1228–1240.

Sachidananda M. and D.S. Zrnić, 1985, "Z_{DR} measurement consideration for a fast-scan capability radar", *Radio Sci.*, **20**, 907-922.
Sachidananda M. and D.S. Zrnić, 1986, "Differential propagation phase shift and rainfall rate estimation", *Radio Sci.*, **21**, 235-247.
Sachidananda M. and D.S. Zrnić, 1987, "Rain rate estimates from differential polarization measurements", *J. Atmos. Ocean. Technol.*, **4**, 588-598.
Sachidananda M. and D.S. Zrnić, 1989, "Efficient processing of alternately polarized radar signals", *J. Atmos. Ocean. Technol.*, **6**, 173-181.
Sachidananda M. and D.S. Zrnić, 1999, "Systematic Phase Codes for Resolving Range Overlaid Signals in a Doppler Weather Radar", *J. Atmos. Ocean. Technol.*, **16**, 1351-1363.
Sachidananda M. and D.S. Zrnić, 2000, " Clutter filtering and spectral moment estimation for Doppler weather radars using staggered pulse repetition time (PRT)", *J. Atmos. Ocean. Technol.*, **17**, 323-331.
Saffle R.E., 1976, "D/RADEX products and field operation ", *Preprints, 17th Conf. Radar Meteor.*, Amer. Meteor. Soc., 555-559.
Saha K.R., 1966, "*Tornadoes and their occurrence in India*", IMD ITM Sci. Report No. **30**, 14 pp + 6 figures.
Saha K.R., 1971, "Tornadoes in India", *Vayu Mandal*, **1**, 84-87.
Sakakibara H., M. Ishihara, and Z. Yanagisawa, 1985, "Structure of a typhoon rainstorm in the middle latitudes observed by Doppler radar", *J.Meteor. Soc. Japan*, Ser II, **63**, 901-922.
Sánchez-Diezma R., D. Sempere-Torres, G. Delrieu and I. Zawadzki, 2001, "An improved methodology for ground clutter substitution based on a pre-classification of precipitation types", *Preprints, 30th Conf. Radar Meteor.*, 271-273.
Sánchez-Diezma R., I. Zawadzki and D. Sempere-Torres, 1999, "Identification of the bright band through the analysis of volumetric radar information", *Preprints, 29th International Radar Meteor. Conf.*, Amer. Meteor. Soc., 834-835.
Sand W. and C. Biter, 1989, "*TDWR display experiences*", AIAA-89-0807, 27th Aerospace Sciences Meeting, January, 1989, Reno, Nevada, American Institute of Aeronautics and Astronautics.

Sarker R.P. and A. Chowdhury, 1988, "A diagnostic structure of monsoon depressions", *Mausam*, **39**, 9-18.
Sarkar S.K. and A.B. Bhattacharya, 1980, "Sea thunderstorms and associated atmospherics", *Mausam*, **31**, 465-468.
Sarkar S.K., P.K. Pasricha, H.N. Dutta, B.M. Reddy and S.M. Kulshrestha, 1985, *"Atlas of tropospheric radio propagation parameters over the Indian subcontinent"*, National Physical Lab., New Delhi, 244 pp.
Sarkar S.K., P.K. Pasricha, A.B. Ghosh, H.N. Dutta, M.V.S.N. Prasad and B.M. Reddy, 1992, "High resolution refractivity measurement and its influence on radar clutter extent", *Ind.J. Radio Space Phys.*, **21**, 5-9.
Sato T. and R.F. Woodman, 1982, "Fine altitude resolution radar observations of upper tropospheric and lower stratospheric winds and waves", *J. Atmos. Sci.*, **39**, 2539-2545.
Sato T. and R.F. Woodman, 1982b, "Fine altitude resolution radar observations of stratospheric turbulent layers by the Arecibo 430 MHz radar", *J. Atmos. Sci.*, **39**, 2546-2552.
Satoh S. and J. Wurman, 2001, "Accuracy of wind fields in convective and stratiform echoes observed by a bistatic Doppler radar network", *Preprints, 30th Conf. Radar Meteor.*, 139-141
Sato T., N. Ao, M. Yamamoto, S. Fukao, T. Tsuda and S. Kato, 1991, "A typhoon observed with the MU radar", *Mon. Wea. Rev.*, **119**, 755-768.
Sato T., H. Doji, H. Iwai, I. Kimura, S. Fukao, M. Yamamoto, T. Tsuda and S. Kato, 1990, "Computer processing for deriving drop-size distribution and vertical air velocities from VHF Doppler radar spectra", *Radio Sci.*, **25**, 961-973.
Sauvageot H., 1991, *"Radar Meteorology"*, Artech House, London, 315 pp.
Sauvageot H., 1996, "Polarimetric radar at attenuated wavelengths as a hydrological sensor", *J. Atmos. Ocean. Technol.*, **13**, 630-637.
Sauvageot H. and G. Despaux, 1996, "The clear-air coastal vespertine radar bands", *Bull. Amer. Meteor. Soc.*, **77**, 673-681.
Saxton J.A. and J.A. Lane, 1946, "The anomalous dispersion of water at very high radio frequencies", in four parts, *Meteorological Factors in Radio Wave Propagation*, Phys. Soc., London, 278-325.
Schaefer J.T., 1986, "The dryline", pp549-572 in *"Mesoscale Meteorology and Forecasting"*, Ed., P.S.Ray, Published by Amer. Meteor. Soc., 793 pp.
Schaffner M., E. Gorgucci, R. Leonardi and G. Scarchilli 1983, "Radar calibration with weather echoes", *Preprints, 21st Conf. Radar Meteor.*, Amer. Meteor. Soc., 182-185.
Schiesser H.H., R.A. Houze Jr., and H. Huntrieser, 1995, "The mesoscale structure of severe precipitation systems in Switzerland", *Mon. Wea. Rev.*, **123**, 2070-2097.
Schlatter T.W., F.H. Carr, R.H. Langland, R.E. Carbone, N.A. Crook, R.W. Daley, J.C. Derber and S.L. Mullen, 1999, *"A five-year plan for research related to the assimilation of meteorological data"*, NCAR Tech. Note NCAR/TN-443., National Center for Atmospheric Research, Boulder, CO., USA, 41pp.
Schmid W. and A. Waldvogel, 1986, "Radar hail profiles in Switzerland", *J. Climate Appl. Meteor.*, **25**, 1002-1011.
Schmidt G., R. Rüster and P. Czechowsky, 1979, "Complementary code and digital filtering for detection of weak VHF radar signals from the mesosphere", *IEEE Trans. Geosci. Electronics*, **GE-17**, 154-161.
Schroeder J.A., E.R.Westwater, P.T. May and L.M. McMillin, 1991, "Prospects for temperature sounding with satellite and ground-based RASS measurements", *J. Atmos. Ocean. Technol.*, **8**, 506-513.
Schroeder M.J. and G.E. Klazura, 1978, "Computer processing of digital radar data gathered during HIPLEX", *J.Appl. Meteor.*, **17**, 498-507.

Schroth A.C., M.S. Chandra, and P.F. Meischner, 1988, "A C-band coherent polarimetric radar for propagation and cloud physics research", *J. Atmos. Ocean. Technol.*, **5**, 803-822.

Schuur T.J., A. V. Ryzhkov, D.S. Zrnić, M. Schönhuber, 2001, "Drop size distributions measured by a 2D video disdrometer: Comparison with dual-polarization radar Data", *J. Appl. Meteor.*, **40**, 1019–1034.

Schwerdt R.F., F.P. Ho and R.R. Watkins, 1979,"*Meteorological criteria for standard project hurricane and probable maximum hurricane windfields, Gulf and East coasts of the United States*", NOAA Tech. Rep., NWS 23, Sept. 1979, Published by NTIS, USA.

Scialom G., 1984, "Retrieval of horizontal wind field in stratiform precipitation by conical scannings from two Doppler radars", *Preprints, 22nd Conf. Radar Meteor.*, Amer. Meteor. Soc., 509-513.

Scialom G. and Y. Lemaître, 1991, "Pressure and temperature perturbations analytical retrieval from dual Doppler radar data", *Preprints, 25th International Conf. Radar Meteor.*, Amer. Meteor. Soc., 147-150.

Scialom G. and Y. Lemaitre, 1994, "QVAD; A method to obtain quadratic winds from conical scans by a Doppler weather radar network", *J. Atmos. Ocean. Technol.*, **11**, 909-926,

Scott, R., P. Krehbiel, and W. Rison, 2001: The use of simultaneous horizontal and vertical transmissions for dual-polarization radar meteorological observations, *J. Atmos. and Oceanic Tech .*, **18**, 629-648.

Sekhon R.S. and R.C. Srivastava, 1970, "Snow size spectra and radar reflectivity", *J. Atmos. Sci.*, **27**, 299-307.

Sekhon R.S. and R.C. Srivastava, 1971, "Doppler radar observations of drop-size distribution in a thunderstorm", *J. Atmos. Sci.*, **28**, 983-994.

Seliga T.A and V.N. Bringi, 1976, "Potential use of of radar differential reflectivity measurements at orthogonal polarizations for measuring precipitation", *J.Appl. Meteor.*, **15**, 69-76.

Seliga T.A, and V.N. Bringi, 1978, "Differential reflectivity and differential phase shift: Applications in radar meteorology", *Radio Sci.*, **13**, 271-275.

Seliga T.A, K. Aydin and H. Direskeneli, 1986, "Disdrometer measurements during an intense rainfall event in Central Illinois: Implications for differential reflectivity radar observations", *J. Climate Appl. Meteor.*, **25**, 835-846.

Seliga T.A, V.N. Bringi and H.H. Al-Khatib, 1981, "A preliminary study of comparative measurements of rainfall rate using the differential reflectivity radar technique and a raingage network", *J. Appl. Meteor.*, **20**, 1362-1368.

Sen S.N., 1931, "Mechanism of Bengal tornadoes in the nor'wester season", *Nature*, **127**, 128-129.

Senn H.V., 1966, "Radar hurricane precipitation patterns as track indicators", *Proc. 12th Conf. Radar Meteor.*, Amer. Meteor. Soc., 436-439.

Senn H.V. and H.W. Hiser, 1959, "On the origin of hurricane spiral bands", *J. Meteor.*, 16, 419-426.

Senn H.V., J.A. Stevens and P.J. Davies, 1965, "*Radar hurricane research, Sept. 1964 to Aug. 1965*", Institute of Marine Science, Univ. of Miami, Final Report.

Seo D-J., J. Breidenbach, R. Fulton, D. Miller and T. O'Bannon, 2000, "Real-Time Adjustment of Range-Dependent Biases in WSR-88D Rainfall Estimates due to Nonuniform Vertical Profile of Reflectivity", *J. Hydrometeor.*, **1**, 222–240.

Serafin R.J. and W.F. Dabberdt, 1990, "Profiling networks and developments in the United States", *Meteorol. Rdsch.*, **42**, 70-83.

Sevruk B., 1989, "Reliability of precipitation measurement", pp 13-19 in *Instrument and Observing Methods Report No. 48*, International Workshop on Precipitation Measurements, WMO/TD-No. 328, WMO, Geneva, 584 pp.

Shackford C.R., 1960, "Radar indication of a precipitation-lightning relationship in New England thunderstorms", *J. Meteor.*, **17**, 15-19.

Shafer M.A., D. R. MacGorman and F. H. Carr, 2000, "Cloud-to-ground lightning throughout the lifetime of a severe storm system in Oklahoma", *Mon. Wea. Rev.*, **128**, 1798–1816.

Shapiro A. and J. J. Mewes, 1999, "New formulations of dual-Doppler wind analysis", *J. Atmos. Ocean. Technol.* **16**, 782-792.

Shapiro, A., S. Ellis, and J. Shaw, 1995, "Single-Doppler velocity retrievals with Phoenix II data: Clear air and microburst wind retrievals in the planetary boundary layer", *J. Atmos. Sci.*, **52,** 1265–1287.

Shapiro L.J., 1983, "The asymmetric boundary layer flow under a translating hurricane", *J.Atmos. Sci.*, **40**, 1984-1998.

Shapiro M.A., T. Hample and D.W. van de Kamp, 1984, "Radar wind profiler observations of fronts and jet streams", *Mon. Wea. Rev.*, **112**, 1263-1266.

Sharma B.L., 1965, "Dry hole in a thunderstorm", *Ind. J. Meteor. Geophys.*, **16**, 134-135.

Sharma R.V., 1999, "Structure of the Kakinada cyclone of 6 November 1996", *Mausam*, **50**, 129-134.

Shea D.J. and W.M. Gray, 1973, "The hurricane's inner core region: I. Symmetric and asymmetric structure", *J. Atmos. Sci.*, **30**, 1544-1564.

Sheets R.C., 1985, "Hurricane tracking using a mass field envelope approach – impacts on forecast tracks", *Extended abstracts, 16th Conf. Hurricanes and Tropical Meteorology*, Amer. Meteor. Soc., 38-39.

Sheppard B.E., 1990, "Measurement of raindrop size distribution using a small Doppler radar", *J. Atmos. Ocean. Technol.*, **7**, 255-268.

Sheppard B.E. and P.I. Joe, 1994, "Comparison of raindrop size distribution measurements by a Joss-Waldvogel disdrometer, a PMS 2DG spectrometer, and a POSS Doppler radar", *J. Atmos. Ocean. Technol.*, **11**, 874-887.

Sheppard B.E., K.K.K. Wu and R.A. Wilson, 1983, "A new precipitation occurrence sensor system", *Preprints, 5th Symposium Meteorological Observations and Instrumentation*, Toronto, Amer. Meteor. Soc., pp 38-41.

Shibuya S., 1987, "*A basic Atlas of radiowave propagation*", Translated by H.Ishizuka, John Wiley & Sons, 778 pp.

Shimazu Y., 1997, "Wide and slow-moving rainbands and narrow fast-moving rainbands observed in typhoon 8913", *J. Meteor. Soc. Japan*, **75**, 67-80.

Short D.A., M. S. Gremllion , C. S. Pinder, and W. P. Roeder, 2000, "Volume scan strategies for the WSR-74C in support of space launch", *Preprints, Ninth Conf. Aviation, Range, and Aerospace Meteor.*, Amer. Meteor. Soc., 551-556.

Shravan Kumar and A.K.Sen Sarma, 1970, "Visual and radarscope observations of a funnel cloud", *Ind. J. Meteor. Geophys.*, **21**, 495-496.

Sikdar D.N., R.E. Schlesinger and C.E. Anderson, 1974, "Severe storm latent heat release: Comparison of radar estimate versus a numerical experiment", *Mon. Wea. Rev.*, **102**, 455-465.

Simpson J., R.F.Adler and G.R. North, 1988, "A proposed Tropical Rainfall Measuring Mission (TRMM) satellite", *Bull. Amer. Meteor. Soc.*, **69**, 278-295.

Simpson R.H., 1956, "Some aspects of tropical cyclone structure", *Proc. Tropical Cyclone Symp.*, Brisbane, Bureau of Meteorology, Australia,139-157.

Simpson R.H., and J.M. Pelissier, 1971, "Atlantic hurricane season of 1970", Mon. Wea. Rev., **99**, 269-277.

Singal S.P. and M. Goel, 1997, "Radio Acoustic Sounding System (RASS) for studying the lower atmosphere", pp142-176 *in 'Acoustic Remote Sensing Applications'*, Ed. S.P. Singal, Narosa Publishing House, New Delhi.

Sirmans D., D. Zrnić and B. Bumgarner, 1976, "Extension of maximum unambiguous Doppler velocity by use of two sampling rates", *Preprints, 17th Conf. Radar Meteor.*, Amer. Meteor. Soc, 23-28.

Sivaramakrishnan T.R. and S.Sridharan, 1989, "Wind observations from Bay cyclone of November 1984", Mausam, **40**, 344-345.

Sivaramakrishnan T.R., B. Ramakrishnan and V. Ramaswamy, 1985, "A study of convective clouds around Madras during the northeast monsoon season", *Mausam*, **36**, 120.

Skoda G., 1987, "Fractal dimension of rainbands over hilly terrain", *Meteorol. Atmos. Phys.*, **36**, 74-82.

Skolnik M.I., 1970, *"Radar Handbook"*, McGraw-Hill.

Skolnik M.I., 1980, "*Introduction to Radar Systems*", 2nd ed., McGraw Hill, 581 pp.

Sloss D.H. and D. Atlas, 1968, "Wind shear and reflectivity gradient effects on Doppler radar spectra", *Preprints, 13th Conf. Radar Meteor.*, Amer. Meteor. Soc, 44-49.

Smalley D.J. and F.I. Harris, 1993, "Wind field characteristics of Hurricane Bob (1991)", *Preprints, 26th International Conf. Radar Meteor.*, Amer. Meteor. Soc., 97-99.

Smith C.J., 1986, "The reduction of errors caused by bright bands in quantitative rainfall measurements made using radar", *J. Atmos. Ocean. Technol.*, **3**, 129-141.

Smith P.L., 1972, "Siting considerations for weather radars", *Preprints, 15th Radar Meteor. Conf.*, Amer. Meteor. Soc., 99-100.

Smith P.L., 1993, "An update on weather radar system sensitivity", *Preprints, 26th International Conf. Radar Meteor.*, Amer. Meteor. Soc., 384-386.

Smith S.D. and R.M. Rabin, 1989, "Considerations in estimating horizontal wind gradients from an individual Doppler radar or a network of wind profilers", *J.Atmos. Ocean. Technol.*, **6**, 446-458.

Smith S.D. and R.M. Rabin, 1989b, "Estimation of divergence in the prestorm boundary layer", *J.Atmos. Ocean. Technol.*, **6**, 459-475.

Smull B.F. and R.A. Houze Jr., 1987, "Dual-Doppler radar analysis of a midlatitude squall line with a trailing region of stratiform rain", *J. Atmos. Sci.*, **44**, 2128-2148.

Smyth T.J. and A.J. Illingworth, 1998, "Radar estimates of rainfall rates at the ground in bright band and non-bright band events", *Quart. J. Roy. Meteor. Soc.*, **124**, 2417-2434.

Smythe G.R. and D.S. Zrnić, 1983, "Correlation analysis of Doppler radar data and retrieval of the horizontal wind", *J. Climate Appl. Meteor.*, **22**, 297-311.

Snell W.L. and E.W. McCaul Jr., 1993, "Doppler signatures of tornadoes spawned by hurricane ANDREW near Montgomery, Alabama", *Preprints, 26th International Conf. Radar Meteor.*, Amer. Meteor. Soc., 80-82.

Snow C.P, 1971, "*Public Affairs*", Macmillan, 324 pp.

Sohoni V.V., 1928, "*Thunderstorms of Calcutta, 1900-1926*", IMD Sc. Notes **I**, 3, 25-36.

Soundararajan K. and S. Raghavan, 1962, "Severe squall at Madras airport on 17 August 1961", *Ind. J. Meteor. Geophys.*, **13**, 548-550.

Spilhaus A.F., 1948, "Drop size, intensity, and radar echo of rain", *J.Meteor.*, **5**, 161-164.

Srivastava G.P., B.B. Huddar and V. Srinivasan, 1966, "Radar observations of monsoon precipitation", *Ind. J. Meteor. Geophys.*, **17**, 249-252.

Srivastava R.C., 1990, "A method for improving rain estimates from vertical-incidence Doppler radar observations", *J. Atmos. Ocean. Technol.*, **7**, 769-773

Srivastava R.C. and D. Atlas, 1974, "Effect of finite radar pulse volume on turbulence measurements", *J. Appl. Meteor.*, **13**, 472-480.

Srivastava R.C., T.J. Matejka and T.J. Lorello, 1986, "Doppler radar study of the trailing anvil region associated with a squall line", *J. Atmos. Sci.*, **43**, 356-377.

Stankov B.B., 1996, "Ground- and space-based temperature and humidity retrievals: Statistical evaluation", *J. Appl. Meteor.*, **35**, 444-463.

Stankov B.B., 1998, "Multi-sensor retrieval of atmospheric properties", *Bull. Amer. Meteor. Soc.*, **79**, 1835-1854.

Stankov B.B., B.E. Martner and M.K. Politovich, 1995, "Moisture profiling of the cloudy winter atmosphere using combined remote sensors", *J. Atmos. Ocean. Technol.*, **12**, 488-510.

Stankov B.B., E.R.Westwater and E.E. Gossard, 1996, "Use of wind profiler estimates of significant moisture gradients to improve humidity profile retrieval", *J. Atmos. Ocean. Technol.*, **13**, 1285-1290.

Steiner M and R.A. Houze Jr., 1993, "Three-dimensional validation at TRMM ground truth sites: some early results from Darwin, Australia", *Preprints, 26th International Conf. Radar Meteor.*, Amer. Meteor. Soc., 417-420.

Steiner M. and J. A. Smith, 2002, "Use of Three-Dimensional Reflectivity Structure for Automated Detection and Removal of Nonprecipitating Echoes in Radar Data" *J. Atmos. Ocean. Technol.* **19**, 673–686.

Steiner M., R.A. Houze Jr., and S.E. Yuter, 1995, "Climatological characterization of three-dimensional storm structure from operational radar and rain gauge data", *J. Appl. Meteor.*, **34**, 1978-2007

Steiner M., A. Waldvogel, J.W.F. Goddard, A.J. Illingworth and I.J. Caylor, 1991, "A relation between mean Doppler velocity, radar reflectivity factor and differential reflectivity", *Preprints, 25th International Conf. Radar Meteor.*, Amer. Meteor. Soc., 662-665.

Stern A.D., R.H. Brady III, P.D. Moore and G.M. Carter, 1994, "Identification of aviation weather hazards based on the integration of radar and lightning data", *Bull. Amer. Meteor. Soc.*, **75**, 2269-2280.

Stewart R.E., J.D. Marwitz, J.C. Pace and R.E. Carbone, 1984, "Characteristics through the melting layer of stratiform clouds", *J. Atmos. Sci.*, **41**, 3227-3237.

Stewart S.R. and S.W. Lyons, 1996, "A WSR-88D radar view of tropical cyclone Ed", *Wea. Forecasting*, **11**, 115-135.

Stout G.E. and F.A. Huff, 1953, "Radar records Illinois tornadogenesis", *Bull. Amer. Meteor. Soc.*, **34**, 281-284.

Stout G.E. and E.A. Mueller, 1968, "Survey of relationships between rainfall rate and radar reflectivity in the measurement of precipitation", *J. Appl. Meteor.*, **7**, 465-474.

Stout J., D.W. Martin and D.N. Sikdar, 1979, "Estimating GATE rainfall from geosynchronous satellite images", *Mon. Wea. Rev.*, **107**, 585-598.

Straka J.M. and D.S Zrnić, 1993, "An algorithm to deduce hydrometeor types and contents from multi-parameter radar data", *Preprints, 26th Imternational Conf. Radar Meteor.*, Amer. Meteor. Soc., 513-515.

Stratton J.A., 1941, "*Electromagnetic theory*", McGraw-Hill, 615 pp.

Strauch R.G., A.S. Frisch and B.L. Weber, 1986, "Wind measurement in the upper troposphere with UHF and VHF radar", *Preprints, 23rd Conf. Radar Meteor.*, Amer. Meteor. Soc, 48-51.

Strauch R.G., D.A. Merritt, K.P. Moran, K.B. Earnshaw and J.D. van de Kamp, 1984, "The Colorado wind profiling network", *J. Atmos. Ocean. Technol.*, **1**, 37-49.

Strauch R.G., B.L. Weber, A.S. Frisch, C.G. Little, D.A. Merritt, K.P. Moran and D.C. Welsh, 1987, "The precision and relative accuracy of of profiler wind measurements", *J. Atmos. Ocean. Technol.*, **4**, 563-571.

Stutzman W.L. and H.C. Ko, 1974, "Measurement of antenna beamwidth using extraterrestrial radio sources", *IEEE Trans. Antennas Propag.*, **AP-22**, 493-495.

Subramanian D.V., 1981, "Radar observations of cyclones in the Arabian sea and the Bay of Bengal", *Mausam*, 32, 189-194.

Subramanian D.V. and U.N. Sehgal, 1967, "Radar study of thunderstorm activity in northeast India during the pre-monsoon season", *Ind. J. Meteor. Geophys.*, **18**, 111-114.

Sulakvelidze G.K., 1968, "VGI studies of hail center detection" pp 46-53, in "*Radar meteorology, Proc. 3rd all-Union conference*", Moskva, 1968, Ed. Kostarev V.V., A.A.Chernikov and A.B. Shupyatskii, Israel Programme for Translations, Jerusalem, 1971.

Sulakvelidze G.K. and Yu. A. Dadali, 1968,"Multi-wavelength radar measurements of precipitation intensity" pp 32-45, in "*Radar meteorology, Proc. 3rd all-Union conference*", Moskva, 1968, Ed. Kostarev V.V., A.A.Chernikov and A.B. Shupyatskii, Israel Programme for Translations, Jerusalem, 1971.

Sulakvelidze G.K., B.I. Kiziriya and V.V. Tsykunov., 1974, "Progress of hail suppression work in the USSR", pp. 410-431 in "*Weather and Climate Modification* ", Ed. W.N. Hess, John Wiley & Sons, 842 pp.

Sun J., 1994, "Fitting a Cartesian prediction model to radial velocity from single-Doppler radar", *J. Atmos. Ocean. Technol.,* **11,** 200–204.

Sun J. and A. Crook, 1994, "Wind and thermodynamic retrieval from single-Doppler measurements of a gust front observed during Phoenix II", *Mon. Wea. Rev.*, **122,** 1075-1091.

Sun J. and N. A. Crook, 1996, "Comparison of thermodynamic retrieval by the adjoint method with the traditional retrieval method". *Mon. Wea. Rev.,* **124,** 308–324.

Sun J. and R.A. Houze Jr., 1991, "Application of a thermodynamic retrieval technique to a numerical model", *Preprints, 25th International Conf. Radar Meteor.*, Amer. Meteor. Soc., 131-134.

Sun J. and R.A. Houze Jr., 1992, "Validation of a thermodynamic retrieval technique by application to a simulated squall line with trailing stratiform precipitation", *Mon.Wea. Rev.*, **120**, 1003-1018.

Sun J., D. W. Flicker, and D. K. Lilly, 1991, "Recovery of three-dimensional wind and temperature fields from simulated single-Doppler radar data", *J. Atmos. Sci.,* **48,** 876–890.

Suomi V.E., R. Fox, S.S. Limaye and W.L. Smith, 1983, "McIDAS III: A modern interactive data access and analysis system", *J. Climate Appl. Meteor*, **22**, 766-778.

Suzuki O., H. Niino, H. Ohno and H. Nirasawa, 2000, "Tornado-Producing Mini Supercells Associated with Typhoon 9019", *Mon. Wea. Rev.*, **128,** 1868–1882.

Szoke E.J., E.J. Zipser and D.P. Jorgensen, 1986, "A radar study of convective cells in mesoscale systems in GATE. Part I: Vertical profile statistics and comparison with hurricanes", *J. Atmos. Sci.*, **43**, 182-197.

Tabary P., G. Scialom, and U. Germann, 2001, "Real-time retrieval of the wind from aliased velocities measured by Doppler radars", *J.Atmos. Ocean. Technol.*, **18**, 875–882.

Tabata A., H. Sakakibara, M. Ishihara, K. Matsuura and Z. Yanagisawa, 1992, "A general view of the structure of typhoon 8514 observed by a Doppler radar", *J. Meteor. Soc. Japan*, **70,** 897-917.

Tabata A., H. Sakakibara, H. Okamura, K. Mori and K. Akaeda, 1993, "Structure of the inner core region of typhoon 9119 (MIREILLE) observed by a Doppler radar", *Preprints, 26th International Conf. Radar Meteor.*, Amer. Meteor. Soc., 88-90.

Tan J., A.R. Holt, A. Hendry and D.H.O. Bebbington, 1991, "Extracting rainfall rates from X-band CDR radar data by using differential propagation phase shift", *J. Atmos. Ocean. Technol.*, **8**, 790-801.

Tang C. -H., 1985, "Typhoon precipitation patterns and their relation to flood areas", in "*Report of the seminar on the application of radar data to tropical cyclone forecasting, Bangkok, 1983*", World Meteorological Organization(WMO), Rep. No. TCP-19, WMO/TD No. 26.

Tang X., 1993, "Radar echo characteristics and precipitation mechanism of cloud system of Mei-yu frontal torrential rain over China", Preprints, 26th International Conf. Radar Meteor., Amer. Meteor. Soc., 300-301.

Tatarskii V.I., 1961, "*Wave propagation in a turbulent medium*", English translation by R.A.Silverman, Dover Publications, New York, 285 pp.

Tatarskii V.I., 1971, "*The effects of the turbulent atmosphere on wave propagation*", Israel Program for Scientific Translations, Jerusalem, 472 pp.

Tatehira R. and T. Shimizu, 1978, "Intensity measurement of precipitation echo superposed on ground clutter - A new automatic technique for ground clutter rejection", *Preprints, 18th Conf. Radar Meteor.*, Amer. Meteor. Soc., 364-369.

Terblanche D.E., 1996, "A simple digital signal processing method to simulate linear and quadratic responses from a radar's logarithmic receiver", *J. Atmos. Ocean. Technol.*, **13**, 533-538.

Testud J. and P. Amayenc, 1989, "Stereoradar meteorology: A promising technique for observation of precipitation from a mobile platform", *J. Atmos. Ocean. Technol.*, **6**, 89-108.

Testud J. and M. Chong, 1983, "Three-dimensional wind field analysis from dual-Doppler radar data. Part I: Filtering interpolating and differentiating the raw data", *J.Climate Appl. Meteor.*, **22**, 1204-1215.

Testud J. and P.H. Hildebrand, 1991, "A technique to correct airborne Doppler data for coordinate transformation errors using surface clutter", *Preprints, 25th International Conf. Radar Meteor.*, Amer. Meteor. Soc., 493-496.

Testud J., P. Amayenc, X. Dou and T. Tani, 1996, "Tests of rain profiling algorithms for a spaceborne radar using raincell models and real data precipitation fields", *J. Atmos. Ocean. Technol.*, **13**, 426-453.

Testud J., P. Amayenc and M. Marzoug, 1991, "The dual-beam radar: A promising instrument to measure rain from space", *Preprints, 25th International Conf. Radar Meteor.*, Amer. Meteor. Soc., 392-395.

Testud J., P. Amayenc and M. Marzoug, 1992, "Rainfall-rate retrieval from a spaceborne radar: comparison between single-frequency, dual frequency, and dual-beam techniques", *J. Atmos. Ocean. Technol.*, **9**, 599-623.

Testud J., P.H. Hildebrand and W-C. Lee, 1995, "A procedure to correct airborne Doppler data for navigation errors using the echo returned from the earth's surface", *J. Atmos. Ocean. Technol.*, **12**, 800-820.

Testud J., E. Le Bouar, E. Obligis, and M. Ali-Mehenni, 2000, "The Rain Profiling Algorithm Applied to Polarimetric Weather Radar", *J. Atmos. Ocean.Technol.*,**17**, 332–356.

Thiele O.W., 1991, "Meteorological radar methods validating space observations of precipitation", *Preprints, 25th International Conf. Radar Meteor.*, Amer. Meteor. Soc., 374-379.

Thomas S.I.T., 1973, "Heights of cumulonimbus cloud tops over the Deccan plateau and adjoining plains of Andhra Pradesh and east Maharashtra - A preliminary radar study", *Ind. J. Meteor. Geophys.*, **24**, 163-164.

Thomas S.I.T. and V.K. Raghavendra, 1977, "Heights of cumulonimbus cloud tops over the Deccan plateau and adjoining plains of Andhra Pradesh and east Maharashtra - A radar study", *Ind. J. Meteor. Hydrol. Geophys.*, **28**, 479-482.

Thomson A.D. and R. List, 1996, "Raindrop spectra and updraft determination by combining Doppler radar and disdrometer", *J. Atmos. Ocean. Technol.*, **13**, 465-476.

Thurai M., H. Kumagai, T. Kozu and J. Awaka, 2001, "Effects of incorporating a brightband model in a downward-looking radar rainfall retrieval algorithm, *J. Atmos. Ocean. Technol.*, **18**, 20-25.

Tijin A.B.C. and H.T. Wu, 2000, "*ASWAN: Assessment of the impact of wind profiler/sodar data on the mesoscale wind analysis in the Netherlands*", Royal Netherlands Meteorological Institute (KNMI) Postbus 201, 3730, A.E. De Bilt, The Netherlands, 48 pp.

Tohsha M. and I. Ichimura, 1961, "Studies of shower and thunderstorm by radar", *Pap. Meteor. Geophys.*, **XII**, 18-29.

Tokay A and D, J. Short, 1996, "Evidence from tropical raindrop spectra of the origin of rain from stratiform versus convective clouds"., *J. Appl. Meteor*, **35,** 355-371.

Tong H., V. Chandrasekar, K.R. Knupp and J. Stalker, 1995, "Evaluation of latent heating estimates from multiparameter radar data and multiple Doppler analysis", *Preprints, 27th Conf. Radar Meteor.*, Amer. Meteor. Soc., 450-452.

Trotter B.L., C.L. Frush and R.G. Strauch, 1980, "An airborne pulse-Doppler radar for meteorological research", *Preprints, 19th Conf. Radar Meteor.*, Amer. Meteor. Soc., 110-114.

Tsuda T., T. Adachi, Y. Masuda, S. Fukao and S. Kato, 1994, "Observations of tropospheric temperature fluctuations with the MU radar-RASS", *J. Atmos. Ocean. Technol.*, **11**, 50-62.

Tsuda T., M. Miyamoto and J. Furumoto, 2001, Estimation of a Humidity Profile Using Turbulence Echo Characteristics, *J. Atmos. Ocean. Technol.*, **18**, 1214-1222.

Turnbull D., J. McCarthy, J. Evans and D. Zrnić, 1989, "The FAA Terminal Doppler Weather Radar (TDWR) program", *3rd International Conf. Aviation Weather System*, Amer. Meteor. Soc..

Tuttle J.D. and G.B. Foote, 1990, "Determination of the boundary layer airflow from a single Doppler radar", *J. Atmos. Ocean. Technol.*, **7**, 218-232.

Ulbrich C.W., 1983, "Natural variations in the analytic form of the raindrop size distribution", *J. Climate Appl. Meteor.*, **22**, 1764-1775.

Ulbrich C.W., 1992, "Algorithms for determination of rainfall integral parameters using reflectivity factor and mean Doppler fall speed at vertical incidence", *J. Atmos. Ocean. Technol.*, **9**, 120-128.

Ulbrich C.W. and D. Atlas, 1978, "The rain parameter diagram: Methods and applications", *J. Geophys. Research*, **83**, No. C3, 1319-1325.

Ulbrich C.W. and D. Atlas, 1982, "Hail parameter relations; A comprehensive digest", *J. Appl. Meteor.*, **21**, 22-43.

Ulbrich C.W. and L. G. Lee, 1999, "Rainfall measurement error by WSR-88D radars due to variations in Z–R law parameters and the radar constant", *J. Atmos. Ocean. Technol.*, **16**, 1017–1024.

Uttal T. and R. A. Kropfli, 2001, "The effect of radar pulse length on cloud reflectivity statistics", *J. Atmos. Ocean. Technol.*, **18**, 947–961.

Uyeda H. and D.S.Zrnić, 1986, "Automatic detection of gust fronts", *J. Atmos.Ocean. Technol.*, **3**, 36-50.

Uyeda H., N. Koizumi, K. Iwanami, R. Shirooka, N. Takahashi and K. Kikuchi, "Structure of wind field and precipitation system around an isolated mountain observed with a Doppler radar during the passage of a typhoon", *Preprints, 25th International Conf. Radar Meteor.*, Amer. Meteor. Soc., 686-689.

Vali G., R. Kelly, A. Pazmany, J. Mead and R. McIntosh, 1993, "Observations of clouds with an airborne, 95 GHz. polarimetric radar", *Preprints, 26th International Conf. Radar Meteor.*, Amer. Meteor. Soc., 723-725.

Van Vleck J.H., 1947a, "Absorption of microwaves by oxygen", *Phys. Rev.*, **71**, 413.

Van Vleck J.H., 1947b, "Absorption of microwaves by uncondensed water vapour", *Phys. Rev.*, **71**, 425.

VanAndel J., C. Kessinger and D. Ecoff, 1999, "APCAT: An AP clutter analysis tool", *29th International Conf. Radar Meteor.*, AMS, 267-269.

VanZandt T.E., K.S. Gage and J.M. Warnock, 1980, "Progress in modelling turbulence parameters from profiles of background wind, temperature and humidity", *Preprints, 19th Conf. Radar Meteor.*, Amer. Meteor. Soc, 613-615.

VanZandt T.E., J.L. Green, K.S. Gage and W.L. Clark, 1978, "Vertical profiles of refractivity turbulence structure constant: Comparison of observations of the Sunset radar with a new theoretical model", *Radio Sci.*, **13**, 819-829.

Vasiloff S.V., 1991, "The TDWR tornadic vortex signature detection algorithm", *Preprints 25th International Conf. Radar Meteor.*, Amer. Meteor. Soc., J43-J48.

Venkataraman K.S. and N.S. Bhaskara Rao, 1966, "Mesoscale study of summer thunderstorms in Delhi area", *Ind. J. Meteor. Geophys.*, **17**, 529-544.

Verma R.K. and Awadhesh Kumar, 1993, "Prediction of movement of radar echoes around Lucknow airport", *Vayu Mandal*, **23**, 82-83.

Vicente G. A., R. A. Scofield and W. P. Menzel , 1998, "The operational GOES infrared rainfall estimation technique", *Bull. Amer. Meteor. Soc.*, **79**, 1883-1898.

Vignal. B and W.F. Krajewski, 2001, "Large-sample evaluation of two methods to correct range-dependent error for WSR-88D rainfall estimates", *J. Hydrometeor.*, **2**, 490–504.

Vignal B., H. Andrieu, and J. D. Creutin, 1999, "Identification of vertical profiles of reflectivity from volume scan radar data", *J. Appl. Meteor.*, **38**, 1214–1228.

Vignal B., G. Galli, J. Joss, and U. Germann, 2000, "Three methods to determine profiles of reflectivity from volumetric radar data to correct precipitation estimates", *J. Appl. Meteor.*, **39**, 1715–1726

Vincent R.A. and I.M. Reid, 1983, "HF Doppler measurements of mesospheric gravity wave momentum fluxes", *J. Atmos. Sci.*, **40**, 1321-1333.

Viswanathan G., P. Kishore, P.B. Rao, A.R. Jain, R. Aravindhan, P. Balamuralidhar and S.H. Damle, 1993, "Preliminary results using ST mode of the Indian MST radar: Measurement of vertical velocity", *Ind. J. Radio Space Phys.*, **22**, 97-102.

Vivekanandan J., D.N. Yates and E.A. Brandes, 1999, "The influence of terrain on rainfall estimates from radar reflectivity and specific propagation phase observations", *J. Atmos. Ocean. Technol.*, **16**, 837–845.

Vivekanandan J., G. Zhang, and M. K. Politovich, 2001, "An Assessment of Droplet Size and Liquid Water Content Derived from Dual-Wavelength Radar Measurements to the Application of Aircraft Icing Detection", *J. Atmos. Ocean. Technol.*, . **18**,1787–1798.

Vivekanandan J., D. S. Zrnić, S. M. Ellis, R. Oye, A. V. Ryzhkov and J. Straka, 1999b, "Cloud microphysics retrieval using S-band dual-polarization radar measurements", *Bull. Amer. Meteor. Soc.*, **80**, 381-388.

Wakasugi K., A. Mizutani, M. Matsuo, S. Fukao and S. Kato, 1986, "A direct method for deriving drop-size distribution and vertical air velocities from VHF Doppler radar spectra", *J. Atmos. Ocean. Technol.*, **3**, 623-629.

Wakasugi K., A. Mizutani, M. Matsuo, S. Fukao and S. Kato, 1987, "Further discussion on deriving drop-size distribution and vertical air velocities directly from VHF Doppler radar spectra", *J. Atmos. Ocean. Technol.*, **4**, 170-179.

Wakimoto R.M., 1982, "The life cycle of thunderstorm gust fronts as viewed by Doppler radar and rawinsonde data", *Mon. Wea. Rev.*, **110**, 1060-1082.

Wakimoto R.M. and P.G. Black, 1994, "Damage survey of hurricane Andrew and its relationship to the eyewall", *Bull.Amer. Meteor. Soc.*, **75**, 189-200.

Wakimoto R.M. and V.N. Bringi, 1988, "Operational detection of microbursts associated with intense convection: The 20th July storm during the MIST project", *Mon. Wea. Rev.*, **116**, 1521-1539.

Waldteufel P., and H. Corbin, 1979, "On the analysis of single-Doppler radar data", *J.Appl. Meteor.*, **18**, 532-542.

Waldvogel A., 1974, "The N_0 jump of raindrop spectra", *J. Atmos. Sci.*, **31**, 1067-1078.

Waldvogel A., B. Federer and P. Grimm, 1979, "Criteria for the detection of hail cells", *J. Appl. Meteor.*, **18**, 1521-1525.

Waldvogel A., B. Federer and W. Schmid, 1978, "The kinetic energy of hailfalls. Part II: Radar and hailpads", J. *Appl. Meteor.*, **17**, 1680-1693.

Waldvogel A., W. Schmid and B. Federer, 1978b, "The kinetic energy of hailfalls. Part I: Hailstone spectra", *J. Appl. Meteor.*, **17**, 515-520.

Walker G.B. and P.S. Ray, 1974, "Precipitation characteristics at vertical incidence from multiple wavelength Doppler radars", *J. Atmos. Sci.*, **31**, 1948-1951.

Walker G.B., P.S. Ray, D. Zrnić and R. Doviak, 1980, "Time, angle and range averaging of radar echoes from distributed targets", *J.Appl. Meteor.,* **19**, 315-323.

Walther C., 1991, "A scientific domain driven radar control windowing system", *Preprints, 25th International Conf. Radar Meteor.*, Amer. Meteor. Soc., 229-232.

Walther C., C. Frush and P. Hildebrand, 1986, "The NCAR airborne Doppler radar, Part III: Overview of radar design details", *Preprints, 23rd Conf. Radar Meteor.*, Amer. Meteor. Soc., 155-158.

Wang K.-C., and H.Y. Tseng, 1989, "A case study of the circulation structure of typhoon Alex by using a single Doppler weather radar", *Preprints, 24th Conf. Radar Meteor.*, Amer. Meteor. Soc., 537-544.

Wark D.Q. and H.E. Fleming, 1966, "Indirect measurements of atmospheric temperature profiles from satellites: 1. Introduction", *Mon. Wea. Rev.*, **94**, 351-362.

Warnock J.M., J.L. Green and W.L. Clark, 1986, "Studies of C_n^2 and its variability measured by the Sunset clear-air radar", *Preprints, 23rd Conf. Radar Meteor.*, 34-37.

Warnock J.M., T.E. VanZandt, W.L. Clark, S.J. Franke, H.S. Kim, G.D. Nastrom, P.E. Johnston, 1994, "Measurement of synoptic-scale vertical velocities by two nearby VHF Doppler radars in very flat terrain", *J. Atmos. Ocean. Technol.*, **11**, 5-13.

Watanabe K, 1963, "The radar spiral bands and typhoon structure", *Proc. Symposium Trop. Meteor.,* Rotarua, New Zealand, 484-491.

Watson A.I., R.L. Holle and R.E. López, 1995, "Lightning from two national detection networks related to Vertically Integrated Liquid and echo-top information from WSR-88D radar", *Wea. Forecasting*, **10**, 592-605.

Wears. D.J., 1982, "Radar Smell", *New Scientist*, 25 Mar., 805.

Weatherford C.L.and W.M. Gray., 1988, "Typhoon structure as revealed by aircraft reconnaissance, Part I: Data analysis and climatology, Part II: Structural variability." *Mon. Wea. Rev.,* **116**, 1032-1043,1044-1056.

Weber B.L., and D.B.Wuertz, 1990, "Comparison of rawinsonde and wind profiler radar measurements", *J. Atmos. Ocean. Technol.*, **7**, 157-174.

Weber B.L., D.B. Wuertz, D.C. Law, A.S. Frisch and J.M. Brown, 1992, "Effect of small-scale vertical motion on radar measurements of wind and temperature profiles", *J. Atmos. Ocean. Technol.*, **9**, 193-209.

Weber B.L., D.B. Wuertz, R.G. Strauch, D.A. Merritt, K.P. Moran, D.C. Law, D. van de Kamp, R.B. Chadwick, M.H. Ackley, M.F. Barth, N.L. Abshire, P.A. Miller, and T.W. Schlatter, 1990, "Preliminary evaluation of the first NOAA demonstration wind-profiler", *J. Atmos. Ocean. Technol.*, **7**, 909-918.

Weber B.L., D.B.Wuertz, D.C. Welsh and R. McPeek, 1993, "Quality control for profiler measurements of winds and RASS temperatures", *J. Atmos. Ocean. Technol.*, **10**, 452-464.

Weber M., M. Stone, C. Primeggia and J. Andersons, 1991, "Airport surveillance radar based wind shear detection", *Preprints, 25th International Conf. Radar Meteor.*, Amer. Meteor. Soc., J11-J16.

Weber M.E., 1989, "Weather sensing with airport surveillance radars", *3rd International Conf. Aviation Weather System,* Amer. Meteor. Soc..

Weber M.E., 2000, FAA surveillance radar data as a complement to the WSR-88D network", *Preprints, 9th Conference on Aviation, Range, and Aerospace Meteorology,* Amer. Meteor. Soc., 35-39.

Weber M.E. and T.A. Noyes, 1988, "*Low altitude wind shear detection with airport surveillance radar: Evaluation of 1987 field measurements*", Lincoln Lab, Massachusetts Institute of Technology (MIT), 107pp.

Weinman J.A., R. Meneghini and K. Nakamura, 1990, "Retrieval of precipitation profiles from airbone radar and passive radiometer measurements: Comparison with dual frequency radar measurements", *J. Appl. Meteor.*, **29**, 981-993.

Weinstock J., 1981, "Energy dissipation rates of turbulence in the stable free atmosphere", *J. Atmos. Sci.*, **38**, 880-883.

Weston K.J., 1972, "The dry-line in northern India and its role in cumulonimbus convection", *Quart. J. Roy. Meteor. Soc.*, **98**, 519-531.

Wexler R., 1951, "Theory and observation of radar storm detection", *Compendium of Meteorology*, Amer. meteor. Soc., 1283-1289.

Weygandt, S., A. Shapiro, and K. K. Droegemeier, 1995, "Adaptation of a single-Doppler velocity retrieval on a deep-convective storm", *Preprints, 27th Conf. Radar Meteor.,* Amer. Meteor. Soc., 264–266.

White A. B., D. J. Gottas, E. T. Strem, F. M. Ralph and P. J. Neiman, 2002, "An Automated Brightband Height Detection Algorithm for Use with Doppler Radar Spectral Moments", *J. Atmos. Ocean. Technol.* **19**, 687–697.

Whiton R.C., P.L. Smith and A.C. Harbuck, 1976, "Calibration of weather radar systems using the sun as a radio source", *Preprints, 17th Conf. Radar Meteor.*, Amer. Meteor. Soc., 60-65.

Wieler J.G., 1986, "Realtime automatic detection of mesocyclones and tornadic vortex signatures", *J. Atmos. Ocean. Technol.* **3**, 98-113.

Wiener G. and S. Yuter, 1991, "A storm cell extraction algorithm for parallel computers", *Preprints, 25th International Conf. Radar Meteor.*, Amer. Meteor. Soc., 193-196.

Wilczak J.M., R.G. Strauch, F.M. Ralph, B.L. Weber, D.A. Merritt, J.R. Jordan, D.E. Wolfe, L.K. Lewis, D.B. Wuertz, J.E. Gaynor, S.A. McLaughlin, R.R. Rogers, A.C. Riddle and T.S. Dye, 1995, "Contamination of wind profiler data by migrating birds: characteristics of corrupted data and potential solutions", *J. Atmos. Ocean. Technol.*, **12**, 449-467.

Wiley R.L., K.A. Browning, J. Joss and A. Waldfogel, 1970, "Measurement of drop size distribution and vertical air motion in widespread rain using pulsed Doppler radar and distrometer", *Preprints, 14th Radar Meteor. Conf.* Amer. Meteor. Soc.., 167-170.

Wilfong T.L., D. A. Merritt, R. J. Lataitis, B. L. Weber, D. B. Wuertz and R. G. Strauch, 1999, "Optimal generation of radar wind profiler spectra", *J. Atmos. Ocean. Technol.*, **16**, 723–733

Wilheit T.T., 1986, "Some comments on passive microwave measurement of rain", *Bull. Amer. Meteor. Soc.*, **67**, 1226-1232.

Wilheit T.T., A.T.C. Chang, M.S.V. Rao, E.B. Rodgers and J.S. Theon, 1977, "A satellite technique for quantitatively mapping rainfall rates over the oceans", *J. Appl. Meteor.*, **16**, 551-560.

Williams C.R., W.L. Ecklund and K.S. Gage, 1995, "Classification of precipitating clouds in the tropics using 915-MHz wind profilers", *J. Atmos. Ocean. Technol.*, **12**, 996-1012.

Williams C.R., W.L. Ecklund, P. E. Johnston and K. S. Gage, 2000, "Cluster analysis techniques to separate air motion and hydrometeors in vertical incident profiler observations", *J. Atmos. Ocean. Technol.*, **17**, 949–962.

Williams E. and S. Geotis, 1986, "Radar diagnostics of the plasma and geometry of lightning", *Preprints, 23rd Conf. Radar Meteor.*, Amer. Meteor. Soc., JP365-372.

Williams E., S. Geotis and A.B. Bhattacharya, 1989, "A radar study of the plasma and geometry of lightning", *J. Atmos. Sci.*, **46**, 1173-1185.

Willis P.T., 1984, "Functional fits to some observed drop size distributions and parametrization of rain", *J. Atmos. Sci.*, **41**, 1648-1661.

Willis P.T. and A.J. Heymsfield, 1989, "Structure of the melting layer in mesoscale convective system stratiform precipitation", *J. Atmos. Sci.*, **46**, 2008-2025.

Willoughby H.E., 1988, "The dynamics of the tropical cyclone core", *Aust. Met. Mag.*, **36**, 183-191.

Willoughby H.E., 1990, "Temporal changes of the primary circulation in tropical cyclones", *J.Atmos. Sci.*, **47**, 242-264.

Willoughby H.E. and P.G. Black, 1996, "Hurricane Andrew in Florida: Dynamics of a disaster", *Bull. Amer. Meteor. Soc.*, **77**, 543-549.

Willoughby H.E., J.A. Clos and M.G. Shoreibah, 1982, "Concentric eyewalls, secondary wind maxima, and the evolution of the hurricane vortex", *J.Atmos. Sci.*, **39**, 395-411.

Willoughby H.E., H-L. Jin, S.J. Lord and J.M. Piotrowicz, 1984b, "Hurricane structure and evolution as simulated by an axisymmetric nonhydrostatic numerical model", *J.Atmos. Sci.*, **41**, 1169-1186.

Willoughby H.E., D.P. Jorgensen, R.A. Black and S.L. Rosenthal, 1985, "Project Stormfury: a scientific chronicle 1962-1983", *Bull. Amer. Meteor. Soc.*, **66**, 505-514.

Willoughby H.E., F.D. Marks Jr. and R.J. Feinberg, 1984, "Stationary and moving convective bands in hurricanes", *J.Atmos. Sci.*, **41**, 3189-3211.

Wilson J.W., 1970, "Integration of radar and raingage data for improved rainfall measurement", *J. Appl. Meteor.*, **9**, 489-497.

Wilson J.W. and E.A. Brandes, 1979, "Radar measurement of rainfall", *Bull. Amer. Meteor. Soc.*, **60**, 1048-1058.

Wilson J.W and C.K. Mueller, 1993, "Nowcasts of thunderstorm initiation and evolution", *Wea. Forecasting*, **8**, 113-131.

Wilson J.W. and D. Reum, 1988, "The flare echo: reflectivity and velocity signature", *J.Atmos. Ocean. Technol*, **5**, 197-205.

Wilson J.W. and W.E. Schreiber, 1986, "Initiation of convective storms at radar-observed boundary layer convergence lines", *Mon. Wea. Rev.*, **114**, 2516-2536.

Wilson J., R. Carbone, H. Baynton and R. Serafin, 1980,"Operational application of meteorological Doppler radar", *Bull. Amer. Meteor. Soc.*, **61**, 1154-1168.

Wilson J.W., J.A. Moore, G. Brant Foote, B. Martner, A. R. Rodi, T. Uttal and J.M. Wilczak, 1988, "Convection Initiation and Downburst Experiment (CINDE), Bull. Amer. Meteor. Soc., **69**, 1328-1348.

Wilson J.W., R.D. Roberts, C. Kessinger and J. McCarthy, 1984, "Microburst wind structure and evaluation of Doppler radar for airport wind shear detection", *J. Climate Appl. Meteor.*, **23**, 898-915.

Wilson J.W., T.M. Weckwerth, J. Vivekanandan, R.M. Wakimoto and R.W. Russell, 1994, "Boundary layer clear-air radar echoes and accuracy of derived winds", *J. Atmos. Ocean. Technol*, **11**, 1184-1206.

Winston H.A. and L.J. Ruthi, 1986, "Evaluation of RADAP-II severe-storm-detection algorithms", *Bull. Amer. Meteor. Soc.*, **67**, 145-150.

Witt A. and J.T. Johnson, 1993, "An enhanced storm cell identification and tracking algorithm", *Preprints, 26th International Conf. Radar Meteor.*, Amer. Meteor. Soc., 141-143.

Witt A. and S.P.Nelson, 1991, "The use of single-Doppler radar for estimating maximum hailstone size", *J.Appl. Meteor.*, **30**,425-431.

WMO (World Meteorological Organization), 1974, *Proc. WMO/IAMAP Scientific Conf. on Weather Modification, Tashkent, 1-7 October 1973*, WMO-No. 399, 1974, 538 pp.

WMO, 1986, "*Abridged final report of the extraordinary session of the Commission for Basic Systems (CBS), Hamburg, 21 October - 1 November, 1985*", WMO No. 654, p158.

WMO, 1988, "*Manual on Codes, Vol. 1, Annex II to WMO Technical Regulations, International Codes 1988 Edn.*", and "*Suppl. 3.*, (VIII, 1991)".

WMO, 1994, "*Guide to hydrological practices*", WMO No. 168, 5th edition, 735 pp.

Wolf D.E., D.J. Hall and R.M. Endlich, 1977, "Experiments in automatic cloud tracking using SMS-GOES data", *J. Appl. Meteor.*, **16**, 1219-1230.

Wood V.T., 1991, "Objective determination of tropical cyclone tracks from land-based Doppler radar observations", *Preprints, 19th Conf. Hurricanes and Trop. Meteor.*, Amer. Meteor. Soc., 198-201.

Wood V.T., 1994, "A technique for detecting a tropical cyclone center using a Doppler radar", *J.Atmos. Ocean. Technol.*, **11**, 1207-1216.

Wood V.T., and R.A. Brown, 1986, "Single-Doppler velocity signature interpretation of non-divergent environmental winds", *J. Atmos. Ocean. Technol.*, **3**, 114-128.

Wood V.T., and R.A. Brown, 1987, "Simulated land-based NEXRAD Doppler measurements within hurricanes", *Extended Abst., 17th Conf. Hurricanes and Trop. Meteor.*, Amer. Meteor. Soc., 174-177.

Wood V.T., and R.A. Brown, 1989, "Time-lapse images of simulated NEXRAD Doppler radar measurements from hurricanes", *Preprints, 24th Conf. Radar Meteor.*, Amer. Meteor. Soc., 255-258.

Wood V.T., and R.A. Brown, 1992, "Effects of radar proximity on single-Doppler velocity signatures of axisymmetric rotation and divergence", *Mon. Wea. Rev.*, **120**, 2798-2807.

Wood V.T., and R.A. Brown, 1997, "Effects of radar sampling on single-Doppler velocity signatures of mesocyclones and tornadoes", *Wea. Forecasting*, **12**, 928–938.

Wood V.T., and R.A. Brown, 2000, "Oscillations in mesocyclone signatures with range owing to azimuthal radar sampling", *J. Atmos. OceanicTechnol.*, **17**, 90–95.

Wood V.T., and F.D. Marks Jr., 1989, "Hurricane Gloria: Simulated land-based Doppler velocities reconstructed from air borne Doppler radar measurements", *Preprints, 18th Conf. Hurricanes and Trop. Meteor.*, Amer. Meteor. Soc., 115-116.

Woodley W.L., A.G. Barnston, J.A. Flueck and R. Biondini, 1983, "The Florida Area Cumulus Experiment's second phase (FACE-2). Part II: Replicated and confirmatory analyses", *J. Climate Appl. Meteor.*, **22**, 1529-1540.

Woodley W.L., J. Jordan, J. Simpson, R. Biondini, J. Flueck and A. Barnston, 1982, "Rainfall results of the Florida Area Cumulus Experiment, 1970-76", *J. Appl. Meteor.*, **21**, 139-164.

Woodley W.L., A.R. Olsen, A. Herndon and V. Wiggert, 1975, "Comparison of gage and radar methods of convective rain measurement", *J. Appl. Meteor.*, **14**, 909-928.

Woodman R.F. and A. Guillen, 1974, "Radar observations of winds and turbulence in the stratosphere and mesosphere", *J. Atmos. Sci.*, **31**, 493-505.

Wuertz D.B., B.L. Weber, R.G. Strauch, A.S. Frisch, C.G. Little, D.A. Merritt, K.P. Moran and D.C. Welsh, 1988, "Effects of precipitation on UHF Wind profiler measurements", *J. Atmos. Ocean. Technol.*, **5**, 450-465.

Wurman J., 1994, "Vector winds from a single-transmitter bistatic dual-Doppler radar network", *Bull. Amer. Meteor. Soc.*, **75**, 983-994.

Wurman J., 2001, "The DOW mobile multiple-Doppler network", *Preprints, 30th Conf. Radar Meteor.*, 95-97.

Wurman J., and M. Randall, 2001, "Bistatic radar networks", *Preprints, 30th Conf. Radar Meteor.*, 130-132.

Wurman J., and M. Randall, 2001b, "An inexpensive, mobile, rapid-scan radar", *Preprints, 30th Conf. Radar Meteor.*, 98-100.

Wurman J., J. Straka, E. Rasmussen, M. Randall and A. Zahrai, 1997, "Design and deployment of a portable, pencil-beam, pulsed, 3-cm Doppler radar", *J. Atmos. Ocean. Technol.*, **14**, 1502-1512.

Xiao R. and V. Chandrasekar, 1997, "Development of a neural network based algorithm for rainfall estimation from radar observations", *IEEE Trans. Geosci. Remote Sens.*, **35**, 160-171.

Xu Q., C.-J. Qiu, and J.-X. Yu, 1994, "Adjoint-method retrievals of low-altitude wind fields from single-Doppler reflectivity measured during Phoenix II", *J. Atmos. Ocean. Technol.*, **11**, 275–288.

Xu Q., H.-D. Gu, and J.-X. Yu, 1995, "Simple adjoint retrievals of microburst winds from single-Doppler radar data", *Mon. Wea. Rev.*, **123**, 1822–1833.

Xu Yumao, T. Xunchang and X. Wenmei, 1988, "An objective method for discrimination and short-term forecasts of severe storms using weather radar", *Chinese J. Atmos. Sci.*, **12**, 467-472.

Yamada Y., and M. Chong, 1999: VAD-based determination of the Nyquist interval number of Doppler velocity aliasing without wind information. *J. Meteor. Soc. Japan*, **77**, 447–457.

Yamasaki M., 1988, "Towards an understanding of the interaction between convection and the larger-scale in the tropics", *Aust. Met. Mag.*, **36**, 171-182.

Yeh T.C., 1950, "The motion of tropical storms under the influence of a superimposed southerly current", *J. Meteor.*, **7**, 108-113.

Yoe J.G. and R. Rüster, 1992, "Doppler radar observations of vertical velocities in the vicinity of the jet stream", Mon. Wea. Rev., **120**, 2378-2382.

Yoe J.G., M.F. Larsen, and E.J. Zipser, 1992, "VHF wind-profiler data quality and comparison of methods for deducing horizontal and vertical air motions in a mesoscale convective storm", *J. Atmos. Ocean. Technol.*, **9**, 713-727.

Young C.B., B.R. Nelson, A.A. Bradley, J.A. Smith, C.D. Peters-Lidard, A. Kruger and M.L. Baeck, 1999, "An evaluation of NEXRAD precipitation estimates in complex terrain", *J. Geophys. Res.*, **104**, 19,691- 19,703.

Zahrai A., and D.S. Zrnić, 1993, "The 10-cm wavelength polarimetric weather radar at NOAA's National Severe Storms Laboratory", *J. Atmos. Ocean. Technol.*, **10**, 649-662.

Zamora R.J., M.A. Shapiro and C.A. Doswell III, 1987, "The diagnosis of upper tropospheric divergence and ageostrophic wind using profiler measurements", *Mon. Wea Rev.*, **115**, 871-884.

Zawadzki I., 1984, "Factors affecting the precision of radar measurements of rain", *Preprints, 22nd Conf. Radar Meteor.*, Amer. Meteor. Soc., 251-256.

Zawadzki I., C. Desrochers, E. Torlaschi and A. Bellon, 1986, "A radar-raingage comparison", *Preprints, 23rd Conf. Radar Meteor.*, Amer. Meteor. Soc., 121-124.

Zhang J., and T. Gal-Chen, 1996, "Single-Doppler wind retrieval in the moving frame of reference". *J. Atmos. Sci.*, **18**, 2609–2623.

Zhang G., J. Vivekanandan and E. Brandes, 2001, "A method for estimating rain rate and drop size distribution from polarimetric radar measurements", *IEEE Trans. Geosci. Remote Sens.*, **39**, 830-841.

Zhou Ducheng, 1985, in "*Report of the seminar on the application of radar data to tropical cyclone forecasting, Bangkok, 1983*", World Meteorological Organization(WMO), Rep. No. TCP-19, WMO/TD No. 26.

Ziegler C.L. 1985, "Retrieval of thermal and microphysical variables in observed convective storms. Part I: Model development and preliminary testing, *J.Atmos. Sci.*, **42**, 1487-1509.

Ziegler C.L. 1988, "Retrieval of thermal and microphysical variables in observed convective storms. Part II: Sensitivity of cloud processes to variation of the microphysical parametrization" *J. Atmos. Sci.* **45**,1072-1090.

Zipser E.J., 1982, "Use of a conceptual model of the life-cycle of mesoscale convective systems to improve very-short-range forecasts", in "*Nowcasting*", Ed. K.A. Browning, Acad. Press, 191-204.

Zipser E.J., R.J. Meitin and M.A. LeMone, 1981, "Mesoscale motion fields associated with a slowly moving GATE convective band", *J. Atmos. Sci.*, **38**, 1725-1750.

Zittel W.D., 1976, "Computer applications and techniques for storm tracking and warning", *Preprints, 17th Conf. Radar Meteor.*, Amer. Meteor. Soc., 514-521.

Zou Jingmeng, 1993, "Monitoring and prediction of heavy rainfalls in China, *Proc. 2nd International Conf. East Asia and Western Pacific Meteorology and Climate,* 7-10 Sept. 1992, Hong Kong, Ed. Kyle W.J. and C.P. Chang, World Scientific, 621 pp.

Zrnić D.S., 1975, "Moments of estimated input power for finite sample averages of radar receiver outputs", *IEEE Trans. Aerospace and Electronic Systems*, **AES-11**, 109-113.

Zrnić D.S., 1979, "Estimation of spectral moments for weather echoes", *IEEE. Trans. Geosci. Electronics.*, **GE-17**, 113-128.

Zrnić D.S., 1987, "Three-body scattering produces precipitation signatures of special diagnostic value", *Radio Sci.*, **22**, 76-86.

Zrnić D.S., and R.J. Doviak, 1975, "Velocity spectra of vortices scanned with a pulse-Doppler radar", *J.Appl. Meteor.*, **14**, 1531-1539.

Zrnić D.S., and M. Istok, 1980, "Wind speeds in two tornadic storms and a tornado, deduced from Doppler spectra", *J. Appl. Meteor.*, **19**, 1405-1415.

Zrnić D.S. and P. Mahapatra, 1985, "Two methods of ambiguity resolution in pulse Doppler weather radars", *IEEE Trans.* Aerosp*ace and Electronic Systems*, **AES-21**, 470-483.

Zrnić D.S., and A.V. Ryzhkov, 1996, "Advantages of rain measurements using specific differential phase", *J. Atmos. Ocean. Technol.*, **13**, 454-464.

Zrnić D. S. and A. V. Ryzhkov, 1999, "Polarimetry for Weather Surveillance Radars", *Bull. Amer. Meteor. Soc.*, **80**, 389–406.

Zrnić D.S., N. Balakrishnan and B. Gordon, 1993, "Co-polar correlation coefficient at vertical incidence", *Preprints, 26th International Conf. Radar Meteor.,* Amer. Meteor. Soc., 742-744.

Zrnić D.S., D.W. Burgess and L. Hennington, 1985, "Doppler spectra and estimated wind speed of a violent tornado", *J. Climate Appl. Meteor.*, **24**, 1068-1081.

Zrnić D.S., D.W. Burgess and L.D. Hennington, 1985b, "Automatic detection of mesocyclonic shear with Doppler radar", *J. Atmos. Ocean. Technol.* **2**, 425-438.

Zrnić D.S., R.J. Doviak, and D.W. Burgess, 1977, "Probing tornadoes with a pulse-Doppler radar", *Quart. J. Royal Meteor. Soc.*, **103**, 707-720.

Zrnić D.S., A. Ryzhkov, J. Straka, Y. Liu and J. Vivekanandan, 2001, "Testing a Procedure for Automatic Classification of Hydrometeor Types", *J. Atmos. Ocean. Technol.*, **18**, 892–913.

ABBREVIATIONS AND ACRONYMS

4DD	Four-dimensional Doppler Dealiasing
ABCD	Analyse de Balayages Coniques Doppler
ADC	Analogue-to-Digital Converter
AFAST	Alternative Fore/Aft Scanning Technique
AGC	Automatic Gain Control
a.g.l.	above ground level
AMS	American Meteorological Society
a.m.s.l.	above mean sea level
AP	Anomalous Propagation
APCAT	AP Clutter Analysis Tool
APE	Areal Precipitation Estimate
ARMAR	Airborne Rain MApping Radar
ARSS	Automated Radar Subsystem (former USSR)
ASOS	Automatic Surface Observing System (USA)
ASR	Airport Surveillance Radar
ASTEX	Atlantic Stratocumulus Transition Experiment
ASTRAIA	Analyse STéréoscopique par RAdar à Impulsions Aéroportes
ATC	Air Traffic Control
ATI	Area Time Integral
AZTI(Z_x)	Area of reflectivity Z Time Integrated (Z above threshold Z_x)
BITE	Built-In Test Equipment
BVF	Brunt-Väisälä Frequency
BVP	Band-Velocity-Processing
BUFR	Binary Universal Form (for the) Representation (of meteorological data)
BWER	Bounded Weak Echo Region
CAPPI	Constant Altitude Plan Position Indicator
CAT	Clear Air Turbulence
CBE	single-Cell-type Bow Echo
CDF	Cumulative Distribution Function
CDR	Circular Depolarisation Ratio
CDR	Cyclone Detection Radar
CEDRIC	(software package for) Cartesian space Editing, synthesis and Display of Radar fields under Interactive Control
CFD	Cumulative Frequency Distribution
C-G	Cloud-to-Ground (lightning)
CII	Cyclone Intensity Indicator
CLOVAR	Canadian (London, Ontario) Vhf Atmospheric Radar
CM	Combined Moment (product)
CMM	Climatological Mature Mesocyclone
CNET	Centre d'études des Environnements Terrestre et Planetaire
COHO	COHerent Oscillator
COPT-81	Convection Profonde Tropicale, (an experiment conducted in West Africa in 1981)
COST	Co-Operation for Science and Technology (in Europe)
CPRS	Cloud Profiling Radar System
CRT	Cathode Ray Tube
CSI	Critical Success Index
CSS	Coherent Signal Source
CSU	Colorado State University
CW	Continuous Wave
DFR	Dual Frequency Ratio
DFT	Discrete Fourier Transform

DMSP Defense Meteorological Satellite Program (USA)
DOW Doppler On Wheels
DPPP Dual Polarisation Pulse Pair
D/RADEX Digital RADar EXperiment
DSD Drop Size Distribution
DSP Digital Signal Processor
DSRT Dual wavelength Surface Reference Technique
DVAD Double VAD
DVIP Digital Video Integrator and Processor
DWR Doppler Weather Radar

ECMWF European Centre for Medium range Weather Forecasting
EDOP ER-2 Doppler radar
EIO Extended Interaction Oscillator
ELDORA ELectra DOppler RAdar
EODD Extended Over-determined Dual-Doppler
ERKE Excess Rotational Kinetic Energy
EVAD Extended VAD
EVTD Extended Velocity Track Display

FAA Federal Aviation Administration (USA)
FACE Florida Area Cumulus Experiment
FAR False Alarm Rate
FAST Fore/Aft Scanning Technique
FFT Fast Fourier Transform
FM Frequency Modulation
FRONTIERS Forecasting Rain Optimised using New Techniques of Interactively Enhanced Radar and Satellite data (UK)

GARP Global Atmospheric Research Programme
GATE GARP Atlantic Tropical Experiment.
GBVTD Ground-Based Velocity Track Display
GFDA Gust Front Detection Algorithm
GIS Geographical Information System
GOES Geostationary Operational Environmental Satellite (USA)
GPCP Global Precipitation Climatology Project
GPS Global Positioning System
GSD Geographical Situation Display
GTS Global Telecommunication System

HART Height-Area Rainfall Threshold
HF High Frequency
HRSA High Radial Shear Area

IAGC Instantaneous Automatic Gain Control
ICAO International Civil Aviation Organization
ICRAD Interactive Color RAdar Display
IEEE Institution of Electrical and Electronic Engineers
IF Intermediate Frequency
IIR Infinite Impulse Response
IJRSP Indian Journal of Radio and Space Physics
IMD India Meteorological Department
INSAT Indian National SATellite
IOS Integrated Observing System
IRR Inner Radar Radius
ISRO Indian Space Research Organisation
IST Indian Standard Time
IWC Ice Water Content

JAWS Joint Airport Weather Studies
JMA Japan Meteorological Agency

LAWP Lower Atmospheric Wind Profiler
LAWS Low Altitude Wind Shear
LCC Launch Commit Criteria
LDR Linear Depolarisation Ratio
LDS Lightning Detection System
LEWP Line Echo Wave Pattern
LHR Latent Heat Release
l.h.s. left hand side
LIDAR LIght Detection and Ranging
LLWAS Low Level Wind shear Alert System
LLWS Low Level Wind Shear
LNA Low Noise Amplifier
LT Lead Time; Local Time
LWC Liquid Water Content

McIDAS Man-computer Interactive Data Access System

MCC Mesoscale Convective Complex
MCS Mesoscale Convective System
MCTEX Maritime Continent Thunderstorm Experiment
MDR Manually Digitised Radar
MDS Minimum Detectable Signal
MDZ_e Minimum Detectable Effective Reflectivity Factor
MONEX MONsoon EXperiment
MOPA Master Oscillator Power Amplifier
MSLP Mean Sea Level Pressure
MSSW Maximum Sustained Surface Wind
MST Mesosphere Stratosphere Troposphere
MTBF Mean Time Between Failures
MTI Moving Target Indicator
MTTR Mean Time To Repair
MU Middle and Upper atmosphere
MUTSI MU Radar Temperature Sheets and Interferometry campaign (Japan).
MVVP Modified Volume Velocity Processing

NASA National Aeronautics and Space Administration (USA)
NCAR National Center for Atmospheric Research (USA)
NCP Normalised Coherent Power
NEXRAD NEXt generation RADar
NHRE National Hail Research Experiment
NMRF National MST Radar Facility (India)
NN Neural Network
NOAA National Oceanic and Atmospheric Administration (USA)
NSSL National Severe Storms Laboratory (USA)
NTIS National Technical Information Service (USA)
NWP Numerical Weather Prediction

OI Optimum Interpolation
OMT OrthoMode Transducer
OSSE Observing System Simulation Experiment
OTH Over-The-Horizon

PDF Probability Distribution Function
PMS Particle Measuring System
POD Probability Of Detection
PPI Plan Position Indicator
POSS Precipitation Occurrence Sensor System
PR Precipitation Radar
p.r.f. Pulse Repetition Frequency
PROUST Prototype de Radar pour l'Observation en Uhf de la Stratosphère et Troposphère.
PRT Pulse Repetition Time
PSD Power Spectral Density
PUP Principal User Processor
PVF Potential Vortex Fit
PDPP Polarisation Diversity Pulse Pairs
QPE Quantitative Precipitation Estimate
QVAD Quadratic VAD

RADAP Radar Data Processor
RADAR RAdio Detection And Ranging
RADARNET RADAR NETwork centre
RAPID Radar data Analysis, Processing and Interactive Display
RASS Radio Acoustic Sounding System
RDA Radar Data Acquisition (NEXRAD)
RDF Radio Direction Finding
REI Range Elevation Indicator
RES Radar-Estimated Size
RF Radio Frequency
RHI Range Height Indicator
r.h.s. right hand side
RI Radar Interferometry
RKE Rotational Kinetic Energy
RMR Radius of Maximum Reflectivity
r.m.s root mean square
RMW Radius of Maximum Winds
RONSARD Recherche sur les Orages et les Nuages par un Système A deux Radars Doppler
RPG Radar Product Generator
rpm revolutions per minute
RTI Range-Time-Intensity (Plot)
RTS Relative Top Shear
RV Rain Volume

SAR	Synthetic Aperture Radar
SBC	Stationary Band Complex
SDD	Synthetic Dual-Doppler
SEVAD	Stepped Elevation VAD
SHARP	SHort term Automatic Radar Prediction
SLBE	Squall-Line-type Bow Echo
SNR	Signal-to-Noise Ratio
SOLFEGE	Système d'Observation Locale Favorisant l'Exploitation des paramètres GEophysiques
SOUSY	SOUnding SYstem
SPAwind	wind determined from movement of Small Precipitation Areas
SRT	Surface Reference Technique
SSI	Storm Strength Indicator
SSM/I	Special Sensor Microwave Imager
ST	Stratosphere Troposphere
STALO	Stable Local Oscillator
STC	Sensitivity Time Control
SWP	Severe Weather Probability
TC	Tropical Cyclone
TDWR	Terminal Doppler Weather Radar
TIROS	Television and Infra-Red Orbiting Satellite
TOGA COARE	Tropical Oceans-Global Atmosphere Coupled Ocean Atmosphere Response Experiment
TOVS	TIROS-N Operational Vertical Sounder
TRACIR	TRacking Air with CIrcular polarisation Radar
TRACON	Terminal RAdar CONtrol
TREC	Tracking of Radar Echo with Correlations
TRMM	Tropical Rainfall Measuring Mission
TSS	Tornado Spectral Signature
TVS	Tornadic Vortex Signature
TWT	Travelling Wave Tube
UHF	Ultra High Frequency
UK	United Kingdom (of Great Britain and northern Ireland)
USA	United States of America
USSR	(former) Union of Socialist Soviet Republics
UTC	Universal Time Coordinated
VAD	Velocity Azimuth Display
VARD	Velocity ARea Display
VHF	Very High Frequency
VID	Vertical Incidence Doppler
VIL	Vertically Integrated Liquid water content
VIP	Video Integrator and Processor
VPR	Vertical Profile of Reflectivity
VSWR	Voltage Standing Wave Ratio
VTD	Velocity Track Display
VVP	Volume Velocity Processing
WER	Weak Echo Region
WMO	World Meteorological Organization
WPDN	Wind Profiler Demonstration Network
WPMM	Window Probability Matching Method
WSA	Wind Shift Algorithm
WSR	Weather Surveillance Radar
WSR-88D	Weather Surveillance Radar-1988-Doppler

INDEX

GPSR Compliance

The European Union's (EU) General Product Safety Regulation (GPSR) is a set of rules that requires consumer products to be safe and our obligations to ensure this.

If you have any concerns about our products, you can contact us on ProductSafety@springernature.com

In case Publisher is established outside the EU, the EU authorized representative is:

Springer Nature Customer Service Center GmbH
Europaplatz 3
69115 Heidelberg, Germany

Batch number: 10755248

Printed by Printforce, the Netherlands